ADVANCED MOLECULAR PLANT BREEDING

ADVANCED MOLECULAR PLANT BREEDING

Meeting the Challenge of Food Security

Edited by
D. N. Bharadwaj, PhD

Apple Academic Press Inc.
3333 Mistwell Crescent
Oakville, ON L6L 0A2 Canada

Apple Academic Press Inc.
9 Spinnaker Way
Waretown, NJ 08758 USA

© 2019 by Apple Academic Press, Inc.
Exclusive worldwide distribution by CRC Press, a member of Taylor & Francis Group
No claim to original U.S. Government works
International Standard Book Number-13: 978-1-77188-664-2 (Hardcover)
International Standard Book Number-13: 978-0-203-71065-4 (eBook)

Library and Archives Canada Cataloguing in Publication

Bharadwaj, D. N., editor
Advanced molecular plant breeding : meeting the challenge of food security / edited by D.N. Bha-radwaj, PhD.
Includes bibliographical references and index.
Issued in print and electronic formats.
ISBN 978-1-77188-664-2 (hardcover).--ISBN 978-0-203-71065-4 (PDF)
1. Plant breeding. 2. Crops--Molecular genetics. 3. Crops--Genetics.
4. Crop improvement. I. Title.

| SB123.B53 2018 | 631.5'233 | C2018-903201-4 | C2018-903202-2 |

Library of Congress Cataloging-in-Publication Data

Names: Bharadwaj, D. N. (Dinesh Narayan) editor.
Title: Advanced molecular plant breeding : meeting the challenge of food security / editor, D.N. Bharadwaj.
Description: Toronto ; New Jersey : Apple Academic Press, 2018. | Includes bibliographical refer-ences and index.
Identifiers: LCCN 2018024230 (print) | LCCN 2018025492 (ebook) | ISBN 9780203710654 (ebook) | ISBN 9781771886642 (hardcover : alk. paper)
Subjects: LCSH: Plant genetic engineering. | Plant breeding. | Crops--Genetic engineering. | MESH: Plant Breeding | DNA Shuffling | Crops, Agricultural--genetics | Crop Production
Classification: LCC SB123.57 (ebook) | LCC SB123.57 .A35 2018 (print) | NLM SB 123.57 | DDC 631.5/233--dc23
LC record available at https://lccn.loc.gov/2018024230

Apple Academic Press also publishes its books in a variety of electronic formats. Some content that appears in print may not be available in electronic format. For information about Apple Academic Press products, visit our website at **www.appleacademicpress.com** and the CRC Press website at **www.crcpress.com**

CONTENTS

ABOUT THE EDITOR

Dinesh Narayan Bharadwaj, PhD
Former Professor of Plant Sciences, Haramaya University, Ethiopia;
Former Head, Department of Genetics and Plant Breeding,
C.S. Azad University of Agriculture & Technology, Kanpur, India
Mobile: +91-7275380950/+91-9956211207
E-mail: prof.bharadwaj52@gmail.com

D. N. Bharadwaj, PhD, is currently an expert and examiner at several Indian universities and a reviewer for various scientific journals in India and abroad. He was formerly Professor of Plant Sciences at Haramaya University, Ethiopia (Africa), under the United Nations Development Program (UNPD) and was also Professor of Biology under a UNDP program at the Eritrea Institute of Technology, Asmara, Eritrea. In addition, he was Head of the Department of Genetics and Plant Breeding at C.S. Azad University of Agriculture and Technology, Kanpur, India.

Other former positions include lecturer at Kanpur University, where he taught several botany subjects to undergraduate and postgraduate students, and Scientist Pool Officer of the CSIR (Council of Scientific and Industrial Research, Govt. of India). He has made significant contributions to several crop breeding programs of ICAR (Indian Council of Agriculture Research) on projects such as cotton, soybean, sorghum, groundnut, wheat, and barley, etc. He also worked as a seed production officer at the National Seed Project of ICAR.

During his stay of more than nine years in foreign countries, he visited over 20 countries and had opportunities to visit various educational institutes, universities, and their laboratories. With over 40 years of teaching and research experience, he has supervised many agriculture students in MSc and PhD programs. Dr. Bharadwaj has participated in several national and international workshops, seminars, and symposia and presented 15 research papers. He has published over 40 refereed research papers in national and international journals, contributed several book chapters, and published a dozen or so textbooks and reference books with Indian and foreign publishers. He also awarded a best scientist award in 1998 by the American Biographical Institute, Inc. USA.

LIST OF CONTRIBUTORS

M. H. Basha
Breeder, Best Crop Science, Guntur–522007, (Andhra Pradesh) India

Dinesh Narayan Bharadwaj
Ex–Professor and Head, Department of Genetics and Plant Breeding, CS Azad University of Agriculture and Technology, Kanpur–208002 (UP) India

Mohd Yasin Bhat
Research Scholar, Department of Plant Sciences, School of Life Sciences, University of Hyderabad, Gachibowli, Hyderabad, 500046, (Telangana) India

Megha Bhatt
Department of Molecular Biology and Genetic Engineering, College of Basic Science and Humanities, GB Pant University of Agriculture and Technology, Pantnagar–263145, (UK) India

Devmani Bind
Senior Research Fellow, ICAR–Indian Institute of Wheat and Barley Research (IIWBR), Karnal–132001, (Haryana) India

Akshaya Kumar Biswal
Postdoctoral Research Associate, Department of Biology, University of North Carolina, Chapel Hill, NC – 27599, USA

Mir Zahoor Gul
Research Associate, Department of Biochemistry, University College of Sciences, Osmania University, Hyderabad, 500007, (Telangana) India

Vikas Gupta
Scientist (Plant Breeding), ICAR–Indian Institute of Wheat and Barley Research (IIWBR), Karnal–132001, (Haryana) India

Anirudh Kumar
Assistant Professor, Department of Botany, Indira Gandhi National Tribal University, Amarkantak–484886, (MP) India

Rakesh Kumar
Senior Research Fellow, ICAR–Indian Institute of Wheat and Barley Research (IIWBR), Karnal–132001, (Haryana) India

Ramesh Kumar
PhD Scholar, Department of Genetics and Plant Breeding, College of Agriculture, Junagadh Agricultural University, Junagadh –362001, (Gujarat) India

Satish Kumar
Scientist (Plant Breeding), ICAR–Indian Institute of Wheat and Barley Research (IIWBR), Karnal–132001, (Haryana) India

Pushpa Lohani
Professor, Department of Molecular Biology and Genetic Engineering, College of Basic Science and Humanities, GB Pant University of Agriculture and Technology, Pantnagar–263145, (UK) India

H. M. Mamrutha
Scientist (Plant Physiology), ICAR–Indian Institute of Wheat and Barley Research (IIWBR), Karnal–132001, (Haryana) India

Satendra Kumar Mangrauthia
Scientist, Crop Improvement Section, Indian Institute of Rice Research, Rajendranagar, Hyderabad–500030, (Telangana) India

Dharmendra Rasiklal Mehta
Associate Professor, Dept of Genetics and Plant Breeding, College of Agriculture, Junagarh Agriculture University, Junagadh–362001, (Gujarat) India

A. K. Mehta
Principal Scientist, Department of Plant Breeding and Genetics, JNKVV, Jabalpur–482004, MP, India

Chandra Nath Mishra
Scientist (Plant Breeding), ICAR–Indian Institute of Wheat and Barley Research (IIWBR), Karnal–132001, (Haryana) India

Nishi Mody
DST Inspire Fellow, Drug Delivery Research Laboratory, Department of Pharmaceutical Sciences, Dr. H S Gour Vishwavidyalaya, Sagar–470003, (MP) India

Senthilkumar K. Muthusamy
Scientist (Agril Biotechnology), ICAR–Indian Institute of Wheat and Barley Research (IIWBR), Karnal–132001, (Haryana) India

Abhijeet K. Nandha
PhD Research Scholar, Dept of Biotechnology, College of Agriculture, Junagarh Agriculture University, Junagadh–362001, Gujarat, India

V. C. Pandey
Executive, Research and Development, Hester Biosciences Limited, Ahmedabad–380006, (Gujarat) India

Suneeta Pandey
Scientist, Executive– Research and Development, Department of Plant Breeding and Genetics, JNKVV, Jabalpur, 482004, (MP) India

Hitendra Kumar Patel
Senior Scientist, CSIR–Centre for Cellular and Molecular Biology (CCMB), Uppal Road, Hyderabad–500007, (Telangana) India

Revathi Ponnusami
Scientist, Hybrid Rice Section, Indian Institute of Rice Research, Rajendranagar, Hyderabad–500030, (Telangana) India

Rahul Priyadarshi
Research Associate, Hybrid Rice, ICAR–Indian Institute of Rice Research, Hyderabad–500030, (Telangana) India

Masochon Raingam
Department of Molecular Biology and Genetic Engineering, College of Basic Science and Humanities, GB Pant University of Agriculture and Technology, Pantnagar–263145, (UK) India

Chet Ram
Scientist, Division of Genomic Resources, ICAR–National Bureau of Plant Genetic Resources, New Delhi–110001, India

Beedu Sashidhar Rao
Professor, Department of Biochemistry, University College of Sciences, Osmania University, Hyderabad, 500007, (Telangana) India

Rajeev Sharma
Senior Research Fellow (ICMR–SRF), Drug Delivery Research Laboratory, Department of Pharmaceutical Sciences, Dr. H. S. Gour Vishwavidyalaya, Sagar–470003, (MP) India

Arun Kumar Singh
PhD Research scholar, Crop Improvement Section, IIRR, (Indian Institute of Rice Research), Rajendranagar, Hyderabad–500030, (Telangana) India

Akash Sinha
Department of Molecular Biology and Genetic Engineering, College of Basic Science and Humanities, GB Pant University of Agriculture and Technology, Pantnagar–263145 (UK) India

Vivek Thakur
Bioinformatics Specialist, C4 Rice Center, International Rice Research Institute, DAPO 7777, Metro Manila, Philippines

Kashi Nath Tiwari
Senior Research Fellow, ICAR–Indian Institute of Wheat and Barley Research (IIWBR), Karnal–132001, (Haryana) India

Karnam Venkatesh
Scientist (Plant Breeding), ICAR–Indian Institute of Wheat and Barley Research (IIWBR), Karnal–132001, (Haryana) India

Ajay Verma
Principal Scientist, ICAR–Indian Institute of Wheat and Barley Research (IIWBR), Karnal–132001, (Haryana) India

Suresh P. Vyas
Professor, Department of Pharmaceutical Sciences, Dr. H S Gour Vishwavidyalaya, Sagar–470003, (MP) India

Dhammaprakash P. Wankhede
Scientist, Division of Genomic Resources, ICAR–National Bureau of Plant Genetic Resources, New Delhi–110001, India

Sheel Yadav
Scientist, Division of Genomic Resources, ICAR–National Bureau of Plant Genetic Resources, New Delhi–110001, India

LIST OF ABBREVIATIONS

AAS	atomic absorption spectrometry
ABA	abscisic acid
ABS	Africa Biofortified Sorghum
ADCS	amino deoxychorismate synthase
AEA	average-tester axis
AFLP	amplified fragment length polymorphisms
AIF	apoptosis-inducing factor
ALO	arabinono-1,4-lactone oxidase
AMMI	additive main effect and multiplicative interaction
ASV	AMMI stability value
BALT	bronchus-associated lymphoid tissues
BAP	benzylaminopurine
BARC	Bhabha Atomic Research Centre
BBTV	banana bunchy top virus
BC	backcross
BCG	bacille calmette-guerin
BMGF	bill and melinda gates foundation
BMV	brome mosaic virus
BOAA	β-oxalylamino-alanine
BRR	bayesian ridge regression
BS	bundle sheath
BT	biotechnological
BYMV	bean yellow mosaic virus
CA	carbonic anhydrase
CAPS	cleaved amplified polymorphism sequence
CBB	common bean blight
CBL	calcineurin B-like
CFIA	Canadian Food Inspection Agency
CHIP	Chicken and Hen Infection Program
CHS	chalcone synthase
CIPK	CBL-interacting protein kinases
CMS	cytoplasmic male sterile
CMV	cowpea mosaic virus

CN	cyst nematode
CP	coat protein
CT	cholera toxin
CTB	cholera toxin B subunit
CTV	citrus tristeza virus
CWR	crop wild relatives
DE	double exponential
DH	doubled haploid
DSB	double-strand breaks
EAHB	East African Highland bananas
EDTA	ethylene diamine tetraacetic acid
EF	edema factor
EMS	ethyl methane sulfonate
EN	elastic net
EST	expressed sequence tag
ETEC	enterotoxigenic E. coli
FDA	Food and Drug Administration
FHB	fusarium head blight
FMD	foot and mouth disease
FMDV	foot-and-mouth disease virus
GA	gibberellic acid
GALT	gut-associated lymphoid tissues
GCHI	GTP-cyclohydrolase I
GDC	glycine decarboxylase complex
GEBV	genomic estimated breeding value
GM	genetically modified
GMO	genetically modified organisms
GPS	global positioning systems
GRP	golden rice project
GS	genomic selection
GURT	Genetic Use Restriction Technology
GWAS	genome-wide-association study
GWS	genome-wide selection
HCN	hydrocyanic acid
HDR	homology-directed repair
HFRS	hemorrhagic fever with renal syndrome
HHP	high hydrostatic pressure

HPV	hepatitis B virus
HR	hypersensitive response
HSV	herpes simplex virus
IAA	indole-3-acetic acid
IARI	Indian Agricultural Research Institute
IBT	ion beam technology
IGKV	Indira Gandhi Krishividyalaya
IITA	International Institute of Tropical Agriculture
IPC	interaction principal component
IPM	integrated pest management
IPR	Intellectual Property Rights
IRRI	International Rice Research Institute
ISEC	Institute for Social and Economic Change
LEA	late embryogenesis activity
LET	linear energy transfer
LF	lethal factor
LL	late lethality
LT	labile toxin
MAB	marker-assisted breeding
MABC	marker-assisted backcrossing
MACC	marker-assisted complex or convergent crossing
MAF	minor allele frequency
MALT	mucosal-associated lymphoid tissues
MARS	marker-assisted recurrent selection
MAS	marker-assisted selection
MAS	molecular assisted selection
MAT	multi-auto-transformation
MBC	map-based cloning
MF	molecular farming
MM	microhomology
MMEJ	microhomology-mediated end joining
MN	meganucleases
MO	moncompu
MS	murashige and skoog
MSP	merozoite surface protein
MSV	maize streak virus
NALT	nasal-associated lymphoid tissues

NARO	National Agricultural Research Organization
NBI	National Botanical Institute
NBRI	National Botanical Research Institute
NGS	next generation sequencing
NHEJ	non-homologous end joining
NIAS	National Institute of Agrobiological Sciences
NMD	nonsense-mediated mRNA decay
NPT	new plant type
OAA	oxaloacetate
ODM	oligonucleotide-directed mutagenesis
OFSP	orange flesh sweet potato
PA	phytic acid
PAM	protospacer adjacent motif
PCA	principal components analysis
PCK	PEP carboxykinase
PCO	principal coordinate analysis of genotypes
PCR	photosynthetic carbon reduction cycle
PDBK	Plant DNA Bank in Korea
PDP	plant-derived pharmaceutical protein
PEG	polyethylene glycol
PEPC	phosphoenolpyruvate carboxylase
PGR	plant genetic resources
PLRV	potato leaf roll virus
PP	Peyer's patches
PP	polyphenol(s)
PPDK	pyruvate pi dikinase
PPV	plum pox virus
PR	pathogenesis-related
PS	phenotypic selection
PTGS	posttranscriptional gene silencing
PTR	peptide transporter
PV	potato virus
PVP	plant variety protection
PVPP	polyvinylpolypyrrolidone
PVX	potato virus X
PVY	potato virus Y
QPM	quality protein maize

QTL	quantitative trait locus
QUT	Queensland University of Technology
RAPD	random amplified polymorphic DNA
RB	reverse breeding
RBE	relative biological effectiveness
RF	random forest
RFLP	restriction fragment length polymorphism
RIP	ribosomal inactivating protein
RISC	RNA-induced silencing complex
RKHS	reproducing Kernel Hilbert spaces
RKN	root-knot nematode
RP	recurrent parent
RTSV	rice tungro spherical virus
RUE	radiation use efficiency
SAGE	serial analysis of gene expression
SCAR	sequence characterized amplified region
SCN	soybean cyst nematode
SDA	engineered stearidonic acid
SDN	site-directed nucleases
SH	synthetic hexaploid
SHMM	shifted multiplicative model
SNP	single nucleotide polymorphism
SPCA	supervised principal component analysis
SRSR	short regularly spaced repeats
SSA	sub-Saharan Africa
SSH	suppression subtractive hybridization
SSR	simple sequence repeats
STMS	sequence-tagged microsatellite sites
SVD	singular value decomposition
TB	tuberculosis
TE	targeting efficiency
TGS	transcriptional gene silencing
THN	tri-hydroxy-naphthalene reductase gene
TMV	tobacco mosaic virus
TNAU	Tamil Nadu Agricultural University
TP	training population
UPOV	Union pour la Protection des Obtentions Vegetales

VAM	vesicular-arbuscular mycorrhizae
WCR	western corn rootworm
WDV	wheat dwarf virus
WGS	whole genome sequencing
WHO	World Health Organization
WTSS	whole transcriptome shotgun sequencing
XRF	x-ray fluorescence
YDV	yellow dwarf virus
YSI	yield stability index
ZF	zinc finger
ZFA	zinc finger transcription activators
ZFM	zinc finger methylases
ZFN	zinc-finger nucleases
ZFR	zinc finger transcription repressors
ZIRC	Zebrafish International Resource Center

FOREWORD

Plants are essential parts of our life. They have provided us with food, feed, shelter, and healthy environment to breathe since their origin. The last five decades have been the most productive period in the world agriculture history due to the 'Green Revolution,' which saved about billion people from hunger and starvation. This increased agriculture production has been based on over-exploitation of natural resources with the implementation of advanced technology. In spite of this, one billion people are still suffering from chronic hunger, and most of them belong to the poor countries; these people are mostly laborers and small-scale farmers.

It is expected that by 2050, the world population will reach a new height of around 10 billion, which will require extra global food production—to nearly 70% higher than today's. The big challenge is that we will have to achieve this goal from a shrinking agricultural land base due to industrialization and urbanization and the limiting factors of global warming, and climatic or environmental changes with limited natural resources. Therefore, agriculture produce and production may be reduced; food prices may go up and will pose further problems for several countries. Although higher food prices affect everyone, but especially the poor, because they are compelled to spend of their most income on food.

Despite these serious threats and challenges, there is hope with new developing science and technology—these are biotechnology and genetic engineering, which have the great potential to solve the world's biggest challenge to cope up with food insecurity. These newer technologies have contributed invaluable new scientific methodologies to produce more productive and value-added agricultural food products, never existed before in nature. The plant breeder's journey is still continues and reaches from the genome to the molecular level of the cell. The genomics-based sequencing methods have enabled breeders to achieve greater precision in selecting and transferring useful genes, not only from distant species but also to exploit genes between unrelated animal to plant and vice versa. Furthermore, this technology has also reduced the time needed to eliminate undesirable genes beyond conventional approaches.

The developments of science and technology made it possible to meet the challenges of future environmental hazards ahead of the 21st century. With the new molecular genetic tools, we are poised for another explosion in agricultural innovation in the field of molecular plant breeding. These new tools are openingup new ways to produce nutritionally enhanced food, coupled with higher production food crops. The newer technology success-fully produces genetically modified crops, plant-based edible vaccines, with the implementation of molecular marker-based selection, RNA interference to turn on-off genes function, OMICS biology, and genome editing, etc.

The present book, *Advanced Molecular Plant Breeding: Meeting the Challenge of Food Security*, edited by Dr. Dinesh Narayan Bharadwaj, Former Head Department of Genetics and Plant Breeding, C.S. Azad University of Agriculture and Technology, Kanpur, has made an outstanding compila-tion of research on genetics and genomics that can drive progress in modern plant breeding. The chapters are well designed by several specialist scientist/breeders from diverse specialized areas in the field of plant breeding. It is a wonderful job in integrating information about traditional and molecular plant breeding approaches. I hope the book will be read widely.

—M. S. Swaminathan

Founder Chairman Ex-Member of Parliament (Rajya Sabha)
3rd Cross Road, Taramani Institutional Area, Chennai (Madras) – 600 113,
India, Phone: +91-44-2254-2790/2254-1698, Fax: +91-44-2254-1319,
E-mail: founder@mssrf.res.in, swami@mssrf.res.in

ACKNOWLEDGMENTS

First and foremost, I am immensely grateful to my parents, brothers, sister, and teachers who helped me to become a successful person. Secondly, I am highly thankful to all the contributing authors of this book for their positive response. I am also grateful to Dr. M.S. Swaminathan, former DG, Indian Council of Agriculture Research, and currently Founder Chairman of MSSRF, Chennai, for kindly agreeing and writing the 'Foreword' for this book. I wish to express my thanks to a number of friends and colleagues for their invaluable suggestions from time-to-time during the manuscript preparation. I also thank to my colleague Dr. Sanjay Kumar Singh for his help in preparing the subject index.

I wish to express my appreciation for help rendered by Ms. Sandra Sickels, Mr. Rakesh Kumar, and Mr. Ashish Kumar, staff of Apple Academic Press, for their cooperation and valuable suggestions, above all their professionalism, which made this book a reality, is greatly appreciated.

At last but not least, I wish to express my most sincere thanks to my life partner, Mrs. Shobha Bharadwaj, and lovely daughter, Ms. Sneha Bharadwaj, for their unending support and encouragement in preparing this manuscript, without which it was not possible to finish this multifaceted task.

—**Dinesh Narayan Bharadwaj, PhD**
Editor

PREFACE

Plants are the natural resource of human sustenance. They provide us with food, feed, shelter, and the natural environment; therefore, better understanding of plant biology has been a subject of biologists since from ancient times and continues today. The past half-century has been very crucial due to the Green Revolution, but within the last 30 years, plant biologists have unraveled several physiological, biochemical, and genetic mysteries of the structure and function of cells and, further, from gene to genome. This journey of the transformation of plant knowledge has been presented in the form of an uncountable number of books over time by the biologists. Continuing this trend for curious readers, the present book, *Advanced Molecular Plant Breeding: Meeting the Challenge of Food Security*, is an effort to contribute, share, and document contemporary knowledge of genetics and plant breeding research between researchers, teachers, and students.

This book has taken almost five years in its preparation. In fact, during my 40 years of teaching and research career, I always felt the need for recent books; therefore, the initial idea for the book was stimulated by several of my colleagues and students during many discussions on molecular plant breeding. I strongly believe that this is publication provides comprehensive knowledge of the relevant fields for a new generation of molecular plant breeders.

This book includes five sections: introduction, role of biodiversity, molecular selection tools, genetic engineering and tissue culture, and most recent techniques employed now to solve the future food problems ahead of the 21st century. Better understanding of molecular plant breeding can be exploited to boost the quality of agriculture produce with high production rates and provide nutritious food for everyone by 2050, when the global population is expected to a new height of about 10 billion people. The problem to provide quality food to everyone by 2050 is a cumbersome task, because along with the increased population comes the depletion of natural resources and also shrinking arable land areas. The gigantic problem can be solved by implementing newer techniques of quantitative trait inheritance to integrate genomics and molecular biology into appropriate tools and methodologies to create genetically engineered plants using biotic and abiotic stress tolerance, molecular markers, '-omics' technology, and genome editing to the benefit of mankind. The milestone of whole genome sequencing from Arabidopsis to rice opened up new evidence to enter in the era of functional genomics, which aims to determine the functions of all the genes identified in their genomes that help in improving crop species qualitatively and quantitatively to meet the ongoing challenges of doubling the world food production by 2050.

The contributors to this book were selected from a wide and diverse range of institutions. They have great understanding, expertise, and experience on these aspects and use the depth of their knowledge by integrating the most recent information from classical to modern sources.

The book aims to provide advanced recent knowledge of plant breeding tools to update researchers, teachers, and graduate and postgraduate students in plant sciences, plant biotechnology, applied botany, agricultural sciences, plant genetics, and molecular biology.

—Dinesh Narayan Bharadwaj, PhD
Editor

PART I

INTRODUCTION TO ADVANCED MOLECULAR PLANT BREEDING

CHAPTER 1

SUSTAINABLE AGRICULTURE AND FOOD SECURITY: 2050 (FUTURE PLANT BREEDING)

D. N. BHARADWAJ

Ex-Professor and Head, Department of Genetics and Plant Breeding, C.S. Azad University of Agriculture and Technology, Kanpur, India, Tel.: 9956211207, E-mail: bharadwajdncsau@gmail.com

CONTENTS

ABSTRACT

According to FAO estimate, by 2050, the world's population will reach about 9.6 billion (34% higher than today). The population increase will occur mainly in developing countries, and urbanization will continue to be about 70% (compared to 49% as of today). Income growth would be many fold higher; so, consumption of food and biofuels would increase by more than 70% (FAO, 2009). Annual cereal production requirement will be one and half times more, i.e., 3 billion tons, from 2.1 billion tons as of today. Similarly, meat requirement will also be around 470 billion tons as compared to 270 billion tons as of today. Agricultural land will shrink, and more food will have to be produced from less land; moreover, water and energy resources will become limiting factors. Agricultural intensification in richer countries and extensification (land clearing) in poor countries will continue; consequently, ~1 billion ha of land would be cleared globally by 2050. The CO_2-C equivalent greenhouse gas emissions will then reach by ~3 Gt y^{-1} and N use by ~250 Mt y^{-1}. The conditions will be absolutely different, and demand of resources will be much higher than today. Meeting this additional demand of food and feed will be a difficult task for several nations.

1.1 INTRODUCTION

During the first green revolution, the development of improved agricultural technology, new high-yielding varieties, mechanization, and enhanced use of chemicals and fertilizers led to maximize the yields (Rosset, 2000), but this development adversely affected agro-ecology and labor demands in agriculture. All these developments could not lead to solve the food safety in future when global population will increase to reach around 9.6 billion (Table 1.1). Foresight is needed to avoid the difficult situation before 2050; the world may run into another food crisis. In the current global scenario,more than 800 million people are periodically hunger or food unsecured (Table 1.2).

To ensure national nutritional security, adequate food production is required to save the population from chronic undernourishment and malnutrition. To meet this additional need, food supply will depend on higher yield and global agriculture expansion. Modern agriculture is essentially based on high-performance varieties under high input system, viz., fertilizers, water, pesticides, and fossil fuels. Currently, the use of these chemical inputs is

TABLE 1.1   Projections of Global Population Growth

Areas	World Population (millions) during year		
	2015	**2030**	**2050**
Asia	4393	4923	5367
Africa	1186	1679	2478
Europe	738	734	707
Latin America and Caribbean	634	721	784
USA	358	396	433
Oceania	39	47	57
World Total	7249	8501	9725

TABLE 1.2   Countries with Global Hunger Index (GHI) extremely alarming (GHI $\geq$ 50) and alarming (GHI between 35.0 and 49.9)

Country	Global Hunger Index			
	1992	**2000**	**2008**	**2017**
Central African Republic	52.2	50.9	47.0	50.9
Chad	62.5	51.9	50.9	43.5
Sierra Leone	57.2	54.7	44.5	38.5
Madagascar	43.9	43.6	36.8	38.3
Zambia	48.5	52.3	45.0	38.2
Yemen	43.5	43.4	36.2	36.1
Sudan	-	-	-	35.5
Liberia	51.2	48.2	38.9	35.3

at their peaks because these varieties do not perform well under low input conditions. On the one hand, the availability of these inputs will not continue forever, and on the other hand, the situation is alarming for degradation of natural resources and ecosystems. Millions of rural poor will depend on agriculture for their livelihoods that may enhance ecosystem degradation and high greenhouse gas emission. The new croplands will be the result of conversion of forests, grasslands, and wetlands, which has high environmental and social impacts and reduced wildlife habitat. This will also induce migration of people from rural to urban areas in different ecosystems, which may give rise to some social conflicts within the communities. Several countries will depend on import of food commodities to ensure their food security. It is estimated that by 2050, net imports of cereals of developing countries will be more than double from today's, i.e., from 140 million metric tons to 300 million metric tons in 2050. The increasing population trend, food production, and future resource demand is presented in Table 1.3. We may encounter shortage of food supply and hunger or undernourishment, if the problem is not addressed well within the time.

Agriculture is the world's largest use of land, occupying about 38% of the Earth's terrestrial surface (Foley et al., 2011). The agricultural community had tremendous success in the world's food production over the past five decades with increasing population. Global grain production has tripled since 1960 (Godfray et al., 2010) due to the green revolution. Modern agriculture took advantage of inexpensive fossil fuels to increase agricultural productivity, but many countries like sub-Saharan Africa have been left behind.

1.2 PROBLEMS AND RISK AHEAD OF FOOD SECURITY TOWARD 2050

As predicted, the world's population is expected to grow about 9.6 billion people by 2050. At the same time, per capita income will also increase many fold in 2050 as compared to today's levels. The world is still away in solving the problem of economic deprivation and malnutrition for large population. The market demand for food would continue to grow for cereals (food and feed); this increase may up by 3 billion tons from today's 2.1 billion tons (Figure 1.1). Higher income groups in the developing countries will have more demand for other food products like meat, dairy products, vegetables, oils, etc., which will grow much faster than cereals. Production of these com-

TABLE 1.3 Increasing trend in population, food production, and future resources

Resources	Demand during years		
	1960	2000	2050
Population (billions)	3	6	9 – 10
Food production (Mt)	1.8 x109	3.5 x 109	6.5 x 109
Agricultural water (km^{-3})	1500	7130	12– 13.500
N fertilizer use (Tg)	12	88	120
P fertilizer use (Tg)	11	40	55 – 60
Pesticide use (Tg, active ingredient)	1.0	3.7	10.1

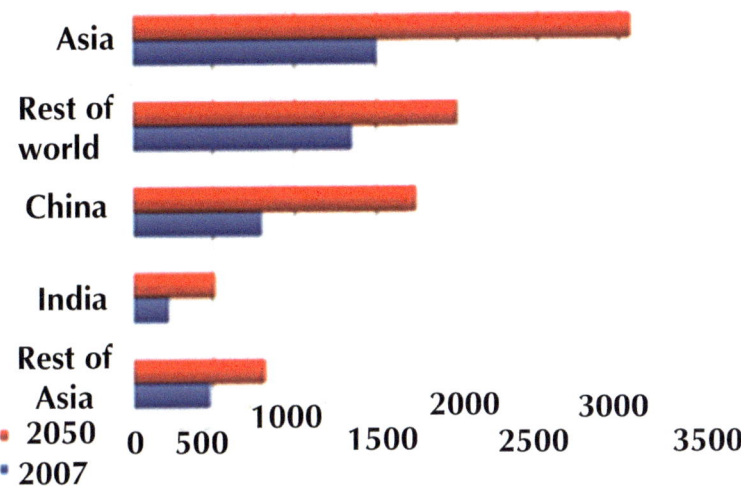

FIGURE 1.1   Increase in consumption of major field crops during 2007–2050.

modities in the developing countries would need to be almost double. This implies a significant increase in the production of several key commodities.

The global ability to meet the demand for food, feed, and fiber through the agricultural system could be severely limited due to a number of risk factors and challenges. The most important risk is hunger and malnutrition that could continue in spite of sufficient food production because of uneven food supplies. The second big challenge is climate change that may affect developing countries at various levels. The third emerging challenge is rapid increase in energy prices, causing additional transport scarcity in markets for food and feed.

Measures should be taken for sustainable food safety in future by minimizing food gap in demand and production before 2050; these includes reduction in excess food consumption, obesity, rural migration, greenhouse gas emissions, and pollution of freshwater; generating benefits to women, and protection of ecosystems. Most of the world's people consume excess milk, dairy products, meat, eggs, beef, and fish than required. The modest consumption of these products could be 20% beneficial to large population. The current fertility rate is roughly 2.1 children per woman in most of the developed, countries, but in Africa and Muslims both are the exception with a rate of >5.4 children per woman; this will increase their population to nearly triple from its 2006 level, i.e., more than 2 billion people by 2050. Global reduction in fertility rates by 2050, especially in hungriest areas, may enhance food security. Some of the challenges in sustainable food security are listed below.

1.2.1 GLOBAL WARMING – THREAT TO AGRICULTURE

In future, climate change due to greenhouse gas emission will pose serious problems for agriculture by burning of fossil fuels that releases billions of tons of CO_2 into the atmosphere. This CO_2 accumulation will increase global temperature by about 2°C to 6°C by the end of the century. Rising temperatures will desiccate forests and lead to forest fires that will release more CO_2 to the atmosphere and further temperature increase. The Arctic region will release vast amounts of additional carbon into the atmosphere to increase the temperature again. The melting of Arctic ice will cause more sunlight to be absorbed by the ocean, increasing its temperature even further. And, if the ice sheets of Greenland and Antarctica melt, the sea level would increase by 67 m, which may submerge coastal areas and reduce the arable land areas. This global warming will also affect rainfall patterns. Some areas may receive greater or lesser rainfall than average; all these can hamper agricultural productivity. Increased temperature will also have various deleterious effects on crop growth, increased evapotranspiration rates, decreased germination rates, inhibition of photosynthesis, alteration in flowering times, inhibition of sexual reproduction, and an increase in plant diseases and herbivore attacks. Every 1°C temperature rise will decline crop yields of wheat and rice by 10% and of maize and soybean yields by 17%. So, because of global warming, the yields of important food, feed, and fiber crops will decline.

1.2.2 CLIMATE CHANGE – GREENHOUSE EMISSION

Increased biofuel production and consumption will represent major risks for climate change and long-term food security. Increased use of petroleum products is already posing a serious threat to the environment. According to FAO (2011) estimate, the aggregate negative impact of climate change on

African agricultural output by 2080–2100 would be 15–30%. Agriculture will have to adapt and mitigate the effects of climate changes. Increased production of food crops coupled with consumption of biofuels could have serious implications for future food security. Excessive uses of these inputs are posing serious threats to ecology, environment, soil health, and ground water. Presently, the production of agricultural crops and animal products releases around 13% of global greenhouse gas emissions, which is equivalent to 6.5 Gt (gigatons) of CO_2 per year; by 2050, this emission could grow up to 9.5 Gt, and if combined with emissions from land use change, it could reach 15 Gt. Further, increase in the global warming below 2° Celsius, the world annual emissions from all sources would reach around 21–22 Gt by 2050; this would severely harm the ecosystem. Therefore, agriculture must reduce its greenhouse gas emissions even after boosting production. The great balancing act requires reducing greenhouse gas emissions from both existing and additional agricultural production of food and consuming less. There is also a need to reduce the use of fossil fuels and its gradual replacement with other nonrenewable resources such as solar energy, hydrogen (water), alcohol, pyrolysis, etc.

1.2.3   FOOD SECURITY AND NUTRITION

In various developing countries, a sizeable amount of food is lost during harvesting, handling, and storage. Reduction of food wastes would reduce the food gap by around 20% by 2050; reduced losses can be an immediate and cost-effective opportunity for food availability in many areas. Increase in global agricultural food production and supply up to 70–100% by 2050 on the existing land is quite complex job; therefore, it is needed to reduce food waste as it is very high in developing countries (Figure 1.2); there is also an urgent need to reduce food losses and improve food distribution system to eradicate hunger and malnutrition by 2030.

1.2.4   CHANGING SOCIO-ECONOMIC ENVIRONMENT

Eliminate extreme poverty and make food affordable to the poor by 2030. Besides this other measures are to increase the income of rural households this will reduce migration ratio of rural peoples to urban areas for their eco-

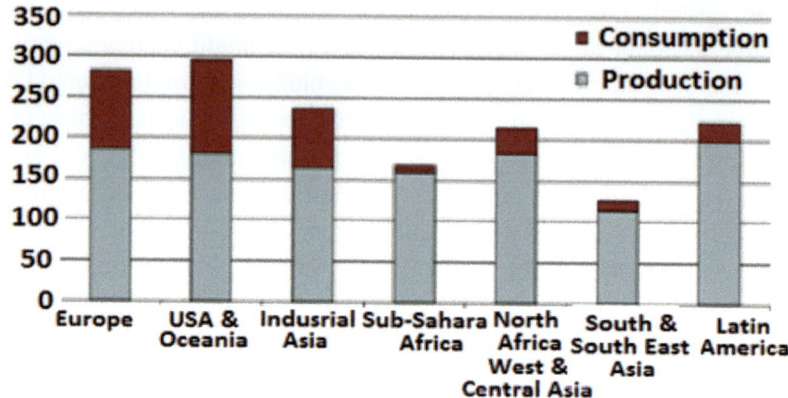

FIGURE 1.2   Per capita food losses at consumption and production (kg/year).

nomic development and further slow down the ratio of increasing population growth which is mainly responsible for excess food demand.

1.2.5   ENVIRONMENTAL DEVELOPMENT

Conserve natural resource ecosystems (water, energy, fertilizer, etc.), lower global warming and reduce water and air pollution, soil degradation and soil nutrient depletion. Protect wildlife, biodiversity, agricultural landscapes, closing yield gaps, increase diversity in number of growing crops and reduction of post-harvest losses.

1.2.6   THE NATURAL RESOURCE BASE

1.2.6.1   Land

Globally, there are still sufficient land resources available to feed the future world population. Much of the potential land is suitable for growing only a few crops, and much of the land yet suffers from chemical and physical constraints, endemic diseases and lack of infrastructure; therefore, it is needed to bring this land into production by reclamation. Future sustainable agriculture resource management and the ability of the world's food system need improved stewardship of natural resources with rising scarcity and degradation of land, water, and biodiversity.

1.2.6.2 Water

Water availability may become constraint in future; thus, there is a need to develop water use-efficient crops. Globally, fresh water resources are sufficient, but they are very unevenly distributed, and water scarcity may reach to alarming levels in several countries or regions within countries, particularly in the near East, North Africa, and South Asia.

1.2.6.3 Air

The most responsible pollutants of air are smoke, dust from tillage, harvest, transportation, spraying of insecticides and pesticide, and emission of other harmful gasses like nitrous oxide from fertilizers. The burying of crop residue into the soil, planting wind breaks, covering of crops, and growth of perennial grasses on open land can reduce the pollution of the air and enhance soil fertility as well.

1.2.6.4 Soil

Healthy soil is one of the most important factors of sustainability; it involves the management of physical, chemical, and biological properties of soil to contain large quantity of organisms to maintain the soil's ecosystem (Kibblewhite et al., 2008). Organic matter and compost are nutrients for beneficial soil microorganisms such as protozoa, bacteria, fungi, and nematodes. Properly managed soil microorganisms perform vital functions that enable to grow healthy and more vigorous crop plants without diseases. Regular additions of organic matter can increase soil fertility by increasing soil nutrients and microbes. Soil erosion caused by heavy storm, wind, and flood results into a severe threat to crop productivity.

1.2.7 CROP GENETIC DIVERSITY

Genetic diversity plays an important role in the survival and adaptability of any species and enough crop genetic diversity expanded by our ancestors. (Doebley et al., 2006) is the best source for sustainable food production against drastic changes in local climate, soil chemistry, and biotic influences. A wide genetic base provides "built-in insurance" (Harlan, 1992) against crop pests, pathogens, and climatic vagaries. We need to conserve the global natural resource base (crop genetic biodiversity) which is under threat and showing degradation or un-sustainability with soil nutrient depletion, erosion, desertification, depletion of freshwater reserves, loss of tropical forest , etc. Marine resources need maintenance and rehabilitation for future use. The land use practices must be more sustainable; otherwise, the productive potential of land, water, and genetic resources may continue to degrade at alarming rates.

Biodiversity is another essential resource for agriculture and food production, which is threatened by urbanization, deforestation, pollution, and conversion of wetlands. Agricultural modernization, changes in population density, and human diet depend on little agricultural and biological diversity (gene pool) for its food supplies. Only a dozen species of animals provide 90% of the animal protein consumed, and just five crops (rice, wheat, maize, pulses, and potato) provide more than half of plant-based calories in the human diet (FAO, 2011).

1.2.8   WILDLIFE

The conversion of wildlife habitat to agricultural land can reduce soil erosion and sedimentation. The support of wildlife diversity will enhance natural ecosystems and agricultural pest management (Green et al., 2005).

1.2.9   HEIRLOOM VARIETIES

Heirloom plant varieties of heritage fruits, vegetables, cereals, etc. are old cultivars maintained by gardeners and farmers, particularly in isolated or ethnic minority communities in countries like Australia, New Zealand, Ireland, and UK. These are not commonly used in modern agriculture. Typically, heirlooms have adapted over time to the climate and soil they have been grown, and because of their genetic diversity they are often resistant to local pests, diseases, and extremes of weather. Therefore, these heirloom genetic variability must be used in further breeding and development of new crop varieties.

1.3   SUSTAINABLE AGRICULTURE

The concerns of sustainability in the agricultural system are based on the need to develop technologies and practices that do not have adverse effects on environment, food quality, and productivity and services to farmer's society. New approaches are required to integrate biological and ecological concerns into food production and minimize the use of nonrenewable inputs that cause harm to the environment or human health. Agricultural sustainability suggests a focus on genotype improvements through complete exploitation of modern biological approaches by understanding the benefits of ecological and agronomic management, manipulation and redesign. The ecological management of agro-ecosystems that addresses energy flows, nutrient cycling, population-regulating mechanisms and system resilience can lead to redesign agriculture at site-specific magnitude (Pretty, 2008). The ideal sustainable agriculture resource model is shown in Figure 1.3.

1.3.1 SUSTAINABLE RENEWABLE ENERGY SOURCES

The nonrenewable energy sources may not be available forever and can cause a big catastrophe in future; therefore, sustainable agricultural systems should not rely upon nonrenewable energy sources, and these must be replaced with renewable, economical, and feasible energy sources in future (Boyle 1996). The sustainable sources are biofuels, hydrogen, wind power,

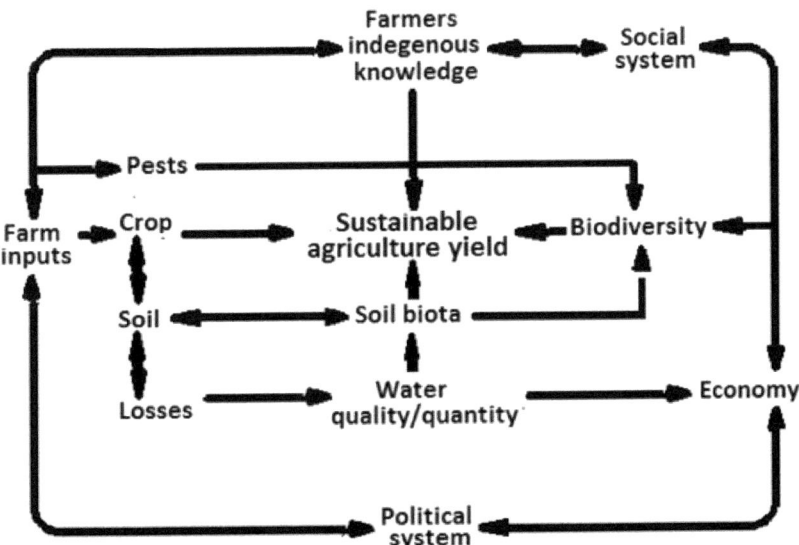

FIGURE 1.3   Sustainable agriculture resource model.

hydropower, geo-thermal energy, ocean energy, solar energy, and bioenergy. The bioenergy from agricultural waste may lead to conservation of food crops. Forage crops are more potential crops as source of renewable biological energy because they capture about two-thirds of the total solar energy used in agriculture, and together with forestry, it can be doubled again. The use of forage crops to produce more renewable energy has two advantages: (a) Most forage crops require less nitrogen fertilizer and pesticides than corn and soybeans, meaning less reliance on fossil energy in minimizing pollution and (b) Most forage crops are perennial and require no annual tillage; thus, they cause less soil erosion which is also a good storehouse of biological energy. The current technologies used to produce hydrogen, ethanol and biodiesel hold more promise to realize the limited use of nonrenewable energy source and can replace petroleum energy significantly. Pyrolysis can be another alternative for renewable energy source.

1.3.2   PYROLYSIS

It is another higher priority area for the sustainability of bioenergy that seems to be one of the most promising and prospective technology of today (Marker et al., 2012). It is a kind of chemical decomposition of organic materials under high temperature in the absence of oxygen; this process has been extensively used in the chemical industry to produce charcoal, activated carbon, and methanol from wood. The technology needs to be adapted more vigorously.

1.4   SUSTAINABLE RESOURCE MANAGEMENT

The growing demands for food seem uncertain because one of our greatest challenges is to increase food production in a sustainable manner with judicious use of natural resources and without over-exploiting them within balanced ecosystems (Cassman et al., 2002), so that everyone can be fed adequately and nutritiously. Sustainable farming must have continued flow of biological energy by recycling of nutrients through crop and livestock residues to the soil, as these organic matters restore healthy and productive organic soils that minimize the use of nonrenewable inputs that are harmful to the environment and the health of farmers and consumers (Goulding et al., 2008).

A better concept of resource management centers on intensification of natural resources, i.e., land, water, biodiversity (flora and fauna), technologies, genotypes, and ecological management practices that minimize harm to the environment. The basic resources for agricultural development are prone to various kinds of degradation. Sustainable resource management can be achieved by the following steps:

1. Recycling of crops and livestock waste to be used as manure.
2. Growing legume crops to enhance bacterial symbioses to sequester atmospheric nitrogen.
3. Enhance production of nitrogen fertilizer from natural gas.
4. Use of renewable energy from hydrogen (water), solar, and windmills.
5. Use of genetically engineered (nonleguminous) crops to fix atmospheric nitrogen.

Options for increasing food production without land expansion

1.4.1 SMARTER FARMS AND FARMERS

The current yield capacity of crops is limited due to higher use of fertilizers and water. Smarter farming will therefore have improved agricultural technology management and input resource-efficient, high-yielding crop seed varieties that are suited to optimum use of fertilizers, nutrients, and other inputs.

1.4.2 BOOST AGRICULTURAL YIELDS

Sub-Saharan Africa has the highest hunger rate, imports 25% of the required grain, and has the world's lowest staple crop yield. About 285 million people live in this dry region and consume only 9% of the world's calories. Soil degradation has harsh effects, particularly loss of soil carbon, which is a challenge to agricultural production in sub-Saharan Africa; here, over 300 mh of dry cropland (agro-forestry) has greater potential to boost yields, while managing water harvesting and small use of fertilizers.

1.4.3   MULTIPLE CROPPING SYSTEM

According to FAO estimation (2013), more than 400 million hectares of cropland is left fallow each year. Reducing fallow or by increasing double cropping could increase crop production without requiring new land. This increased cropping intensity can be an option to meet increased future food demand, still only 150 mh land is used for crops planting twice or more times every year.

1.4.4   CROP CONSERVATION ON LOW CARBON-DEGRADED LAND

Several million hectares of low carbon-degraded land are available in countries like Indonesia and Malaysia, which are suitable for palm oil production. Abandoned cropland areas are capable of supporting trees and forest that sequester carbon and play an important role to avoid climate change. Grazing lands produce valuable forage, tropical savannas, and sparse woodlands that have high carbon storage and biodiversity value.

1.4.5   INTENSIFY PASTURE PRODUCTIVITY

Wet pastures have large opportunities to intensify the production of milk and meat. These pastures can be promoted into cattle farms by growing legume crops, adding small amount of fertilizer, and promoting crop rotation periodically to develop into small grazing areas.

1.4.6   AVOID CONVERSION OF FORESTS INTO AGRICULTURAL LANDS

It is important to avoid conversion of forests into agricultural land that causes millions of hectares of deforestation due to new agricultural expansion. This causes losses in carbon storage and other ecosystem services due to new deforestation.

1.4.7 AGRICULTURE BIODIVERSITY AND FOOD SECURITY

Agricultural biodiversity is the backbone of food security since ancient period of civilization throughout the world; biodiversity of species and farming systems provides valuable ecosystem and functions for agricultural production (FAO, 1996). The sustainable use of agro-biodiversity, conservation, and its enhancement are precious for sustaining food security because there are several serious threats to loss of agro-biodiversity.

1.4.8 BREEDING BETTER SEEDS

Improved breeding techniques and genetic engineering can play a vital role to produce transgenic crops suitable to specific areas (higher yield and resistant/tolerant to insects, pests, diseases, cold, and drought) with improved nutrients and water uptake. Conventional breeding approaches may also be improved through modern biological tools like molecular assisted selection (MAS), which will help in faster identification to select desired gene combinations.

1.4.9 FOOD ALGAE AND SEAWEEDS

Algae can provide a vital solution to the world's most complex problem during food shortages, but they are at the bottom of our food chain. They can feed both humans and animals and can be grown in fresh water and ocean. Many types of seaweeds are used as staple food in Asia. There are 10,000 types of seaweeds found in the world; out of them, 145 species of red, brown, or green seaweeds are used as food. Algae farming could become the world's biggest cropping industry. Asian countries and Japan have huge farms where it could be cultivated at a large scale. Scientists at Sheffield Hallam University, UK, used seaweed granules to replace salt in bread and processed foods. The granules provide a strong flavor with low salt and could be used to replace salt in ready processed food such as sausages and cheese and can be helpful for patients with high blood pressure.

1.4.10 INCREASED PRODUCTIVITY OF AQUACULTURE

Aquaculture has expanded rapidly to produce nearly half of all the fish protein consumption. Fish farms are very efficient to convert feed to food, and chicken can be an attractive source of animal protein, if produced sustainably. The aquaculture production can be more than double by 2050 and can make more efficient feeding. Future fish production growth will require increased fish per hectare of pond.

1.4.11   CARBON SEQUESTRATION

Carbon sequestration strategies have received much of the attention specially for agricultural soils. The increase in soil carbon changes can be caused by carbon movement from one location to another or divert carbon biomass such as crop residues in the form of animal feed. Increased soil carbon accumulation can be a significant part of a strategy to boost long-term crop production to enhance productivity for making better use of productive resources.

1.4.12   INCREASED INPUT USE EFFICIENCY

Strategies must be developed for increased food production with agricultural climate mitigation that reduces greenhouse gas emissions. Increasing production efficiency provides the strongest opportunity to reduce emissions from agricultural production and include:

1.4.12.1   Balanced Fertilizer Use

Overdoses of fertilizers are used in several countries of Asia, USA, and Europe; this leads to high emissions as well as unnecessary expenses, although nitrogen fertilizer is under used in Africa and some other countries.

1.4.12.2   Reduce Emissions from Paddy rice

Removal of rice straw from rice paddies can cut emissions by more than half compared to those farms that do not employ these measures.

1.4.12.3   Avoid Competition from Bioenergy

Heavy fuel consumption required during farm mechanization and food transportation can be minimized by mixing 10% biofuels by 2020. Strategic goals should be used for bioenergy by 20% mixing of alcohol from biomass. This would require biomass from harvest of crops, grasses, crop residues, and trees, and it will improve soil carbon health and ecosystems. The potential solutions can help to close the food gap, thus reducing waste and saving greenhouse gas emissions.

1.4.12.4   Improve the Feeding and Health of Livestock

Plant debris generates nearly half of agricultural emissions, but its feeding improves the health of cattle and may reduce two-thirds of farm emissions per kilogram of milk or meat. Livestock farms that mix crop residue provide promising and efficient feeding opportunities.

1.4.13   INCREASE SOILLESS CROP PRODUCTION

Hydroponics is an old method of growing plants using mineral nutrient solutions in water, without soil (Santos et al., 2013). Terrestrial plants may be grown with their roots in the mineral solution or in an inert medium (sand and gravel). In this technique, plants are grown in containers of nutrient solution, such as glass jars, plastic buckets, tubs, metal tanks, vegetable solids, and wood. The nutrients in hydroponics can be waste from fish waste and duck manure and nutrients from coir peat or rice husk, etc. Aeroponics is another method that can be used for some specific soilless culture.

1.5   FUTURE CHALLENGES

Agricultural farms are complex biological systems, and the integrated knowledge of farming components related to agritech such as human capital (skilled labor), genetics, soil science, ecology, etc. is necessary for its true potential to be realized.

1.5.1   HUMAN CAPITAL

It is the key to sustainable growth in crop production, as the smallholder farmers produce up to 80% of the world's food supply in terms of value (FAO, 2014). The technology can be more efficient and environmentally sound as compared to that used in industrial operations. The crucial need of the agriculture sector is the development of skilled labor in food production. In many areas, agricultural industry is facing shortage of skilled labor due to rural–urban migration (Luo and Escalante, 2015). The future of food production is dependent on the skill and expertise of the farmers (Puri, 2012). In order to overcome skill constraints, there is a need to educate and train the agricultural workers around the globe in farming, forestry, and fishing (ILO, 2014). Skilled labor must know the negative impacts of climate change, adaptation strategies, water conservation, salinization, organic farming, sustainable biofuel production, etc.

1.6   PRECISION AGRICULTURE

Traditional agriculture follows agronomical practices in a schedule of planting to harvesting, but if farmer has real-time data on weather, soil, air quality, crop maturity, weather patterns, soil temperature, humidity, growth factors, crop rotation, crop diversity, handling of equipment, and economy data analysis, then he can make profitable smarter decisions. This is known as site-specific precision agriculture or modern farming practices. Farmers may use global positioning systems (GPS), satellite pictures of their field, or robotic drones that can make production more efficiently. They can also use recent practices to economize input of nutrients, water, seed, and other agricultural inputs to grow more crops on their farms. Thus, farmers are able to produce more food at lesser input cost. Consequently, farmer may conserve soil for sustainable food production, and precision agriculture results in a stable economical food supply at reduced cost for strong community development.

By 2050, increase in population and food production will be met by developing countries like Brazil, which has the largest area and arable land for agriculture. With growing demands on the world's food supply chain, it is crucial to maximize agriculture resources in a sustainable manner. Farmers should also be aware of management of low-leaching cropping systems, soil biological processes, maximum recirculation, crop intensity, nutrient imbalance caused by regional farm specialization, and the development of an agricultural quality assessment system.

1.7 SUSTAINABLE AGRICULTURE AND MODERN MOLECULAR BREEDING TECHNIQUES

Modern breeding techniques provide key opportunities to make agriculture more sustainable, because crop improvements through conventional techniques are very time-consuming. In recent years, the knowledge of genes and their function at the molecular level has increased considerably; these techniques have proved to be very useful in successful breeding programs. For example, gene editing in rice breeds has enabled to switch off one particular gene via a specific targeted mutation to create resistance to certain pathogens. Now, breeders are expected to develop more resistant varieties against various abiotic stresses such as drought and heat without allergens. Plant breeders also developing crop varieties with higher yields, improved resource use efficiency, and reduced environmental impact to achieve goals of sustainability in agriculture.

1.8 FUTURE PLANT BREEDING

To extend further agricultural production and yield potentials of crops, we need a new green revolution, better to say a "Gene Revolution" (Pingali and Raney, 2005), which could boost higher crop yields. Genetic engineering techniques at the molecular level are under development to more precisely target the DNA of plants and organisms. During the last decade, several new plant breeding techniques (NPBTs) have been developed and have made it possible to accurately perform genome modifications in plants. There have been ongoing debates on whether these genetically modified organisms (GMO) are harmful to humans and biodiversity or not. GMO bio-safety legislation would severely hamper the use of NPBT. Therefore, genetically modified plants must pass through a costly and time-consuming GMO approval procedure. Crop plants modified by genetic engineering techniques must be evaluated for bio-safety issues.

1.8.1 GRAFTING

Grafting is an old technique where the stem of one plant species or variety is grafted onto the root/stem of another species or variety. This technique is adapted in horticulture and tree nursery such as roses and grape vines. Now, this technique has come again under scrutiny as part of the GMO debate, with the question that what should be the regulatory level if a non-GM scion is grafted onto a GM rootstock? Do the fruits of the non-GM scion come under the GMO regulations? The answer may be so simple that the DNA of the scion-bearing fruits is not modified; therefore, the fruits should not be regulated according to the GMO legislation. But scientifically, there is exchange of water, sugars, and other metabolites and small molecules (such as RNA molecules) derived from the GM rootstock, which will be transferred to scion from the rootstock. The GM plant that is used as a rootstock can interfere or silence the expression of one or more genes through the exchange of RNA molecules when transported to the scion, which may influence the expression of specific genes. So, even though the DNA of the scion is unchanged by the rootstock, the production of certain proteins in the scion can still be modified by the rootstock.

1.8.2 REVERSE BREEDING (RB) AND DOUBLED HAPLOID (DH)

Reverse breeding (RB) is a novel plant breeding technique designed to directly produce parental lines for any heterozygous plant. RB produces perfectly complimentary homozygous parental lines through genetically modified meiosis by eliminating meiotic crossing over in the heterozygote. In this process, meiotic crossing over is knockdown by RNA interference (Higgins et al., 2004; Siaud et al., 2004;); this results in predominantly post-transcriptional gene silencing (PTGS) of required genes in the selected hybrid plant. Gametes obtained from such heterozygote plants contain combinations of nonrecombinant parental chromosomes. Now, these non-GM (achiasmatic) gametes are selected, cultured *in vitro*, and used to generate doubled haploids (DHs). RB meets the challenge of fixation of complex heterozygous genomes by constructing complementary homozygous lines (Dirks et al., 2003). Because the genotype of unknown heterozygous could not be fixed in traditional plant breeding, RB could change the future plant breeding.

The DH lines can then be used to construct the elite heterozygote on a commercial scale.

1.8.3   MARKER-ASSISTED BREEDING

A molecular marker is a genetic tag that identifies a particular location within a plant's DNA sequences. Marker-assisted plant breeding involves the application of molecular marker techniques and statistical and bioinformatics tools to achieve plant breeding objectives in a cost-effective and time-efficient manner. Molecular markers are used to track the genetic makeup of plants during the varietal development, which provides efficiency in selection with desirable combination of specific genes where a trait of interest is linked for productivity, disease resistance, abiotic stress tolerance, quality, etc.

1.8.4   MUTATION BREEDING

Scientific knowledge of genes and their functioning has increased considerably over recent years. To apply this knowledge in a useful way, new breeding techniques have been developed to achieve very specific results in the breeding process. The switching off of genes via mutation was until recently only possible in a non-targeted way. Now, modern breeding techniques such as gene editing have been developed to allow more specific breeding. Plant breeders will be using these techniques over the coming years to develop resistant crop varieties against various abiotic stress like drought and heat.

1.8.5   HYBRID GENETICS

Various gene combinations have improved the spectacular yield in several crops, and the progress still continues. Breeders are using new molecular technologies such as marker-assisted breeding, biotechnology, genomics, transcriptomics, proteomics, phenomics, genome sequencing, etc. Better understanding of DNA alterations and function in precision can achieve better changes. Several seed development programs have promising potential for reducing water requirements and improving nitrogen utilization in established crops such as corn and cotton. Modern genetics exploit modifications to plant DNA; these modifications are results of mutations in the wild form that results into increased crop yields in the form of more food, feed, fiber, and fuels with efficient use of land, water, or fossil fuel. The present breeding techniques include jumping of DNA sequences from one part of a plant's genome to another, inducing random genetic changes, or inserting known gene sequences. In all these cases, plant genes are moved within or across species, thereby creating novel combinations.

Advances in molecular genetics have been made possible through DNA sequencing and database of genomes, transcriptomes, expression profiles, small RNAs, and epigenomes. These DNA and RNA databases can be correlated with proteomes, metabolomes, phenotypic data, QTLs, and expression level of Quantitative Traits Loci(QTLs), which can help to infer gene function. This will help in better understanding of gene function; potentially genetically engineered traits give rise increased crop yields and even to thrive them under the adverse conditions.

1.8.6 AGRO-INFILTRATION

Agro-infiltration uses soil bacteria *Agrobacterium tumefaciens* (Bt) as a vector that transfers DNA to plant cells during the development of a genetically modified plant. In this technique, *Agrobacterium* is injected into the plant tissue or leaves so that the bacteria can spread into the plant cells. Then, *Agrobacterium* introduces the gene of interest to one or more plant cells, leading to transient expression and production of the desired protein. The gene can be inserted into the DNA of some plant cells for the production of complex molecules such as antibodies.

1.9 FUTURE CROPS

Genomic understanding of plant molecular biology on the inheritance of photosynthetic activity of the chloroplast reveal how chloroplasts absorb sunlight and use its energy to split water into hydrogen and oxygen. This hydrogen then combines with carbon dioxide to form small intermediate molecules and finally into sugars; this form of photosynthesis is known as C3, because the intermediates contain three carbon atoms. Another way of photosynthesis using a four-carbon intermediate known as C4 photosynthesis, which is more efficient than C3, occurs especially in crops of tropical climates such as maize, millet, sorghum, sugarcane, etc. Rice is the second most important crop after wheat. Rice is a tropical plant that would produce yields around 50% higher than the present if it used the C4 pathway. Scientists at the International Rice Research Institute in Los Banos (Manila) are conducting experiments under the "C4 Rice Project," and the project is headed by Paul Quick and his co-researchers from Asia, Europe, North America, and Australia. They added five alien enzymes to rice for an extra biochemical pathway like C4. The team has already created rice strains that contain genes from maize plants for the extra enzymes to improve their efficacy. Other scientists are working on crops for resistance to drought, heat, cold, salt, immunity to infection, infestation, improving nutritional value, ability to fix atmospheric nitrogen, and efficient resource utilization of water and nutrients.

Other successful genetically engineered crops have been produced, such as brinjal, maize, soybeans and cotton, which contain two types of genes from bacteria (Bt). One protects its host from the insect larvae and the other for specific herbicide tolerance. Cotton is nonedible, but soybean and maize are used mainly as food and fodder. Therefore, the anti-GM lobby is raising voice against transgenic edible crops on bio-safety issues.

In 1994. the first GMO crop, the "Flavr-Savr" tomato with adequate shelf-life was transformed with the antisense gene for polygalacturonase, which inhibited cell wall softening; this allowed the tomato to delay ripening. But Flavr-Savr tomatoes were softened during shipping; so, in 1996, it was withdrawn from the market. In 1996, the first glyphosate-resistant crops soybean and maize were introduced, which were quickly adopted by many countries around the world. In 2001, Potrykus et al. published a paper on "Golden Rice" in *Science*; this plant is a source of vitamin A, the deficiency of which causes blindness in thousands of Asian children every year. In 1995, the first Bt-maize crop was introduced in the market. By 1999, 12 million hectares of GM crops such as Bt maize, potato, and cotton were grown globally, which reduced the pesticide use. Later, a number of virus-resistant crops such as potato, tomato, papaya, squash, and bean were produced. Currently, GM crops occupy a global area of 179.7 mh (Source: Clive James, 2014).

The next decade of molecular genetic development will create combinations of new desirable traits and combinations of polygenic traits such as nutritional improvement, resistance to viruses and pests, drought tolerance, and salt tolerance. In the coming years, we would have succeeded in creating nitrogen-fixing cereals, perennial cereals, and crops with increased photosynthetic efficiency like C4 plants.

Domesticated maize evolved from one of the teosintes: *Zea mays* ssp. *parviglumis*. The transformation of teosinte into maize was one of the most dramatic changes in the history of plant domestication. The major change was the fasciation of the teosinte inflorescence, giving rise to the maize cob with its numerous rows of kernels. Recent studies by Bommert et al. (2013) have identified the gene "*FASCIATED EAR2*" locus, which regulates the number of rows of kernels in maize cobs. Genes that regulate fasciation could potentially increase the yields of cereal like cob of wheat and legumes or perennial legumes in future.

1.10 INTRODUCTION OF TISSUE CULTURE TECHNIQUES FOR CLONAL PROPAGATION

The first use of chemical selection in plant breeding led to the creation of Canola oil by eliminating erucic acid from *Brassica napus* seeds; and last but not least, the discovery of the Ti-plasmid of *Agrobacterium tumefaciens* by Schell, Montegu in 1977 set the stage for the creation of an ever-expanding list of GMO crops for clonal selection with tissue culture techniques.

1.11 CONCLUDING REMARKS

To feed the 9.6 billion people living on the planet Earth, the production of high-quality food must be increased with reduced inputs, which will be quite challenging in global environmental change. Plant breeders need to focus on desired quality traits with the greatest potential to increase yield. CO_2 enrichment is likely to increase yields of most crops by approximately 13% in C3 crops but not in yields of C4 crops. In many places, increased temperature will provide opportunities to manipulate agronomy to improve crop performance. Ozone concentration increase will decrease yields by 5% or more. Plant breeders will probably be able to increase yields considerably in the CO_2-enriched environment in the future. Hence, new technologies must be developed to accelerate breeding through improved genotyping and phenotyping methods and by increasingly using genetic diversity in breeding programs. Our challenge is to increase agricultural yields while decreasing the use of fertilizers, water, fossil fuels, and other negative environmental impacts of inputs. To keep pace with both population growth and potential economic improvement, we will have to double the current rate of food production in order to feed every person by 2050.

KEYWORDS

- **carbon sequestration**
- **gene editing**
- **genome**
- **genotypic**
- **marker assisted breeding**
- **metabolites**
- **metabolomics**
- **phenotypic**

REFERENCES

Bommert, P., Nagasawa, N. S., & Jackson, D., (2013). Quantitative variation in maize kernel row number is controlled by the *FASCIATED EAR2* locus. *Nature Genetics, 45,* 334–337.

Boyle, G. (ed.), (1996) *Renewable Energy–Power for a Sustainable Future,* Open University, UK.

Cassman, K. G., Doberman, A., & Walters, D. T., (2002). Agroecosystems, nitrogen use efficiency and nitrogen management. *Ambio., 31,* 132–140.

Dirks, R., Van Dun, K., De Snoo., C. B., et al., (2009). Reverse breeding: a www.intechopen. com Haploids and doubled haploids in plant breeding 103 novel breeding approach based on engineered meiosis. *Plant Biotechnology Journal, 7*(9), 837–845, ISSN 1467–7652.

Doebley, J., Gaut, B. S., & Smith, B. D., (2006). The molecular genetics of crop domestication. *Cell, 127,* 1309–1321.

Evans, L. T., (1998). Feeding the ten billion: *Plants and Population Growth,* Cambridge University Press.

FAO, (1996). Global plan of action for the conservation and sustainable utilization of plant genetic resources for food and agriculture. *Adopted by the International Technical Conference on Plant Genetic Resources,* Leipzig, Germany, 17–23.

FAO, (2011). *The state of the world's land and water resources for food and agriculture: Managing systems at risk.* Rome and Earthscan, London.

FAO, (2013). IFAD, and WFP, The state of food insecurity in the world 2013: *The Multiple Dimensions of Food Security*, Rome, UN Food and Agriculture Organization.

Foley, J. A., et al., (2011). Solutions for a cultivated planet. *Nature, 478,* 337–342.

Food and Agriculture Organization of the United Nations, (2009). *How to Feed the World in 2050*, Rome, FAO.

Food and Agriculture Organization of the United Nations, (2014). *Building a Common Vision for Sustainable Food and Agriculture: Principles and Approaches*, Rome, Italy, FAO.

Godfray, H. C. J., Beddington, J. R., Crute, I. R., et al., (2010). Food Security: the challenge of feeding 9 billion people. *Science, 327,* 812–818, [DOI: 10. 1126/science. 1185383].

Godfray, H. C. J., Crute, I. R., Haddad, L., et al., (2010). The future of the global food system. *Philosophical Transactions of the Royal Society, B., 365,* 2769–2778.[DOI: 10. 1098/ rstb. 2010. 0180].

Goulding, K., Jarvis, S., & Whitmore, A., (2008). Optimizing nutrient management for farm systems. *Phil. Trans. R. Soc., B., 363,* 667–680.

Green, R. E., Cornell, S. J., Scharlemann, J. P. W., et al., (2005). Farming and the fate of wild nature. *Science, 307,* 550–555.

Harlan, J. R., (1992). 'Crops and man', *American Society of Agronomy – Crop Science Society, 2nd Edition*, Madison WI., USA.

Higgins, J. D., Armstrong, S. J., Franklin, F. C. H., et al., (2004). The *Arabidopsis* MutS homolog AtMSH4 functions at an early step in recombination: evidence for two classes of recombination in *Arabidopsis. Genes. Dev., 18,* 2557–2570.

ILO, (2014). Greening the economies of least developed countries: *The Role of Skills and Training. Skills for Employment Policy Brief.*

Kibblewhite, M. G., Ritz, K., & Swift, M. J., (2008). Soil health in agricultural systems. *Phil. Trans. R. Soc., B., 363,* 685–701.

Luo, T., & Escalante, C., (2015). Would more extensive out-migration of rural farmers expedite farm mechanization? *Evidence from a Changing Chinese Agriculture Sector.* (Paper submitted for the southern agricultural economics association's 2015 annual meeting, 2015, Atlanta, Georgia.

Marker, T. L., Felix, L. G., Linck, M. B., et al., (2012). Integrated hydropyrolysis and hydroconversion (IH2) for the direct production of gasoline and diesel fuels or blending components from biomass, part 1: Proof of principle testing. *Environmental Progress & Sustainable Energy, 31*(2), 191. doi: 10.1002/ep. 10629.

Pingali, P., & Raney, T., (2005). From the green revolution to the gene revolution: *How Will the Poor Fare?* (ESA, Working Paper No. 05–09. November 2005 Poorest? Food and Agriculture Organization of The United Nations, Rome, Italy. http://www.worldbank. org/content/dam/Worldbank/document/State of the poor paper April17. pdf.

Potrykus, I., (2001). Golden rice and beyond. *Plant Physiology, 125,* 1157–1161.

Pretty, J., (2008). Agricultural sustainability: concepts, principles and evidence. *Philos. Trans. R. Soc. Lond. B. Biol. Sci., 363,* 447–465.

Puri, V., (2012). Global networks, global perspectives and global talent. *Discussions on the Development of Human Capital in Agribusiness.* IFAMA 2012 Forum in Shanghai, China.

Rosset, P., (2000). Lessons from the green revolution. Oakland: *Food First*, [online]. Available from: http://www.foodfirst.org/media/opeds/2000/4-greenrev.html [Accessed 2 August 2011].

Santos, J. D., et al., (2013). Development of a vinasse nutritive solutions for hydroponics. *Journal of Environmental Management, 114,* 8–12.

Schell, J., & Van Montagu, M., (1977). The Ti-plasmid of *Agrobacterium tumefaciens*, a natural vector for the introduction of nif genes in plants? *Basic Life Sci., 9,* 159–79.

Siaud, N., Dray, E., Gy, I., et al., (2004). Brca2 is involved in meiosis in *Arabidopsis thaliana* as suggested by its interaction with Dmc1. *EMBO J., 233,* 1392–1401.

Susanne von, C., Paul, Q. W., Robert, T., & Furbank, P. N., (2016). Research school of biology, Australian national university, Canberra, ACT 0200, Australia and International Rice Research Institute, Los Banos, Philippines, and University of Sheffield, Sheffield S10 2TN, UK.

Targeted knockdown of GDCH in rice leads to a photorespiratory deficient phenotype useful as a building block for C4 rice, (2016). *Plant and Cell Physiology, 57(*5), pcw033.

CHAPTER 2

APPLICATION OF GENOTYPE AND ENVIRONMENTAL INTERACTIONS IN PLANT BREEDING

CHANDRA NATH MISHRA, SATISH KUMAR, K. VENKATESH, VIKAS GUPTA, DEVMANI BIND, and AJAY VERMA

ICAR–Indian Institute of Wheat and Barley Research (IIWBR), Karnal – 132001, India, E-mail: kumarsatish227@gmail.com

CONTENTS

2.1 INTRODUCTION

Genotype × Environment interaction (G×E) is a common phenomenon in agricultural research and plays a key role in developing strategies for crop improvement. G×E interaction (GE) is a differential genotypic expression across environments. Various statistical techniques for quantifying genotypic adaptation and for characterizing environments are focused on analyses of two-way tables of G×E data. Genotype estimates for different environments in the same G×E table may be derived using different methods of trial design and analyses. Means across environments are adequate indicators of genotypic performance in trials with nonsignificant G×E. However, for significant G×E, the genotype means often mask subsets of environments where genotypes differ markedly in relative performance. Selections from one environment may often perform poorly in another.

Phenotype (P) is the function of a genotype (G) and environment (E) as genotype is subjected to environment to express full potential. Moreover, the phenotypic value usually varies with the environmental conditions as observed for the same set of genotypes evaluated in different kinds of environment. The phenotype expression in multienvironment testing is function of P = G + E + (G×E) and total phenotypic variance explained as follows:

$$\sigma_P^2 = \sigma_P^2 + \sigma_E^2 + \sigma_{GE}^2$$

The above equation holds true for the genotypes that behave differently in different environments due to the presence of GE interaction.

2.1.1 GENOTYPES (G)

Genotype is the particular assemblage of genes possessed by the individuals. It might be noted that the genes are transferred from the parents to the offsprings, and the genotypes are constituted afresh in each generation until all the genes are fixed. In plant breeding terminology, the genotypes consist of cultivars, genetic stocks, breeding lines, etc.

2.1.2   ENVIRONMENT (E)

All the nongenetic factors that influence the phenotypic value are termed as environment, and a favorable environment is required for genotypes to express their potential yield. The factors, i.e., locations, growing seasons, years, precipitation, temperature etc. has positive or negative influence on genotypes. The environment is highly variable in nature and might vary over space and time.

The environments are classified into two broad categories:

1. Microenvironment: soil type, solar radiation, disease and pests, and weather fluctuations represent the microenvironment.
2. Macroenvironment: this environment comprises geographical area latitude, longitude, temperature humidity, rainfall, etc., which collectively determine the different climatic zones like temperate, tropical, subtropical, etc.

Variations in the environment are divided into two classes as follows: predictable and unpredictable. The predictable variation includes all permanent characteristics of the environment, such as general features of the climate and soil type, as well as those characteristics of the environment that fluctuate in a systematic manner, such as day length. It also includes those aspects of environment that are determined by man and can therefore be fixed more or less at will, such as planting date, sowing density, methods of harvest, and other agronomic practices.

The unpredictable category includes fluctuations in weather, such as amount and distribution of rainfall and temperature, and other factors such as established density of the crop. In an inefficient agricultural system, it may also include variations in agronomic practice, which in more advanced agricultures might be maintained reasonably constant.

2.2   GENOTYPE × ENVIRONMENT INTERACTION

The phenotypic expression of genotype is rarely constant across the environments; some genotypes do well in some environments, while show poor performance in other environments, which clearly indicate the existence of GE interactions. Interaction may be due to heterogeneity of genotypic variance across environments and imperfect correlation of genotypic performance across environments.

2.2.1 CLASSIFICATION OF G×E INTERACTIONS

G×E interactions can be broadly classified into three categories: no G×E interaction, noncrossover interactions, and crossover interactions (Figure 2.1).

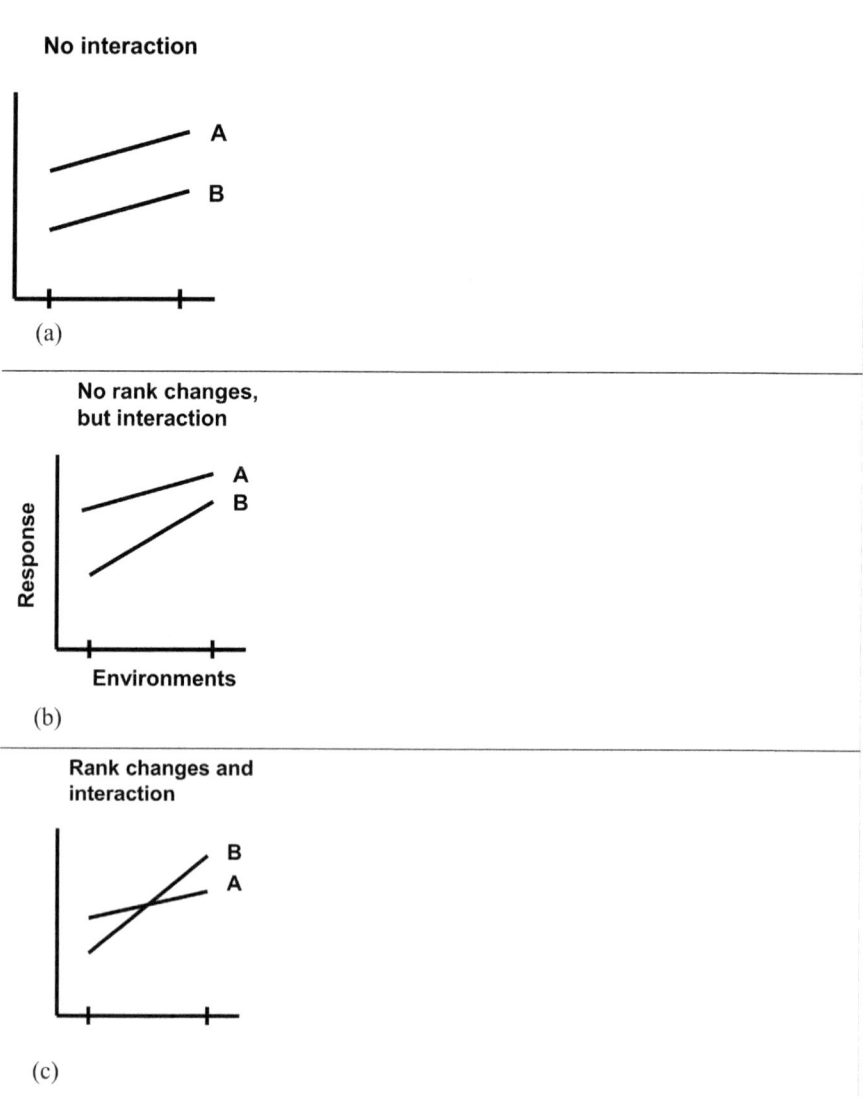

FIGURE 2.1 (a) The performance of two genotypes in two environments showing no
GE interactions. (b) The performance of two genotypes in two environments showing
quantitative GE interactions. (c) The performance of two genotypes in two environments
showing qualitative GE interactions.

2.2.2   NO G×E INTERACTIONS

Consistent performance of one genotype over the other genotype by the same amount in the tested environments indicates the absence of interaction. The two genotypes form parallel lines as the inter-genotypic variance remains unchanged in various environments.

2.2.3   NONCROSSOVER (QUANTITATIVE) G×E INTERACTIONS

A noncrossover or quantitative GE interactions occurs when one genotype performs better than the other genotype, but the differential performance is not the same in all the environments. These are also known as noncrossover type of interactions. The ranking of genotypes does not change in such interactions.

2.2.4   CROSSOVER (QUALITATIVE) G×E INTERACTIONS

The differential and unstable response of genotypes to different environment is referred to as a crossover interaction, and the ranking of genotypes changes from one environment to other environment. Crossover interactions or qualitative interaction shows that no genotype is best across all the environments and genotypes are specifically adapted to an environment.

2.3   ADAPTATION OF GENOTYPES ON THE BASIS OF G×E INTERACTIONS

2.3.1   BROAD ADAPTATION OR GENERALLY ADAPTED GENOTYPES

Some genotypes are better adapted in all the environments as compared to other genotypes, implying that they show consistent performance under both favorable and unfavorable environments. The main breeding objective is to develop a variety that performs consistently well across a range of environments (high mean across environments). The broad adaptation is equivalent to selection for multiple traits that may reduce the rate of progress from selection. The approach will not necessarily identify the best genotype for a specific environment.

2.3.1 SPECIFIC ADAPTATION OR SPECIFICALLY ADAPTED GENOTYPES

Some genotypes may be specifically adapted to specific environment, and they perform well either in rich environment or poor environment. In this approach, the environments are subdivided into groups so that there is little GE interaction within each group. The breeding objective is to breed varieties that perform consistently well in each environment. In this approach, either there are multiple breeding programs for each environment or a common set of material is evaluated across the environment, and specific recommendations are made for each environment.

2.4 ASSESSMENT OF STABILITY OF GENOTYPES

Genotype performance changes due to environmental pressures or stresses (due to population heterogeneity or population buffering and changes in the genetic makeup taking place over generations) and differences in their ability to adapt to stress factors (short-term acclimatization). A number of statistical models to study genotypic adaptation based on phenotypic performance have been discussed in literature. Three concepts, namely stability, adaptability, and predictability, are discussed as follows for completeness.

Phenotypic stability refers to the ability of a genotype to maintain a near constant phenotype for the characteristic of interest over variable environments. Such a genotype would be regarded as having wide adaptation. But certain genotypes may also show predictably superior performance in particular types of environments, indicating that broad adaptation inevitably involves sacrifice of performance in specific environments. The strategies of crop improvements for broad adaptation (minimizing G×E interaction) and specific adaptation (emphasizing favorable interaction) are in direct conflict.

Predictability refers to the extent to which the response is systematic. Responsiveness is the ability of a genotype to respond in a particular manner to a general change in the environmental potential.

Sensitivity (also stability) refers to the extent of unpredictable variation in response. Some researchers relate stability to variability of performance over time (temporal variation) at a location, while adaptability refers to variability in performance across locations (spatial variation).

Stability of genotypes to environmental fluctuations is important for the stabilization of crop production over the regions and years. Estimation of phenotypic stability has proved to be a valuable technique in the assessment of varietal stability. Stability analysis is useful in the identification of adoptable genotypes and in predicting the response of various genotypes over changing environments. There are different stability analysis models.

2.5   MODELS FOR STABILITY ANALYSIS

2.5.1   CONVENTIONAL MODELS

A number of researchers have made significant contributions in the estimation of stability of genotypes for yield over environments and some of the widely used models are described here:

1. Stability factor model (1954): Lewis (1954) introduced the term stability factor for measuring the phenotypic stability, which is estimated as follows:

SF = Mean of genotype under high environment/Mean of that genotype under low environment

    The stable genotypes should have the similar performance in both high- and low-yielding environments, i.e., SF=1. The greater deviation from the stability factor from unity indicates the lesser stability of the genotypes.

2. Ecovalence model (Wi^2) (1962): Wricke's ecovalence model evaluates the stability of genotypes on the basis of contribution of each genotype to the total G×E sum of squares. The genotypes with low value Wi^2 values are considered as stable as these have smaller deviations from mean across the environments.

$$W_i^2 = \sum_{j=1}^{q} \left(y_{ij} - \dot{y}_{i.} - \dot{y}_{.j} + \dot{y}_{..} \right)^2$$

This G×E interaction effects for genotype *i*, squared and summed across all environments, is the stability measure for genotype *i*.

3. Stability variance model – Shukla (1972): The stability statistic of Shukla (1972) is a measure of the share of each particular variety in the G×E interaction.

$$\sigma_i^2 \left(p\Sigma\left(y_{ij} - \dot{y}_{i.} - \dot{y}_{.j} \dot{y}..\right)^2 - \Sigma\Sigma\left(y_{ij} - \dot{y}_{i.} - \dot{y}_{.j} + y..\right)^2 / (p-1)/(p-2)/(q-1)\right)$$

Based on residuals in a two-way classification, the variance of a genotype across environments is the stability measure.

4. Lin and Binn's cultivar performance measure (Pi), (Linn and Binn, 1988): *Pi* is obtained by the expression $P_i = \Sigma (X_{ij} - M_j)^2/2n$, where *Pi*= superiority index of the ith cultivar, X_{ij} = yield of the ith cultivar in the jth environment, M_j = maximum response obtained among all the cultivars in the jth environment, and n = number of environments. (Pi) is estimated by the square of differences between a genotype's and the maximum genotype mean at location, summed and divided by twice the number of locations. The genotypes with the lowest values are considered the most stable.

2.5.2   REGRESSION COEFFICIENT MODELS

Linear regression provides a conceptual model for genotypic stability. The slope, β_1, of the regression of an individual genotype's values against an environmental index estimate genotypic stability. The index is usually estimated as the mean of all genotypes in a trial. The G×E term may be partitioned between the heterogeneity of regressions (differences in slopes, β) and deviations from linear regressions.

2.5.2.1   Finley and Wilkinson Model (1963)

The observed values are regressed on environmental indices defined as the difference between the marginal mean of the environments and the overall mean. The regression coefficient for each genotype is then taken as its stability parameter.

$$b_i \sum_j \left(y_{ij} - \dot{y}\right)\left(\dot{y}_{.j} - \dot{y}..\right) / \Sigma\left(\dot{y}_{.j} - \dot{y}..\right)^2$$

2.5.2.2   Eberhart and Russell Model (1966)

Eberhart and Russell deviation parameter is slight improvement over Finlay and Wilkinson model. This model measures the stability of genotypes in terms of three parameters, namely.

- Genotypic Mean
- Regression or linear response
- Deviation from linearity

$$Y_{ij} = \mu + g_i + b_i t_j + \delta_{ij} + e_{ij}$$

gi: effect of ith genotype; bi: regression of ith genotype; tj: environmental index; Delta ij: deviation from regression

$$\delta_i^2 = \sum_j \left(y_{ij} \dot{y}_{i.} - \dot{y}_{.j} + y.. \right)^2 - \beta_i^2 \sum \left(\dot{y}_{.j} - \dot{y}.. \right)^2 / (q-2)$$

2.6   MULTIVARIATE EXPLORATION OF RELATIONSHIP AMONG ENVIRONMENT AND GENOTYPES

The relationship among environments is also critically important in plant breeding, especially in determining domains for selection and recommendation of genotypes. Multivariate techniques can describe relationships among environments and among genotypes. A genotype performance can be described in a multidimensional space, with each dimension representing a test environment the coordinates for which are the measured yield (one other traits). Conversely, environments can be considered in multidimensional space, with each dimension a genotype yield data from genotypes × location tables commonly generated by breeding programs. The pattern analysis describes the parallel use of classification and ordination techniques to represent the essential variation from GE tables in a few dimensions. Classification techniques such as cluster analysis presume discontinuities within the data, while principal components analysis (PCA) and other methods of ordination assume samples (genotypes and environments). An increasing trend has been noted for using multivariate stability procedures by breeders owing to good returns relative to stability parameters in agricultural trials.

Ordination techniques are used to simplify multivariate data for a set of individuals by summarizing relationships among individuals or among attributes describing them. This is done by producing a simple visual representation of the individuals as points that can be plotted to portray their relationships acceptably free of distortion. The ordination techniques try to reduce the dimensionality of the multivariate systems efficiently to preserve the relationships among individuals as far as possible, and to provide a simplified view of those relationships in fewer dimensions than specified by original variables. There are two methods of ordination.

2.7 PRINCIPAL COMPONENT ANALYSIS (PCA)

PCA considers finding a new set of coordinate axes that account more effectively for the variation among individuals than do those based on original variables. PCA represent a transformation of data from one set of coordinates to another. This may not necessarily lead to reduction of dimensionality. However, when only (first) few principal components account for most of the variation, then it becomes effectively useful. Algebraically, the principal axes are determined by the latent vectors from the matrix of corrected sums of squares and products among variables. Elements of each vector specify the linear combinations of original variables necessary to give the corresponding PC, and the associated latent root give the variation attributable to the component. PCA can also be applied on environment in the same way as it could be done to genotypes.

2.8 PRINCIPAL COORDINATE ANALYSIS OF GENOTYPES (PCO)

PCO analysis requires finding a set of rectangular coordinate axes that which account as efficiently as possible for variation among individuals and may lead subsequently to a reduction in dimensionality for simplification. These objectives are similar to PCA, but PCO is based on a much more general approach. It does not automatically assume that original variables define a multidimensional Euclidian space, in which relationships between pairs of individuals are indicated by Euclidian distance. Many similarity measures (e.g., correlation coefficients) or dissimilarity measures (distance) could be used. PCO involves two steps for computation:

1. Presentation of the set of individuals as points in a coordinate space derived from the original matrix of measures. Gower (1966) showed that the interpoint Euclidian distances in this space are a simple function of the original measures of relationship between individuals. The significance of this method is that it refers individuals to Euclidian coordinate axes even when an initial coordinate framework is unavailable, and it represents original measures of relationships as Euclidian distances even if they are non-Euclidian.
2. Carrying out a PCA on the data derived in the first step.

The only requirement to guarantee a distortion-free representation by PCO is that the original matrix of measures must be symmetric (so that no negative latent roots are obtained). In general, principal axes will not be a linear combination of original variables as in usual PCA. However, it is possible to investigate the relationship of original variables to each principal axis by correlating the set of principal coordinate scores for each axis with each of the original variables. A correlation of large magnitude for a particular variable implies that it is strongly reflected in the axis concerned.

2.9   CLUSTER ANALYSIS

The analysis is based on differences in genotypes' responses across test environments and commonly used multivariate methods. The two major types of the multivariate methods are classification and ordination techniques. Cluster analysis involves grouping similar entities in clusters and is effective for summarizing redundancy in data. Identifying those genotypes with similar responses to environmental changes but different from genotypes in other groups can be profitable.

Several cluster methods have been suggested, some of which classify individuals for similarity according to the one-way method and others classify individuals for similarity of interactions based on the two-way method. There are two major procedures for grouping genotypes according to genotype response to environmental changes in which genotype is a vector of n attributes indicated by m environments using the distance coefficient. Mungomery et al. (1974) used squared distance as a similarity index for clustering. Lin and Thompson (1975) used the deviation MS from the linear regression model of the G×E interaction as a dissimilarity index for clustering. Lin (1982) used the G×E interaction mean square as a dissimilarity index for genotype classification through a slight adjustment of the distance coefficient of the Abou-El-Fittouh procedure. An important aspect of cluster analysis is having a well-defined stopping criterion or cutoff point. A cut-off point can be determined if the dissimilarity index has some relationship with the deviation mean square from a regression model or the G×E inter-action MS in ANOVA. For dissimilarity indices, some F-tests for stopping the clustering procedure are very well defined. There are various other ways of scaling and standardizing data, including environment-centered, environ-ment-standardized, environment heritability-weighted, and environment-ranked methods, as well as these above-mentioned clustering procedures. The basic similarity of all clustering methods is that they use some similarity or distance measurements to classify items into groups.

2.10   FACTOR ANALYSIS

This analysis is used to explain the variability in terms of a lower number of artificial variables or factors. It explores for linked variations in relation to artificial factors, and the variables are modeled as linear combinations of potential factors as well as error term. The FA is an ordination method related to PCA, the factors of the former being similar to the PCA of the latter. A large number of related variables are reduced to a small number of factors, and variation is described in terms of these factors. These general factors are common to all studied variables, and in terms of factors, they are unique to each variable. The FA has been used to grasp interrelationships between dif-ferent yield components of crops as well morphological properties of plants. Factors are conceptualized as real entities such as yield, but the components of PCA are abstractions that may not map easily onto real phenomena. PCA analyses total variance, but FA shares variances that are analyzed. Fritsche-Neto et al. (2010) applied FA to GE interaction stratification in maize and reported that stratification of the test environment by FA was more selective in joining similarities according to a genotype's yield performance.

2.11   THE ADDITIVE MAIN EFFECT AND MULTIPLICATIVE INTERACTION

The essential statistical background for the additive main effect and multiplica-tive interaction (AMMI) models was developed by Williams (1952) after the invention of PCA and ANOVA procedures. The AMMI model consists of fitting an additive model (ANOVA) for producing general means, genotypes' means, and environments' means, and then fitting a multiplicative model (PCA) for the residual of an additive model or a GE interaction. The AMMI model became widely used in different scientific fields. The AMMI model is usually referred to as biplot analysis, even though this term was actually intended to refer to a graph or plot containing two kinds of points. This model is an appropriate choice when both main effects and GE interaction are important.

The AMMI model is an effective tool for (i) understanding GE interaction, (ii) identifying mega-environment patterns, (iii) improving the accuracy of yield estimates, (iv) imputing missing data, and (v) increasing the flexibility of experimental designs.

The AMMI model for the main and G×E interaction effects is defined as follows:

$$Y_{ij} = \frac{1}{4} + gi. + e.j + \sum_{k=1}^{n} \lambda_k \gamma_{ik} \delta_{jk} + \acute{A}_{ij}$$

where Y_{ij} is the yield of the ith genotype in the jth environment; μ is the grand mean; $gi.$ and $e.j$ are the genotype and environment deviations from the grand mean, respectively; λ_k is the eigen value of the principal component analysis axis k; γ_{ik} and δ_{jk} are the genotype and environment principal component scores (eigenvectors) for axis k; n is the number of principal components retained in the model and ρ_{ij} is the error term.

The AMMI model increases the probability of successfully selecting genotypes with the highest yields. The results of the AMMI model can be used to construct a biplot with a point for each genotype and for each environment, located in a graph indicating the main effects on the abscissa and the G×E interaction scores on the ordinate. Such a graph as AMMI-1 biplot indicates, at a glance, both the main effects and the GE interaction effects for both genotypes and environments. Another useful biplot as AMMI-2 biplot indicates interaction PCA1 scores on the abscissa and interaction PCA2 scores on the ordinate. Biplots can readily provide deep insights into a large, complex experiment. Mega-environment analysis is included for the AMMI-1 model through biplots. One of the main objectives in the evaluation of multienvironment trials is to identify superior genotypes for a target area and to determine if this area can be subdivided into different mega-environments to better guide breeding strategies. The AMMI-2 biplot is an efficient means for detecting the possible mega-environments in multienvironment trials. Mega-environments are identified in the investigation of the annually repeatable GE interaction (Figure 2.2).

Purchase et al. (2000) developed the AMMI stability value (ASV) based on the AMMI model's IPC1 and IPC2 scores for each genotype as follows. The lowest ASV value associated with stable performance of genotypes.

$$\text{AMMI Stability Value (ASV)} = \sqrt{\left[\tfrac{SSIPCA1}{SSIPCA2} \times IPCA1score\right]^2 + IPCA2score^2}$$

where SSIPCA1 and SSIPCA2 are sum of squares by the IPCA1 and IPCA2, respectively

The AMMI distance statistic coefficient (D) (Zang et al., 1998) was calculated where D is the distance of the interaction principal component (IPC) point from the origin in space, n is the number of significant IPCs, and γ_{is} is the score of genotype i in IPC. The genotype with the lowest value of D statistic considered as the most stable.

$$\text{AMMI Distance } (D_i) = \sqrt{\sum_{i=1}^{n} \gamma_{is}^2} \ (i = 1,2,3,.. \text{n})$$

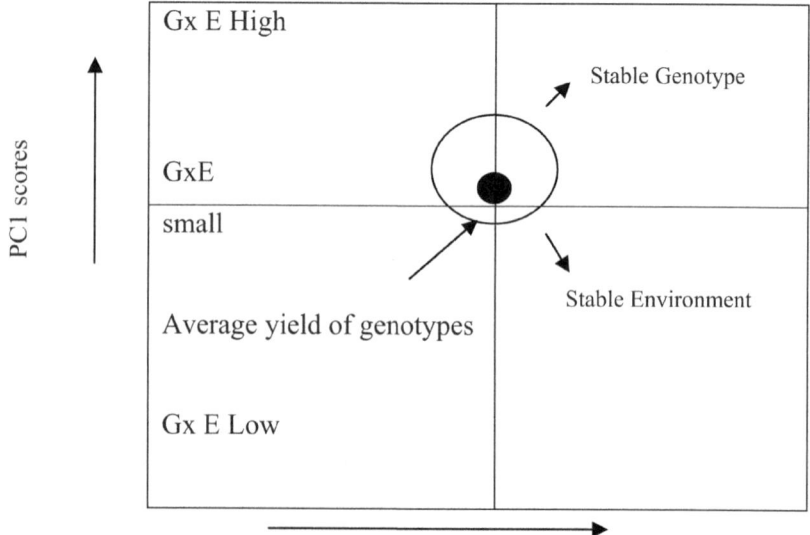

FIGURE 2.2 GE interaction.

Yield stability index (YSI) considered the rank of yield of genotypes across environments and rank of AMMI stability value. This index incorporates mean and stability index in a single criterion and is calculated as (Farshadfar et al., 2011):

$$YSI = RASV + RY,$$

where RASV is the rank of AMMI stability value and RY is the rank of mean yield of genotypes (RY) across environments. Low values show desirable genotypes with high mean yield and stability.

2.12 SHIFTED MULTIPLICATIVE MODEL

The shifted multiplicative model (SHMM) grouped genotypes into classes within which crossover interactions do not exist, and within such groups, the genotype with the best mean would be the best. Multiplicative models for multienvironment trials have been used for studying GE interactions and for developing methods for grouping test environments and genotypes into groups with negligible crossover interaction. These models have an additive component (such as interception of linear regression, main effects of environments and genotypes) and a multiplicative component (G×E interaction effect). The singular vectors on effects for genotypes and test environments for the ordered components are primary, secondary, and so forth.

In SHMM models, differences among genotypes in a special test environment are proportional to genotype differences in any other environment, but differences among environments with respect to the performance of a special genotype are proportional to environmental differences with respect to performance of any other genotype. When an SHMM model is fitted to the dataset of multienvironment trials, secondary and perhaps even higher-order effects must be included if a sufficient fit is to be achieved. In clustering via the SHMM model, the measurement of distance between two test environments is taken as the residual mean square after fitting SHMM1 to the data from the two test environments subject to an additive constraint.

2.13 SITE REGRESSION BIPLOT

Environment is the predominant source of yield variation in most multienvironment trials and genotype, and G×E interaction contributions are relatively small.

The genotypes main effect and G×E interaction effects are relevant to genotype evaluation in comparison to large environmental effect. It is essential to remove the environment effect from data and to focus on the other variation sources (G+GE). The biplot method was expanded by Kempton (1984) and Zobel et al. (1988), highlighting the usefulness of G+GE biplots. This method has attained much attention of plant breeders and agronomists. The GGE model deals with analysis of multienvironment trials' data and identifies (i) mega-environment for understanding the target environment, (ii) genotype evaluation for each mega-environment, and (iii) understanding the causes of GE interaction.

The GGE model describes what is called genotype main effect in terms of GE interaction by defining a constant value for a genotype across test environments. The genotypic PCA-1 score of GGE model indicates a tendency of the genotypes to respond to environmental factors represented by the environmental PCA-1 scores. The yield of genotype relative to PCA-1 of the GGE model is not the same in all environments; rather, it is proportional to the location of PCA-1 scores. Thus, the GGE model emphasizes the fact that the genotype main effect not only has a genotypic basis but is also dependent on environmental conditions. The PCA-1 scores not only detect genotypes with better overall performance but also suggest environmental conditions that facilitate identification of these genotypes.

The basic model for a GGE biplot based on singular value decomposition (SVD) of the first two principal components as follows:

$$Yij - Yj = \lambda_1 \xi_{i1} \eta_{j1} + \lambda \xi_{i2} \eta_{j2} + \varepsilon_{ij}$$

where Yij is the measured mean of genotype i in environment j; Yj is the mean across all genotypes in environment j; λ_1 and λ_2 are the singular values for PC1 and PC2; ξ_{i1} and ξ_{i1} are the PC1 and PC2 scores, for genotype I; η_{j1} and η_{j2} are the PC1 and PC2 scores, for environment j; ε_{ij} is the residual of the model associated with the genotype i in environment j.

The first is the polygon ("which-won-where" mean when we are testing number of genotypes in number of environment which genotype is performing better in which environment, this term is commonly used) view of the GGE biplot consists of an irregular polygon and a set of lines drawn from the biplot origin to intersect the sides of the polygon at right angles. The vertices of the polygon are the genotype markers located away from the biplot origin in various directions, in such a manner to accommodate all genotype markers. A line that starts from the biplot origin and perpendicularly intersects a polygon side represents the set of environments in which the two cultivars defining that side perform equally; the relative ranking of the two cultivars would be reversed in environments on opposite sides of the line (the so-called "crossover GE"). The perpendicular lines to the polygon sides divide the biplot into sectors, each having its own winning cultivar. The winning cultivar for a sector is the vertex cultivar at the intersection of the two polygon sides. It is positioned usually, within its winning sector. The environment markers fall into different sectors, indicating different cultivars won in different sectors contrary to the situation wherein all environment markers fall into a single sector; this indicates that a single cultivar would have the maximum yield in all environments.

Next is the average tester coordination view; this method ranks entries along the average-tester axis (ATC abscissa), with an arrow pointing to the entry with greater average performance across all testers. A vertical line (also called average-tester coordinate [ATC] y axis) divides the abscissa into two halves in order to separate entries with below and above average means. The average performance of an entry is approximated by the projection of its entry's marker on the abscissa while stability is measured by its projection onto the double-arrow line. The less stable entry shows largest projection length.

Finally, the third view displays the discriminating ability and representativeness of the tester. The discriminating power of a tester, which refers to its ability to differentiate among entries being tested, is displayed by the length of the vector, which proportionate the standard deviation of the tester. Testers with shorter vectors provide little or no information about the entries evaluated as compared to those with longer vectors. The small circle located on the abscissa represents the average-tester axis (AEA). The cosine of the angle between two testers approximates the correlation coefficient. Testers that have small angles with AEA are more representative than those that have large angles with AEA.

KEYWORDS

- **developing strategies**
- **environmental (E)**
- **factor analysis**
- **genotype(G)**
- **interactions**
- **phenotype (P)**
- **stability analysis**

REFERENCES

Eberhart, S. A., & Russell, W. A., (1966). Stability parameters for comparing varieties. *Crop Sci.*, *6*, 36–40.

Farshadfar, E., Mahmodi, N., & Yaghotipoor, A., (2011). AMMI stability value and simultaneous estimation of yield and yield stability in bread wheat (*Triticum aestivum* L.). *Australasian Journal of Crop Science, 5*, 1837–1844.

Finlay, K. W., & Wilkinson, G. N., (1963). The analysis of adaptation in a plant-breeding programme. *Aust. J. Agric. Res.*, *14*, 742–754.

Fritsche-Neto, R., Miranda, G. V., DeLima, R. O., De Souza, L. V., & Da Silva, J., (2010). Inheritance of traits associated with phosphorus utilization efficiency in maize. *Pesq. Agropec. Bras.*, *45*(5), 465–471.

Gower, J. C., (1966). Some distance properties of latent root and vector methods used in multivariate analysis. *Biometrika.*, *53*, 325–338.

Kempton, R. A., (1984). The use of biplots in interpreting variety by environment interactions. *J. Agric. Sci.*, *103*, 123–135.

Lewis, L. B., (1954). Gene-environment interaction. *Heredity, 8*, 333–356.

Lin, C. S., & Binns, M. R., (1988). A superiority measure of cultivar performance for cultivar x location data. *Canad. J. Plant Sci.*, *68*, 193–198.

Lin, C. S., (1982). Grouping genotypes by a cluster method directly related to genotype–environment interaction mean square. *Theor. Appl. Genet.*, *62*, 277–280.

Lin, Chuang-Sheng, & Thompson, B., (1975). An empirical method of grouping genotypes based on a linear function of the genotype-environment interaction. *Heredity, 34*, 255–263.

Mungomery, V. E., Shorter, R., & Byth, D. E., (1974). Genotype × environment interactions and environmental adaptation. I. Pattern analysis – Application to soybean populations. *Aust. J. Agric. Res.*, *25*, 59–72.

Purchase, J. L., Hatting, H., & Vandeventer, C. S., (2000). Genotype × environment interaction of winter wheat (*Triticum aestivum* L.) in South Africa: II. Stability analysis of yield performance. *South African Journal of Plant Soil*, *17*, 101–107.

Shukla, G. K., (1972). Some statistical aspects of partitioning genotype-environmental components of variability. *Heredity, 29*, 237–245.

Williams, E. J., (1952). The Interpretation of interactions in factorial experiments. *Biometrika.*, *39*, 65–81.

Wricke, G., (1962). About a method for the detection of Okogische Streubretein field tests. *Plant Breeding, 47*, 92–96.

Zhang, et al., (1998). Mechanism of phospholipids binding by the C2A-domain of synaptotagmin *I Biochemistry*, *37*, 12395–12403.

Zobel, R. W., Wright, M. J., & Gauch, H. G., (1988). Statistical analysis of a yield trial. *Agron. J.*, *80*, 388–393.

CHAPTER 3

FORWARD AND REVERSE GENETICS IN PLANT BREEDING

MASOCHON RAINGAM, MEGHA BHATT, AKASH SINHA, and PUSHPA LOHANI

Department of Molecular Biology and Genetic Engineering, College of Basic Science and Humanities, GB Pant University of Agriculture and Technology, Pantnagar, India, E-mail: pushpa.lohani@yahoo.com

CONTENTS

3.1   INTRODUCTION

The world of genetics began with the simple experiments of the Austrian monk Gregor J. Mendel on garden pea plants, which suggested the existence of genes. Later, the discovery of structure of double-stranded DNA, DNA sequencing technique, and PCR technique revolutionized the world of genetics. Technological advances in sequencing have greatly accelerated the accumulation of genetic sequence data to the point that now whole genome sequences are publicly available for a large number of organisms, including many crop plants. The tool and techniques for carrying out research in genetics are expanding at a very fast rate. Genome sequencing, in combination with various computational and empirical approaches to sequence annotation, has made it possible to identify genes in various plants. Increasingly sophisticated genetic tools are being developed to understand how the coordinated activities of these genes give rise to a complex organism. To understand the activity of the genes, two approaches can be followed; forward and reverse genetics. These two approaches together have made molecular analysis of both known and previously unknown genes controlling numerous traits in plants possible. One of the most important ways to identify the function of each and every gene in the genome is by the study of a mutant in which a particular gene has been inactivated. The characterization of the function of a protein domain in one organism often provides a hint to its function in another organism. It is therefore necessary to identify as many genes as possible in major model organisms like *Arabidopsis thaliana* and characterize their function (Alonso et al., 2000). Mutants can be generated by classical "forward" genetic methods and by more modern "reverse" genetic approaches. In other words, there are basically two ways to link the sequence and function of a specific gene, i.e., by studying forward genetics and reverse genetics.

Forward genetics refers to the identification and characterization of the gene that is responsible for the mutant phenotype. It starts with the identification of interesting mutants with the aim to discover the function of genes defective in mutants. This technique is based on "phenotype to gene approach." In forward genetics, the researcher selects a mutant plant with a modified phenotype and finds out the gene responsible for that phenotype. Mutagenesis can be used to create random mutations followed by searching for the genotypes that underlie the resulting phenotypes. A plant population is mutagenized by ethyl methane sulfonate (EMS), other mutagens, radiation, or T-DNA insertion, and the mutants are screened for a particular phenotype by using mutant screening. The selected mutants are analyzed by PCR for the identification of changes in the gene that result in an altered phenotype. The advantage of this technique is that it works very well in the phenotype that can be easily distinguished, for e.g., color of the flower or grain, shape of the plant, fruit size, etc. It also works in an unbiased manner because it is random in nature (Alonso et al., 2006).

The term reverse genetics is used because the scientists already know the sequence of the gene and want to find its function by working backward in the direction in which geneticists worked before the era of genome sequencing. Reverse genetics is based on "gene to phenotype" approach (Bouche et al., 2001). It starts with a known gene sequence. The function of the gene is altered to determine the role of the gene based on its effects produced in the plants. The aim of reverse genetics is to examine the effect of induced mutation or altered expression of a particular gene sequence and understand the function of the gene. This strategy works on the single gene for which the function is to be determined. A gene sequence is selected for which the function or the phenotype it creates is not known. The gene is targeted for mutagenesis and tested for its *in vivo* function. This results in a phenotype that can be easily assigned to the gene under study (Brown et al., 2005).

Forward genetics is more time consuming than reverse genetics because it involves the process called "genetic mapping." Forward genetics permits the discovery of gene function by identifying the mutation that causes variant phenotype. This process is tedious and time-consuming. New sequencing technologies have enabled rapid and cost effective whole genome sequencing which in turn has dramatically accelerated the progress in forward genetics. This method has allowed rapid identification of mutant alleles in model organisms. In the near future, mapping by sequencing is expected to become a center piece in efforts to discover the genes responsible for quantitative traits like increased disease resistance, higher yields, tolerance to abiotic stress like drought and salinity, improved nutritional quality, etc. (Schneeberger and Weigel, 2011). In forward genetics, mutations are generated either naturally or by mutagens, followed by the selection of phenotype. Finally, the sequence of the gene is identified, and its function is elucidated. However, in reverse genetics, the sequence of the gene is known, but the function of the gene is unknown. Mutation is induced or the expression of the gene is altered to understand its function (Zakhrabekovaa et al., 2013) (Figure 3.1). Although the reverse genetic approach can yield deep understanding of the function of individual genes, it is limited by hypotheses about the phenotypic outcome of the targeted genetic alteration. The forward genetic approach begins with a particular biological phenomenon or characteristic to find out the genes responsible for a particular trait. Mutant lines are created that show variant phenotypes. The mutational cause can be determined by mapping and positional cloning. Differences in the expression of phenotypes that result from mutations in different genes may indicate the importance of some genes in the process of expression of a phenotype (Lukowitz et al., 2000; Janny et al., 2003). Forward genetics does not rely on any previous knowledge. Because it is an unbiased approach, there is ample possibility to find new regu-

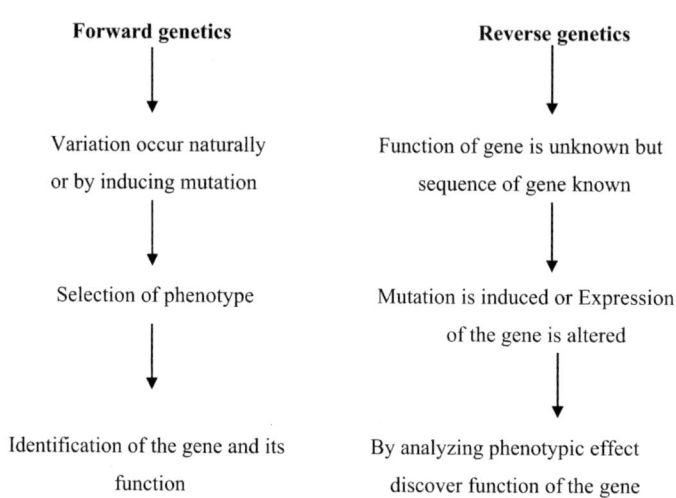

FIGURE 3.1   Comparison of strategies used in the forward and reverse genetics approach.

lators for certain physiological processes in the plant cell and to identify a large number of genes involved in the same process.

Reverse genetics is an important complement as compared to forward genetics, because in reverse genetics, the researcher can investigate function of all genes in the gene family, which is not easily done with forward genetics. A vast majority of genes have not yet been mutated in most of the plants, and their study can be made possible by using the reverse genetics approach. The availability of the complete genome sequence is a boon in the reverse genetics approach and has allowed to study every single gene. The reverse genetic approach can yield deep understanding of the function of individual genes but is limited by hypotheses about the phenotypic outcome of the targeted genetic alteration (Nordborg and Weigel, 2008; Hardy et al., 2010).

3.2   TOOLS FOR STUDYING FORWARD GENETICS

The aim of forward genetics is to determine the genetic basis of observed phenotypic variation. Various approaches are exploited to generate random mutations in an organism, for example X-rays, ultraviolet irradiation, and chemical treatment. These gene disruptions are followed by selection of aberrant phenotypes associated with various traits used in plant breeding, such as high yield, early maturity, lodging resistance, disease resistance, drought tolerance, cold tolerance, toxic metal resistance, etc. After mutants are identified, they need to be classified. The aim is to collect mutants into complementation groups by using allelism tests. Such groups of multiple independent mutant alleles can efficiently be used to validate a candidate gene. There are many tools to study forward genetics, for example, genome-wide mutagenesis, genetic screening, mutation or genetic mapping, expression analyses with microarray, candidate gene approach, etc.

3.2.1   GENERATION OF MUTANTS

Mutant phenotypes provide clues about genetically controlled functions, and co-inherited genetic markers indicate the region of the genome containing the responsible gene. This information is used for isolating and cloning the gene. The cloned gene is sequenced, and its base sequence is determined. In many cases, the observable variations are induced using a DNA damaging agent or a mutagen, but the variation can also occur naturally. Various methods of mutagenesis have been suggested. These include irradiation, chemical mutagenesis, insertional mutagenesis, etc. (Kutscher et al., 2014).

3.2.1.1   Chemical Mutagenesis

EMS, TMP/UV (trimethyl psoralen with ultraviolet light), ENU (N-ethyl-N-nitrosourea), formaldehyde, NTG (nitrosoguanidine), DES (diethyl sulfate), acetaldehyde, DEO (diepoxyoctane), and DEB (diepoxybutane) have been shown to cause mutations. Chemical mutagenesis is one of the easiest and most straight-forward methods used to create germ line and somatic mutations with high frequency. For example, the chemical EMS causes point mutations, which are changes at a single nucleotide position. Mutations may be nonsense or mis-sense. They may also be in the non-coding sequence, thereby affecting splicing signals, or in regulatory elements that control gene expression (Denver, 2009). This approach allows for many different mutations within the gene regions, but they are very difficult to map.

3.2.1.2 Radiation Mutagenesis

X-rays, UV rays, and γ-radiations can cause break in DNA or may cause formation of dimmers in DNA. X-rays and γ-rays cause breaks in double-stranded DNA, resulting in large deletions of pieces of chromosome or chromosomal re-arrangements. These mutations are typically easy to map by cytological examination of chromosomes, but they may not be limited to a single gene. It is also not good for fine-scale mutagenesis. X-rays typically induce large-scale genomic alterations, including translocations, inversions, and deletions and may be better suited for the analysis of regulatory regions.

3.2.1.3 Transposon Insertional Mutagenesis

Transposable elements (TEs) containing a marker gene(s) are mobilized in the genome. The TE can be insert within a coding region and disrupt the amino acid sequence, or they may be inserted into neighboring non-coding DNA and affect intron splicing or gene expression. The major advantage is that the TE insertion can easily be mapped and the region of the affected genome can be cloned.

3.2.1.4 Genome-wide Mutagenesis

Genome-wide mutagenesis is commonly used to create nondirected mutation for forward genetic screening. Mutagenesis can be accomplished with a variety of chemical agents, each offering a specific array of favored lesions and mutation frequencies. Ionizing radiation mutagenesis is used to generate large deletions or complex rearrangements. Alternatively, mutagenesis is achieved with transposon insertion, using either the endogenous transposons or an exogenous transposon. It has been applied extensively and with great success to *Arabidopsis thaliana* (Drenkard, 2000).

3.2.2 FORWARD GENETIC SCREENS

Screening methods to identify the mutants must be designed before doing genome-wide mutagenesis. In this process, genetic variability is induced artificially. However, natural variations are also considered. Identification of mutants is followed by isolation of the affected gene. Mutation may be natural, but in many cases, the observed variation has been induced using a DNA damaging agent or mutagen (Colbert et al., 2001). Screening is done for phenotypes of interest, like morphology, lethality, specific embryonic phenotype, color of the seed, regulatory elements etc. The study is carried out to isolate the mutated gene that causes altered phenotype. Genetic screens or screening through chromosomes (often one at a time) is mostly utilized to identify mutant alleles of interest. Genetic selection allows only the mutant of interest to grow (Forsburg, 2001). Screening on defined media allows recovery of mutants that are defective in biosynthetic pathways like lipid biosynthesis, amino acid biosynthesis, etc. All the undesired mutant alleles are eliminated. Genetic selection is easier and less laborious than genetic screening but is often not possible and is potentially more biased. Genetic screen is an invaluable tool in the identification of gene required for specific cellular and biological process. Screening under specific biotic and abiotic conditions may allow understanding the process of growth and development in plants, photo-morphogenesis, plant-pathogen interaction, etc. For those genes that are difficult to isolate and identify, a two-step screening process for mutant is performed.

    i.   Screening for an easily scorable trait and for secondary phenotype that is characteristic of the mutant targeted, which are difficult to assay but can be easily observed and scored, i.e., simple screens.

    ii.   Secondary screening to screen for mutants with defects in particular developmental or metabolic pathway, i.e., modifier screens.

A second site mutation screen or the modifier screen can either enhance or suppress the phenotype to be screened. Classical genetics approach uses wild type as background for recovering mutants with striking phenotypes. It is not suitable for detecting redundant genes or mutant genes involved in a process but do not immediately lead to a morphological change. These genes do not translate into easily scorable biochemical phenotypes. Transgenic approach uses a drug or herbicide resistance gene for selection. Sometimes, reporter genes, e.g., green florescent protein gene, β-glucuronidase gene, etc., are also used to construct a visible phenotype. Identifying mutants from the wild type through marked phenotypic (qualitative) changes is carried out after mutagenesis. In the case of *Arabidopsis,* mutants with altered leaf shape, altered flowering time and flower morphology, and altered embryo or seedling lethality were grouped and screened for recovery of mutants defective in phytohormone synthesis and perception, and other biosynthetic pathways (Page and Grossniklaus, 2002). An enhancer mutation identifies genes (redundant genes) that act redundantly with the primary mutation or possibly those that interact physically with the mutant gene product. Screening is performed to identify the mutation that enhances the phenotype of the primary mutant. It begins with a mutant individual with defects in a particular developmental or metabolic pathway with a known gene mutation. The double mutant phenotype should be more prominent than either of the single mutant. Suppressor is a second site mutation that suppresses the phenotype of the original mutation. Mutation is either an intragenic suppression where the mutation is in the same gene as the original mutation or intergenic suppression where the mutation is located somewhere other than the original mutation. Suppressor mutations are used for identifying interacting proteins or alternative pathways that become activated by the second site suppressor defining the functions of biochemical pathways within a cell and the relationships between different biochemical pathways. Thus, identification of suitable genetic background and efficient screening procedure for the identification of mutants of interest is the key for success of forward genetic screening.

3.2.3 GENETIC MAPPING

A genetic map is a collection of genetic elements arranged in an ordered manner according to their segregation patterns. Genetic mapping also known as linkage mapping or meiotic mapping refers to the determination of the relative position and distances between markers along chromosomes. Genetic map is a key tool in both classical and modern research (Cheema and Dicks, 2009).The principle of genetic mapping is based on recombination frequency between markers during crossover of homologous chromosomes at the time of meiosis. Genetic mapping is based on the principle that genes, markers or loci segregate via chromosome recombination during meiosis (i.e., sexual reproduction), thus allowing their analysis in the progeny. It is an important tool in marker-assisted plant breeding programs. It enables plant breeders to develop new plant varieties for different traits like increased yield, drought tolerance, resistance to pest and pathogens, etc., in a target-oriented manner.

Genetic mapping is a widely applicable approach to scan the genetic information of an organism for those genes that are responsible for a particular trait. Genome mapping facilitates the isolation of a gene based on the measurement of their effect on a particular phenotype with no requirement of prior knowledge about the biochemical function performed by that gene. There are basically two broad categories of genome mapping that offers different levels of resolution at which a particular genome can be studied, namely genetic mapping and physical mapping. Genetic mapping is based on recombination frequency, i.e., the naturally occurring "breaking and rejoining" of the chromosome to determine the relative proximity of DNA landmarks to one another based on the frequency at which they co-occur on the same chromosome segment. Physical mapping involves determination of the relative proximity of DNA landmarks by direct measurement of physical quantity of DNA that lies between them. This is often similar to genetic mapping because it also involves chromosome breakage but artificially by using radiation, DNA modification enzymes, or shear forces (Paterson, 1996). Genetic linkage map construction requires appropriate mapping population and sample size, molecular marker(s) for genotyping the mapping population, screening parents for marker polymorphism, genotyping of the mapping population, linkage analyses by calculating pair-wise recombination frequencies between markers, establishing linkage groups, estimating map distances, and determining map order using statistical programs (Semagn et al., 2006; Paterson, 2009).

Genetic mapping involves out-crossing of the affected organism to a different inbred strain (the mapping strain) and backcrossing or intercrossing the F1 hybrid progeny. For recessive mutation, phenotype is tested in the F2 progeny, which is then genotyped to distinguish the mutant and mapping strains. Linkage mapping (linkage analysis) of the mutant strain is done with the help of markers by identifying a marker tightly linked to the gene in a "large" mapping population. Mutation is localized to a smaller chromosomal region by genotyping of additional markers within the region of linkage. Thus, a unified genetic map provides new opportunities for molecular interpretation of simple and complex phenotypes. Physical maps may help in cloning of agriculturally important genes or quantitative trait loci from major crops. Thousands of genetically mapped mutants of rice, maize, and other taxa might be united into a central tool to carry out comparative study of plant development (Paterson, 1996).

3.2.4 POSITIONAL CLONING OR MAP-BASED CLONING

Rapid identification of the gene of interest is theoretically possible in T-DNA or transposon insertion-induced mutation by locating the sequence tag and analyzing its neighboring sequences. Mutation induced by chemicals or radiation requires a very laborious map-based cloning (MBC) approach in which markers linked to the mutated gene are used to delimit the proximal and distal boundary of the critical region containing the gene of interest. In spite of this, there are several advantages to screen for mutants in chemical- or radiation-mutagenized populations. The method is applicable to any species of interest. It has a wider spectrum than tagging approaches. Mutagenesis is usually very efficient, and second-site mutations are easy to obtain in any plant system.

The first step in positional cloning is identification of a molecular marker that lies close to the gene of interest. Mutant phenotype and homozygosity for one or more markers of the mutant strain suggest linkage of a particular chromosomal region with the mutant phenotype. The genotyping of additional markers within the region of linkage has made it possible to localize the mutation to a smaller chromosomal region. The proximal and distal boundaries of the critical region defined by fine mapping are amplified by PCR and sequenced directly. Quantitative trait loci (QTLs) underlying natural variation simply require map-based cloning for their sequence identification (Peters et al., 2003).

In the early days, progress in mapping was inefficient and tedious due to lack of sufficient markers that did not exhibit epistatic interactions. MBC was a tedious, time-consuming approach before the availability of saturating marker systems. Development in sequencing technology and DNA-based markers such as Random Fragment Length Polymorphism (RFLP) and PCR-based markers like random amplified polymorphic DNA (RAPD), simple sequence repeats (SSR), SNPs, and amplified fragment length polymorphisms (AFLP) marker are no longer a limiting factor in map-based cloning. To identify genes within un-annotated DNA sequences, exon trapping is performed by cloning fragments of genomic DNA into vectors that would be expressed in mammalian cells, given the presence of functional splice junctions.

cDNA cloning is used to retrieve and identify the expressed fragments, which are then sequenced and designated as putative exons. The hybridization selection technique is used to find out the expressed gene within the genome.

The major drawback of positional cloning, however, is the difficulty of narrowing down the field of candidate clones to a manageable number for complementation testing. New sequencing technologies have dramatically accelerated the progress in forward genetics, thereby allowing rapid identification of mutant alleles. Thus, map-based cloning can be considered as an excellent approach for gene functional analysis.

3.2.5 MAPPING BY SEQUENCING

It includes the following techniques.

3.2.5.1 Massively Parallel Sequencing

Massively parallel sequencing refers to the simultaneous sequencing of large numbers of DNA fragments. With the wide availability of massively parallel sequencing technologies, genetic mapping has become the rate limiting step in the study of forward genetics of mammalian and plant systems. The development of massively parallel sequencing has shifted the focus in mutation identification from mapping to sequencing. The advent of massively parallel sequencing techniques has given rise to more rapid "mapping-by sequencing" methods in which genome-wide marker genotyping and DNA sequencing are combined into a single step applied to either individual or pooled groups of organisms (Wang et al., 2015). This technique can also be used to identify genes in which mutations have been induced (Zhang, 2009)

3.2.5.2 Whole Genome Sequencing

New sequencing technologies have accelerated the progress of forward genetics. Genome sequencing has advanced to a great extent and will be soon used to find out the mutant alleles in the genome. Application of whole genome sequencing (WGS) for simultaneous mapping and identification of the mutant gene was first reported in *Arabidopsis thaliana*. EMS-induced F2 mutant strains with growth defect was out-crossed to a genetically diverged line, followed by one round of self-crossing. A total of 500 mutant recombinants were pooled and sequenced to a depth of up to 22 times the genome size. The short-read data were aligned against the reference sequence, and more than 80,000 marker sites were used to estimate the putative center of the region harboring the causal mutation, which was fixed for mutant alleles in the pooled genomes. Less than 5 kb away from this estimated center was a nonsynonymous nucleotide substitution. As the second closest mutation was located ~250 kb away, this mutation was the undisputed candidate and later verified as a causal mutation for the slow-growth phenotype (Metzker, 2010; Schneeberger, 2014).

3.2.6   EXPRESSION ANALYSIS USING MICROARRAY

DNA microarray is one of the most efficient methods for gene expression analysis. Microarray is a very promising technology for the identification of genes in transcription-deficient mutants. DNA microarrays are solid supports, usually of glass or silicon, upon which DNA is attached in an organized grid fashion. Each spot of DNA, called a probe, represents a single gene. Microarray technology evolved from Southern blotting. The use of microarrays for gene expression profiling was first reported in 1995 (Schena et al., 1995), and a complete eukaryotic genome (*Saccharomyces cerevisiae*) on a microarray was published in 1997 (Lashkari et al., 1997). Microarray is a high density and high throughput approach that allows quantitative analysis of thousands of genes, expression patterns, and gene network in parallel (Guindalini et al., 2007). Microarray provides information about primary DNA sequence in coding and regulatory regions, polymorphic variation within a species or subgroup, time and place of expression of RNAs during development environmental stress and disease, etc. The information that can be obtained using microarray method depends on the type of sequences immobilized on the array, nature of the target sequences used for hybridization and the hybridization conditions. Probes for hybridization are designed using either known gene sequences or predicted gene sequences (Kuo et al., 2004). Microarray can be spotted cDNA arrays or oligonucleotide arrays.

3.2.6.1   Spotted Microarray

In spotted microarrays, the probes are oligonucleotides, cDNA, or small fragments of PCR products that correspond to mRNAs. A common approach utilizes an array of fine pins or needles controlled by a robotic arm that is dipped into wells containing DNA probes and then depositing each probe at designated locations on the array surface. Microarray "spotters" are high-precision robots with metal pins that dip into the DNA solution and tap down on the glass slide. The metal pins work like a fountain pen.

3.2.6.2   Oligonucleotide microarray

Oligonucleotide arrays are produced by printing short oligonucleotide sequences designed to represent a single gene or family of gene splice-variants. In oligonucleotide microarray, printing is done by synthesizing the sequence directly onto the array surface. In the microarray approach, the first step is to convert RNA into c-DNA through RT-PCR, and the c-DNA is then labeled with florescent dyes, i.e., cy3 and cy5. Probe preparation can be done either by spotted or by affimatrix technique based on direct synthesis of probe onto the glass slide. After probe preparation labeled target DNA is hybridized with the probe. Scanning is done using different wavelength of light, and data are analyzed for differential expression of genes (Galbraith et al., 2010). The approach of using phenotypically similar mutants minimizes the number of candidate genes for sequencing, due to the reduction of genes that are secondarily affected by the mutation. cDNA and affymetrix microarray platforms are able to successfully pinpoint the gene that is down- or upregulated due to induced or naturally occurred mutation events.

Unbiased information about the transcriptional activity of the entire genome can be obtained by carrying out microarray studies using high probe density. This technique facilitates not only identification of novel transcription units but also validation of existing gene models (Kapranov et al., 2007). Microarray technique is mostly used for transcription analysis rather than for exploring genome structure. The most popular application of microarray is in analysis of targets produced from transcripts, which may provide information about alterations in the sites of transcriptional activity, promoter binding, chromatin state, and polymorphism across genotypes.

Gene expression profiling with microarray is done by pair-wise hybridization using targets labeled with fluorochromes. Comparison is done among different treatments of the samples, like different developmental stages, different genotypes, organs, tissues, etc. It may represent different environmental conditions like biotic and abiotic stress, etc. Treatments can be sampled at one time or at different time intervals. Analysis of the data is done to arrive at a conclusion (Rensink et al., 2005). Such studies have resulted in the identification of candidate genes involved in cellulose biosynthesis (Persson et al., 2005), glucosinolate synthesis (Hirai et al., 2007), and pollen development (d'Erfurth et al., 2008).

3.2.7   CANDIDATE GENE APPROACH

The RNA-sequencing method is also called "whole transcriptome shotgun sequencing" (WTSS). This is a high-throughput sequencing technology to sequence cDNA in order to get information about the RNA content of the cells. The RNA sequencing method was developed in 2008 and was first applied to the analysis of yeast, mouse, and *Arabidopsis* transcriptomes (Mortazavi et al., 2008). RNA sequencing was the first sequencing-based method that allowed the entire transcriptome to be surveyed in a very high-throughput and quantitative manner. A population of RNA (total or fractionated, such as poly (A)+) was converted to a library of cDNA fragments with adaptors attached to one or both ends. Each molecule, with or without amplification, was then sequenced in a high-throughput manner to obtain short sequences from one end (single-end sequencing) or both ends (pair-end sequencing). Any high-throughput sequencing technology can be used, for example: Illumina, Applied Biosystems SOLiD and 454 pyro-sequencing for sequencing the gene of interest. Because converting RNA into cDNA by using reverse transcriptase might introduce mutations, single-molecule direct RNA sequencing technology has been developed. The sequences of all RNA in the mutant population are finally compared with the wild-type to discriminate mutant candidates (Wang et al., 2009; Martin et al., 2013). The RNA sequencing technique has increased our understanding of plant gene expression and evolution. It is accurate and facilitates comparison of even those samples that may not be generated at the same time as a part of the same experiment. It provides accuracy in resolution of splice junctions and alternative splicing events. As a result, this technology has a great potential for molecular breeding wherein multiple cultivars with variation in traits of interest are sequenced and genetic variation is identified. It also facilitates identification of molecular markers for the selection of genotypes in molecular genetics research. It has been successfully used in the discovery of more than 50 unknown drought-responsive genes in sorghum subjected to drought stress in conjunction with published transcriptome analysis for Arabidopsis, maize, and rice (Dugas et al., 2011; Casassola et al., 2013).

3.3 TOOLS USED TO STUDY REVERSE GENETICS

The reverse genetic approach can be used for determining the function of the gene by making use of various tools and techniques like gene silencing, TILLING, CRISPR, T-DNA, transposons, etc.

3.3.1 GENE SILENCING BY RNA INTERFERENCE

RNA interference is one of the most exciting breakthroughs of the past decade in functional genomics and promises to be a very useful instrument for determining the function of gene. The phenomenon of RNAi was first discovered during experiments associated with changes in pigmentation in the petunia plants. Introducing extra copies of a pigment biosynthesis gene did not increase the color intensity in the flower as expected; instead, the flowers became less colorful than the wild flowers. The term RNAi was first used when it was discovered that the injection of dsRNA into the nematode worm *Caenorhabditis elegans* caused the specific silencing of genes highly homologous to the supplied sequence (Fleenor et al., 1999). A sense strand is a 5′ to 3′ mRNA molecule or DNA molecule. The complementary strand or mirror strand to the sense is called an antisense strand. Antisense technology is the process in which the antisense strand is linked to the targeted sense strand by hydrogen bonds. When an antisense strand binds to an mRNA sense strand, a cell will recognize the double helix as foreign to the cell and proceed to degrade the faulty mRNA molecule, thus preventing the production of undesired protein. Double-stranded RNA (dsRNA) triggers the RNAi process or the gene silencing process. The interfering RNAs (microRNA (miRNA) or small interfering RNA (siRNA)) can be introduced in the cell either endogenously or exogenously. The basis of the RNAi process, production of functionally similar endogenously produced siRNAs, is quite similar in many organisms, and the enzymes required for this process show high interspecies homology. Processing of dsRNA precursors into siRNAs is mediated by special dsRNA-specific RNase-III endonucleases, known as Dicer. This results in the formation of 21- to 25-nucleotide dsRNA duplexes with symmetric 2–3 nucleotide 3′ overhangs, which are called siRNAs. These siRNAs are afterwards incorporated into the RNA-induced silencing complex (RISC), where an RNA helicase unwinds the inactive double-stranded siRNA, converting it to an active single-stranded form. Only one strand, known as the guide strand, is stabilized in

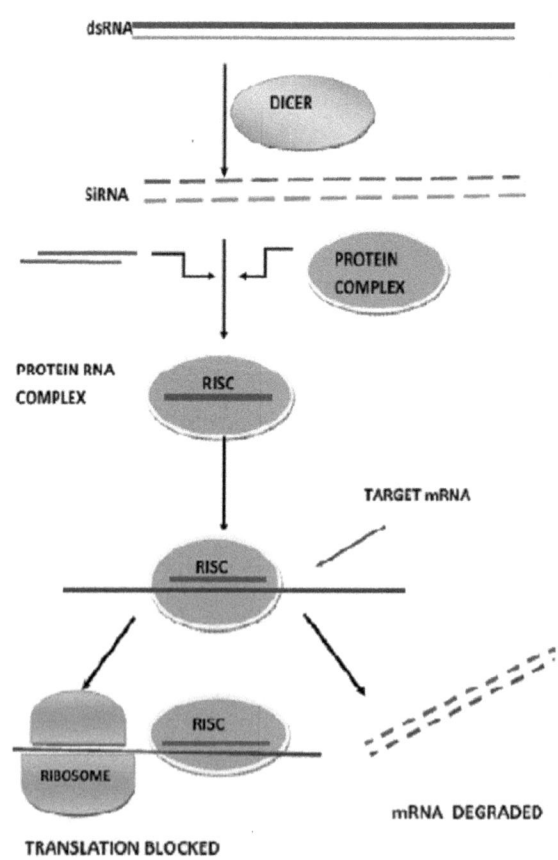

FIGURE 3.2   The process of gene silencing.

the RISC complex, while the passenger siRNA strand is degraded. An active RISC complex uses the guide siRNA to find and destroy the complementary sequence of mRNA, resulting in gene silencing (Figure 3.2).

In plants, in contrast to other organisms, miRNAs have perfect or near perfect complementarity to their targets and mediate gene silencing through mRNA cleavage. It is also easy to design plant siRNAs. In plants, like other eukaryotes, the small noncoding RNAs are utilized for posttranscriptional gene silencing (PTGS) to regulate gene expression and to direct epigenetic modifications. siRNAs (21-24 nucleotide long) and miRNAs (21-22 nucleotide long) are the two types of RNAs involved in PTGS, with important difference in their biogenesis and target genes. siRNAs originate commonly from invading or aberrant nucleic acids and produce cis-acting silencing by targeting the same molecule from which they were derived. miRNAs instead originate from the endogenous genes (Batel et al., 2004). miRNA technique of gene silencing offer fewer biosafety and environmental problems than other strategies. The advantages of this technique are strand-specific silencing, tissue differential expression, and possibility of complementation with other noncleavable targets. Identification and functional assessment of various genes responsible for crop improvement can be done by RNAi mechanism. This approach has opened new avenues for crop improvement. This novel tool has been applied in crop improvement extensively for resistance against both abiotic and biotic stresses. Artificial microRNAs (amiRNAs) represent a recently developed miRNA-based strategy to silence endogenous genes (Sablok et al., 2011). Histone deacetylase genes were studied in rice using miRNAs. These genes are important for overall gene expression regulation. The target genes were downregulated by artificial miRNAs, and it was discovered that the expression of rice histone deacetylase genes is specific for tissue/organ/treatment (Hu et al., 2009). In rice, OsSSI2 codes for fatty acid desaturase activity, which attracts diseases like leaf blight and blast. RNAi can knockdown OsSSI2, which increases resistance against bacterial pathogen of leaf blight (*Xanthomonas oryzae* pv. *Oryzae*) and blast fungus (*Magnaporthe grisea*) (Jiang et al., 2009).

3.3.2   TILLING

TILLING (Targeting Induced Local Lesions in Genomes) is a reverse genetic technique that is suitable for most plants. It provides a simple and innovative way to study induced or natural variation in the genomes of plants and animals. Most important breakthrough in the history of genetics was the discovery that mutations can be induced. Ionizing radiations and certain chemicals provide an easy and cost-effective way to mutate genes in the genome. This is coupled with PCR-based mutation detection in the TILLING project. It has made possible to perform genetic studies that were not feasible when only spontaneous mutations were available. With the expansion of sequence data banks, reverse genetics approaches have become popular for phenotypic screening and functional analysis of genes. Development of automated genome-scale reverse genetics approaches are required to enable to create a wide range of mutant alleles needed for functional analysis (Henikoff et al., 2004). Generation of mutated lines by transposons, T-DNA or RNA interference is technically difficult in some organisms. The difficulty comes due to the lack of an efficient transformation system and due to the large genome for some organisms like barley. Variation in the breeding process can be brought using radiation or chemical mutagens such as EMS. The mutagenic substance EMS preferentially alkylates guanine bases. The resulting O-6-ethyl guanine pairs with cytosine that is misread by the DNA-replicating polymerase, which inserts a thymine residue instead of a cytosine residue. This causes mutation of GC to AT. Mutations in the coding regions can be silent, miss-sense, or nonsense, and mutations in the noncoding promoter or intron regions can result in up- or downregulation of transcription (Banik et al., 2007; Sabetta et al., 2011).

TILLING is a recently developed reverse genetics technique, based on the use of a mismatch-specific endonuclease (*Cel*I), which finds mutations in a target gene containing a heteroduplex formation. If the mutation frequency is high and the population size is large enough, mutated alleles of most, if not all, genes will be present in the population. The technique involves PCR amplification of the target gene using fluorescent-labeled primers, formation of a DNA heteroduplex between wild-type and mutant alleles, followed by endonuclease digestion specifically cleaving at the site of an EMS-induced mismatch. The sizes of the amplicon cleavage fragments are often analyzed by a Li-COR or MALDI-TOF (matrix-assisted laser desorption/ionization-time-of-flight mass spectrometry). An allelic series of miss-sense mutations can be discovered by TILLING. It can also be used to detect insertions/deletions in genes. The success of the TILLING approach lies in the construction of good-quality DNA mutant libraries.

Purity and uniformity of the seed stock used in the TILLING project can be established by genotyping plant material using microsatellite markers. One of the greatest benefits of the TILLING approach is that it does not involve genetic manipulations, which results in genetically modified organisms (GMO), which are not legal for agricultural applications in many countries. It also has the potential to increase the power of QTL and association mapping (Okabe et al., 2011).

3.3.3   CRISPR

Plant biotechnology is evidencing a new phase where tools for random mutagenesis are being superseded by sequence-specific genome editing technologies. These are the molecular tools capable of creating DNA double-stranded breaks in a precise sequence in a specific manner. Although genome editing technologies using zinc finger nucleases (ZFNs)1 and transcription activator-like effectors nucleases (TALENs) can generate genome modifications, advances in the study of the prokaryotic adaptive immune system, involving type II clustered, regularly interspaced, short palindromic repeats (CRISPR), provide an alternative genome editing strategy. Type II CRISPR systems are widespread in bacteria; they use a single endonuclease, a CRISPR-associated protein Cas9, which provides defense against invading viral and plasmid DNAs. Cas9 forms a complex with a synthetic single-guide RNA (sgRNA), consisting of a fusion of CRISPR RNA (crRNA) and trans-activating crRNA. Cas9 is guided by sgRNA for recognizing and cleaving the target DNA. Cas9 consists of an HNH nuclease domain and a RuvC-like domain; each cleaves one strand of a dsDNA. The system is based on the Cas9 nuclease and a designed guide RNA that is specific for a target DNA sequence. Because a single RNA is sufficient to generate target specificity, the CRISPR/Cas system is convenient to use for genome editing as compared to ZFNs and TALENs. The average length of the guide sequence is nearly 20 bp. Therefore, the length of the target DNA is also 20 bp followed by a PAM sequence and consensus NGG. Many studies have shown that the 3′ end of the guide sequence within the sgRNA confers specificity of the CRISPR/Cas system. It is also observed that if there are mismatches between the target DNA and the guide RNA sequence located within the last 8–10 bp of the 20-bp target sequence, then target recognition will not be done by Cas9. However, if these mismatches are toward the 5′ region of the target sequences, they are better tolerated (Belhaj et al., 2013).

The CRISPR system is simple to design, cheap, and versatile for many biological applications and cell types, which have made it the most popular among the programmable nucleases (Shen et al., 2013; Harms et al., 2014). Other sequence-specific nucleases like ZFNs and TALENs are bipartite enzymes. The designing and construction of such large molecular proteins demand huge labor and cost. Moreover, ZFN recognition and cleavage of the desired DNA sequence involve high rate of failure. By using the CRISPR system that requires a single Cas 9 nuclease that can be programmed by engineering the sgRNA, one can produce knock out in many model organisms (Jiang et al., 2016). We can also create point mutations without any gene disruption, multiple genetic mutations in one experiment, large-scale genomic INDELS, chromosomal translocations, or replace several hundred kilobase stretches of DNA (Fujii et al., 2013).

In plants, this technology can be used for making targeted single and multiple gene knockouts, introducing SNPs into the gene of interest, expressing proteins with affinity tags at their native loci in genomes, etc. Sequence-specific editing can be done by delivering the guide RNA and Cas9 nuclease into the cell. A study was carried out to target inositol oxygenase (inox) and phytoene desaturase (pds) genes using cell suspension culture of wheat (Upadhyay et al., 2013). INDEL mutations were detected in all tested samples. It was also demonstrated that two genes at the same time could be targeted if the chimeric guide RNA (cgRNA) was multiplexed. They also carried out a cgRNA target specificity analysis and showed that the system showed high specificity toward target sequences. It was demonstrated that CRISPR–Cas-guided genome editing was achievable using wheat suspension cells. All these advantages make this technique applicable in crop improvement programs (Miao et al., 2013). Scientists have demonstrated that mutations in specific genes of rice could be generated by the CRISPR–Cas technology with high efficiency of targeted mutagenesis in different genes. They mutated the chlorophyll A oxygenase 1 (CAO1) gene and transformed it to a japonica variety with short life cycle. They obtained seedlings with a pale green leaf phenotype due to defective synthesis of chlorophyll b. Genotyping analysis of the transformed lines using gene-specific primers was conducted, and in the sequencing chromatogram, they obtained either loss of peak/gain of peak or overlapping peak around the target site, which confirmed mutations in the CAO1 gene. About 83.3% of the lines of the T1 transgenic rice showed mutations in the CAO1 gene, thus exhibiting the high mutagenesis efficiency of the CRISPR–Cas system (Shan et al., 2013). Rice and wheat, the two most important food crops, have been found to be amenable to this gene editing approach.

CRISPR technology has become a new plant breeding technique and promises to modify plant genome in a similar manner as done by conventional breeding technique (Qiwei et al., 2013). As a result, plant varieties developed using this technique will be classified as non-genetically modified crops. This will boost development of plant biotechnology and plant breeding.

3.3.4 T-DNA AS A REVERSE GENETIC TOOL

The frequency of homologous recombination is very low in higher plants. Therefore, insertional mutagenesis is a more appropriate option for reverse genetics studies in plants. Of the two systems of insertional mutagenesis, those involving the use of t-DNA and those involving the use of plant transposons, it is the T-DNA insertions that can generate stable insertions in the primary transformants (Primrose et al., 2013). Moreover, because T-DNA insertions unlike transposons are not biased toward a particular gene group, thus near saturation mutagenesis of the genome is possible (An et al., 2005).

T-DNA is a segment of Ti-plasmid of the soil bacterium *Agrobacterium tumefaciens* that is used for insertional mutagenesis as it randomly integrates into the plant genome causing disruption or activation of the gene of interest depending on the construct used. This technique involves use of two separate plasmids; one carrying the insert DNA flanked by the left and right border sequences of T-DNA and other carrying virulence genes from the Ti plasmid. Mutagenesis by T-DNA is not only used for identifying knockout mutants but also used to study promoter and activator tagging. Gene knockout for more than 70% of *Arabidopsis* genes was obtained by T-DNA mutagenesis. Generating a reasonable number of knockout lines is not too difficult in *Arabidopsis* because it can be easily transformed, but it is difficult to obtain many thousand requisite transgenics in other species as the transformation efficiency is not high. In a study to evaluate the effectiveness of T-DNA insertion in knocking out a gene, 1084 published *A. thaliana* insertion mutants representing 755 genes were evaluated. It was found that insertion in the protein-coding region of a gene generated a knockout at least 90% of the time. It was also inferred that T-DNA insertion can also result in deletion and translocation (Wang et al., 2008). It was also observed in an experiment to generate a library of T-DNA enhancer trap lines in rice cv. Nipponbare that T-DNA appeared less prone to hot spots and cold spots of integration and T-DNA inserts rarely integrated into repetitive sequences (Sallaud et al., 2004).

3.3.5   TRANSPOSONS

Insertional mutagenesis can be carried out in plants by using mobile genetic elements known as transposons. Transposon-based mutagenesis has been used in reverse genetic screens to identify disruptions in target gene or gene of interest. It involves two components: an autonomous element containing a transposase gene and one or more non-autonomous elements that are active only when the transposase expressed by the autonomous element is active. It can be used for generating multiple insertional events within one region of the genome. Transposon-generated populations are advantageous over T-DNA insertion lines because many unique insertion lines can be produced from a few initial plant lines using transposons, and these lines lack epigenetic changes as shown by T-DNA insertion lines. Transposons have been widely used by researchers for reverse genetics experiments. Several transposons from plants like maize (Ac/Ds: Activator/dissociation, Spm: suppressor mutator, Mu: mutator), petunia (Tph1), and *Antirrhinum majus* (Tam3) have been used for genome-scale mutagenesis projects. Maize transposon Ac/Ds shows local transposition, and therefore, saturation is difficult to obtain; however, the Mu element does not show preferential local transposition but shows insertion most of the time into transcription units, which makes it a preferred tool for mutagenesis experiments (Primrose et al., 2013). High activity level of Mu transposon often leads to deleterious mutation; therefore, it is difficult to use in heterologous system. The Spm transposons of maize have been used in heterologous plants like *Arabidopsis*, and a large number of mutant lines have been generated (Speulman et al., 1999). This has helped in the identification of interrupted genes though searchable databases. The first endogenous transposon demonstrated to be active in rice were Tos reterotransposons. There are various other transposons that have been used to generate tagged population in rice (Upadhyay et al., 2011).

3.4   CONCLUDING REMARKS

One of the major task of forward genetics is to associate with the corresponding mutated genes and one of the best way to definitively identify a function for a gene is to find out what happens when it is inactivated, which leads to the concept of reverse genetics. The advent of TILLING and EcoTILL-ING technologies has provided a powerful tool to scientists for carrying out reverse genetics research for any plant because TILLING can provide allelic series of mutations, including knockouts. In addition, TILLING can be applied even if genomic sequencing is limited to selected target genes. The high density of chemically induced point mutations makes TILLING suitable for targeting small genes, and it allows a researcher to focus on single protein domain when targeting larger genes (Henikoff et al., 2004). TILLING and EcoTILLING can use pooled DNA samples to increase throughput and DNA polymorphism that can be detected by nonsophisticated instruments or low cost-labeled primers (Raghavan et al., 2007). These features make these techniques more efficient and affordable to many laboratories, particularly in developing countries.

KEYWORDS

- CRISPR
- forward genetics
- gene mapping
- genetic marker
- microarray
- reverse genetics
- RNA sequencing
- RNAi
- TILLING

REFERENCES

Alonso, J. M., & Ecke, J. R., (2006). Moving forward in reverse: genetic technologies to enable genome-wide phenomic screens in *Arabidopsis. Nature Reviews Genetics, 9,* 38–45.

Alonso, B. C., & Koornneef, M., (2000). Naturally occurring variation in Arabidopsis: an underexploited resource for plant genetics. *Trends Plant Sci., 5,* 22–29.

An G., Lee S., et al., (2005). Molecular genetics using T-DNA in rice. *Plant Cell Physiology., 46*(11), 14–22.

Banik, M., Shuyu, L., Kangfu, Y., Poysa, V., & Soon, J., (2007). Molecular tilling and ecotilling: Effective tool for mutant gene detection in plants. *Gene Genome and Genomics, 1*(2), 123–131.

Batel, D. P., (2004). MicroRNAs: genomics, biogenesis, mechanism and function. *Cell, 116,* 281–297.

Belhaj, K., et al., (2013). Plant genome editing made easy: Targeted mutagenesis in model and crop plant using the CRISPR/Cas system. *Plant Methods, 9,* 39–48.

Bouche, N., & Bouchez, D., (2001). Arabidopsis gene knockout: phenotypes wanted. *Curr. Opin. Plant Biol., 4,* 111–117.

Brown, D. M., Zeef, L. A. H., Ellis, J., Goodacr, R., & Turne, S. R., (2005). Identification of novel genes in arabidopsis involved in secondary cell wall formation using expression profiling and reverse genetics. *The Plant Cell, 17,* 2281–2295.

Casassola, A., Brammer, S. P., Chaves, M. S., José, A. I., Grando, M. F., & Denardin, N. D., (2013). Gene expression: A review on methods for the study of defense-related gene differential expression in plants. *American Journal of Plant Sciences, 4,* 64–73.

Cheema, J., & Dicks, J., (2009). Computational approaches and software tools for genetic linkage map estimation in plants. *Briefings in Bioinformatics, 10*(6), 595–608.

Colbert, T. G., et al.,(2001). High-throughput screening for induced point mutations. *Plant Physiol., 126,* 480–484.

D' Erfurth, I., Jolivet, S., Froger, N., Catrice, O., Novatchkova, M., Simon, M., Jenczewski, E., & Mercier, R., (2008). Mutations in AtPS1 (*Arabidopsis thaliana* parallel spindle 1) lead to the production of diploid pollen grains. *PLoS Genetics, 4,* e1000274.

Denver, D. R., (2009). A genome-wide view of *Caenorhabditis elegans* base-substitution mutation processes. *Proc. Natl. Acad. Sci., 106,* 16310–16314.

Drenkard, E., (2000). A simple procedure for the analysis of single nucleotide polymorphisms facilitates map-based cloning in Arabidopsis. *Plant Physiol., 124,* 1483–1492.

Dugas, D., et al., (2011). Functional annotation of the transcriptome of sorghum in response to osmotic stress and abscisic acid. *BMC Genomics, 12,* 512. Doi: 10.1186/11471-2164-12-514.

Fleenor, J., Grishok, A., Fire, A., & Mello, C. C., (1999). The RDE-1 gene, RNA interference and transposon silencing in C. elegans. *Cell, 99*(2), 123–132.

Forsburg, S. L., (2001). The art and design of genetic screens: yeast. *Nature Rev. Genet., 2,* 659–668.

Fujii, W., Kawasaki, K., Sugiura, K., & Naito, K., (2013). Efficient generation of large-scale genome-modified mice using gRNA and CAS9 endonuclease. *Nucleic Acids Res., 41,* e187.

Galbraith, D. W., & Edwards, J., (2010). Applications of microarrays for crop improvement: Here, there, and everywhere. *Bio. Science, 60,* 337–348.

Guindalini, S., & Tufik, S., (2007). Use of microarrays in the search of gene expression patterns—Application to the study of complex phenotypes. *Revista Brasileira de Psiquiatria.*, 370–374.

Hardy, S., Legagneux, V., Audic, Y., & Paillard, L., (2010). Reverse genetics in eukaryotes. *Biol. Cell*, *102*, 561–580.

Harms, D. W., et al., (2014). Mouse genome editing using the CRISPR/Cas system. *Current Protocols in Human Genetics*, *15*(7), 1–27.

Henikoff, S., Till, B. J., & Comai, L., (2004). TILLING traditional mutagenesis meets functional genomics. *Plant Physiology*, *135*, 630–636.

Hirai, M. Y., et al., (2007). Omics based identification of *Arabidopsis* Myb transcription factors regulating aliphatic glucosinolate biosynthesis. *Proceedings of the National Academy of Sciences*, *104*, 6478–6483.

Hu, Y., Qin, F., Huang, L., Sun, Q., Li, C., Zhao, Y., & Zhou, D., (2009). Rice histone deacetylase genes display specific expression patterns and developmental functions. *Biochem. Biophys. Res. Commun.*, *388*, 2266–2271.

Janny, L., Peters, C. F., & Gerats, T., (2003). Forward genetics and map-based cloning approaches. *Trends Plant Sci.*, *8*(10), 484–491.

Jiang, C. J., Shimono, M., Maeda, S., Inoue, H., Mori, M., Hasegawa, M., Sugano, S., & Takatsuji, H., (2009). Suppression of the rice fatty-acid desaturase gene OsSSI2 enhances resistance to blast and leaf blight diseases in rice. *Mol Plant-Microbe In.*, *22*, 820–829.

Jiang, F., Taylor, D. W., Thompson, A. J., Nagales, E., & Doudna, J. A., (2016). Structures of a CRISPR-Cas9R loop complex primed for DNA cleavage. *Science*, *351*(6275), 867–871.

Kapranov, P., et al., (2007). Genome wide transcription and implications for genomic organization. *Nature Reviews Genetics*, *8*, 413–423.

Kuo, W. P., Kim, E. Y., Trimarchi, J., Trimarchi, T. K., Vinterbo, S. A., & Ohno-Machado, L., (2004). A primer on gene expression and microarrays for machine learning researchers. *Journal of Biomedical Informatics*, *37(4)*, 293–303.

Kutscher, L. M., & Shaha, S., (2014). Forward and reverse mutagenesis in *C. elegans*. *Worm Book*, (ed.). *The C. Elegans Research Community*, 1–26.

Lashkari, D. A., et al., (1997). Yeast microarrays for genome wide parallel genetic and gene expression analysis. *PNAS*, *94*(24), 13057–13062.

Lukowitz, W., et al., (2000). Positional cloning in Arabidopsis. Why it feels good to have a genome initiative working for you. *Plant Physiol.*, *123*, 795–805.

Martin, L. B. B., Fei, Z., Giovannoni, J., Rose, J., & Jocelyn, K. C., (2013). Catalyzing plant science research with RNA-seq. *Frontiers in Plant Science*, *4*, 1–10.

Metzker, M. L., (2010). Sequencing technologies – the next generation. *Nat. Rev. Genet.*, *11*, 31–46.

Miao, J., Guo, D., Zhang, J., Huang, Q., Qin, G., Zhang, X., Wan, J., Gu, H., Qu, Li, J., (2013). Targeted mutagenesis in rice using CRISPR-Cas system. Letter to the Editor. *Cell Research*, *23*, 1233–1236.

Mortazavi, A., Williams, B. A., McCue, K., Schaeffer, L., & Wold, B., (2008). Mapping and quantifying mammalian transcriptomes by RNA-Seq. *Nature Methods*, *5*, 621–628.

Nordborg, M., & Weigel, D., (2008). Next-generation genetics in plants. *Nature*, *456*, 720–723.

Okabe, Y., Asamizu, E., Saito, T., Matsukura, C., Ariizumi, T., Brès, C., Rothan, C., Mizogu-chi, T., & Ezura, H., (2011). Tomato TILLING technology: Development of a reverse genetics tool for the efficient isolation of mutants from micro-tom mutant libraries. *Plant Cell Physiol., 52*(11), 1994–2005.

Page, D. R., & Grossniklaus, U., (2002). The art and design of genetic screens: Arabidopsis thaliana. *Nature Reviews Genetics, 3,* 124–136.

Paterson, A. H., Lan, T. H., Reischmann, K. P., & Chang, C., (1996). Towards a unified genetic map of higher plants, transcending the monocot-dicot divergence. *Nature Genetics, 14,* 380–382.

Paterson, A. H., (2009). Plant genome mapping : Strategies and application. *Biotechnology, 7,* 1–9.

Persson, S., Wei, H., Paige, G. P., & Somerville, C., (2005). Identification of genes required for cellulose synthesis by regression analysis of public microarray data sets. *Proceedings of the National Academy of Sciences, 102,* 8633–8638.

Peters, J. L., Cnudde, F., & Gerats, T., (2003). Forward genetics and map based cloning approaches. *Trends in Plant Science, 8*(10), 484–491.

Primrose, S. B., & Twyman, R., (2013). *Principles of Gene Manipulation and Genomics.* John Wiley & Sons.

Qiwei, S., Yanpeng, W., Jun, L., Yi, Z., Kunling, C., Zhen, L., Kang, Z., Jinxing, L., Jian-zhong, J. X., Jin-Long, Q., & Caixia, G., (2013). Targeted genome modification of crop plants using a CRISPR-Cas system. *Nature Biotechnology, 31,* 686–688.

Raghavan, C., Naredo, M. E. B., Wang, H., & leung, H., (2007). Rapid method for detecting SNPs on agarose gels and its application in candidate gene mapping. *Mol. Breeding, 19,* 87–101.

Rensink, W. A., & Buel, R. C., (2005). Microarray expression profiling resources for plant genomics *TRENDS in Plant Science,* doi 10. 1016/j.tplants. 2005. 10. 003.

Sabetta, W., Alba, V., & Blanco, A., (2011). Montemurro, C., sunTILL: a TILLING resource for gene function analysis in sunflower. *Plant Methods, 7,* 20–25.

Sablok, G., Perez-Quintero, A. L., Hassan, M., Tatarinova, V. T., & Lopez, C., (2011). Artificial microRNAs (amiRNAs) engineering – On how microRNA-based silencing methods have affected current plant silencing research. *Biochemical and Biophysical Research Communications, 406,* 315–319.

Sallaud, C., Gay, C., Larmande, P., Bes, M., Piffanelli, P., Piegu, B., Droc, G., Regad, F., Bourgeois, E., Meynard, D., Perin, C., Sabau, X., Ghesquiere, A., Glaszmann, J. C., Delseny, M., & Guiderdoni, E., (2004). High throughput T-DNA insertion mutagenesis in rice: A first step towards in silico reverse genetics. *The Plant J., 39*(3), 450–464.

Schena, M., et al., (1995). Quantitative monitoring of gene expression patterns with a complementary DNA microarray. *Science, 270,* 467–470.

Schneeberger, K., (2014). Using next-generation sequencing to isolate mutant genes from forward genetic screens. *Nat. Rev. Genet., 15*(10), 662–676.

Schneeberger, K., & Weigel, D., (2011). Fast-forward genetics enabled by new sequencing technologies. *Trends Plant Sci., 16,* 282–288.

Semagn. K., & Bjornstad, A., (2006). Principles, requirements and prospects of genetic mapping in plants. *African Journal of Biotechnology, 5*(25), 2569–2587.

Shan, Q., et al.,(2013). Targeted genome modification of crop plants using a CRISPR/Cas system. *Nature Biotechnology, 31*(8), 686–688.

Shen, B., Zhang, J., et al., (2013). Generation of gene-modified mice via Cas9/RNA mediated gene targeting. *Cell Re., 23,* 720–723.

Speulman, E., Metz, P. J., Gert van Arkel, Bas te Lintel Hekkert, Stiekema, W., & Pereira, A., (1999). A two-component enhancer-inhibitor transposon mutagenesis system for functional. Analysis of the arabidopsis genome *The Plant Cell, 11,* 1853–1866.

Upadhyay, S. K., et al., (2013). RNA guided genome editing for target gene mutations in wheat. *Genetics, 3,* 2233–2238.

Upadhyaya, N. M., Zhu, Q. H., & Bhat, R. S., (2011). Transposon insertional mutagenesis in rice. In: Pereira A (ed) *Plant Reverse Genetics: Methods and Protocols, Methods in Molecular Biology, 678,* 147–177.

Wang, Y. H., et al., (2008). How effective is T-DNA insertional mutagenesis in Arabidopsis ? *Journal of Biochemical Technique., 1*(1), 11–20.

Wang, Z., Gerstein, M., & Gerstein, M., (2009). RNA-Seq: a revolutionary tool for transcriptomics, *Nat. Rev. Genet., 10*(1), 57–63.

Wang, T., et al., (2015). Real time resolution of point mutations that cause phenovariance in mice. *PNAS,* E440–E449.

Zakhrabekovaa, S. M., Gougha, S., Lundhb, L., & Hanssona, M., (2013). Functional genomics and forward and reverse genetics approaches for identification of important QTLs in plants. *News of AMEA (Biological and Medical Sciences),68*(2)*,* 23–28.

Zhang, Z., (2009). Massively parallel sequencing identifies the gene Megf8 with ENU-induced mutation causing heterotaxy. *Proc. Natl. Acad. Sci. USA, 106*(9), 3219–3224.

PART II

THE ROLE OF BIODIVERSITY IN PLANT BREEDING

CHAPTER 4

BIODIVERSITY AND CONSERVATION OF PLANT GENETIC RESOURCES (*IN-SITU* AND *EX-SITU*)

SHEEL YADAV, CHET RAM, and
DHAMMAPRAKASH PANDHARI WANKHEDE

Division of Genomic Resources, ICAR–National Bureau of Plant Genetic Resources, Pusa Campus, New Delhi–110012, India, E-mail: d.wankhede@icar.gov.in

CONTENTS

4.1   INTRODUCTION

Plants are a vital component of life and drivers of global sustainability. The biodiversity of plant kingdom can be best gauged by the fact that there are about 32,212 different plant species on the Earth, and 87% of these are flowering plants (Nautiyal, 2011). A total of 30,000 plant species are edible, and some 7,000 species have been cultivated since the time agriculture came into practice (IUCN, 2012). India alone accounts for having 11.9% of the world flora, with a total of 2,50,000 plant species found in the Hindu Kush-Himalayan belt (IUCN, 2012). This biodiversity has long served humanity by acting as a source of food, fiber, shelter, and medicine. Life, as we know today, would not have been possible and will not be possible without plants, thus rendering their conservation essential. The loss of biodiversity globally is one of the most eminent pressing crisis. About 22–47% of the plant species in the world are under the endangered category (Graham, 2002), and we have already lost 112 species of plants. This loss can be attributed to factors like rapid land degradation and conversion of natural areas to agricultural farms and other development projects, the introduction of invasive alien species, and artificial selection of a few superior varieties that causes "genetic erosion" and overexploitation of soil and water resources. At the rate at which the world population is believed to be growing, it is estimated that by the year 2050, the world would need double the amount of food to be produced, which was produced in 2000 (FAO, 2006). This would not only entail a manifold increase in food production but also a sustainable increase in food production. This task becomes daunting, given the fact that the resource base in the form of land and water will continue to dwindle further in times to come. The impending challenge of climate change further exacerbates the situation at hand. Conserving our plant biodiversity to meet the demand for increased food grains and sustain the global population, is, therefore, imperative. The vast array of genetic resources we have, at our disposal, together with new techniques, offers the potential for meeting the world's future food requirements. A meticulous and long-term conservation strategy for the conservation of our plant genetic resources (PGR) is thus the need of the hour.

4.2   BIODIVERSITY

Biodiversity or biological diversity refers to the variability that exists among all living organisms. According to the definition of UN Convention on Biological Diversity, biodiversity is "the variability among living organisms from all sources, including, inter alia, terrestrial, marine, and other aquatic ecosystems and the ecological complexes of which they are part; this includes diversity within species, between species, and of ecosystems" (UNEP, 1992). Humans receive several directs and indirect benefits from biodiversity, starting from basic necessities such as food, fiber, and medicine to luxuries. However, with the over-exploitation of biodiversity without the substantial conservation measures, the negative impacts on the ecosystems at local to global scales are evident. The situation warrants an urgent need to frame strategies and implementation in order to protect biodiversity to meet the goals of poverty reduction and eco-friendly sustainable development.

There are three major levels of biodiversity: (a) Ecosystem diversity (variety of interactive systems of living components such as plants, animals, and microbes and nonliving components such as water, earth forms, soil, rocks, etc.), (b) Species diversity (variety among the species or different types of living organisms), and (c) Genetic diversity (variations in the genetic constitution of individuals within a species).

This chapter is focused on genetic diversity in agriculturally important plant genetic resources and its conservation.

4.3   IMPORTANCE OF GENETIC DIVERSITY

Since the dawn of agriculture, humans collected seeds and vegetative propagules for their consumption as well as for planting in subsequent seasons. Since that time, early farmers have utilized genetic variability within crop species to meet the requirement of food, clothing, and other necessities. From the early to the present day modern agriculture, the diversity in PGR bestowed opportunities for farmers and plant breeders to breed for new and improved cultivars with the desirable traits.

Post-green revolution (mid of the 1060s), there has been a predominance of improved and uniform cultivars in major crops in agriculture. Further, several improved cultivars have been developed using limited germplasm, thereby making genetic base of the widely grown cultivars very narrow. The narrow genetic base of cultivars increases its vulnerability to pest and diseases. The most laudable examples of narrow genetic base leading to catastrophic consequences are Irish potato famine caused by the late blight of potato (*Phytophthora infestans*) and Southern leaf blight epidemic of maize in USA caused by cytoplasmic uniformity in 1970 (Cooper et al., 2001). In the backdrop of such incidences, it is very important to broaden the genetic base of crop plants by utilizing a greater amount of genetic diversity.

Traditionally the landraces that farmers were growing are gradually disappearing. Landraces are primitive or antique varieties usually associated with traditional agriculture and are often highly adapted to local conditions (FAO, 2003). It has been observed that the landraces possess many useful characteristics and could be of a great importance in crop improvement programs. Further, several crop wild relatives could form an important source in present breeding programs (Paroda and Arora, 1991).

4.4 THREATS TO GENETIC DIVERSITY

There are three major threats to the genetic diversity of currently growing crops, namely genetic erosion, genetic vulnerability, and genetic wipe-outs (Paroda and Arora, 1991; Wilkes, 1984).

4.4.1 GENETIC EROSION

Genetic erosion is the loss of alleles, the loss of specific allele combinations, and the loss of locally adapted varieties and breeds. In simple words, it is the loss of variation in crops owing to the modernization of agriculture. Genetic erosion occurs at two stages: first with the replacement of landraces with improved modern cultivars and second in diversity as a result of modern breeding practices. Genetic erosion could also occur at the crop, variety, and allele level (Rogers, 2004). It could either be a loss of alleles/genes or the loss of entire genotypes and landraces. It can also occur at the level of germplasm collections and genebanks due to improper management and inadequate regeneration procedures (Mathur, 2011). Anthropogenic activities have accelerated the process of genetic erosion. Habitat loss and fragmentation and a narrow genetic base could also cause genetic erosion, especially in native populations. The improved variety often eliminates the genetic resources from which it is bred (Paroda and Arora, 1991). Plant breeders have been taking refuge in landraces and folk varieties that are considered to be a reservoir of several abiotic and biotic stress-tolerant genes/alleles to be used in breeding programs. However, in modern agriculture, this genetic diversity is replaced with a substantially small number of improved varieties often with similar genetic base. Further owing to increased anthropogenic activities (urbanization, industrialization, etc.), the natural habitat of wild progenitors and weedy forms of crop plants is being diminished, which ultimately further reduces genetic diversity (Paroda and Arora, 1991). Further, a reduction in population size without an increase in recombination is one of the principal causes of genetic erosion (Brown, 2008; Mathur, 2011).

4.4.2   GENETIC VULNERABILITY

it is defined as "the condition that results when a widely planted crop is uniformly susceptible to a pest, pathogen, or environmental hazard as a result of its genetic constitution, thereby creating a potential for widespread crop losses" (NRC, 1972). Low level of genetic diversity in crops grown over large acreages and/or monoculture of a crop variety is a typical example of genetic vulnerability. The narrow genetic base makes crop varieties at much higher risk of complete failures as have been observed in historical cases such as the Irish potato famine in the 1840s, wheat crop failure in 1954 due to stem rust, and southern corn blight in 1970. A viable strategy for the better deployment of diversity is required to address genetic vulnerability in farming systems. Farmers should be encouraged to use diverse crop species through crop rotation, crop mixtures, agroforestry, and promotion of underutilized species. Further, efforts should be made to broaden genetic base crop varieties by using diverse germplasm in breeding programs (NRC, 1972).

4.4.3   GENETIC WIPEOUT

In contrary to genetic erosion that is gradual and slow, genetic wipeout is rapid, one stroke destruction. Thus, genetic wipe out is defined as wholesale loss of plant genetic resources (Harlan, 1975). Consequences of institutional failures, social causes (political instability), famines, floods, and crop failures could lead to the destruction of plant genetic resources rapidly. A few institutional decisions such as discarding of a genetic collection after the retirement of a curator or if the material is of no use to the institution could also lead to the elimination of important plant genetic resource (Paroda and

Arora, 1991). Therefore, it has been recommended that safeguarding the germplasm should be given priority in case of arising of any such eventualities. Further streamlining of the institutional germplasm collection-conservation-utilization and coordination at the regional, national and international level have been suggested (Paroda and Arora, 1991).

It has been well established that an effective PGR conservation should have clear knowledge about the concerned species with respective to degree of genetic diversity, its structure, distribution in nature, and in the material conserved (Allard, 1988; Hamrick and Godt, 1990, 1997; Hamrick, 1993; Hamrick et al., 1993). Ramanatha Rao and Hodgkin (2002) have well elaborated the factors affecting the distribution of genetic diversity and what information on genetic diversity is required in the conservation and utilization of plant genetic resources.

4.5 THE DEGREE AND DISTRIBUTION OF GENETIC DIVERSITY

In the perspective of genetic resources, genetic diversity is defined as "the range of the gene pool, the amount of genetic variation present in a population or species as consequences of its evolutionary pathways" (IBPGR, 1991). It is a measure of the variety of versions (alleles) of the same gene within individual species reflected in phenotypes. Genetic diversity could be distinguished in four components: a) variety of alleles present in populations, b) distribution of alleles in and among populations, c) impact of different alleles on phenotype/performance, and d) distinctness between different populations of a species. Variations at the nucleic acid level, which is the basis of genetic diversity, arises with mutation and recombination. Further, variation in diversity between different populations arises due to selection (natural and/or artificial), genetic drift, and gene flow (Suneson, 1960; Frankel, 1977; Nevo et al., 1984; Brown, 1988; Hamrick et al., 1992). In order to answer, what, how, and where to conserve, the knowledge of the extent and distribution of the various aspects of genetic diversity in a species is an essential prerequisite (Ramanatha Rao and Hodgkin, 2002).

Factors crucial for proper conservation of genetic diversity need to be identified and considered before conservation efforts. For these requisites, the following factors need to be addressed:

4.5.1 ECO-GEOGRAPHIC FACTORS

Geographic differences play an important role in the distribution of genetic diversity. Populations often show a difference over several aspects of diversity, number, and identity of alleles and their effect on the characteristics in the population. The habitats in which plants grow have a direct relation to morpho-physiological traits of plants and, therefore, the geographic variation in the distribution of genetic diversity is not appropriate to separate from ecologically determined variation. Different geographic locations often differ with respect to several ecological characteristics (Ramanatha Rao and Hodgkin, 2002).

4.5.2 BREEDING SYSTEM

The breeding system of a species has a significant effect on the distribution of alleles. The mating system, floral morphology, and mode of reproduction have a profound effect on the degree and distribution of genetic diversity (Loveless and Hamrick, 1984). Breeding systems of wild species are often different from those of their cultivated crop relatives and, therefore, need to be treated differently in genebanks for maintenance and regenerations. Therefore, information about the mating system of a species gives significant information on genotypic distribution in natural populations. Outcrossing naturally helps plant populations maintain a high degree of genetic diversity (Ramanatha Rao and Hodgkin, 2002).

4.5.3 POPULATION SIZE, A BOTTLENECK CONCEPT

A population bottleneck is an event (natural or artificial disasters such as earthquakes, flood, fire, diseases, etc.) that leads to sharp decline in the size of a population. The population bottleneck results in a reduced gene pool of the population owing to loss of several alleles from the original population. As a result of a population bottleneck, the remaining population is left with a relatively lower level of genetic diversity and is at higher risk of susceptibility to pests and diseases, thereby increasing the chances of its extinction. Further, with the loss of genetic variation, the remaining population has a high chance of becoming genetically distinct from the earlier parent population and, therefore, can lead to the evolution of new species. According to genetic models, the proportion of initial heterozygosity maintained after each generation is $[1-(1/2Ne)]$ where Ne is the effective population size, generally $< N$, the actual population size. Hence, it is estimated that a population with 10 individuals loses 5% of its heterozygosity per generation. This suggests that bottlenecks degrade heterozygosity as well as genetic diversity (Pimm et al., 1989; Govindaraj et al., 2015). Accordingly, conservation biologists are expected to maintain the appropriate population size for effective conservation genetic diversity without any bias.

In the past decade, there have been several efforts on adopting a holistic view of biodiversity including agrobiodiversity, conservation as well as sustainable utilization and development (Arora, 1997; Ramanatha Rao and Hodgkin, 2002). A few significant advances have also been made in conserving plant genetic resources that are the valuable natural resources and form the basis for food and nutritional security (Watanabe et al., 1998; Ramanatha Rao et al., 1999).

4.6 PLANT GENETIC RESOURCES

According to the International Undertaking on Plant Genetic Resources (FAO, 1983), PGR can be defined as the reproductive or vegetative propagating material of (i) cultivated varieties (cultivars) in current use and newly developed varieties; (ii) obsolete cultivars; (iii) primitive cultivars (landraces); (iv) wild and weed species, and (v) special genetic stocks . According to the Convention for Biological Diversity (CBD, 1992), PGR has been defined as any living material of present and potential value for humans. To this effect, plant genes, DNA fragments, and RNA also come under the ambit of plant genetic resources as a new concept. Broadly speaking, there are five basic kinds of plant genetic resources (Rubenstein et al., 2005).

4.6.1 *WILD OR WEEDY RELATIVES*

These are undomesticated plants that share a common ancestry with a crop species. It is estimated that about 9000 wild plant species are threatened (Kumar, http://www.preservearticles.com). These plants hold great potential in facilitating crop improvement as they are found to be a rich source of resistance traits.

4.6.2   LANDRACES

These are varieties of crops that have been selected, raised, and improved by farmers over many generations. Because they are grown at specific regions, these have become extremely specialized to a specific environment and owe unique features. The landraces are usually found to be very diverse within species. In a modern breeding program, they are sometimes used as a source for agronomical traits.

4.6.3   IMPROVED GERMPLASM

This refers to any plant material that possesses one or more traits of interest that have been incorporated by scientific selection or planned crossing.

4.6.4   ADVANCED (OR ELITE) GERMPLASM

This includes "cultivars" or cultivated varieties, suitable for planting by farmers, and advanced breeding material that breeders combine to produce new cultivars.

4.6.5   GENETIC STOCKS

These are mutants or other germplasm with chromosomal abnormalities that may be used by plant breeders, often for sophisticated breeding and basic research.

4.7   NEED FOR CONSERVATION OF GENETIC DIVERSITY/PLANT GENETIC RESOURCES

It is estimated that human beings have used about 7,000 plant species for consumption purpose. At present, about 95% of human food energy needs comes from about 30 crops only, and among them, rice, wheat, maize, and potato constitute about 60% share (FAO, http://www.fao.org/biodiversity/components/plants/en/). With changing climatic conditions and prevalence of several biotic and abiotic stress conditions daunting agriculture production and productivity, the dependence of human beings on such small number of plants necessitates to maintain a high genetic diversity within these crops. Such situation stresses the importance of conservation of diversity in plant genetic resources, which would form the basis for researchers to breed varieties that can be cultivated under unfavorable conditions, such as drought, salinity, flooding, poor soils, and heat and cold conditions.

The conservation of our plant genetic resources is extremely important to support livelihoods of millions of people inhabiting the world. The maintenance of ecosystem functionality and environmental stability also requires that the plant biodiversity should be conserved. The importance of conservation of PGR can be explained by the following points

- The diverse plant genetic resources harbor a vast array of valuable traits, which are needed for breeding of new crop varieties and for sustainable growth in productivity. These resources are a reservoir of agronomically beneficial genes and alleles needed for imparting resistance and tolerance to plants against diseases, pests, and harsh environments, found in their natural habitats. An individual genotype that is not cultivated or used by the farmer today may suddenly become essential tomorrow due to changing climatic conditions or outbreaks of disease. Although it would be in theory perfect to conserve all gene resources, it would be impractical to conserve all the available gene resources; therefore, the primary goal should be to conserve as many representative samples of existing germplasm as human resources permit. Priority should be given to those being threatened by extinction or displacement and genetic erosion (Paroda and Arora, 1991). The classic example of the revival of cassava cultivation in Uganda, after massive loss incurred by the incidence of cassava mosaic disease in 1989, through identification and selection of resistant varieties, conserved in the genebanks around the world is a case in the point (Hahn and Howland, 1972; Hahn et al., 1980; Fregene et al., 2000; Legg and Thresh, 2000).
- Loss of plant biodiversity has profound consequences on the functioning of the ecosystem. It has been proven that a reduced biodiversity

results in a loss of its functional diversity and leads to a reduction in ecosystem productivity (Rainforest Conservation Fund). Reduced biodiversity refers to a reduction in species richness (number of species), reduction in the composition of species or reduced/changed specific interactions between species. A change in any of above factors has been shown to adversely affect the ecosystem stability. Thus, it can be concluded that plants play a key role in maintaining environmental balance on the Earth by stabilizing its functional unit, the ecosystem.

- Modern intensive agriculture utilizes only a select few varieties for cultivation. These varieties have been artificially selected over a period of time by farmers and breeders. This has led to the creation of genetic uniformity and narrowing of genetic base, with wide stretches of arable land under cultivation of the same variety. In the event of an epidemic or a natural calamity, due to this genetic uniformity, the entire harvest is lost. The 1970 incidence of southern corn leaf blight in the U.S. led to a 15% reduction in corn production due to genetic uniformity. Genetic diversity, in the form of traditional landraces and cultivars, provides insurance against future, unpredictable adverse conditions such as extreme and variable environments, and evolving pests and diseases.
- Our native plant genetic resources hold enormous cultural and historical value and are our national heritage. These were grown and maintained by our ancestors through their painstaking efforts. It is, therefore, imperative for us to conserve this wealth for our future generations.

4.8 CONSERVATION OF PGR

Conservation of plant genetic resources refers to the protection and prevention from extinction, of genetic diversity of crop plants. Based on the size of sample, nature, and breeding behavior of the species and the objectives of conservation, the conservation strategies can be broadly classified into two categories:

4.8.1 IN SITU CONSERVATION

It refers to the conservation of germplasm or genetic diversity under natural conditions, within their ecosystem or natural habitat. Developments of natural parks, biosphere reserves, or gene sanctuaries are methods for achieving in situ conservation of natural biodiversity. Anthropogenic activities are largely curtailed to various degrees in these protected areas. The conservation of plant genetic resources in their natural ecosystems provides for their continuous evolution and their adaptation to changing environment. Farmers and local communities play a critical role in the conservation of plant genetic diversity in situ, especially on-farm conservation (FAO, http://www.fao.org/agriculture/crops/thematic-sitemap/theme/seeds-pgr/gpa/priority-areas/en/).

NBPGR, New Delhi, established gene sanctuaries in Meghalaya for citrus and in northeastern regions for *Musa, Citrus, Oryza, and Saccharum.*

4.8.2 EX-SITU CONSERVATION

Because in situ conservation is often not possible for each species, it becomes essential to conserve germplasm or diversity outside its natural habitat. According to CBD, the *ex situ* conservation is "the conservation of components of biological diversity outside their natural habitats" (UNCED, 1992). The *ex situ* conservation involves collection and storage of seeds and another plant reproductive material of cultivated plants and their wild relatives in genebanks. Genebanks ensure a secure conservation of plant genetic resources under suitable conditions and their easy access to breeders and researchers. A genebank is responsible for registering, studying, describing, and documenting its collection and making this information and plant material available to researchers and other interested users (Rasmussen and NordGen, http://cropgenebank.sgrp.cgiar.org). Around the world, there are about 1,750 individual genebanks, of which 130 genebanks hold 10,000 plus germplasm accessions (FAO, 2010; Jacob et al., 2015). A list of major international (genebank of the consultative group on international agricultural research) and national genebanks have given below.

- International Crop Research Institute for the Semi-Arid Tropics, India
- International Rice Research Institute, The Philippines
- International Centre for Agricultural Research in Dry Areas, Syria
- International Institute of Tropical Agriculture, Nigeria
- International Livestock Research Institute, Ethiopia
- Centro Internacional de Agricultura Tropical, Colombia
- Centro Internacional de Mejoramiento de Maíz y Trigo, Mexico

- West African Rice Development Association, Ivory Coast
- Musa International Transit Centre, Bioversity International, Belgium
- Centro Internacional de la Papa, Peru
- Information and Communication Division, International Center for Research in Agroforestry, Kenya
- National Centre for Genetic Resources Preservation (NCGRP) in the United States of America
- Institute of Crop Germplasm Resources, Chinese Academy of Agricultural Sciences (ICGR-CAAS) in China,
- ICAR-National Bureau of Plant Genetic Resources, New Delhi, India
- N.I. Vavilov All Russian Scientific Research Institute of Plant Industry (VIR) in the Russian Federation
- N.I. Vavilov All-Russian Scientific Research Institute of Plant Industry, Russia
- Asian Vegetable Research and Development Center, Taiwan
- Leibniz Institute of Plant Genetics and Crop Plant Research, Germany
- Department of Applied Genetics, John Innes Centre, Norwich Research Park, UK
- Plant Breeding and Acclimatization Institute, Poland
- Millennium Seed Bank Project, Seed Conservation Department, Royal Botanic Garden, Kew, UK
- Division of Genetics and Plant Breeding, Research Institute of Crop Production, Czech Republic

The *ex situ* conservation of plant genetic resources has been achieved in the form of seed genebanks, *in vitro* genebanks, field genebanks, cryogenebanks, and DNA genebanks. The choice of method of plant germplasm conservation depends on several factors and amenability of the target species to the particular technique in question (Thormann et al., 2005).

4.8.2.1   Seed Genebanks

Conservation of seeds in controlled conditions is the most effective and widespread method of *ex situ* conservation. Seeds are the most popular tissue of choice for conservation, as, by virtue of their small size, they occupy less space and, therefore, are convenient to store. They also represent wide genetic variability. Seed banking, therefore, offers considerable advantages over the other *ex situ* conservation methods in terms of ease of storage, relatively low labor demands, and the ability to maintain large samples at an economically viable cost. On the basis of their storability, seeds are classified into three groups

- Orthodox seeds: These are seeds that can be dried to low moisture content (5% to 10%) and stored at low temperature for long periods of time, without them losing their viability (Roberts, 1973).
- Recalcitrant seeds: These seeds are desiccation sensitive and show very drastic loss in viability with a decrease in moisture content below a value, which may range from 12% to 40%, depending on the species (Roberts, 1973). Out of a total of 600 species whose storage behavior has been studied and described, 49 were classified as recalcitrant (Ellis, 1984).
- Intermediate seeds: These seeds survive desiccation to low moisture content (~10%), but the dried seeds are injured by low temperature (Ellis et al., 1990, 1991; Hong and Ellis, 1996).

4.8.2.1.1 Conservation Strategies for Storage of Orthodox, Intermediate, and Recalcitrant Seeds

Before the seed is conserved, it is essential to know the seed storage behavior of the seeds to be conserved. Correct diagnosis and identification are crucial as improper storage conditions can lead to a complete loss of seed viability. Only after this step, the seeds can be optimally stored either for long term (e.g., at –20°C with 5±1% moisture content), medium term (e.g., at +10°C at moisture content in equilibrium with 40–50% RH, i.e., about 7–11% moisture content, depending upon species), or short term (e.g., imbibed seeds at +15°C) (Hong and Ellis, 1996). While success has been achieved to varying degrees in the conservation of seeds, the successful conservation of recalcitrant seeds still poses tremendous challenges. Researchers working on recalcitrant seed storage often encounter unexplained results that can be attributed to the great variation in seed traits between individual species-specific seeds (Chin al., 1989). Recalcitrant seeds differ in the degree of their desiccation sensitivity, in that variable degrees of dehydration are tolerated depending on the species (Berjak and Pammenter, 2008). A classification based on the degree of recalcitrance measured in terms of desiccation and chilling sensitivity is suggested by researchers. A gradient of desiccation sensitivity and low temperature tolerance seems to exist in seed populations ranging from highly recalcitrant, to moderately recalcitrant, to intermediate, and finally to orthodox-type seeds. Successful conservation of recalcitrant seeds thus requires dedicated species-specific research that aims to standardize the storage conditions in terms of moisture content, temperature, and oxygen availability. For orthodox seeds, further classification into hard seed coated and not hard seed coated has been suggested by some researchers (FAO, http://www.fao.org/docrep/006/ad232e/ad232e07.htm). The conservation of orthodox seeds appears to be far less challenging than recalcitrant species and, therefore, researchers have been able to conserve them for a considerable period of time without a loss of their viability.

4.8.2.1.2 Seed Genebank Procedures

The seed samples obtained for the purpose of conservation are added to the seed gene bank only if they prescribe to specific standards, which are as follows:

- Untreated seeds.
- Quality seeds with high viability >85%.
- Minimum 2000 seeds in self-pollinated crops and 4000 seeds in cross-pollinated crops.
- Seed moisture content 3–8%.
- Seed storage temperature at −18°C for long-term storage and +4°C for medium-term storage.
- Monitoring of viability after 10 years in long-term storage and 5 years in medium-term storage.
- Regeneration to be carried out for 100 or more plants to avoid loss of alleles.

4.8.2.2 Field Gene Bank

Field gene bank is an *ex situ* approach of plant genetic resource conservation where PGR are conserved as live undergoing normal growth and development with proper maintenance. This approach is used for plant species where conservation of its seed is difficult or impossible (owing to recalcitrant nature of seeds, difficulty in seed setting or long period of seed production time).

For a few plant species, conservation in field gene bank is essentially owing to recalcitrant seeds, e.g., coconut, cocoa, oil palm, rubber, mango, jackfruit etc. (Roberts et al., 1984). This approach is also a major method of PGR conservation in vegetatively propagated plants species such as sugarcane, sweet potato, yam, taro, cultivated banana, potato, cassava, and pineapple . In few other plant species like fruit trees, in spite of the ability to set fertile seeds, clonal propagation is preferred to maintain the genotype. Like the activities of seed gene bank, the activities of field gene bank include field collection, conservation, evaluation, and utilization of PGR.

4.8.2.2.1   Collection of PGR for Field Gene Bank

The objective of field collection is to obtain maximum diversity from minimum sample size and number following random and/or non-random sampling method (Hawkes, 1987). In crops which are propagated by seeds, collection of both cultivated and wild plant species is done based on the prevalent environmental diversity. In areas which are uniform in terms of soil type, altitude, crop cultivars, climate, farming practices etc. collection is done at large distance intervals, about 20-50 km to avoid redundancy in the material collected. On the other hand, in diverse areas, frequent sampling is done (about every km or less) to capture maximum diversity in the material collected. However, if any interesting variants are spotted (and not included in random sampling), non-random samples are also taken (Saad and Ramanatha Rao, 2001; Ramanatha Rao, 2001). For long-lived plants (trees and shrubs), woody cuttings are preferred, and in case recalcitrant seeds are collected, they should be sown within a few weeks (Yunus, 2001). In case of asexually propagated and short-lived plants, roots, tubers, bulbs, corms, etc. are collected and wherever, reproduction is by both, seeds as well as vegetatively, the germplasm collection should are similar to those of seed crops.

4.8.2.2.2   Documentation

The information to be recorded at the time of collection varies from species to species. In recent years, the amount of information at the time of collection has increased (Yunus, 2001). According to Hawkes (1980), information such as the title of the expedition, plant identification, the collector and the collector's number, date and site of collection, status of material, frequency, provenance (field, farm store, market), etc. should be recorded.

4.8.2.2.3 Conservation

Conservation strategy depends on the purpose of PGR collection and how has it been sampled in the field (Frankel, 1970; Yunus, 2001). For the germplasm collected for research purpose (evolution, systematics, cytogenetics, etc.), care is taken to retain the integrity of its components. For plant breeding activity such as for a pool of broad-based variability, it can be maintained as mass reservoirs to render the locally adapted variability required for crop improvement programs (Simmonds, 1962; Yunus, 2001).

Hawkes (1980) proposed that 8000 to 20,000 seeds should be stored for seed crops, depending on the variability of the population collected with 4000–12,000 for base collection, 1000–3000 for a duplicate of base collection, and 3000–5000 for active collection. To achieve these recommended numbers for conservation, multiplication is often necessary as during field collection the number of seeds collected are far less (Yunus, 2001). For asexually propagated crop, special care is required to control viruses and other diseases and to take strict quarantine measures (Hawkes, 1970).

4.8.2.2.4 Evaluation and Utilization

The advantage with the field gene bank is its complete compatibility evaluation and utilization as the material is available for observation all the time. Identification of the core collection (Frankel and Brown, 1984) in such case would enhance utilization of germplasm by breeders. Field genebank, therefore, in spite of many limitations, scores well among other approaches of conservation in providing all the time easy and ready access for evaluation, characterization, and utilization of the germplasm (Saad and Ramanatha Rao, 2001; Yunus, 2001).

4.8.2.3 *In Vitro* Genebanks

Although seed conservation is the most efficient and economical means of germplasm preservation, this kind of storage is not always possible. For vegetatively propagated plants that do not produce seeds like potato, sweet potato, yam, and cassava, *in vitro* approaches including tissue culture maintenance and cryopreservation are useful tools for medium- to long-term conservation. Also, in cases where seeds are recalcitrant or seeds are heterozygous and not suitable for maintaining true to type genotypes or when seeds are infected with seedborne pathogens, seed genebanking is not the most viable way of conservation (Johnson, 2002). *In vitro* conservation can be defined as conservation of germplasm under defined nutrient conditions in artificially controlled and standardized environment, in the form of *in vitro* cultures (Tyagi and Agrawal, 2014). Adopting *in vitro* techniques for conservation allows for a rapid multiplication of germplasm facilitating an international exchange of plant material. *In vitro* conservation also entails usage of minimum space for the establishment of extensive collections and helps in preventing germplasm loss caused due to pathogen or pest infestation and irregularities of climate. This method is also known as *in vitro* genebank.

The major activities of an *in vitro* genebank are (i) germplasm acquisition; (ii) disease eradication, indexing, and quarantine; (iii) culture establishment and multiplication; (iv) *in vivo* establishment and germplasm exchange; (v) storage by slow growth or cryopreservation; (vi) monitoring stability; (vii) characterisation and evaluation; and ultimately, (viii) utilization through *in vitro* breeding (Withers, 1989).

4.8.2.4 Cryogene Bank

As its name indicates, cryopreservation is conservation of tissues at ultra-low temperature (at −196°C) by using liquid nitrogen. In this method, a wide range of tissues can be used such as shoot tips, buds, embryos, protoplast, etc. This method hold greater promise in long-term conservation, as in theory, the ultra-low temperature effectively reduces or suspends metabolic activities of the conserved tissues. This approach is relatively cheaper, efficient, and relatively less cumbersome *in vitro* conservation as there is limited need for viability indexing or subculturing (Luan, 2001; Saad and Ramanatha Rao, 2001).

Cryopreservation requires more technical expertise, and care needs to taken for a few important factors for successful cryopreservation. Moisture in tissue is the vital component for successful cryopreservation as moisture above the critical level proves to be lethal as it causes ice crystallization. Similarly, desiccation injuries have to be minimized for higher survival rate, especially for recalcitrant seeds and vegetative plant parts such as meristem, shoot tips, and axillary buds. There are several desiccation techniques available for *in vitro* conservation and cryopreservation. The classical methods involve freezing to cause tissue desiccation. Several cryoprotectants such as glycerol, proline, ethylene glycol, and dimethyl sulfoxide (DMSO) are used either alone or in combination before exposing the tissues to LN (Engelmann, 1999; Luan, 2001). For regeneration, the frozen tissues are rapidly thawed in a water bath at 38–40°C and then regenerated on an agar medium. Recent desiccation techniques induce tissue water to vitrify into an amorphous glass and thus preventing injuries caused by ice crystallization (Engelmann, 1999; Luan, 2001). There are a few new promising cryopreservation techniques developed through vitrification and are enlisted below (Luan, 2001).

- **Direct desiccation of naked tissues**: The method is relatively simple where orthodox seeds are cryopreserved by desiccating to low moisture and then direct plunging into LN. For recalcitrant/semi-recalcitrant seeds, embryos are excised and cryopreserved after desiccation as they are more resistant to desiccation and low-temperature injuries. This technique has been successfully employed in a number of recalcitrant or semi-recalcitrant seeds including oil palm (Ginibun, 1997) and longan (Fu et al., 1990).
- **Desiccation of encapsulated tissues**: in this method, tissues are encapsulated in alginate and then desiccated to a suitable moisture level for cryopreservation. This is method is suitable for cryopreservation of even direct desiccation sensitive tissues such as meristem and root tips. Wide ranges of crops such as cassava (Benson et al., 1992), carrot (Dereuddre et al., 1991), oil palm (Ginibun, 1997), and rubber (Yap et al., 1999) have been cryopreserved following this technique.
- **Pre-growth in cryoprotectant prior to freezing:** here, tissues are pre-grown directly in cryoprotectants and then plunged into LN. This method was found successful for banana (Panis, 1995).

- **Pre-growth-desiccation technique**: In this technique, embryos are pre-grown in sucrose, and then, the naked embryos are desiccated to optimum moisture and then plunged into LN (Khoo, 1999; Yap et al., 1999).

4.8.2.5 DNA Genebanks

The conservation approaches mentioned earlier in this chapter have their own inherent advantages and limitations. Therefore, experts are of opinion to have a complementary approach for safe conservation of the genetic diversity available in the gene pool of the crops (Bowen, 1999). Along with collecting and conserving germplasm, PGR conservation also includes efficient management and utilization of PGR. For enhancing utilization of diverse germplasm and wild relatives of crop in breeding program, it is essential to document, characterize, and evaluate the genetic variation present in gene pools of target crops as well as respective wild relatives (FAO, 1996). In addition to conventional approaches of PGR documentation, characterization, and evaluation, molecular biology and biotechnology can offer important tools and techniques (such as DNA-based markers) which could facilitate effective conservation and utilization of PGR. The application, usefulness, and efficiency of DNA-based markers in the management of genebanks has been well recognized, such as characterization of genotypes, assessment of genetic diversity, genetic relationships studies within collections, duplicate identification, setting of core collections, and assessing genetic stability and integrity (e.g., Anderson and Fairbanks, 1990; Hodgkin and Debouck, 1992; Karp et al., 1997; Karp, 2002; de Vicente, 2004; de Vicente et al., 2004, 2006). DNA-based molecular markers employ DNA analysis and use total DNA or DNA libraries (cDNA or genomic DNA libraries) as a target. In recent times, there has been the practice of conserving DNA and libraries even after completion of the project as they constitute valuable resources. DNA-based resources can be maintained up to 2 years at –20°C and for longer periods at –70°C or in liquid nitrogen. Recently, there have been efforts to establish DNA banks with planned objectives as evident from several DNA banks across the world. It is important to note that conserving complete genome, a small portion genome, or a gene is different from conserving living organism as an entire genotype for future use. Therefore, DNA banks are not means to replace traditional ways of conserving genetic resources/germplasm (de Vicente et al., 2006).

Genetic analyses are necessary for the systematic and strategic conservation of diversity. Conservation of the species can be enhanced with the access of its DNA and molecular analysis of the same. The studies of genetic variation within plant populations have had a significant impact on the conservation genetics of the species (Byrne, 2005; Rice et al., 2006). Therefore, it has been recognized that documented DNA samples serve a crucial infrastructure need in conservation science and complements efforts to preserve biological diversity both *in situ* and *ex situ* (Dulloo et al., 2006).

4.8.2.5.1 DNA Extraction and Storage Practices

The quality of the DNA (molecular weight and authenticity of sequences) is a prime determinant before the sample is deposited in the DNA bank. There is a need to maintain standards for sample collection, extraction, characterization, and distribution. Further, the effect of different storage platforms on DNA stability over the course of time is also important to be mentioned. Models developed from degradation kinetics predict that fully hydrated DNA will be depolymerized into smaller fragments in around 10,000 years at room temperature (Lindahl, 1993). However, as observed practically, DNA tends to degrade faster if not protected from nucleases, ionizing radiation, or activated oxygen. The condition of tissue/specimen used for DNA extraction before storage, during storage, and the storage environment are crucial for determining the quality of DNA (Walters and Hanner, 2006).

A survey of DNA banks across the world gives an insight into the practice/protocols for processing and storage of DNA (Andersson et al., 2006; de Vicente and Andersson, 2006). In general, the DNA to be stored in the DNA bank is extracted using standard manual extraction protocols with/without modifications. However, the use of commercial kits for DNA extraction is also prevalent largely in developed countries than in developing countries. In the DNA banks, DNA is generally stored for medium term (6 months to 2 years, at –20°C or –70°C) and long term (more than 2 years, at –70°C or in liquid nitrogen). Mostly, total DNA is stored in the DNA banks; DNA fragments are also stored (39%).

Several institutions that store DNA also supply the DNA samples to other individuals for scientific purposes. Few DNA banks also supply DNA samples to private institutions for commercial purposes. The majority of the DNA banks provides DNA samples without any charges. DNA samples to be stored in DNA banks should have along with passport information, additional information such as quantity and quality of the DNA, the extraction protocol, marker data and sequence information, references to works or publications on the respective species, original donor, pedigree information, links to genebank data, names of other institutions that have requested the respective accessions, etc. The list of the major DNA banks worldwide is given below.

- The Royal Botanic Gardens, Kew, UK
- The US Missouri Botanical Garden
- The Australian Plant DNA Bank, Southern Cross University
- The DNA bank, Leslie Hill Molecular Systematics Laboratory of the National Botanical Institute (NBI), Kirstenbosch, South Africa
- The DNA bank, the National Institute of Agrobiological Sciences (NIAS), Ibaraki, Japan
- Plant DNA Bank in Korea (PDBK)
- DNA Bank at Kirstenbosch
- DNA Bank Brazilian Flora Species
- National Plant Genomic Resources Repository, ICAR-National Bureau of Plant Genetic Resources, New Delhi, India
- National Plant Gene Repository, National Institute of Plant Genome Research, New Delhi, India

KEYWORDS

- **genetic diversity**
- **plant genetic resources**
- ***in situ* and *ex situ* conservation**
- **gene banks**
- **orthodox**
- **recalcitrant**

REFERENCES

Allard, R. W., (1988). Genetic changes associated with the evolution of adaptedness in cultivated plants and their wild progenitors. *J. Heredity*, *79*, 225–238.

Anderson, W. R., & Fairbanks, D. J., (1990). Molecular markers: Important tools for plant genetic resources characterization. *Diversity*, *6*, 51–53.

Andersson, M. S., Fuquen, E. M., & De Vicente, M. C., (2006). State of the art of DNA storage: results of a worldwide survey. In: *DNA Banks—Providing Novel Options to Gene Banks*, De Vicente, M. C., & Andersson, M. S. (eds.), 6–10.

Arora, R. K., (1997). Biodiversity convention, global plan of action and the national programmes. In: Hossain, M. G., Arora, R. K., & Mathur, P. N., (eds.) Plant genetic resources–Bangladesh perspective, *Proceedings of a National Workshop on Plant Genetic Resources*, 26–29 August 1997, Bangladesh Agricultural Research Council, BARC-IPGRI, Dhaka, Bangladesh, pp. 28–35.

Benson, E. E., Chabrillange, N., & Engelmann, F., (1992). A comparison of cryopreservation methods for the long term *in vitro* conservation of cassava. *Proceedings of Autumn Meeting of the Society of Low Temperature Biologists*, Stirling, UK, 8–9.

Berjak, P., & Pammenter, N. W., (2008). From avicennia to zizania: Seed recalcitrance in perspective. *Annals of Botany*, *101*, 213–228.

Bowen, B. W., (1999). Preserving genes, species, or ecosystems? Healing the fractured foundations of conservation policy. *Molecular Ecology*, *8*, S5–S10.

Brown, A. H. D., (1988). The genetic diversity of germplasm collections. In: Fraleigh, B., (ed.) *Proceedings of a Workshop on the Genetic Evaluation of Plant Genetic Resources*, Toronto, Canada, Research Branch, Agriculture Canada, Toronto, pp. 9.

Brown, A. H. D., (2008). Thematic background study on *"Indicators of Genetic Diversity, Genetic Erosion and Genetic Vulnerability for Plant Genetic Resources for Food and Agric*ulture." A report submitted to FAO. Food and Agriculture Organization of the United Nations, Rome.

Byrne, C. A., (2005). Genetic variation in plant populations: assessing cause and pattern. In: Henry, R. J., editor. *Plant Diversity and Evolution*. CABI Publishing, Wallingford, UK. pp. 139–165.

CBD, (1992). Convention on Biological Diversity. *Secretariat of the Convention on Biological Diversity*, Montreal, Quebec, Canada.

Chin, H. F., & Krishnapillay, B. P. C., (1989). Stanwood seed moisture: Recalcitrant vs. orthodox seeds crop science society of America, 677 S. Segoe Rd., Madison, WI 53711, USA. Seed moisture, CSSA special publication no. 14. *Seed Storage FAO Corporate Document Repository (Forestry Department)* http://www.fao.org.

Cooper, H. D., Spillane, C., & Hodgkin, T., (2001). Broadening the genetic base of crops: an overview. In: *Broadening the Genetic Base of Crop Production*. Cooper, H. D., Spillane, C., Hodgkin T (eds.), FAO, Rome (Italy).

Day, R. K., Heisey, P., Shoemaker, R., Sullivan, J., & Frisvold, G., (2005). Crop genetic resources: An economic appraisal. Washington, DC, US. Department of Agriculture, *Economic Information Bulletin Number*, *2*, 47.

De Vicente, M. C., & Andersson, M. S., (eds.), (2006). DNA banks—providing novel options for gene banks? *Topical Reviews in Agricultural Biodiversity*. International plant genetic resources institute, Rome, Italy.

De Vicente, M. C., (2004). The evolving role of gene banks in the fast-developing field of molecular genetics. *Issues in Genetic Resources No., 11*. International Plant Genetic Resources Institute, Rome, Italy, 26–32.

De Vicente, M. C., Metz, T., & Alercia, A., (2004). *Descriptors for Genetic Markers Technologies.* International Plant Genetic Resources Institute (IPGRI), Rome, Italy, 1–22.

Dereuddre, J., Blandin, S., & Hassen, H., (1991). Resistance of alginate-coated somatic embryos of carrot (*Daucus carota* L.) to desiccation and freezing in liquid nitrogen. 1. Effects of preculture. *Cryo-Letters, 12,* 125–134.

Dulloo, E., Nagamura, Y., & Ryder, O., (2006). DNA storage as a complementary conservation strategy. In: *DNA banks-Providing Novel Options to Gene banks.* De Vicente, M. C., & Andersson, M. S., (eds.), 11–24.

Ellis, R. H., (1984). Revised table of seed storage characteristics. *Plant Genetic Resources Newsletter, 58,* 16–33.

Ellis, R. H., Hong, T. D., & Roberts, E. H., (1990). An intermediate category of seed storage behaviour? I. Coffee. *J. Exp. Bot., 41,* 1167–1174.

Ellis, R. H., Hong, T. D., & Roberts, E. H., (1991). An intermediate category of seed storage behaviour? II. Effects of provenance, immaturity, and imbibition on desiccation-tolerance in coffee. *J. Exp. Bot., 42,* 653–657.

Engelmann, F., (1999). Management of field and *in vitro* germplasm collections. *Proceedings of a Consultation Meeting,* 15–20 January 1996, CIAT, Cali, Colombia. International Plant Genetic Resources Institute (IPGRI), Rome, Italy, pp. 165.

FAO, (1983), *Commission on Plant Genetic Resources.* Resolution 8/83 of the 22nd session of the FAO conference. Rome, Italy.

FAO, (2003), *Glossary on Forest Genetic Resources* (English version). Development. Rome, Italy.

FAO, (2006) World agriculture: Towards 2030/2050–Interim report. Rome, Italy. (http://www.fao.org/fileadmin/user_upload/esag/docs/Interim_report_AT2050web.pdf).

FAO, (2010), The second report on the state of the world's plant genetic resources. Rome, Italy, FAO, pp. 370.

FAO, http://www.fao.org/docrep/006/ad232e/ad232e07.htm (Accessed on February, 2016).

FAO, (1996) Report on the state of the world's plant genetic resources for food and agriculture. Food and Agriculture Organization of the United Nations (FAO), Rome, Italy, pp. 75.

FAO, http://www.fao.org/biodiversity/components/plants/en/ (accessed on February, 2016).

Frankel, O. H., (1970). Genetic conservation in perspective. In: *Genetic Resources in Plants– Their Exploration and Conservation* (Frankel, O. H., & Bennett, E., eds.). IBP Handbook No. *11*, Blackwell Scientific Publications, Oxford, pp. 469–489.

Frankel, O. H., (1977). Natural variation and its conservation. In: *Genetic Diversity in Plants,* Muhammed, A., Aksel, R., & Von Borstel, R. C., (eds.), Plenum Press, New York, pp. 21–44.

Frankel, O. H., & Brown, A. H. D., (1984). Plant genetic resources today – a critical appraisal. In: *Crop Genetic Resources – Conservation and Evaluation* (Holden, J. H. W., & Williams, J. T., eds.). George Allen and Unwin, London, pp. 249–259.

Fregene, M., Bernal, A., Duque, M., Dixon, A., & Tohme, J., (2000). AFLP analysis of African cassava (Manihot esculenta Crantz) germplasm resistant to the cassava mosaic disease (CMD). *Theoretical and Applied Genetics, 100,* 678–685. doi: 10.1007/s001220051339.

Fu, J. R., Zhang, B. Z., Wang, X. P., Qiao, Y. Z., & Huang, X. L., (1990). Physiological studies on desiccation, wet storage and cryopreservation of recalcitrant seeds of three fruit species and their excised embryonic axes. *Seed Science and Technology, 18,* 743–754.

Ginibun, F. C., (1997). Effects of speed of desiccation and prefreezing on survival of naked and encapsulated oil palm (Elaeis guineensis Jacq.) embryos in liquid nitrogen. *Graduation Project,* Fakulti Pertanian, Universiti Pertanian Malaysia.

Govindaraj, M., Vetriventhan, M., & Srinivasan, M., (2015). Importance of genetic diversity assessment in crop plants and its recent advances: an overview of its analytical perspectives. *Genet. Res. Int. 431487.* doi: 10.1155/2015/431487.

Graham, S., (2002). *Global Estimate of Endangered Plant Species Triples.* Scientific American November 1, 2002. http://www.scientificamerican.com/article.cfm?id=global-estimate-of-endang&print=true).

Hahn, S. K., & Howland, A. K., (1972). Breeding for resistance to cassava mosaic. In: Hahn, S. K., (ed.) IITA International Institute of Tropical Agriculture, Proceed*ings of the Cassava Mosaic Workshop,* Ibadan, Nigeria, pp. 4–7.

Hahn, S. K., Terry, E. R., & Leuschner, K., (1980). *Cassava Breeding for Resistance to Cassava Mosaic Disease.* Euphytica, *29,* 673–683.

Hamrick, J. L., & Godt, M. J. W., (1990). Allozyme diversity in plant species. In: Brown, A. H. D., Clegg, M. T., Kahler, A. L., & Weir, B. S., (eds.) *Plant Population Genetics, Breeding and Genetic Resources.* Sinauer Associates Inc., Sunderland, pp. 43–63.

Hamrick, J., L., Godt, M. J. W, & Sherman-Broyles, S. L., (1992). Factors influencing levels of genetic diversity in woody plant species. *New Forests, 6,* 95–124.

Hamrick, J. L., (1993). Genetic diversity and conservation of tropical trees. In: Drysdale, R. M., John, S. E. T., & Yapa, A. C., (eds.) Proceed*ings an International Symposium on Genetic Conservation and Production of Tropical Tree Seed,* 14–16 June, Chiang Mai, Thailand, ASEAN-Canada Forest Tree Centre, Thailand, pp. 1–10.

Hamrick, J. L., Murawski, D. A., & Nason, J. D., (1993). The influence of seed dispersal mechanisms on the genetic structure of tropical tree populations. *Vegetation, 107/108,* 281–297.

Hamrick, J. L., & Godt, M. J. W., (1997). Allozyme diversity in cultivated crops. *Crop Sci., 37,* 26–30.

Harlan, J. R., (1975). *Crops and Man.* American Society of Agronomy, Madison, Wisconsin, 295.

Hawkes, J. G., (1970). The conservation of short-lived asexually propagated plants. In: *Genetic Resources in Plants – Their Exploration and Conservation* (Frankel, O. H., & Bennett, E., eds.). IBP Handbook No. *11,* Blackwell Scientific Publications, Oxford, pp. 495–499.

Hawkes, J. G., (1980). *Crop Genetic Resources Field Collector's Manual.* IBPGR/Eucarpia, pp. 7.

Hawkes, J. G., (1987). World strategies for collecting, preserving and using genetic resources. In: *Improving Vegetatively Propagated Crops* (Abbott, A. J., & Atkin, R. K., eds.). Academic Press, London, pp. 285–301.

Hodgkin, T., & Debouck, D. G., (1992). Some possible applications of molecular genetics in the conservation of wild species for crop improvement. In: *Conservation of Plant Genes,* Adams, R. P., & Adams, J. E., editors. DNA banking and *in vitro* biotechnology. Academic Press, San Diego, California, USA. pp. 135–152.

Hong, T. D., & Ellis, R. H., (1996). A protocol to determine seed storage behaviour. IPGRI *Technical Bulletin No. 1,* (Engels, J. M. M., & Toll, J., eds.) International Plant Genetic Resources Institute, Rome, Italy.

IBPGR, (1991) *Elsevier's Dictionary of Plant Genetic Resources.* Elsevier Science Publishers, B. V., Amsterdam, The Netherlands.

IUCN, (2012) The IUCN Red List of threatened species 2012. 2, www.iucnredlist.org.

Jacob, S. R., Tyagi, V., Agrawal, A., Chakrabarty, S. K., & Tyagi, R. K., (2015). Indian Plant Germplasm on the Global Platter: *An Analysis. PLoS One, 10,* 5: e0126634, doi: 10.1371/journal.pone. 0126634.

Johnson, K., (2002). *In vitro* conservation including rare and endangered plants, heritage plants and important agricultural plants. in: Taji, A., Williams, R., (eds.) *The Importance of Plant Tissue Culture and Biotechnology in Plant Sciences,* Armidale, Australia, University of New England, pp. 79–90.

Karp, A., Kresovich, S., Bhat, K. V., Ayad, W. G., & Hodgkin, T., (1997). *Molecular tools in Plant Genetic Resources Conservation*: A guide to the technologies. *Technical Bulletin No. 2.* International Plant Genetic Resources Institute (IPGRI), Rome, Italy, ISBN: 92-9043-323-X.

Karp, A., (2002). The new genetic era: will it help us in managing genetic diversity? In: Engels, J. M. M., Rao, V. R., Brown, A. H. D., & Jackson, M. T., editors. *Managing Plant Genetic Diversity.* CABI Publishing and International Plant Genetic Resources Institute (IPGRI), Wallingford, UK and Rome, Italy. pp. 43–56.

Khoo, S. F., (1999). Effects of sucrose concentration and desiccation on survival of naked oil palm (*Elaeis guineensis* Jacq.) embryos in liquid nitrogen. *Graduation Project, Fakulti Pertanian,* Universiti Pertanian Malaysia.

Kumar, S., (2016). *What is in Vitro Germ Plasm Conservation?* http://www.preservearticles. com/2012042631174/what-is-in-vitro-germplasm-conservation.html.

Legg, J. P., & Thresh, J. M., (2000). Cassava mosaic virus disease in East Africa: a dynamic disease in a changing environment. *Virus Res., 71,* 135–149. doi: 10.1016/s0168–1702(00)00194–5.

Lindahl, T., (1993). Instability and decay of the primary structure of DNA. *Nature, 362,* 709–715.

Lorenz, J. G., Jackson, W. E., Beck, J. C., & Hanner, R., (2005). The problems and promise of DNA barcodes for species diagnosis of primate biomaterials. *Phil. Trans. R. Soc. B., 360,* 1869–1877.

Loveless, M. D., & Hamrick, J. L., (1984). Ecological determinants of genetic structure in plant populations. *Ann. Rev. Ecol. Syst. 15,* 65–96.

Luan, H. Y., (2001). *In vitro* conservation and cryopreservation of plant genetic resources. In: *Establishment and Management of Field Gene bank, a Training Manual.* Saad, M. S., Ramanatha Rao,V., (eds.). IPGRI-APO, Serdang., 54–58.

Mathur, P., (2011). Assessing the threat of genetic erosion. In: *Collecting Plant Genetic Diversity: Technical Guidelines – 2011 Update* Guarino, L., Ramanatha Rao, V., Goldberg, E., eds.), Bioversity International, Rome, Italy, 1–7.

Nautiyal, S., (2011). Plant biodiversity and its conservation in Institute for Social and Economic Change (ISEC) Campus, Bangalore: A case study. *Journal of Biodiversity, 2,* 9–26.

Nevo, E., Beiles, A., & Ben-Shlomo, R., (1984). The evolutionary significance of genetic diversity: ecological, demographic and life history correlates. *Lect. Notes Biomath., 53,* 13–21.

NRC, (1972) *Genetic Vulnerability of Major Crops*. National Research Council, National Academy of Sciences, Washington, DC.

Panis, B., (1995). Cryopreservation of banana (*Musa* spp.) germplasm. *Dissertationes de Agricultura*, Katholieke Universiteit Leuven, Belgium, 201.

Paroda, R. S., & Arora, R. K., (1991). Plant genetic resources: general perspective (English) In: *Plant Genetic Resources: Conservation and Management*. Concepts and approaches/ International Board for Plant Genetic Resources, New Delhi (India). Regional Office for South and Southeast Asia, 1–23.

Pimm, S. L., Gittlaman, J. L. G., McCracken, F., & Gilpin, M., (1989). Genetic bottlenecks: alternative explanations for low genetic variability. *Trends in Ecology and Evolution, 4,* 176–177.

Ramanatha, R. V., Quek, P., Bhag, M., & Zhou Ming-De, (1999). Role of IPGRI in promoting research on PGR conservation and use, and GPA implementation, with a focus on Asia and the Pacific. In: Zhou Ming-De, Zhang Zongwen, & Ramanatha Rao, V., (ed) *Proceedings of a National Workshop on Conservation and Utilization of Plant Genetic Resources in China*, Beijing, October 1999 (in press) IPGRI Office for East Asia, Beijing, 25–27.

Ramanatha, R. V., (2001). Principles and concepts in plant genetic resources conservation and use. In: *Establishment and Management of Field Gene bank, a Training Manual*. Saad, M. S., Ramanatha Rao,V., (eds.). IPGRI-APO, Serdang, 1–16.

Ramanatha, R. V., & Hodgkin, T., (2002). Genetic diversity and conservation and utilization of plant genetic resources. *Plant Cell, Tissue and Organ Culture, 68,* 1–19, 2002.

Rasmussen, M., (2017) NordGen–plants the role of gene banks in finding and conserving our cultural heritage. *In vitro* conservation of cassava genetic resourceshttps://www.nordgen.org/en/plants/.

Rice, N., Henry, R., & Rossetto, M., (2006). DNA banks: a primary resource for conservation research. In: *DNA Banks—Providing Novel Options to Gene banks*. De Vicente, M. C., & Andersson, M. S., (editors), ISBN: 9789290437024.

Roberts, E. H., (1973). Predicting the storage life of seeds. *Seed Science and Technology, 1,* 499–514.

Roberts, E. H., King, M. W., & Ellis, R. H., (1984). Recalcitrant seeds: their recognition and storage. In: *Crop Genetic Resources: Conservation and Evaluation* (Holden, J. H. W., & Williams, J. T., eds.). George Allen and Unwin, London, pp. 38–52.

Rogers, D. L., (2004). Genetic erosion: No longer just an agricultural issue. *Native Plants J., 5,* 112–122.

Saad, M. S., & Ramanatha, R. V., (eds), (2001). *Establishment and Management of Field Gene Bank, a Training Manual*. IPGRI-APO, Serdang, 1–107.

Simmonds, N. W., (1962). Variability in crop plants, its use and conservation. *Biol. Rev., 37,* 442–465.

Suneson, C. A., (1960). Genetic diversity–a protection against diseases and insects. *Agron. J., 52,* 319–321.

Thormann, I., Dulloo, M. E., & Engels, J. M. M., (2005). Techniques of *ex situ* plant conservation. In: *Plant Conservation Genetics*, Henry, R., editor. Centre for Plant Conservation Genetics, Southern Cross University, Lismore, Australia. The Haworth Press Inc., Binghamton, New York, USA, 9–12.

Tyagi, R. K., & Agrawal, A., (2014). *Genetic Resources Conservation, Ex situ* conservation strategies" during International Training on "*In vitro* Conservation and Cryopreservation.

UNEP, (1992) *Convention on Biological Diversity*, NA 92–7807, New York.

Walters, C., & Hanner, R., (2006). Platforms for DNA banking, In: *DNA banks—Providing Novel Options to Gene Banks*. de Vicente, M. C., & Andersson, M. S., (eds).

Watanabe, K. N., Ramanatha, R. V., & Iwanaga, M., (1998). International trends on the conservation and use of plant genetic resources. Pl. *Biotechnology, 15*(3), 115–122.

Wilkes, G., (1984). Germplasm conservation towards the year 2000. Potential for new crops and enhancement of present crops. In: *Plant Genetic Resources: a Conservation Imperative*. Yeatman, C., Kefton, D., & Wilkes, G., (eds.). *American Assoc. for the Advancement of Sciences*. Washington, D. C., USA.

Yap, L. V., Hor, Y. L., & Normah, M. N., (1999). Effects of sucrose preculture and subsequent desiccation on cryopreservation of alginate encapsulated *(Hevea brasiliensis* Muel.- Arg) embryos. In: *Proc. IUFRO Seed Symposium -Recalcitrant Seeds*. FRIM, Malaysia, pp. 140–145.

Yunus, A. G., (2001). Introduction to field gene bank. In: *Establishment and Management of Field Gene bank, a Training Manual*. Saad, M. S., Ramanatha Rao, V., (eds.). IPGRI-APO, Serdang, 59–63.

CHAPTER 5

USES AND IMPORTANCE OF WIDE HYBRIDIZATION IN CROP IMPROVEMENT

VIKAS GUPTA, CHANDRA NATH MISHRA, and SATISH KUMAR

ICAR–Indian Institute of Wheat and Barley Research (IIWBR), Karnal – 132001, India, E-mail: kumarsatish227@gmail.com

CONTENTS

5.1   INTRODUCTION

Improvement in any crop involves crossing among the different varieties or genotypes to create variability, followed by selecting the desirable genotype to be released as a variety. Genetic variation has been eroded in most of the crops under modern agriculture systems. Genetic erosion not only limits further improvement in traits like yield and quality but also makes the crop vulnerable to biotic and abiotic stresses. The realization of yield increment during green revolution resulted in the narrow genetic base because of utilization of common dwarfing systems in crop breeding mainly in wheat and rice. The introduction of genetic variation available in the related or distant relatives of crop plants provides a means to enhance variability to be available to breeders. The term hybridization is used to refer to cross-fertilization. When individuals belonging to different species but having the same genus are crossed, it is termed as interspecific hybridization. When individuals belonging to two different genera are crossed, it is referred to as intergeneric hybridization. Wide hybridization refers to the mating between individuals of different species or genera, which are reproductively iso-lated from one another to various extents. Wide hybridization provides a means to break species barriers for gene transfer, thus making it possible to combine genomes of different species into one nucleus. It is difficult to establish when actual plant breeding was started and practiced, but hybrid-ization, which led to the significant establishment of plant breeding, was first practiced as early as the 18th century. The first documented record of hybridization was the development of first hybrid between Sweet William and carnation in 1717, popularly known as Fairchilds mule. But practical application of such crosses became obvious when Kolreuter (1733–1806) produced hybrids through artificial crossing in tobacco (Chahal and Gosal, 2002). Thomas Knight produced varieties of horticultural crops using arti-ficial hybridization during 1811–1833. Subsequently, interspecific hybrid-ization was started in different crops, and in some crops, many commercial varieties have been released, particularly in ornamentals. The first success-ful intergeneric hybrid between wheat and rye resulted in the development of a new crop Triticale by Rimpau (1890). Triticale is the first manmade crop that took a long time for its improvement so that it is acceptable to be grown as a crop. Now, Triticale varieties have been released and are culti-vated in Canada and India. Similarly, another classical example of interge-neric hybrid Raphanobrassica was developed by Karpachenko (1924). He crossed *Raphanus sativus* and *Brassica oleracea* and obtained F_1 hybrid that was completely sterile, but upon spontaneous chromosome doubling were fertile. Raphanobrassica does not occur naturally, as far as it is known. Intergeneric crossing is an integral part of most of the breeding programs of field crops. In intergeneric crosses, sterility is a common problem due to the different genomes; it is therefore important to go for repeated backcross-ing of wide hybrids to their parental species for attaining stability of the introgressed chromosomes. In some crops, wide hybridization is common for further improvement such as sugarcane, but in most of the crops, wide hybridization is practised for transferring few traits of economic importance, particularly resistance traits.

5.2   WHY THERE IS A NEED FOR WIDE HYBRIDIZATION

Plant breeding employs the use of genetic variation available within a crop species as well as creating variability by transferring genes from related and distant species for developing improved varieties. The target trait that breeder incorporates to improve depends upon the available variability for that trait in the available germplasm to be used as a donor. If the trait is available in related germplasm, then it is easy to transfer compared to that if available in an unrelated species. Gene transfer within a species is easy as problem of linkage drag is not associated as compared to gene transfer across species. The modern day cultivars in most of the cultivated crop species derived from only a small fraction of germplasm in the primary gene pool, having a narrow genetic base leading to limited buffering potential. The monoculture of these varieties with high yield potential occupying larger areas generally eroded genetic variability. The use of wild species that have contributed to the evolution of crop species will be beneficial for adding new variability in the breeding germplasm. The use of wild relatives of wheat and rice is well documented by Jiang et al. (1993) and Brar and Khush (1997), respectively, for the transfer of genes of economic importance.

5.3   BARRIERS IN THE PRODUCTION OF WIDE HYBRIDS

Problems in the production of hybrids vary greatly depending upon the choice of species, particularly those belonging to primary, secondary, or tertiary gene pool. The hybrids are easily produced in some crop species The hybrids are easily produced in some crops like tomato (cultivated tomato with *Lycopersicon pimpinellifolium*) and rice (cultivated rice with wild *O. rufipogon*). But, in the vast majority of the cases, F_1s are obtained with variable degree of difficulty, and in most cases, it is difficult to have a distant hybrid through conventional methods. Further, they require specific techniques like embryo rescue to have fertile F_1. Several reproductive barriers are known to limit wide hybridization. These are pre- and postfertilization barriers. Prefertilization barriers are failure of pollen germination, slow pollen tube growth, or arrest of the growth of pollen tube in style or ovary tissue, thus preventing it to reach ovule. The postzygotic barriers mainly include chromosome elimination, hybrid embryo breakdown, and failure of hybrid seedling development. These mechanisms are discussed in detail in the following sections.

5.3.1 PREZYGOTIC BARRIERS

Lack of crossability in the interspecific crosses due to incompatibility in the fertilization process was termed as interspecific incompatibility by Nettancourt in 1977. In normal and in interspecific crosses, pollen pistil interaction must pass through certain stages for successful hybridization, which include pollen germination on stigma due to hydration, pollen tube growth and penetration, and finally entry into female gametophyte for fusion to form zygote (Heslop-Harrison, 1987). In the interspecific crosses, inhibition of pollen tube growth was reported in *Lycopersicon, Linum, Brassica,* and *Trifolium.* Similarly, failure of pollen germination and penetration of pollen tube into the stylar tissues in interspecific crosses between *Cicer arietinum* and *C. soongaricum* was observed by Mercy and Kakkar (1975). Evans (1962) observed slower pollen tube growth in the interspecific crosses in *Trifolium,* but there was no failure of pollen germination. Incompatibility of the pollen tube and stylar tissue proved to be an important prefertilization barrier to compatibility. In the normal course of processes in hybridization, the pollen tube delivers two sperm nuclei to the embryo sac, one fertilizes the egg cell and the other fertilizes the two polar nuclei to form triploid endosperm nucleus. In wide crosses, these events are disturbed as observed in wheat × Maize crosses; only embryos were formed without endosperm (Laurie and Bennet, 1988). These embryos abort if left to develop on plants.

5.3.2 POSTZYGOTIC BARRIERS

The processes leading to postzygotic barriers in distant hybridizations are poorly known. The prevailing view is that postzygotic isolation is usually attributed to the negative epistatic interactions of parental genomes in artificially produced interspecific hybrids (Burke and Arnold, 2001). In some of the interspecific crosses, hybridization take place to form zygote, but the development of zygote is affected at different stages (Hermsen, 1967). Besides hybrid embryo lethality, which results in seed abortion (Sears, 1944), distant hybridizations often come with seedling lethality and morphological abnormalities like hybrid weakness and death at later developmental stages (Hermsen, 1967). Postzygotic barriers are discussed in the following subsections.

5.3.2.1 Chromosome Elimination

After zygote formation, regular mitosis should follow for the proper development of embryo for the continuation of generation. Chromosome elimination of one parent takes place during the early mitotic divisions as reported by Subrahmanyam and Kasha (1973) in barley. In the crosses between *Hordeum vulgare* and *H. bulbosum*, the chromosomes of *H. bulbosum* were invariably eliminated regardless of the ploidy level and direction of cross, resulting in the production of haploid barley. Similarly, in the intergeneric crosses between *T. aestivum* and *H. bulbosum*, *bulbosum* chromosomes were eliminated producing wheat haploids (Barclay, 1975). An interesting discovery of chromosome elimination was observed in *H. vulgare* and *H. marinum* crosses (Finch and Bennet, 1983). In these crosses, dual chromosome elimination occurs in which *H. marinum* chromosomes eliminated in the embryo, while the *H. vulgare* chromosomes were lost in the endosperm. Chromosome elimination in wide crosses has been exploited for the production of haploids for genetic studies.

5.3.2.2 Lethal Genes

In some wide crosses, fertilization may occur, but the developing zygote dies. Sears (1944) observed that the intergeneric hybrids between einkorn wheat and *Aegilops umbellulata* either died in the seedling stage or reached maturity depending on whether the lethal gene 1 or its non-lethal allele L was present. In the former species, the author could distinguish between three alleles for early lethality (Le), late lethality (LL), and viability (I). Seedling lethality in *T. aestivum* × *Secale cereale* hybrids results from the new combination of rye and wheat chromatin as reported by May and Appels (1984). Wheat plants disomic for a 2RS/2BL translocation chromosome substituting for chromosome 2B show seedling lethality. Hybrid lethality at the seedling stage has been most thoroughly studied in the genus *Nicotiana* (Kobori and Marubashi, 2004). In crosses between *T. durum* and rye (Raina, 1984), hybrid lethality is characterized by abnormal development of embryo and endosperm and seeds developed failed to germinate.

5.3.2.3 Hybrid Breakdown

Hybrid sterility and hybrid breakdown are commonly encountered during wide hybridization studies. Hybrid breakdown is a difficult problem to solve because of its complex mode of inheritance. Stebbins (1950) proposed that hybrid breakdown is due to disharmonious interaction between combinations of the genes of parental species. In crosses between two *indica* and *japonica*, subspecies of *Oryza sativa*, varying degrees of hybrid sterility and hybrid breakdown were observed (Oka, 1988). Certain rice varieties produce fertile F_1 hybrids when crossed to either *ssp. indica* or *ssp. Japonica*, and thus, they are called widely compatible varieties (Ikehashi and Araki, 1986). Hybrid sterility and hybrid breakdown often concur in *indica/japonica* crosses, but hybrid breakdown is sometimes found in advanced generation progenies from completely fertile F_1 hybrids. Oka (1974) found a pair of complementary recessive sterility genes that caused hybrid breakdown in *indica/japonica* crosses. Many F_1 sterility genes have been identified, and some of them have already been localized on the rice genetic map (Yanagihara et al., 1995). *Triticum boeoticum, T. urartu, Ae. squarrosa,* and *Ae. speltoides* exhibited reciprocal differences in hybrid seed morphology, endosperm development, and embryo viability (Gill and Waines, 1978). *T. urartu* and *Ae. squarrosa* as females in crosses with *T. boeotiaum* and *Ae. speltoides* lead to shriveled inviable seed. *T. boeoticum* accessions as female with *Ae. speltoides* also lead to shriveled seeds, whereas the reciprocal crosses produced plump seeds that either resembled the maternal parent or showed size differences. Genetic experiments involving hybrids of *T. boeoticum, T. urartu, and T. monococcum* showed that a factor is present in pollen or male gametes, which shows dosage effect and by interacting with the maternal genome, leads to endosperm abortion.

5.3.2.4 Hybrid Necrosis

Hybrid necrosis is the premature death of the foliage in some hybrids. Hybrid necrosis occurs in *Triticum, Aegilopes, Lycopersicon, Oryza,* and *Gossypium.* In *Triticum,* this characteristic is studied extensively. Hybrid necrosis, characterized by necrosis of leaf and sheath tissues on hybrid plants, has been frequently observed in F_1 hybrids between genotypes of common wheat (*T. aestivum* L., 2n=6x=42, AABBDD genomes) and between common wheat and tetraploid wheat (*T. turgidium* L., 2n=4x=28, AABB) (Tsunewaki, 1992). Hybrid necrosis is usually lethal or semi-lethal, resulting in gradual death or loss of productivity (Tomar and Singh, 1998). Hybrid necrosis is controlled by the complementary dominant genes *Ne1* and *Ne2*, which are located on chromosome arms 5BL and 2BS, respectively (Nishikawa et al., 1974). The *Ne1* and *Ne2* genes are widely distributed among different subspecies, varieties, and commercial cultivars of common wheat (Tsunewaki, 1992).

5.4 OVERCOMING THE BARRIERS IN WIDE HYBRIDIZATION

Wild relatives of crop plants play an important role in plant breeding by providing a reservoir of genetic diversity for different agronomic traits. Crossing barriers occur frequently in intra and interspecific crosses, which hamper the development of hybrid embryos. The nature of the barrier determines the method to be used to overcome the specific barrier. A range of techniques such as bud pollination, stump pollination, use of mentor pollen, and grafting of the style have been applied successfully to overcome prefertilization barriers. *In vitro* methods employed to prevent embryo abortion like ovule and embryo culture are being used to overcome postfertilization barriers. A combination of both, *in vitro* fertilization followed by embryo rescue, has been applied in many crosses. If sterility is caused due to lack of chromosome pairing during meiosis, fertility may be restored by polyploidization, thereby enabling pairing of homologous chromosomes in the allopolyploid hybrid. Hybrid breakdown occurs in F_2 or later generations. Techniques developed for overcoming these pre- and postfertilization barriers are discussed below.

5.4.1 *TECHNIQUES FOR OVERCOMING PREFERTILIZATION BARRIERS*

5.4.1.1 Manipulation of Chromosome Number

When species with different ploidy levels are crossed, hybridization between them is relatively difficult than when both species with the same ploidy level are crossed. Several crop plants like Wheat, potato, cotton, tobacco, oats, brassicas etc. differ in ploidy levels with their progenitors/ wild relatives. Many of the wild relatives of these crops are diploids or having lower chromosome number than the cultivated crop species. On the other hand, crops like rice, maize, barley, etc. have wild species with higher ploidy level than the cultivated species. For successful hybridization, the chromosome number of either of the parent is doubled. In potato, many wild species have been crossed with the cultivated potato species after doubling the chromosome number. Livermore and Johnstone (1940) doubled the chromosome number of *Solanum chacoense* to increase its crossability with *S. tuberosum*. Similarly, frost and nematode resistance genes have been transferred to *Nicotiana tabaccum* from *Nicotiana* species. Knight (1954) doubled the chromosome number of *Gossypium arboretum* and then crossed with *G. hirsutum* to transfer black arm resistance. Sometimes, the parent with higher ploidy level can be used as female to cross with diploid species.

5.4.1.2   Bridging Species

When two parents are incompatible, a third parent that is compatible with both the parents can be used for bridge crosses, and thus, it becomes possible to perform cross between the original parents. This technique was successfully used in wheat, tobacco, and lettuce to make wide crosses. Many diploid species of wheat do not cross directly with hexaploid wheat. In this case, tetraploid wheat is used as a bridging species to transfer gene of agronomic importance. An example of bridge cross is the transfer of eyespot resistance in wheat from *Ae. ventricosa* by Doussinault et al. (1983). *T. turgidium* was used as a bridge species by crossing it first with *Ae. ventricosa*, and the resulting amphiploid was crossed with hexaploid wheat to obtain hybrid. Crosses between *N. repanada* (2n = 48) with *N. tabacum* (2n = 48) are difficult to make. Burk (1967) made use of *N. sylvestris* (2n = 24 as a bridging species to transfer nematode resistance from *N. repananda* to *N. tabacum*.

5.4.1.3   Use of Mentor Pollen

The use of mixed pollen, i.e., mixture of compatible and incompatible pollen and mentor pollen, in which compatible pollen is genetically inactivated by irradiation but is still capable to germinate, is reported to overcome inhibition in the style in many plant species, when used together with incongruous pollen. Recognition pollen has been used for overcoming self-incompatibility in *Seasmum indicum* × *S. mulayanum* cross by Sastri and Shivanna in 1976.

5.4.1.4   Shortening the Style

Several species differ in their floral morphology that affects the hybridization process. The style of one species is too long or short for the pollen tubes of the other species to reach the ovary. This type of barrier is encountered in hybridization between *Zea mays* and *Tripsacum dactyloides*. In order to achieve hybridization, the part of stigma of maize is cut off when it is crossed as female with *Tripsacum*.

5.4.1.5   Use of Growth Hormones

Pollen tube growth and embryo development were found to be promoted by certain hormones. In addition to this, hormones promote the stigma receptivity as well as prevent the early abscission of flowers. GA3 has been reported to increase the frequency of zygote formation in wheat × rye crosses (Larter and Chaubey, 1965). Similarly, the application of 2,4-D followed by GA3 before pollination allowed intergeneric crosses between *H. vulgare* and species of *Avena*, *Triticum*, and *Lolium*. The application of EACA to flower buds of *Vigna radiata before* pollination accelerated the production of hybrids with *V. umbellate*.

5.4.2   *TECHNIQUES FOR OVERCOMING POSTZYGOTIC BARRIERS*

5.4.2.1   Embryo Rescue

Among the postzygotic barriers, abortion of embryos is the most common, which may occur during different stages of embryo development. Those embryos that are likely to be aborted can be artificially cultured on a nutrient medium and can be grown as mature plants. The first successful example of artificially culturing hybrid embryos in a cross between *Linum perenne* and *L. austriacum* was by Laibach (1929). Embryo rescue technique involves excision of immature embryos and culturing them generally on Murashige and Skoogs artificial media. For few days, the embryos are grown in dark but later shifted to diffused light. This technique has been used in a wide variety of crops for producing interspecific hybrids *Solanum, Lycopersicon, Oryza, Brassica, Hordeum, Agropyron, Triticum,* and *Gossypium.* Resistance to potato leaf roll virus (PLRV) has successfully been introgressed from *S. etuberosum* to *S. tuberosum* through embryo rescue (Chavez et al., 1988). Interspecific hybrids have also been obtained using embryo rescue technique in leguminous crops namely *Phaseolus* (Kuboyama et al., 1991) and *Vigna* (Fatokun and Singh, 1987). Crosses of *Brassica napus* were made with an intergeneric hybrid *Raphanobrassica* (*Raphanus sativus* × *B. oleracea*) to transfer shattering resistance into *B. napus* (Agnihotri et al., 1990). Embryo culture has also been employed to resynthesize *B. napus* from its two progenitor species, *viz., B. campestris* and *B. oleracea, B. alboglabra* (regarded as a form of *B. oleracea*), *B. campestris* and *B. oleracea* and *B. rapa* (Ozminkowski and Jourdan, 1994).

Embryo rescue has been successfully used to produce hybrids involving intergeneric and interspecific crosses in cereals. The important examples of intergeneric hybrids obtained via embryo rescue are barley × rye, wheat × barley, wheat × rye, oat × maize, wheat × maize, *Hordeum* × *Elymus*, and *Elymus* × *Triticum.* Embryo culture has also been used to produce hybrids of wheat with *Agropyron* spp. (Sharma and Ohm, 1990). Crosses of *Thinopyrum scirpeum* were made with different varieties of hexaploid wheat with the objective of transferring salt tolerance into wheat. F_1 hybrids, regenerated through embryo culture, were found to be salt tolerant (Farooq et al., 1993). Monosomic alien addition lines of rice were produced using embryo rescue (Jena and Khush, 1989). *Oryza minuta* (wild rice) is a tetraploid rice with resistance to several insect pests and diseases including blast (caused by *Pyricularia grisea* Syn. *Magnaporthe grisea*) and bacterial blight (caused by *Xanthomonas oryzae* Syn. *X. campestris* pv. *oryzae*). Resistance to these diseases has been transferred from wild rice (*O. minuta*) to cultivated rice (*O. sativa*) by using embryo rescue. Disease resistance has been successfully introduced from *T. timopheevi* to bread wheat via the embryo rescue technique (Bhowal et al., 1993).

5.4.2.2   Ovary Culture

Sometimes, embryo abortion occurs very early in the developmental stages where it is difficult to excise embryos to rescue. To overcome this, ovary culture is practised. In the interspecific crosses between *B campestris* and *B. oleracea*, the ovaries were excised and cultured on artificial media containing minerals and vitamins, and after few weeks, the embryos were excised and cultured on the same media (Inomata, 1978). Production rate of hybrid plants was different in different subspecies of *B. campestris* used as the female parent.

5.4.2.3   Reciprocal Crosses

Sometimes, nuclear cytoplasmic interactions results in sterility or degeneration of wide hybrids. Reciprocal crosses were found to be effective in a cross between *N. debneyi* and *N. tabacum*. The F_1 from the cross *N. debneyi* × *N. tabacum* was male sterile, whereas in reciprocal cross, the male was fertile. Similarly, wheat with *T. timopheevi* cytoplasm and the nucleus of *T. aestivum* are male sterile, whereas the reciprocal cross between these are fertile.

5.4.2.4   Altering Genomic Ratios

Chromosome elimination occurs in wide crosses like *H. vulgare* × *H. bulbosum* and *T. aestivum* × *H. bulbosum*. This chromosome elimination can be avoided by crossing diploid *H. vulgare* with tetraploid bulbosum (Kasha, 1974).

5.4.2.5   Chromosome Doubling

Wide crosses often show sterility in the resulting F_1s. Sterility is overcome by doubling the chromosome number by giving colchicine treatment. Amphiploids were produced using colchicine treatment. Triticale is the classical example produced through chromosome doubling of wheat × rye hybrids. *Tritordeum* is another example of amphiploid produced by chromosome doubling of F_1s resulting from the cross *Hordeum chilense* × *T. turgidum* (Martin and Sanchez-Monge, 1982). In some cases, backcrossing is practised to restore fertility in wide hybrids. Khush (1977) transferred grassy stunt virus resistance gene from *O. minuta* to *O. sativa* through backcrossing. In addition to this, genes for white backed plant hopper were transferred from *O. officinalis* to *O. sativa* through repeated backcrossing to their sterile F_1s.

5.4.2.6 Hybrid Breakdown

In wide crosses, the F_1 is fertile, but in later generations, the recombinants were lethal, semilethal, or weaker and are eventually got eliminated from the population, resulting in only parental types. Tetraploid *G. barbadense* and *G. tomentosum* cross readily with *G. hirsutum*. The F_1 plants are vigorous and fertile. However, F_2 progenies contain weak, dwarf, and inviable plants. Stephens (1949) reported a strong selection in the favor of parental types and against recombinant progenies. The reasons for hybrid breakdown are not well understood. By growing a large F_2 population, it may be possible to isolate desirable recombinants.

5.4.2.7 Limited Recombination

Effective utilization of wild species relies on the transfer of specific chromosome segments of agronomic importance. Reduced recombination or lack of recombination among the genomes of alien species poses problems in their utilization. It would be ideal to transfer the desired gene from the alien species, because the gene under transfer may be linked with other undesirable gene. Generally, in F_1, there is poor pairing among the chromosomes of cultivated and wild species. Reduce homology is a common feature in distant hybridization, which entails uneven chromosome pairing and reduce recombination. Lack of recombination can be overcome by the irradiation treatment of the resulting F_1 seed or pollen to induce chromosomal translocations and by the manipulating the chromosome pairing control system. Sears (1956) used ionizing radiations to induce chromosomal breaks and thereby transferred a gene controlling resistance to leaf rust from *Ae. umbellulata* Zhuk to cultivated wheat. Chromosome pairing and recombination in polyploids species like wheat, oats, and cotton are genetically governed. In wheat, three genomes in one nucleus without intergenomic recombination is a classical example of pairing control mechanism, which is controlled by the *Ph1* gene located on long arm of chromosome 5B. Three methods have been used in wheat to manipulate recombination and achieve homologous pairing: 1) crossing alien addition lines with wild species, 2) crossing nullisomic 5B stocks of wheat with alien addition lines, and 3) crossing a ph mutant of wheat with alien addition lines. Riley et al. (1968) pioneered the transfer of small, nonhomologous alien chromosomal segments by disrupting the normal chromosomal pairing using a high pairing line *Ae. speltoides* Taush in order to transfer stripe rust resistance from *Ae. comosa* ssp. *comosa*. Sharma and Knot (1966) transferred stem rust resistance gene from *Agropyron elongatum* to common wheat. Radiation-induced translocations for transferring alien chromosomal segments have been also used in oats for transferring powdery mildew resistance. Thomas et al. (1980) used *Avena longiglumis* to transfer powdery mildew resistance from *A. barbata* to the cultivated oats through induced pairing. When a nullisomic for 5B wheat is crossed with alien species, the F_1 hybrid slack chromosome 5B and the chromosomes of the alien species pair with the homologous chromosomes of wheat. Using nullisomic 5B, wheat lines carrying rust resistance from *Agropyron elongatum* were developed.

5.4.2.7.1   *Alien Addition Lines*

An alien addition line carries one chromosome pair from an alien species in addition to the normal somatic chromosome number of the parent species. When only one chromosome is present, it is referred to as alien addition monosome. Alien addition lines have been produced in wheat, oats, tobacco, rice, and few other crop species.

5.4.2.7.2   *Alien Substitution Lines*

An alien substitution line has one chromosome pair of a different species in place of a chromosome pair of recipient species. Alien substitution lines have been developed in wheat, cotton, oats etc.

Alien addition and substitution lines have been developed in polyploid species like wheat, oats, cotton etc. as they are having buffering capacity to be viable.

5.5   APPLICATION OF WIDE HYBRIDIZATION IN CROP IMPROVEMENT

Crop wild relatives, which include the progenitors of crops as well as other species more or less closely related to them, have been beneficial to modern agriculture, providing plant breeders with a broad pool of potentially useful genetic resources. Wild species have been exploited as a source of biotic and abiotic stress resistance. Several varieties of crop plants having genes introgressed from distant species and related genera have been developed in the past. Wild relatives were used in crop improvement in sugarcane in the first half of the 20th century. Their utility was recognized in breeding programs of major crops in the 1940s and 1950s (Plucknett et al., 1987), and wild gene use in crop improvement gained prominence by the 1970s and 1980s, with their use being investigated in an increasing wide range of crops. Tanksley and McCouch (1997) pointed the potential role of genome mapping in efficiently utilizing the genetic diversity of wild relatives and suggested that the continued sampling of wild germplasm would result in new gene discoveries and use. The application of wide hybridization for improving different agronomic traits, *viz.,* pest and/or disease resistance or abiotic stress tolerance, increasing yield, providing cytoplasmic male sterility, or fertility restorers for use in producing hybrids, or improving quality traits of the crop plants are described below.

5.5.1   RICE

The genus *Oryza,* to which cultivated rice (*Oryza sativa* L.) belongs, has 22 wild species ($2n = 24$, 48 chromosomes) representing 10 genomes: AA, BB, BBCC, CC, CCDD, EE, FF, GG, HHJJ, and HHKK (Vaughan, 2003). These wild species are an important reservoir of useful genes for rice improvement. Hybrids between *O. sativa* and *O. rufipogon* are partially fertile; however, *O. sativa* × *O. barthii* and *O. sativa* × *O. longistaminata* F_1s are highly sterile. Among the classical examples are the introgression of a gene for grassy stunt virus resistance from *O. nivara* to cultivated rice varieties (Khush, 1977) and the transfer of a cytoplasmic male sterile (CMS) source from wild rice, *O. sativa* f. *spontanea*, to develop CMS lines for commercial hybrid rice production (Lin and Yuan, 1980). Of the 6,000 accessions of cultivated rice and several wild species screened, only one accession of

O. nivara (accession 101508) was found to be resistant (Ling et al., 1970). A single dominant gene *Gs* confers resistance in *O. nivara.* It segregates independently of *Bph1.* Following four backcrosses with improved rice varieties, the gene for grassy stunt resistance was transferred into cultivated germplasm (Khush, 1977). The first set of grassy stunt-resistant varieties, IR28, IR29, and IR30, was released for cultivation in 1974. Subsequently, many grassy stunt resistance varieties, e.g., IR34, IR36, IR38, IR40, IR48, IR50, IR56, and IR58 were released. Another classical example of bacterial blight resistance gene (*Xa*21) transferred from *O. longistaminata,* with a wide spectrum of resistance into rice through backcrossing (Khush et al., 1990). In subsequent studies, *Xa21* has been transferred through MAS and cloned using map-based strategy. *Xa21* was transferred through MAS in several other indica lines, such as IR64 and PR106, Pusa Basmati, and Sambha Mahsuri, including elite breeding lines of new plant type (NPT) rice (Singh et al., 2001). Zhang et al. (1998) transferred bacterial blight resistance from *O. rufipogon* (RBB16) into another rice cultivar. Many genes for bacterial blight resistance, viz., *Xa23, Xa27, Xa29,* and *Xa38*, have been derived from wild species. The "A" genome wild species have been an important source of CMS, the major tool to breed commercial rice hybrids. A number of CMS sources have been developed in rice. However, the most commonly used CMS source in hybrid rice breeding is derived from the wild species *O. sativa* f. *spontanea* (Lin and Yuan, 1980). A new CMS source from *O. perennis* (accession 104823) was transferred into indica rice. More recently, genes for tungro virus tolerance and tolerance to acid sulfate soil conditions have been transferred from *O. rufipogon* into indica rice cultivar. Two new genes for bacterial blight resistance *Xa23* from *O. rufipogon* and *Xa38* from *O. nivara* (Cheema et al., 2008) have been transferred into rice. Three varieties (Matatag 9, AS996, NSIC Rc112) have been released from crosses of *O. sativa* × *O. rufipogon* and *O. sativa* × *O. longistaminata.* Genes have been exploited for tolerance of soils with high acidic-sulfate content in Vietnam and *O. longistaminata* A. Chev. & Roehrich genes for drought tolerance (Brar, 2005) in cultivars in the Philippines, allowing the spread of rice production to previously unusable lands. Yield-enhancing quantitative trait loci (QTLs) have been identified in populations derived from crosses with *O. rufipogon* (Moncada et al., 2001). With rice, 95% of hybrids grown in China are derived from crosses using CMS from wild *O. sativa* f. *spontanea* L. (Virmani and Shinjyo, 1988). Released in 1976, these hybrids are currently planted on approximately 45% of China's rice-planting area.

5.5.2 WHEAT

Breeders continue to isolate and introgress genes from wheat wild relatives for resistance to leaf and stem rust, yellow dwarf virus, root lesion nematode, powdery mildew, and wheat streak mosaic virus. McFadden (1930) was the first to transfer stem rust resistance from a distant species. Most of the high-yielding wheat cultivars carry portions of alien chromosomal introgression from related weedy species. The alien introgression T1BL·1RS, for example, resulted from the breakage of wheat chromosome 1BL·1BS (the long and short arms of chromosome 1 of the B genome) at the centromere and the replacement of the 1BS arm of wheat by the 1RS arm of rye. The 1RS arm in T1BL·1RS carries a sequence of resistance genes specifying resistance to leaf rust (*Lr26*), stem rust (*Sr31*), stripe rust (*Yr9*), powdery mildew (*Pm8*) (Friebe et al., 1996) and genes for adaptation to abiotic stresses, including a robust drought-tolerant root system (Sharma et al., 2011). Another example of a whole-arm alien introgression with a high impact in agriculture is from the distant wheat relative *Dasyprum villosum* (L.) Candargy (syn.) *Haynaldia villosa*). The *Pm21* gene, imparting resistance to powdery mildew fungus *Blumeria graminis*. f. sp. *tritici* (Bgt) of wheat, was transferred from *D. villosum*. In further evaluations, the 6VS arm was shown to harbor other beneficial genes, including resistance to wheat curl mite, stripe rust, *Fusarium* head scab, and soil-borne mosaic virus (De Pace et al., 2011), and all the genes on the 6VS arm are inherited as a single linkage block. Because of linkage on the 1RS arm, the genes are inherited as a single supergene linkage block. Spring wheat germplasm lines derived from *Aegilops tauschii* Coss. for resistance to Hessian fly, a major insect pest causing multimillion dollar crop losses in the US, have recently become available to breeders (Suszkiw, 2005). Stem and leaf rust resistance was transferred from *Agropyron elongatum* Host ex. P. Beauv and *Aegilops umbellulata* Zhuck., respectively, to wheat (Prescott-Allen, 1986). Synthesized amphiploids involving *T. monococcum*, *T. bioeticum*, *Ae. squarrossa*, and *T. durum* were developed and evaluated for bunt resistance (Dhaliwal et al., 1988). Eyespot resistance was transferred from *Ae. ventricosa* to Roazon cultivar. There are an increasing number of cases of high-yielding derivatives of hybrids created with the use of wild relatives, of which synthetic hexaploid wheats are a good example. Produced by the International Maize and Wheat Improvement Center (CIM-MYT), synthetic hexaploid (SH) wheats are a cross between durum wheat and the wild relative *Ae. tauschii* that has undergone artificial chromosome doubling to produce a hexaploid with A, B, and D genomes (Mujeeb-Kazi et al., 1996). These lines are then back-crossed to elite bread wheat cultivars to produce wheat with superior quality, disease resistance, and yield. In 2003, Chuanmai 42, a cross between an SH wheat and a local cultivar, was released in China, producing 20–35% higher yields (CIMMYT, 2004). A total of 600 SH lines with a wide range of positive traits are currently being tested and crossed, including water-logging tolerance (Villareal et al., 2001), spot blotch (*Cochliobolus sativus*) resistance, or karnal bunt (*Neovossia indica*) resistance (Mujeeb-Kazi et al., 2001). Increased protein content in wheat cultivars was derived from *T. dicoccoides* (Kornicke) G. Schweinfurt. Better grain quality synthetic hexaploid wheat cultivar "Carmona" has been released in Spain, and future releases of synthetic hexaploid wheats with higher content of essential minerals such as iron or zinc are expected (CIMMYT, 2004).

5.5.3 MAIZE

Teosintes, the closest wild relative of maize, is considered to be a useful reservoir of genes for maize improvement. *Teosintes* were initially hybridized with maize, and the hybrids were then backcrossed three times with maize while selecting for particular phenotypic traits of teosinte within a maize background. After several generations of selfing to ensure homozygosity, a linkage analysis was undertaken (Mangelsdorf, 1974). The hybrids of cultivated maize and *Z. mays* ssp. *mexicana* exhibit normal chromosomal pairing and general fertility. The crossing-over between maize-teosinte chromosomes occurs at frequencies similar to those observed in maize–maize hybrids (Doebley, 2004). Cohen and Galinat (1984) evaluated the potential use of the alien germplasm for maize improvement and concluded that the alien germplasm was effective on quantitative traits in broadening the genetic variation of maize and increasing heterosis when the germplasm was in a heterozygous state. The annual *mexicana* is a valuable source for traits like strong growth vigor, high protein content in the kernel, and resistance to multiple fungal diseases (Wilkes, 1977). The use of wild relatives of maize

(*Teosintes* and *Tripsacum dactyloides*) for developing striga resistance, a weed particularly prevalent in Africa, was well illustrated by Rich and Ejeta (2008). Some sources have been identified by Kim et al. (1999) in perennial *Teosintes* (*Zea diploperennis*) and *Tripsacum dactyloides* with higher levels of resistance (Gurney et al., 2003). Through a long-term breeding effort, researchers from the International Institute of Tropical Agriculture (IITA) developed a *Striga hermonthica*-resistant inbred, ZD05; this inbred has in its pedigree a *Zea diploperennis* accession as well as tropical maize germplasm (Amusan et al., 2008).

5.5.4 BARLEY

Successful gene transfer from wild species into barley was restricted to crosses *between H. vulgare* and *H. spontaneum* (Lehmann, 1991). A more distantly related species of barley *H. bulbosum* was found to harbor desirable disease-resistance genes, but problems are associated with this species for hybridization with barley. Powdery mildew resistance was transferred from *H. bulbosum* to cultivated barley by Xu and Kasha (1992) from *H. bulbosum*. Also, in 2004, six barley cultivars with drought tolerance derived from *H. spontaneum* K. Koch were released for use in Syria by the International Center for Agricultural Research in Dry Areas.

5.5.5 POTATO

Distant species like *S. andigena*, *S. demissum*, *S. phureja*, *S. acuale*, etc. of *Solanum* had played a major role in improvement of potato, providing resistance to biotic and abiotic stresses. *Solanum demissum* and *S. stoloniferum* Schltdl were used to transfer resistance to potato late blight, which continues to be effective in some areas, and currently 40% of the total area of the most popular potato cultivars in the United States has *S. demissum* in their ancestry (Hajjar, et al., 2007). Along with these wild relatives, *S. chacoense* Bitt., *S. acaule* Bitt., *S. vernei* Bitt. and Wittm. and *S. spegazzinii* Bitt. have provided resistance to several viruses and pests (Love, 1999). *S. berthaultii* is sexually compatible with potato and a valuable source for transfer of insect resistance into potato. This wild diploid species has been used as a male parent in crosses with tetraploid forms to produce fertile tetraploid progenies (Tingey, 1982). *S.*

acuale contributed frost resistance genes to several varieties. Frost resistance was contributed to cultivated polyploid *S. juzepczukii* and *S. curtilobum* from *S. acuale* (Hawkes, 1962). Recently, favorable combinations of good insect resistance traits with high yields were obtained by passing *S. tuberosum* × S. *berthaultii* hybrids through a cycle of tissue culture (Spooner, 2007).

5.5.6 TOMATO

Tomato crosses with its wild relatives with varying degrees of difficulty; thus, wild relatives have been used as sources of genes for crop improvement. Wild species are interesting resources of genetic variation for introgression breeding and comprise exclusive sources of many resistance genes for cultivated tomatoes. One of the first examples was the exploitation of *Cladosporium fulvum* resistance from *S. pimpinellifolium* in 1934 (Walter, 1967). De Ponti et al. (1975) observed high levels of resistance in *L. hirsutum*, *L. hirsutum glabratum* and *S. pennellii* against the glasshouse whitefly (*Trialeurodes vaporariorum*). Many tomato disease resistance genes were introgressed from wild species, mostly from *Lycopersicon pimpinellifolium* Mill. Over 40 resistance genes have been derived from *L. peruvianum* (L.) Mill., *L. cheesmanii* Riley, *L. pennellii* (Correll) D'Arcy, and several other wild relatives (Rick and Chetelat, 1995). In tomatoes, *L. chilense* and *L. pennellii* genes were used to increase drought and salinity tolerance. In a recent study, pyramiding of three independent yield-promoting genomic regions introduced from *Solanum pennellii*, a green-fruited wild relative of tomato, has led to hybrids with a 50% increased yield over a leading variety (Gur and Zamir, 2004). Tomatoes have provided many classic examples of improved quality traits from wild genes, from increased soluble solid content, fruit color, and adaptation to harvesting. Since then, QTL mapping and analysis have aided the discovery of useful quality control genes, such as fruit size, in unlikely candidates such as the small-fruited tomato ancestor *L. pimpinellifolium* (Tanksley and McCouch, 1997).

5.5.7 SUNFLOWER

Crosses between cultivated and wild species of sunflower have been performed with ease, but the resulting progeny was female sterile due to translocations (Chandler et al., 1986). Disease resistances in wild sunflowers have been exploited for decades, with multiple sources of genetic resistance to all the known races of downy mildew, as well as rust, *verticillium* wilt, and broomrape continually being transferred from wild *Helianthus annuus* L. and *H. Praecox* Engelm. & A. Gray into new sunflower hybrids. Wild sunflower species have been a plentiful source of genes for downy mildew resistance. Complete resistance to downy mildew has been reported in annual species *H. annuus, H. argophyllus, H. debilis, and H. petiolaris, and perennial H. decapetalus, H. divaricatus, H. eggertii, H. giganteus, H. x laetiflorus, H. mollis, H. nuttallii, H. scaberrimus, H. pauciflorus, H. salicifolius*, and *H. tuberosus* (Christov et al., 1999). Diploid perennial species *H. divaricatus, H. giganteus, H. glaucophyllus, H. grosseserratus, H. mollis, H. nuttallii,* and *H. smithii* and their interspecific hybrids were resistant to downy mildew (Nikolova et al., 1998). Interspecific hybrids based on *H. eggertii* and *H. smithii* were resistant to downy mildew in Bulgaria. Annual *H. debilis* ssp. *silvestris, H. praecox* ssp. *praecox*, and *H. bolanderi*, and 14 perennial species were tolerant of powdery mildew in both field and greenhouse tests (Saliman et al., 1982). Not all populations of perennial species are resistant; populations of *H. grosseserratus* and *H. maximiliani* showed differential reactions. Skoric (1984) reported that interspecific hybrids with *H. giganteus, H. hirsutus, II. divaricatus*, and *H. salicifolius* had no powdery mildew symptoms. Thirty-six accessions of Jerusalem artichoke (*H. tuberosus*) were field-screened to determine their reaction to powdery mildew, with three accessions showing high levels of tolerance (McCarter, 1993). The most recent trait from wild *H. annuus* L. is herbicide resistance to imidazolinone and sulfonylurea chemicals used to control broomrape. CMS based on using wild *H. annuus* and *H. petiolaris* Nutt. has been used in high-yielding commercial sunflower hybrids since 1972, thereby significantly expanding the sunflower industry (Prescott-Allen, 1986). Currently, 100% of sunflower production in the US and approximately 60–70% of production worldwide is estimated to be from these hybrids.

5.5.8 BRASSICAS

Interspecific hybridization can be easily forced in the family of *Brassicaceae*, whereas under natural conditions, the gene flow is very limited. Spontaneous hybridization of *Brassica rapa* L. and *Brassica oleracea* L., resulted in the production of an amphidiploid oilseed rape (*Brassica napus* L.) and contains the complete diploid chromosome sets of A genome of *B. rapa* and the C genome of *B. oleracea*. Until the 1980s, alien gene introgression lines had been produced by hybridization mainly between the *Brassicaceae* crops, resulting in the development of novel cultivars with desirable agronomic traits. Examples include Chinese cabbage (*B. rapa*) cultivars resistant to bacterial soft rot (Shimizu et al., 1962), early-maturing cultivars of *B. napus* (Namai, 1987), and virus-resistant cultivars of *B. oleracea*. Warwick (1993) described wild genera of the tribe *Brassiceae* as the sources of various agronomically interesting traits such as hairiness, resistance to pod shattering, photosynthesis, soil adaptation, and disease resistance. Introgression of useful traits was successfully achieved by hybridizations with the species of the B-genome *B. juncea*, *B. carinata* and *B. nigra*. These intrageneric hybridizations with *B. napus* have been used to transfer resistance genes to *Leptosphaeria maculans* into the gene pool of oilseed rape. Different resistance traits, viz., nematode resistance to *Heterodera schachtii*, fungal resistance to *Alternaria* ssp. and *L. maculans* and protist resistance to *Plasmodiophora brassicae* has been transferred through intergeneric hybridizations. Successful hybridizations with *B. napus* have been achieved using *B. elongata*, *B. fruticulosa*, *B. souliei*, *Diplotaxis tenuifolia*, *Hirschfeldia incana*, *Coincya monensis*, and *Sinapis arvensis*. The hybrids *B. napus* × *Hirschfeldia incana*, *B. napus* × *Sinapis arvensis*, and *B. napus* × *Coincya monensis* has been successfully backcrossed to *B. napus*. *S. arvensis* was used as a resistance source for blackleg (*L. maculans*) to oilseed rape (*B. napus*). Alternaria resistance was introduced by a backcross program using the hybrids B. napus × *Diplotaxis erucoides*. A spontaneously occurring male sterility inducing cytoplasm from *Brassica* was discovered in a wild population of radish by Ogura (1968). Later, the cytoplasm was transferred to *B. oleracea* and *B. napus* (Bannerot et al., 1974). Male sterility conferred by Ogura cytoplasm was transferred to *B. Juncea* through repeated backcrossing and selection. Cytoplasmic male sterility was artificially transferred from radish to rape (*B. napus*). Resistance to the beet cyst nematode and clubroot was transferred from *R. sativus* to *B. napus* using *R. sativus* MAALs of *B. napus* (Budahn et al., 2008). Prakash and Chopra (1990) carried out interspecific hybridization between *B. juncea* and *B. napus* and were able to isolate a reconstituted *B. napus* plant with complete non-dehiscent fruits. Agnihotri et al. (1990) attempted to transfer shatter resistance from *Raphanus* into

B. napus using *Raphanobrassica* as the bridging material. This resulted in genetic material with variable fertility. Examples of introgression through somatic hybridization include beet cyst nematode resistance (*Heterodera schachtii*) from *Sinapis alba* and *Raphanus sativus* to *B. napus*; alternaria leaf spot resistance from *S. alba* to *B. napus*; clubroot resistance (*Plasmodiophora brassicae*) from *Raphanus sativus* to *B. oleracea* var. botrytis, and blackleg resistance from *B. nigra* to *B. napus*.

5.5.9 COTTON

Gossypium hirsutum, G. barbadense, G. herbaceum, and *G. arboreum* are the cultivated species of cotton. Apart from that, there are about 35 wild species that possess desirable attributes such as fiber quality and insect-pest resistance. The release of interspecific hybrids between *G. hirsutum* and *G. barbadense,* for example DCH- 32, in India had enabled to realize the contribution of *G. barbadense* to extra long staple cotton production. Of late, these hybrids were found to be vulnerable to sucking pests and pink bollworm and resulting in huge yield losses (Krishnamurthy et al., 1975). The wild species of cotton are potential source of resistance to biotic (insects and diseases) and abiotic stresses like salinity, cold, drought, and heat (Narayanan and Singh, 1994). Some wild species, namely, *G. darwinii, G. stocksii, G. harkensii,* and *G. aridum,* are good donors for drought tolerance, whereas *G. thurberii* has resistant genes for frost tolerance (Rooney et al., 1991). Due to their potential, these wild species (diploids and tetraploids) are being utilized in various hybridization programs (Mehetre et al., 2004). Blank and Leathers (1963) transferred resistant genes against cotton rust caused by *Puccinia cacabata* from *G. anomalum* and *G. arboreum* into *G. hirsutum* through interspecific hybridization. Genes for resistance against diseases and drought have been transferred between *G. hirsutum* and *G. arboreum.* Similarly, resistant genes against bacterial blight of cotton present in *G. arboreum* were introgressed into *G. barbadense* (Brinkerhoff, 1970). Sacks and Robinson (2009) introgressed resistance against *Rotylenchulus reniformis* from diploid to tetraploid cotton. Introgression of cotton leaf curl virus-resistant genes from the autotetraploid of the diploid donor species *G. arboreum* into allotetraploid *G. hirsutum* was performed through conventional breeding (Ahmad et al., 2011).

5.5.10 MILLETS

In millets, rust and *Pyricularia grisea* resistances were introgressed from wild relatives. Although the rust resistance was overcome quickly, *Pyricularia* resistance is still effective (Wilson and Gates, 1993). Striga resistance has been identified in millets primary gene pool (Wilson et al., 2000), but the work remains at the early stages of gene transfer. In sorghum, recent success in hybridization between *Sorghum macrospermum* and *S. bicolor* promises to help in introgressing several pest and disease resistance traits to the cultivars (Price et al., 2005). More recently, CMS found in wild millet has been used to produce popular high-yielding and disease resistant hybrids. CMS and fertility restorer lines derived from *Pennisetum purpureum* Schum. were used in the first pearl millet grain hybrids and commercial forage hybrids (Hanna, 1989) and the Tifleaf series derived from these are mostly widely cultivated in North America, and are popular as far down as Brazil. The latest in the Tifleaf series, Tifleaf 3, released in 1997 produces 20% more forage than the disease susceptible hybrids it replaced.

5.5.11 PULSES

The cultivated chickpea is easily crossable with its wild relative *Cicer reticulatum*. *C. judaicum* is reported to have resistance genes for ascochyta blight, fusarium wilt, and botrytis gray mold (Van der Maesen and Pundir, 1984). *C. bijugum, C. pinnatifidum*, and *C. reticulatum* have been identified as suitable donors for resistance to cyst nematode (Greco and Di Vito, 1993). Kaur et al., (1999) reported significantly lower larval density of helicoverpa pod borer on some of the accessions of *C. echinospermum, C. judaicum, C. pinnatifidum*, and *C. reticulatum*.

Drought tolerance has been identified in the wild genepool of *Lens esculenta* (Gupta and Sharma, 2006) and cold tolerance and earliness in *Lens culinaris* ssp. *orientalis* (Hamdi et al., 1996). Some of the wild accessions of *Lens* showing combined resistance to ascochyta blight, fusarium wilt and anthracnose disease have also been identified (Tullu et al., 2006). Gupta and Sharma (2006) evaluated 70 accessions representing four wild species/subspecies (*L. culinaris* ssp. *orientalis, L. odemensis, L. ervoides*, and *Lens nigricans*) for yield attributes and biotic and abiotic stresses.

Wilkinson (1983) reported *Phaseolus coccineus* as a potential source of high yield for common bean. Resistance to angular leaf spot, anthracnose, ascochyta blight, bean yellow mosaic virus (BYMV), common bean blight (CBB), white mold, and cold is found in the secondary gene pool. Some sources of resistance have also been identified in the tertiary gene pool. BG1103, a leading cultivar in Northern India developed by the India Agricultural Research Institute, is tolerant drought and temperature and is derived from *Cicer reticulatum*.

5.5.12 SUGARCANE

Modern sugarcane cultivars are derived from the trispecific hybrids of *Saccharum officinarum*, *S. barberi/sinense*, and *S. spontaneum*. The cane and sugar were contributed by *S. offinarum*, while disease resistance, hardiness, and ability to withstand adverse conditions were derived from *S. spontaneum* (Parthasarathy, 1946). The first Indian commercial hybrid cane Co 205 was an interspecific hybrid between *S. officinarum* Vellai and *S. spontaneum* Coimbatore (Barber, 1920).

5.6 FUTURE PROSPECTS

In this chapter, we have described the continuing increase in the use of crop wild relatives (CWR) for the production of new cultivars over the last few decades. Nearly 10 years ago, Tanksley and McCouch (1997) argued that breeders were relying on the old paradigm of searching for beneficial traits associated with wild relatives, rather than searching for beneficial genes. Considering that phenotypically most wild relatives have poor yield or quality traits, they are rarely sought for contributions of these traits. However, in crop breeding, there are some striking examples of contributions from wild relatives that have led to improvement of domestic types. Some of these can be mentioned as crop-saving instances where one or a few genes introgressed from wild relatives have saved the crop from deadly plant pathogens. In the present scenario of climatic variability, crop plants will require more diversity to tackle changing environments. Thus, a better understanding of the cytogenetic and molecular approaches is need of the future crop breeding programs. Understanding the genetics of each trait to be introgressed and devising feasible methodologies for their transfer into cultivated types are going to be the basis of future plant breeding. Despite a growth in the literature on traits controlled by quantitative trait loci in wild relatives, the potential of new molecular technologies has yet to be fully realized in ways that result in a significant increase in the rate of production of new cultivars carrying genes from the wild type.

KEYWORDS

- **bridging species**
- **embryo rescue**
- **necrosis**
- **sterility**
- **wide hybridization**

REFERENCES

Agnihotri, A., et al., (1990). Production of *Brassica napus* x *Raphanobrassica* hybrids by embryo rescue: an attempt to introduce shattering resistance into *B. Napus*. *Plant Breeding, 105,* 292–299.

Ahmad, S., et al., (2011). Introgression of cotton leaf curl virus-resistant genes from Asiatic cotton (*Gossypium arboreum*) into upland cotton (*G. hirsutum*). *Genet. Mol. Res., 10*(4), 2404–2414.

Amusan, I. O., et al., (2008). Resistance to *Striga hermonthica* in a maize inbred line derived from *Zea diploperennis*. *New Phytol., 178,* 157–166.

Barber, C. A., (1920). The origin of sugarcane. *International Sugar Journal, 22,* 249–251.

Barclay, I. R., (1975). High frequencies of haploid production in wheat (*Triticum aestivum*) by chromosome elimination. *Nature, 256,* 410–411.

Bhowal, J. G., Guha, G., & Vari, A. K., (1993). A cytogenetic study to transfer disease resistance genes from *Triticum timopheevi* to bread wheat. *J. Genet. Breed, 47,* 71–76.

Blank, L. M., & Leathers, C. R., (1963). Environmental and other factors influencing development of south western cotton rust. *Phytopathology, 53,* 921–928.

Brar, D., (2005). Broadening the genepool and exploiting heterosis in cultivated rice, In: *Rice is Life: Scientific Perspectives for the 21st Century*, Toriyama, K., Heong, K. L., Hardy, B., (eds), Proceedings of the World Rice Research Conference held in Tokyo and Tsukuba, Japan, 4-7 November 2004. Los Baños (Philippines): International Rice Research Institute, and Tsukuba (Japan): Japan International Research Center for Agricultural Sciences. CD-ROM. 590 p.

Brar, D. S., & Khush, G. S., (1997). Alien introgression in rice. *Plant Molecular Biology*, *35*, 35–47.

Brinkerhoff, L. A., (1970). Variation in *Xanthomonas malvacearum* and its relation to control. *Ann. Rev. Phytopathol.*, *8*, 85–110.

Budahn, H., Schrader, O., & Peterka, H., (2008). Development of a complete set of disomic rape-radish chromosome-addition lines. *Euphytica.*, *162*, 117–128.

Burk, L. G., (1967). An interspecific bridge cross *Nicotiana repanda* through *N. sylvestris* to *N. tabacum. J. Hered.*, *58*, 215–218.

Burke, J. M., & Arnold, M. L., (2001). Genetics and fitness of hybrids. *Ann. Rev. Genet.*, *35*, 31–52.

Chahal, G. S., & Gosal, S. S., (2002). *Principles and Procedures of Plant Breeding*: Biotechnological and conventional approaches. Alpha Science International Ltd. Harrow, UK, pp. 597.

Chandler, J. M., Chao-Chien, J., & Beard, B. H., (1986). Chromosomal differentiation among the annual *Helianthus* species. *Systematic Botany*, *11*(2), 354–371.

Chavez, R., Brown, C. R., & Iwanaga, M., (1988). Application of interspecific sesquiploidy to introgression of PLRV resistance from non-tuber-bearing *Solanum etuberosum* to cultivated potato germplasm. *Theor. Appl. Genet.*, *76*, 497–500.

Cheema, K. K., et al., (2008). A novel bacterial blight resistance gene from *Oryza nivara* mapped to 38 Kbp region on chromosome 4L and transferred to *O. sativa* L. *Genet. Res.*, *90*, 397–407.

Christov, M., (1999). Production of new CMS sources in sunflower. *Helia.*, *22*(31), 1–12.

CIMMYT, (2004). *Wild Wheat Relatives Help Boost Genetic Diversity*, Mexico City, http://www.cimmyt.org/wild-wheat-relatives-help-boost-genetic-diversity/.

Cohen, J. I., & Galinat, W. C., (1984). Potential use of alien Germplasm for maize improvement. *Crop. Sci.*, *24*, 1011–1015.

De Pace, C., et al., (2011). *Dasypyrum. Wild Crop Relatives: Genomic and Breeding Resources*, Cereals, Kole, C., (ed.). (Springer, Berlin), XXIII, pp. 497

De Ponti, O. M. B., Pet, G., & Hogenboom, N. G., (1975). Resistance to the glasshouse whitefly (*Trialeurodes vaporariorum* Westw.) in tomato (*Lycopersicon esculentum* Mill.) and related species. *Euphytica.*, *24*(3), 645–649.

Dhaliwal, H. S., Navarete, M. R., & Valdez, J. C., (1988). Scanning electron microscope studies of penetration mechanism of *Tilletia indica* in wheat spikes. *Review Mexican Phytopathology*, *7*, 150–155.

Doebley, J. F., (2004). The genetics of maize evolution. *Annu. Rev. Genet.*, *38*, 37–59.

Doussinault, G., et al., (1983). Transfer of a dominant gene for resistance to eyespot disease from a wild grass to hexaploid wheat. *Nature*, *303*, 698–700.

Evans, A. M., (1962). Species hybridization in *Trifolium* II. Investigating the pre-fertilization barriers to compatibility. *Euphytica.*, *11*(3), 256–262.

Farooq, S., et al., (1993). Intergeneric hybridization for wheat improvement. VII. Transfer of salt tolerance from *Thinopyrum scirpeum* into wheat. *J. Genet. Breed*, *47*, 191–198.

Fatokun, C. A., & Singh, B. B., (1987). Interspecific hybridization between *Vigna pubescens* and *V. unguiculata* (L.) Walp through embryo rescue. *Plant Cell Tissue and Organ Culture*, *9*, 229–233.

Finch, R. A., & Bennet, M. D., (1983). The mechanism of somatic chrommosome elimination in *Hordeum*. In: Brandham, P. E., & Bennet, M. D., (eds.) 2nd *Kew Chromosmeee Conf.*, Allen & Unwin London, II, 147–154

Friebe, B., et al., (1996). Characterization of wheat-alien translocations conferring resistance to diseases and pests: Current status. *Euphytica.*, *91*, 59–87.

Gibson, R. W., (1974). Aphid trapping glandular hairs on *Solanum tuberosum* and *Solanum berthaultii*. *Potato Research*, *17*, 152–154.

Gill, B. S., & Waines, J. G., (1978). Paternal regulation of seed development in wheat hybrids. *Theor. App. Genet.*, *51*(6), 265–270.

Greco, N., & Di Vito, M., (1993). Selection for nematode resistance in cool season food legumes. In: Singh, K. B., & Saxena, M. C., (eds.) *Breeding for Stress Tolerance in Cool Season Food Legumes*. John Wiley & Sons/ICARDA, Chichester, UK, pp. 157–166.

Gupta, D., & Sharma, S. K., (2006). Evaluation of wild *Lens* taxa for agro-morphological traits, fungal diseases and moisture stress in northwestern Indian hills. *Gen. Res. and Crop Evol.*, *53*, 1233–1241.

Gur, A., & Zamir, D., (2004). Unused natural variation can lift yield barriers in plant breeding. *PLoS Biol.*, *2*, 1610–1615.

Gurney, A. L., (2003). Novel sources of resistance to *Striga hermonthica* in *Tripsacum dactyloides*, a wild relative of maize. *New Phytol.*, *160*, 557–568.

Hajjar, R., & Hodgkin, T., (2007). The use of wild relatives in crop improvement: A survey of developments over the last 20 years. *Euphytica*, *156*, 1–13.

Hamdi, A., Küsmenoglu, I., & Erskine, W., (1996). Sources of winter hardiness in wild lentil. *Genet. Res. and Crop Evol.*, *43*, 63–67.

Hanna, W. W., (1989). Characteristics and stability of a new cytoplasmic-nuclear male sterile source in pearl millet. *Crop. Sci.*, *29*, 1457–1459.

Hawkes, J., (1962). Introgression in certain wild potato species. *Euphytica*, *11*, 26–35.

Hermsen, J. G. T., (1967). Hybrid dwarfness in wheat. *Euphytica.*, *16*, 134–162.

Heslop-Harrison, J., (1987). Pollen germination and pollen tube growth. *Int. Rev. Cytol.*, *107*, 1–78.

Ikehashi, H., & Araiu, H., (1986). Genetics of F_1 sterility in the remote crosses of rice. *Proceedings of the International Rice Genetics Symposium*, International Rice Research Institute, Manila Philippines, pp. 119–130.

Inomata, N., (1978). Production of interspecific hybrids in *Brassica campestris* x *B. oleracea* by culture *in vitro* of excised ovaries. Development of excised ovaries in the crosses of various cultivars. *The Jap. J. Gen.*, *53*(3), 161–173.

Jena, K. K., & Khush, G. S., (1989). Monosomic alien addition lines of rice: production, morphology, cytology, and breeding behaviour. *Genome.*, *32*(3), 449–455.

Jiang, J., Friebe, B., & Gill, B. S., (1993). Recent advances in alien gene transfer in wheat. *Euphytica.*, *73*(3), 199–212.

Kasha, K. J., (1974). Haploids in higher plants. Advances and potential. Pr*oceedings of the 1st International Symposium*. June 10–14, (1974). University of Guelph. Canada, pp. 3–9.

Kaur, S., Chhabra, K. S., & Arora, B. S., (1999). Incidence of *Helicoverpa armigera* (Hubner) on wild and cultivated species of chickpea. *Inter Chickpea and Pigeon Pea Newsletter*, *6*, 18–19.

Khush, G. S., (1977). Disease and insect resistance in rice. *Adv. Agron.*, *29*, 265–341.

Khush, G. S., Bacalangco, E., & Ogawa, T., (1990). A new gene for resistance to bacterial blight from O. longistaminata. *Rice Genet Newsl.*, *7*, 121–122.

Kim, S. K., Akintunde, A. Y., & Walker, P., (1999). Responses of maize inbreds during development of *Striga hermonthica* infestation. *Maydica*, *44*, 333–339.

Knight, R. L., (1954). The genetics of blackarm resistance. 11. *Gossypium anomalum*. *Journal of Genetics*, *52*, 466–472.

Kobori, S., & Marubashi, W., (2004). Programmed cell death detected in interspecific hybrids of *Nicotiana repanda* x *N. tomentosiformis* expressing hybrid lethality. *Breed Sci.*, *54*, 347–350.

Krishnamurthy, R., et al., (1975). Development of extra long staple *G. barbadense* varieties in India. *Cotton. Dev.*, *5*, 3–6.

Kuboyama, T., Shintaku, Y., & Takeda, G., (1991). Hybrid plant of *Phaseolus vulgaris* L. and *P. lunatus* L. obtained by means of embryo rescue and confirmed by restriction endonuclease analysis of rDNA. *Euphytica.*, *54*, 177–182.

Laibach, F., (1929). Ectogenesis in plants. *J. Hered.*, *20*, 201–208.

Larter, E., & Chaubet, C., (1965). Use of exogenous growth substances in promoting pollen tube growth and fertilization in barley-rye crosses. *Can. J. Genet. and Cytol.*, *7*(3), 511–518.

Laurie, D. A., & Bennett, M. D., (1988). The production of haploid wheat plants from wheat x maize crosses. *Theor. Appl. Genet.*, *76*(3), 393–397.

Lehmann, L. C., (1991). The use of genetic resources for isolating disease resistance for barley cultivar development. In: *Barley genetics VI. Proc 6th Int Barley Genet Symp*, Munck, L., (ed.), *vol I.* Munksgaard International Publishers Ltd., Copenhagen, Denmark, pp. 650–652.

Lin, S. C., & Yuan, L. P., (1980). Hybrid rice breeding in China. In: *Innovative Approaches to Rice Breeding*. International Rice Research Institute, Manila, Philippines, pp. 35–51.

Ling, K. C., Aguiero, V. M., & Lee, S. H., (1970). A mass screening method for testing resistance to grassy stunt disease of rice. *Plant Dis. Rep.*, *56*, 565–569.

Livermore, J. R., & Johnstone, F. E., (1940). The effect of chromosome doubling on the crossability of *Solanum chacoense*, *S. jamesii*, and *S. bulbocastanum* with *S. tuberosum*. *Am. Potato Jour.*, *17*(7), 170–173.

Love, S., (1999). Founding clones, major contributing ancestors, and exotic progenitors of prominent North American potato cultivars. *Am. J. Potato Res.*, *76*, 263–272.

Mangelsdorf, P. C., (1974). *Corn: its Origin Evolution and Improvement*. Harvard University Press, Cambridge, M. A, XIV, pp. 262.

Martin, A., & Sanchez-Monge, L. E., (1982). Cytology and morphology of the amphiploid *Hordeum chilense* x *Triticum turgidum* conv. *durum*. *Euphytica.*, *31*, 261–267.

McCarter, S. M., (1993). Reaction of Jerusalem artichoke genotypes to two rusts and powdery mildew. *Plant Dis.*, *77*, 242–245.

McFadden, E. S., (1930). A successful transfer of emmer characters to *vulgare* wheat. *Am. Soc. Agron.*, *22*, 1020–1034.

Mehetre, S. S., (2004). Ovulo embryo cultured hybrid between amphidiploid (*Gossypium arboreum* x *Gossypium anomalum*) and *Gossypium hirsutum*. *Curr. Sci.*, *87*(3), 286–289.

Mercy, S. T., & Kakkar, S. N., (1975). Barriers to interspecific crosses in *cicer*. *Proc. Ind. Natl. Sci. Acad. B.*, *41*, 78–82.

Moncada, P., et al., (2001). Quantitative trait loci for yield and yield components in an *Oryza sativa Oryza rufipogon* BC_2F_2 population evaluated in an upland environment. *Theor. Appl. Genet.*, *102*, 41–52.

Mujeeb-Kazi, A., et al., (2001). Registration of 10 synthetic hexaploid wheat and six bread wheat germplasms resistant to Karnal bunt. *Crop Sci.*, *41*, 1652–1653.

Mujeeb-Kazi, A., Rosas, V., & Roldan, S., (1996). Conservation of the genetic variation of Triticum tauschii (Coss.) Schmalh (*Aegilops squarrosa* auct. non. L) in synthetic hexaploid wheats (*T. turgidum* L. s. lat. x *T.tauschii*, 2n = 6x = 42, AABBDD) and its potential utilization for wheat improvement. *Genet. Resour. Crop. Evol.*, *43*, 129–134.

Namai, H., (1987). Inducing cytogenetical alterations by means of interspecific and intergeneric hybridization in brassica crops. *Gamma Field Symp.*, *26*, 41–89.

Narayanan, S. S., & Singh, P., (1994). Resistance to *Heliothus* and other serious insect pests in *Gossypium* species- a review. *J. Indian Soc. Cotton Improv.*, *19*, 10–24.

Nettancourt, D., (1977). Interspecific incompatibility. *Monographs on Theoretical and Applied Genetics*, *3*, 141–180.

Nikolova, L. M., Christov, M., & Shindrova, P., (1998). New sunflower forms resistant to *orobanche cumana* wallr. Originating from interspecific hybridization. Current problems of *orobanche* researchers. *Proc. of the 4th Int. Workshop on Orobanche*, Albena, Bulgaria, pp. 295–299.

Nishikawa, K., et al., (1974). Mapping of progressive necrosis gene Ne1 and Ne2 of common wheat by the telocentric method. *Japan J. Breed*, *24*, 277–281.

Ogura, H., (1968). *Studies on the New Male Sterility in Japanese Radish, with Special References to Utilization of this Sterility Towards the Practical Raising of Hybrid Seeds.* Memoirs of the Faculty of Agriculture of Kagoshima University, *6*, 39–78.

Oka, H. I., (1974). Analysis of genes controlling F1 sterility in rice by the use of isogenic lines. *Genetics*, *77*, 531–534.

Oka, H. I., (1988). Functions and genetic bases of reproductive barriers. In: *Origin of Cultivated Rice* edited by. Elsevier, pp. 181–205

Ozminkowski, R. H., & Jourdan, P., (1994). Comparing the resynthesis of Brassica napus L. by interspecific somatic and sexual hybridization. 11. Hybrid morphology and identifying organelle genomes. *J. Am. Soc. Hort. Sci.*, *119*, 816–823.

Parthasarathy, N., (1946). The probable origin of North Indian Sugarcanes. *Journal of Indian Botanical Society*, (M. O. P. Iyengar Commemorative volume), pp. 133–150.

Plucknett, D., et al., (1987). *Gene banks and the World's Food*. Princeton University Press, Princeton, N. J, pp. 233.

Prakash, S., & Chopra, V. L., (1990). Reconstruction of allopolyploid *Brassicas* through nonhomologous recombination: Introgression of resistance to pod shatter in *Brassica napus*. *Genet Res*, Cambridge, *56*, 1–2.

Prescott-Allen, C., & Prescott-Allen, R., (1986). *The First Resource: Wild Species in the North American Economy.* Yale University, New Haven, pp. XV, pp. 529.

Price, H. J., et al., (2005). A Sorghum bicolor x S. macrospermum hybrid recovered by embryo rescue and culture. *Aust. J. Botany.*, *53*, 579–582.

Raina, S. K., (1984). Crossability and *in vitro* development of hybrid embryos of *Triticum durum* x *Secale cereale*. *Indian J. Genet.*, *44*(3), 429–437.

Rich, P. J., & Ejeta, G., (2008). Towards effective resistance to Striga in African maize. *Plant Signal. Behav.*, *3*, 618–621.

Rick, C., & Chetelat, R., (1995). Utilization of related wild species for tomato improvement, first international symposium on solanacea for fresh market. *Acta. Hortic.*, *412*, 21–38.

Riley, R., Chapman, V., & Johnson, R., (1968). Introduction of yellow rust resistance of Aegilops comosa into wheat by genetically induced homeologous recombination. *Nature*, *217*, 383–384.

Sacks, E. J., & Robinson, A. F., (2009). Introgression of resistance to reniform nematode (*Rotylenchulus reniformis*) into upland cotton (*Gossypium hirsutum* L.) from *G. arboreum* L. and a *G. hirsutum* L./*G. aridium* L. bridging line. *Field Crop Res.*, *112*, 1–6.

Saliman, M., Yang, S. M., & Wilson, L., (1982). Reaction of *Helianthus* species to *Erysiphe cichoracearum*. *Plant Dis.*, *66*, 572–573.

Satry, D. C., & Shivanna, K. R., (1976). Attempts to overcome interspecific incompatibility in Seasmum by using recognition pollen. *Ann. Bot.* (London), *40*, 891–893.

Sears, E. R., (1944). Inviability of intergeneric hybrids involving *Triticum monococcum* and *T. aegilopoides*. *Genetics*, *29*, 113–127.

Sears, E. R., (1956). The transfer of leaf rust resistance from *Aegilops umbellulata* to wheat. In: *Genetics in Plant Breeding, Brookhaven Symposium in Biology*, *9*, 1–22.

Sharma, S., et al., (2011). Dissection of QTL effects for root traits using a chromosome arm-specific mapping population in bread wheat. *Theor. Appl. Genet.*, *122*, 759–769.

Sharma, D., & Knott, D. R., (1966). The transfer of leaf rust resistance from Agropyron to Triticum by irradiation. *Can. J. Genet. Cytol.*, *8*, 137–143.

Sharma, H. C., & Ohm, H. W., (1990). Crossability and embryo rescue enhancement in wide crosses between wheat and three *Agropyron* species. *Euphytica.*, *49*, 209–214.

Shimizu, S., Kanazawa, K., & Kobayashi, T., (1962). Studies on the breeding of Chinese cabbage for resistance to soft rot. Part III. The breeding of the resistant variety "Hiratsuka No. 1" by interspecific crossing. *Bull. Hort. Res. Sta., Japan, Ser., A.*, *1*, 157–174.

Singh, S., et al., (2001). Pyramiding three bacterial blight resistance genes (*xa5*, *xa13* and *Xa21*) using marker assisted selection into indica rice cultivar PR106. *Theor. Appl. Genet.*, *102*, 1011–1015.

Skoric, D., (1984). Genetic resources in the *Helianthus* genus // *Proc. of the Int. Symp. on Science and Biotechnology for an Integral Sunflower Utilization*, Bari, Italy, October, *25*, 37–73.

Spooner, D. M., Fajardo, D., & Bryan, G. J., (2007). Species limits of Solanum berthaultii Hawkes and S. tarijense Hawkes and the implications for species boundaries in Solanum sect. *Petota Taxon*, *56*, 987–999.

Stebbins, G. L., (1950). Isolation and the origin of species. In: *Variation and Evolution in Plants*, Stebbins, G. L. Jr., (ed.), Columbia University Press, New York, pp. 643.

Stephens, S. G., (1949). The cytogenetics of speciation in *Gossypium*. I. Selective elimination of the donor parent genotype in interspecific backcrosses. *Genetics*, *34*, 627–637.

Subrahmanyam, N. C., & Kasha, K. J., (1973). Selective chromosomal elimination during haploid formation in barley following interspecific hybridization. *Chromosoma.*, *42*(2), 111–125.

Suszkiw, J., (2005). Hessian fly-resistant wheat germ plasm available agricultural research service, *News and Events*, United States Department of Agriculture, https://www.plant-managementnetwork.org/pub/php/news/2005/hessian/default.asp.

Tanksley, S., & McCouch, S., (1997). Seed banks and molecular maps: unlocking genetic potential from the wild. *Science*, *277*, 1063–1066.

Thomas, H., Powell, W., & Aung, T., (1980). Interfering with regular meiotic behavior in avenasativa as a method of incorporating the gene for mildew resistance from *Avena barbata*. *Euphytica.*, *29*, 635–640.

Tingey, W. M., & Sinden, S. L., (1982). Glandular pubescence glycoalkaloid composition, and resistance to the green peach aphid potato leafhopper, and potato flea beetle in Solanum berthaultii. *Am. Potato J.*, *59*, 22–51.

Tomar, S. M. S., & Singh, B., (1998). Hybrid chlorosis in wheat x rye crosses. *Euphytica.*, *99*, 1–4.

Tsunewaki, K., (1992). Aneuploid analysis of hybrid necrosis and hybrid chlorosis in tetraploid wheats using the D-genome chromosome substitution lines of durum wheat. *Genome.*, *35*, 594–601.

Tullu, A., et al., (2006). Sources of resistance to anthracnose (*Colletotrichum truncatum*) in wild *Lens* species. *Genetic Resources and Crop Evolution*, *53*, 111–119.

Van der Maesen, L. L. G., & Pundir, R. P. S., (1984). Availability and use of wild *Cicer* germplasm. *Plant Genetic Resources Newsletter*, *57*, 19–24.

Vaughan, D. A., (2003). Genepools of genus *Oryza*. In: *Monograph on Genus Oryza. Science Publ.*, Nanda, J. S., Sharma, S. D., (eds.), Enfield, N. H., USA, pp 113–138.

Villareal, R., et al., (2001). Registration of four synthetic hexaploid wheat (*Triticum turgidum/Aegilops tauschii*) germplasm lines tolerant ot waterlogging. *Crop Sci.*, *41*, 274.

Virmani, S., & Shinjyo, C., (1988). Current status of analysis and symbols for male sterile cytoplasms and fertility restoring genes. *Rice Genet Newsl.*, *5*, 9–15.

Walter, J. M., (1967). Heredity resistance to disease in tomato. *Annual Reviews Phytopathol.*, *5*, 131–160.

Warwick, S. I., (1993). Guide to the wild germplasm of Brassica and allied crops. Part IV. Wild species in the tribe Brassiceae (Cruciferae) as sources of agronomic traits. centre for land and biological resources research, research branch, agriculture Canada, Ottawa, *Ont. Technical Bulletin–17E*, 1–19.

Wilkes, H. G., (1977). Hybridization of maize and teosinte, in Mexico and Guatemala and the improvement of maize. *Econ. Bot.*, *31*, 254–293.

Wilkinson, R. E., (1983). Incorporation of *Phaseolus coccineus* germplasm may facilitate *production* of high yielding *P. vulgaris* lines. *Ann. Report of the Bean Improvement Cooperation*, *26*, 28–29.

Wilson, J. P., & Gates, R. N., (1993). Forage yield losses in hybrid pearl millet due to leaf blight caused primarily by Pyricularia grisea. *Phytopathology*, *83*, 739–743.

Wilson, J. P., Hess, D. E., & Hanna, W. W., (2000). Resistance to Striga hermonthica in wild accessions of the primary gene pool of Pennisetum glaucum. *Phytopathology*, *90*, 1169–1172.

Xu, J., & Kasha, K. J., (1992). Transfer of a dominant gene for powdery mildew resistance and DNA from *Hordeum bulbosum* into cultivated barley (*H.* vulgare). *Theor. Appl. Genet.*, *84*, 771–777.

Yanagihara, S., et al., (1995). Molecular analysis for the inheritance of the S-5 locus, conferring wide compatibility in Indica/Japonica hybrids of rice (O. sativa L.) *Theor. Appl. Genet.*, *90*, 182–188.

Zhang, Q., et al., (1998). Identification and tagging of a new gene for resistance to bacterial blight (*Xanthomonas oryzae* pv. *oryzae*) from O. rufipogon. *Rice Genet. Newsl.*, *15*, 138–142.

CHAPTER 6

APPLICATION OF MUTATION BREEDING IN PLANTS

A. K. MEHTA and M. H. BASHA

Department of Plant Breeding and Genetics, Jawaharlal Nehru Krishi Vishwavidyalaya, College of Agriculture, Jabalpur, Madhya Pradesh–482004, India

CONTENTS

ABSTRACT

In recent years, alacrity has rekindled in mutation research, as induced mutagenesis is considered as the tool for mining of new genes/alleles for the production of new varieties. It also studies the nature of genes and the way of controlling the biochemical pathways with the help of genomics and biotechnology. The use of ionizing radiation such as X-rays, gamma rays, and neutrons and chemical mutagens for inducing variation is well established. Induced mutations have been used to improve major crops such as wheat, rice, barley, cotton, peanuts, and beans, which are seed propagated. These studies will definitely have a major impact on the future crop improvement programs. India has become the 3rd largest contributor of mutant varieties in the world. In this chapter, various advanced aspects of mutation induction, applications, and examples of successful use of induced mutants in crop improvement programs are discussed.

6.1   INTRODUCTION

Mutation breeding is simply a new device in the breeder's toolkit that was initiated about seven decades ago, immediately after the discovery of mutagenic actions of X-rays on *Drosophila* by Muller (1927) and in barley by Stadler (1928). It does not involve gene modification and is also known as gene splicing. Mutation breeding uses a plant's own genetic resources as a process of spontaneous mutations, that's under way in nature all the time, the basis of evolution, and more importantly, it broadens biodiversity. It is a means of accelerating the process of developing different traits for selection, such as disease resistance, tolerance to abiotic stress conditions, and other valuable agronomic and nutritional quality traits (Table 6.1).

Nearly 3200 mutant varieties have been officially released for commercial use in more than 214 plant species from more than 60 countries. Over 1,000 mutant varieties of major staple crops, cultivated on tens of millions of hectares, enhance rural income, improve human nutrition, and contribute to environmentally sustainable food security in the world. The vast majority of released mutant varieties consist of cereals (1541) followed by flowers (649), legumes and pulses (432), and edible oil plants (96). Most mutant varieties were released in China (25.00%), Japan (14.85%), India (10.18%), Russia (6.66%), Netherlands (5.43%), Germany (5.27%), and US (4.29%). Many induced mutants were released directly as new varieties; others were used as parents to derive new varieties. For example, of the 3,227 varieties, 2,013 (62.37%) were released as direct mutants, crossing with one mutant (455), crossing with one mutant variety (294), crossing with two mutants (158), treatment of F_1 and F_2 seeds (168), and hybrid F_1 (a cross with a mutant line) (62). Mutation induction with physical mutagens was the most frequently used method to develop direct mutant varieties (77.19%). The use of chemical mutagens (11.4%) was relatively infrequent, and combination of physical and chemical mutagens was very low (0.01%). Among physical mutagens, gamma rays were employed to develop 67.48% of the radiation-induced mutant varieties, followed by X-rays (23.0%). In chemi-

TABLE 6.1   Improved Characteristics of Released Mutant Varieties

S. No.	Improved characters	Number of mutants
I.	**Yield contributing traits**	
	Plant architecture	667
	Growth habits	321
	Maturity	674
	Growing and harvesting	156
	Seed production characters	152
	Dry matter yield	15
	Yield	992
II.	**Resistance to Biotic stresses**	
	Disease resistance	474
	Pest resistance	30
III.	**Tolerance to Abiotic stresses**	
	Frost resistance	92
	Drought & High temperature	94
	Salinity tolerance	27
IV.	**Quality traits**	
	Marketing,food processing and Industrial quality	703
	Cooking quality	276
	Fruit traits	2
V.	**Nutritional quality traits**	
	Specialty and anti-nutrients	121
	Micro & Macro Nutrients	33
VI.	**Agronomic and botanical traits**	772

cal mutagens, ethyl methyl sulfonate (EMS) (28.80%) is the most prominent chemical mutagen, followed by nitroso-ethyl urea (15.44%), colchicine (12.5%), and nitroso-methyl urea (12.5%), as presented in Table 6.2.

TABLE 6.2   Mutant Varieties Developed by Mutagenic Intervention

Mutagenic Sources						Combined
Physical		Chemical				
Gamma, x-rays, Neutrons	Others	Alkalting agents	Colchicine	Sodium-azide	Others	Physical & Chemical
2358	133	278	46	11	79	37

6.2   HISTORY OF MUTATION BREEDING AND ITS APPLICATION

Period I: Observation and documentation of early spontaneous mutants

300 BC	The ancient Chinese book "Lulan" provides the first documentation of mutant election in plant breeding: maturity and other trait in cereals in China (Huang and Liang, 1980).
1590	The first verifiable (spontaneous) plant mutant described, "incisa" mutant of *Chelidonium majus*.
1667	The first known description of a graft–chimera; Bizarria-orange, Florence, Italy.
1672	One of the oldest publications describing variability in trees, shrubs, and herbaceous plants; Waare Oefeninge der Planten by A. Munting (see van Harten (1998).
17th century	"Imperial Rice" in China: a spontaneous mutant?Late 17th century Spontaneous mutant for the ornamental "morning glory" (*Ipomoea nil*) in Japan.
1741– onwards	Description of various examples of heritable variations in both wild and cultivated plants by Linnaeus (reported by Shamel and Pomeroy (1936)).
1859	"The Origin of Species" published by Charles Darwin.
1865	E. A. Carriére published on spontaneous mutations in various groups of crop plants, fruit trees, and potato in his book "Production et fixation des variétes dans les végétaux."

Period II: Induced mutations; the mutation breeding concepts

1895–1900	The discovery of various kinds of radiation (X-rays, α, β, and γ radiation).
1901	Hugo de Vries in "Die Mutations theorie" coined the term "mutation" for sudden, shock-like changes of existing traits.
1901	Korschinsky presented his theory of "Heterogenesis" and gave many examples of mutation-like changes in different plant species and crop plants.

1901 and 1911	First proof of mutations induced by chemicals.
1904 and 1905	Hugo de Vries suggests artificial induction of mutations by radiation.
1907	P.J.S. Cramer's work on bud variations.
1909–1913	W. Johannsen describes spontaneous drastic mutations and slight mutations affecting seed index.
1910	Thomas Hunt Morgan: first mutation experiments with *Drosophila melanogaster*.
Early 1920	From quantitative to quantitative radiation biology.
1920	N. I. Vavilov's "law of homologous series of variation."

Period III: Proof of induced mutations; theoretical background and the first commercial mutants

1926	N. I. Vavilov's theory on gene centers or "centers of origin."
1927	C. Stuart Gager and A. F. Blakeslee reported first proof of induced mutations in plants; radium ray treatment of *Datura stramonium*.
1927	Definite proof of mutation induction by X-rays by H.J. Muller (CIB-Method), indicating the possibility of obtaining genetically superior plants, animals and man by applying X-radiation.
1927	Efficient mass screening techniques for spontaneous mutants in lupine developed by von Sengbusch.
1928	Early practical projects on mutation induction in crop plants.
1928	Stadler publishes the first results of mutation induction in crop plants, barley, maize, wheat, and oat, but is skeptical about the use of induced mutation for crop improvement.
1928–1934	Continued studies on mutation theory and practical applicability.
1930s	Start of the Swedish mutation research program with studies on chlorophyll mutations by Gustafsson.
1934	The physical mutation theory–Target or "Hit and Target" theory was established by N. W. Timoféeff-Ressovsky and co-workers.

1936	The first commercial mutant variety "chlorina" obtained after X-ray radiation in tobacco by D. Tollenar released in Indonesia.
1937	The chromosome doubling effect of colchicine on plant chromosomes.
1941	Chemical mutagenesis: C. Auerbach, I. A. Rapoport, F. Oehlkers, and others.
1942	First report of induced disease resistance in a crop plant; X-ray-induced mildew resistance in barley (Freisleben and Lein).
1944	The term mutation breeding ("Mutationszüchtung") was first coined by Freisleben and Lein.
1949	First plant mutation experiments using 60Co gamma ray installations. Cobalt-60 was chosen as a suitable radioisotope for continuous gamma irradiation Advantages included a long half-life, was relatively cheap, abundant, and available. 60Co became a standard tool in mutation induction of crop plants.
Early 1950s	B. McClintock reported controlling elements (later established as transposable genetic elements or transposons).
1953	The Watson-Crick Model of the gene.
1954	The first release of a mutant variety in a vegetatively propagated crop: tulip var. Faraday with an improved flower color and pattern.
1956	E. R. Sears transfers resistance from *Aegilops* to wheat (*Triticum aestivum*) by radiation-induced translocation.
1958	Application of chemical mutagens on higher plants; Ethyl methane sulfonate EMS.

Period IV: Extensive application of mutation breeding for crop species; Appraisal of prospects and constraints; International cooperative activities developed by the joint IAEA/FAO division in Vienna

1960s	Numerous national research institutes specialized on nuclear techniques in food and agriculture were established.

1964	Establishment of internationally coordinated mutation breeding research programs by the Joint IAEA (International Atomic Energy Agency)/FAO (Food and Agriculture Organization) Division of Nuclear Techniques in Food and Agriculture at Vienna, Austria.
1966	First chemically induced mutant variety of barley, Luther, was released in the USA.
1969	The Pullman Symposium on Induced Mutations in Plant Breeding: The first classified list of mutant varieties published.
1981	First major symposium on the "Use of Induced Mutations as a Tool in Plant Breeding" organized by IAEA / FAO at Vienna, Austria.

Period V: Plant mutation breeding in genomics era

1983	Four groups independently reported the production of first transgenic plants, laying down the basis of T-DNA insertion mutagenesis.
1983	The transposable controlling elements Ac and Ds were isolated, laying down the basis of transposon mutagenesis using Ac-Ds and modified genetic systems.
1990	Joint IAEA/FAO symposium in Vienna, Austria, to assess the results of 25 years of applied mutation breeding.
1997	Retrotransposons were re-activated via tissue culture in rice, which prompted the establishment of the large Tos 17 mutant collection.
2000–2009	Development of high-throughput genotyping and phenotyping using automated, robotic and computerized systems.
2000	The first plant genome (*Arabidopsis* genome) was sequenced. Methodology for targeted screening of induced mutations established, which is now widely known as TILLING (Targeting Induced Local Lesion IN Genomes).

2002–2005	Genomes of the Indica and Japonica rice subspecies were sequenced.
2005	Establishment of mutant populations for functional genomics studies, including TILLING and T-DNA insertion mutant populations in crop plants
2008	International Symposium on Induced Mutations in Plants in Vienna, Austria, to assess applications of induced mutation in plant mutation research and breeding in the genomics era.
2008	Endonucleolytic Mutation Analysis by Internal Labeling (EMAIL) technique developed by Cross et al.
2014	Gurmukh Johal and Brian Dilkes, researchers at Purdue University, developed a new technique called Next GEM for stacking mutations to increasethe mutation density per individual

6.2.1 MUTATION BREEDING IN INDIA

In India, mutation breeding is carried out under the concept of "Atoms for Food and Feed" in several national/state universities/institutes like Bhabha Atomic Research Centre (BARC), Mumbai; Indian Agricultural Research Institute (IARI), New Delhi; National Botanical Research Institute (NBRI), Lucknow; Tamil Nadu Agricultural University (TNAU), Coimbatore, etc. in different crops. Some agricultural universities like Jawaharlal Nehru Krishi Vishwavidyalaya (JNKVV), Jabalpur; Acharya N G Ranga Agricultural University (ANGRAU), Hyderabad; and Indira Gandhi Krishividyalaya (IGKV), Raipur collaborated with BARC in some of the mutation breeding programs. India has become the 3rd largest contributor of mutant varieties in the world and is actively engaged in genetic enhancement of its germplasm through induced mutation techniques. A total of 418 mutant cultivars of crops, belonging to 71 plant species and one sporeless mutant of mushroom were approved and/or released in India. An updated list of 418 mutant cultivars released in India is given in Table 6.3. Summary of mutants, mutagen source, released/identified institutions, and improved traits are given in Table 6.4.

Ahloowalia et al. (2004) provided other examples of the benefits accruing from the cultivation of other notable mutant crop varieties; these included:

TABLE 6.3 Summary of Released/Approved/Registered Mutant Varieties of Different Crop Species in India

S.No.	Crop	Mutant variety	Parent Variety	Year	Mutagenic source	Induced/improved characters		Resistance
						Quantitative	Qualitative	
I. AC&RI, Madurai								
1.		MCU 7	L 1143 EE	1971	X-rays (800 Gy)	Early maturity, higher yield, and improved spinning capacity		Tolerance to drought, resistant to black arm, root rot and leaf blight
		MCU 10	MCU 4	1982	Gamma rays (300 Gy)		Long staple fiber	Drought tolerance, disease resistance
II. APAU, Lam								
1.	Mungbean (*Vigna radiata* L. Wiczeck)	LGG-407	Mutant of Pant M 2	1993	-	High yield, early		Resistant to YMV
		LGG-450	Mutant of Pant M 2	1993	-	High yield, early		Tolerance or resistance to YMV
III. ARS, Bhawanisagar								
1.	Turmeric (*Curcuma domestica* Val.)	BSR-1	Erode local	1986	X-rays	-	Rhizome color	-
IV. ARS, Badnapur								
1.	Mungbean (*Vigna radiata* L. Wiczeck)	BM 4	Mutant of T 44	1992	-	High yield, early maturity	-	Resistant to YMV

TABLE 6.3 (Continued)

S.No.	Crop	Mutant variety	Parent Variety	Year	Mutagenic source	Induced/improved characters			Resistance
						Quantitative	Qualitative		

V. ARS, Durgapur, Jaipur

S.No.	Crop	Mutant variety	Parent Variety	Year	Mutagenic source	Quantitative	Qualitative	Resistance
1.	Cluster bean (*Cyamopsis tetragonoloba* L.)	Kanchan Bahar	Durga Bahar	1996	EMS (0.6%)	Yield	-	-
2.	Barley (*Hordeum vulgare* L.)	RDB-1	R.S.-17	1972	Neutrons (4.5 x 1012 NP/cm2)	Dwarf, early maturity, high grain yield,	-	Resistant to lodging and less water requirements
		RD-103	Crossing of K-18 with mutant variety RDB-1 and	1978	Neutrons (4.5 x 1012 NP/cm2)	Dwarf, erect leaf, higher tillering, high yield under irrigated conditions	-	-
		RD-137	Crossing of EB-795 with mutant variety RDB-1	1981	Neutrons (4.5 x 1012 NP/cm2)	Medium tall plant type, high yield and less water requirements	-	-
		Rejkiran (RD-387)	Crossing of Marocaine-079 (CI-8334) with one mutant variety RDB-1	1982	Neutrons (4.5 x 1012 NP/cm2)	Dwarf, erect leaf, high tillering	-	Nematode resistance

S.No.	Crop	Mutant variety	Parent Variety	Year	Mutagenic source	Induced/improved characters		
						Quantitative	Qualitative	Resistance
		RD-2035	Crossing of PL-101 with mutant variety RD-137	1988	Neutrons (4.5 x 1012 NP/cm2)	Shortness, good tillering ability, wide adaptation, early maturity	-	CNN resistance

VI. ARS, Niphad, Nasik

S.No.	Crop	Mutant variety	Parent Variety	Year	Mutagenic source	Induced/improved characters		
1.	Wheat (*Triticum aestivum* L.)	NI-5643	New Thatch x NI-284-S	1975	Radiation	High yield, early maturity	Amber grain color	Resistant to rust

VII. BARC, Mumbai

S.No.	Crop	Mutant variety	Parent Variety	Year	Mutagenic source	Induced/improved characters		
1.	Groundnut (*Arachis hypogaea* L.)	TG-1 (Vikram)	Spanish Improved	1973	X-rays (750 Gy)	Large seed size, more branches, 50 days seed dormancy	High oil content	TMV resistance
		TG-3	Spanish Improved	1973	X-rays (150 Gy)	Increased pod number, high yield and good adaptability	-	-
		TG-4 (TG-14)	Spanish Improved	1976	X-rays (150 Gy)	Uniform maturity, high yield and good adaptability	-	-
		TG-17	Spanish Improved x dark green	1977	X-rays (150 Gy)	High yield and high harvest index	Desirable plant type (no secondary branches)	-

TABLE 6.3 (Continued)

| S.No. | Crop | Mutant variety | Parent Variety | Year | Mutagenic source | Induced/improved characters | | Resistance |
						Quantitative	Qualitative	
		TGS-1 (Somnath)	Crossing with one mutant TG-18A (TG 18 mut) x M-13	1989	Gamma rays (200Gy)	Early maturity, large seed size, high yield	Sequential flowering, high oil content	-
		TAG-24	Crossing with two mutants (TGS-2 X TGE-1)	1991	Gamma rays (200Gy)	Earliness, semidwarf habit, yield stability, high harvest index	Shorter internodes, dark green, small leaves and high water use efficiency	-
		TKG-19A	Crossing with two mutants (TG-17 X TG-1)	1993	X-rays (750Gy)	High yield, large seed size	Slightly reduced oil content with increased protein percentage	Tolerance to acidic soils
		TG-22	Crossing with one mutant (Robut-33-1 X TG 17)	1994	X-rays (150 Gy)	High yield, medium large seed size	Seed dormancy	Tolerance to acidic soils
		TG-26	Crossing with one mutant BARCG-1 x TG-23	1996	Gamma rays (300 Gy)	High yield, semidwarfness	Fresh seed dormancy, compact pod setting	Tolerance to peanut bud necrosis disease

S.No.	Crop	Mutant variety	Parent Variety	Year	Mutagenic source	Induced/improved characters		Resistance
						Quantitative	Qualitative	
		TPG-41	Crossing with one mutant (TG 28A x TG 22)	2002	X-rays (150Gy)	High yield, large seed, maturity 120 days	25 days seed dormancy, for post-rainy situation under irrigated conditions	-
		TG-37A	Crossing with one mutant variety (TG-25 X variety TG-26)	2004	Gamma rays (300 Gy)	Semi-dwarf, compact pod setting, high yield	Smooth pod surface	-
		TG-38	F_1 seeds of cross Girnar-1 X TG-26	2006	Gamma rays (300 Gy)	Erect growth habit with sequential branching, semi-dwarf	Medium-size dark green leaflets, compact pod setting with smooth pod surface	Tolerance to stem rot and dry root rot
		TLG-45	Crossing with two mutants (TG 19 x TAG 24)	2007	Gamma rays (200 Gy)	High pod yield, semi-dwarf plant	-	-

TABLE 6.3 (Continued)

S.No.	Crop	Mutant variety	Parent Variety	Year	Mutagenic source	Induced/improved characters		Resistance
						Quantitative	Qualitative	
		TG-39	Mutagenic treatment of breeding material F_1 and F_2 seeds (TAG 24 X TG 19)	2008	Gamma rays (200 Gy)	Semi-erect with medium height	Alternate flowering, seed contains50% Oil, 26.5% protein, 12.6% carbohydrate and 4.5% sucrose. Its oil contains 59% oleic acid and 23% linoleic acid.	-
		TG-51	Mutagenic treatment of breeding material F_1 and F_2 seeds (TG 26 X Chico)	2008	Gamma rays (200-350Gy)	Erect with semi-dwarf height	Sequential flowering, seed contains 49% oil with 43% oleic acid and 37% linoleic acid,	-
		TDG 39	-	2009	-	Large seeds, 120 days maturity	High content of oleic acid	-

S.No.	Crop	Mutant variety	Parent Variety	Year	Mutagenic source	Induced/improved characters		
						Quantitative	Qualitative	Resistance
		TG-47 (RARS T1)	Crossing with two mutants (TAG 24 x TG 19)	2011	Gamma rays (200Gy)	Large seed maturity 115 days,	More 3 seeded pods	-
		TGM-167	Mutant of TFDRG 5	2013	-	-	Gibberellin insensitive, dominant, dwarf	-
		TGM-38	Mutant of TAG 24	2013	-	-	Sub-orbicular leaflet, erect, compact & dwarf plant type	-
		TGM-51	Mutant of TAG 24	2013	-	Dwarf plant type	Funnel leaflet	-
2.	Wild Bougainvillea (*Bougainvillea spectabilis*)	Silver Top	Versicolor	1978	Gamma + Colchicine (25 Gy + 0.5% 6h)	-	Ornamental novelty, very vigorous and hardy, medium flowering in winter and summer	-
		TM-4	Crossing with one mutant variety (TM-1 X Varuna)	1978	Beta rays	High yield (34% more) and oil content	Altered seed color	-
		Lady Hudson of C.V.	Lady Hudson of Ceylon	1979	Gamma + Colchicine (10 Gy + 0.5% 6h)	-	Ornamental novelty, variegated leaves, pale yellow	-

TABLE 6.3 (Continued)

S.No.	Crop	Mutant variety	Parent Variety	Year	Mutagenic source	Induced/improved characters		
						Quantitative	Qualitative	Resistance
		Poultoni Variegata	Poultoni	1981	Gamma rays (15 Gy)	-	Variegated leaves	-
		Suvarna	Lady Hudson/Ceylon	1981	Gamma + Colchicine (10 Gy + 0.05% 6h)	-	Altered flower color (young bracts golden color turning to pinkish lemon when older)	-
		TM-2	RL-9	1987	X-rays (750 Gy)	High yield (25%)	Altered pod morphology (black/brown seeds) and	-
		TPM-1	-	2007	-	-	Yellow seed coat	Tolerance to powdery mildew
4.	Pigeon pea (*Cajanus cajan* Milsp.)	Trombay Vishakha-1	T-21	1983	Fast neutrons	Increased seed size with all desirable traits of parent variety T-21	-	-
		TAT 5	T-21	1984	Fast neutrons	Increased seed size (50%), high TGW and early maturity (140 days)	-	-

S.No.	Crop	Mutant variety	Parent Variety	Year	Mutagenic source	Induced/improved characters		
						Quantitative	Qualitative	Resistance
		TAT 10	Crossing with two mutants (TT-2 x TT-8)	1984	Fast neutrons (25 Gy)	Medium large size grain and extra early maturity (115-120)	-	-
		TT-401	-	2007	-	High yield	-	Tolerance to pod borer and pod fly damage
		TJT-501	-	2009	-	High yielding	Early maturity	Tolerance to Phytophthora blight
5.	Tosa jute (Corchorus olitorius L.)	TKJ-40 (Mahadev)	Crossing with 2 mutants (cross Virescent x Involute leathery)	1983	Thermal neutrons	High fiber yield	-	-
6.	Soybean (Glycine max L.)	TAMS-38	-	2005	-	High yielding	-	Resistant to bacterial pustules, Myrothecium leaf spot and SMV
		TAMS-98-21	JS-80-21	2007	Gamma rays	Early maturity	-	Resistant to bacterial pustules, Myrothecium leaf spot.

TABLE 6.3 (Continued)

S.No.	Crop	Mutant variety	Parent Variety	Year	Mutagenic source	Induced/improved characters		
						Quantitative	Qualitative	Resistance
7.	Sunflower (*Helianthus annus* L.)	TAS-82	-	2007	-	-	Black seed coat	Tolerance to drought
8.	Rice *Oryza sativa* L.	Hari (TR-RNR-21)	Dwarf mutant TR-5 of a salt tolerant variety SR-26-B x IR-8	1987	Fast neutrons	Higher grain yield	Short plant height	-
9.	Sesbania (*Sesbaniar rostrata*)	TSR-1	-	1994	-	Higher biomass	Late flowering mutant and photoperiod insensitive mutant	-
10.	Blackgram (*Vigna mungo* L. Hepper)	TAU-5	Mutant of EC-168200	-	Gamma rays	High yield	-	-
		UM-196	Mutant of No. 55	-	Gamma rays	-	Large seeds	-
		UM-201	Mutant of No. 55	-	Gamma rays	-	Large seeds	-

S.No.	Crop	Mutant variety	Parent Variety	Year	Mutagenic source	Induced/improved characters		
						Quantitative	Qualitative	Resistance
		TAU 1	T-9 x UM-196 (Mutant of No. 55)	1985	Gamma rays	High yield	Large seeds	Resistant to powdery mildew
		TAU-2	T-9 x UM-196 (mut.#55)	1992	Gamma rays	-	Large seeds	-
		TPU-4	UM-201 (Mutant of No. 55) x T-9	1992	Gamma rays	High yield and high seed weight	-	-
		TU-94-2	TPU-3 x TAU-5 (Mutant of EC 168200)	1999	Gamma rays	High yield	-	Resistant to yellow mosaic virus disease
11.	Mungbean *Vigna radiata* L. Wiczeck	TAP-7	S-8	1983	Gamma rays	Higher yield (23 %), early maturity (5-7 days)	-	Resistant to mildew and leaf spot
		TARM-2	RUM 5	1993	Gamma rays	High yield, medium late maturity	-	Resistant to powdery mildew disease
		TARM-1	RUM 5	1996	Gamma rays	High yield, and medium maturity	-	Resistant to powdery mildew disease
		TARM-18	PDM-54 x TARM-2	1996	Gamma rays	High yield	-	Resistant to powdery mildew disease

TABLE 6.3 (Continued)

S.No.	Crop	Mutant variety	Parent Variety	Year	Mutagenic source	Induced/improved characters		Resistance
						Quantitative	Qualitative	
		TMB-37	Kopergaon x TARM-2	2005	Gamma rays	Early and high yielding	Large seed size	Resistant to Yellow mosaic virus
		TJM-3	Kopergaon x TARM-1	2007	Gamma rays	Early maturity	-	Resistant to Yellow mosaic virus, powdery mildew, Rhizoctonia root-rot disease
		TM-96-2#	Kopergaon x TARM-2	2007	Gamma rays	Early maturity and suitable for rice fallow cultivation	-	Resistant to Powdery mildew, Corynespora leaf spot
		TM 2000-2 (TJT-501)	-	2010	-	-	Suitable for rice fallows	Resistant to powdery mildew
12.	Cowpea (*Vigna unguiculata* L. Walp.)	TCM 148-1	-	1998	-	-	-	Multiple disease resistance against yellow mosaic virus, root rot, leaf curl and leaf blight diseases
		TRC77-4 (Kalleshwari)	V-130	2007	Gamma rays (200 Gy)	Dwarf compact plant type suitable for rice fallows and yield	-	-

S.No.	Crop	Mutant variety	Parent Variety	Year	Mutagenic source	Induced/improved characters		
						Quantitative	Qualitative	Resistance
VIII. BAU, Ranchi								
1.	Groundnut (*Arachis hypogaea* L.)	BG-1	41-C	1979	Gamma rays	Higher yield, semi-erect habit, gigas type	Large seed	-
		BG-2	41-C	1979	x-rays	Higher yield, and semi erect habit	Large seed	-
		BP-1	41-C	1979	Gamma rays (450Gy)	Semi-erect habit and high yield	Improved seed size	-
		BP-2	41-C	1979	Gamma rays (450Gy)	Semi-errect gigas type and high yield	Improved seed size	-
2.	Soybean (*Glycine max* L.)	Birsa Soybean-1	Sepaya Black	1983	Spontaneous mutant	High yield	-	-
IX. BHU, Varanasi								
1.	Lentil (*Lens culinaris* L. Medik.)	HUL 57	Mutant of HUL–11	2005	-	Small seed		Resistant to rust
2.	Rice (*Oryza sativa* L.)	HUR-36	Mahsuri	1990	Gamma rays (15 kr) + EMS (0.04M)	Early maturity	Longer and shining grains	-

TABLE 6.3 (Continued)

S.No.	Crop	Mutant variety	Parent Variety	Year	Mutagenic source	Induced/improved characters			Resistance
						Quantitative	Qualitative		
X. CAZRI, Jodhpur									
1.	Moth bean (*Vigna aconitifolia* Jacq. M.)	Maru Moth-1 (IJMM-259)	Jadia	-	Gamma rays (300 Gy)	Early maturity	-		Drought resistant
		RMO 40	Jwala	1994	Gamma rays (400 Gy)	Extra early maturing, short stature, non-spreading variety with synchronous maturity	-		Tolerance to drought
		RMO-257	Jadia	1997	Gamma rays (300 Gy) + EMS (0.6%)	Yield and earli-ness	-		-
		CAZRI Moth-1	Jadia	1999	Gamma rays (300 Gy)	High yield, and semi-erect plant	High seed protein		Resistant to YMV
		RMO-225	Jadia	1999	Gamma rays (400 Gy) + EMS (0.6%)	Earliness	-		Resistant to diseases
XI. CCSU, Meerut									
1.	Mungbean (*Vigna radiata* L. Wiczeck)	MUM-2	K-851	1992	EMS (0.2% for 6 hr)	High yield	-		Resistant to diseases

S.No.	Crop	Mutant variety	Parent Variety	Year	Mutagenic source	Induced/improved characters		Resistance
						Quantitative	Qualitative	
XII. CIMAP, Lucknow								
1.	**Citronella** *(Cymbopogon winterianus Jowitt.)*	Manjari	Manjusha M3-8	1998	Spontaneous mutant	-	High oil content	-
		Jalparllavi	-	1998	-	High oil content	-	-
2.	**Indian henbane** *(Hyocyamus niger)*	Aekela	Balck hen-bane	1996	Gamma rays (20-40 kR)	High tropane alkaloid	Unbranched mutant	-
		Aela	-	1996	Gamma rays (20-60 kR)	high yield of tropane alkaloid and biomass	Yellow flowered mutant	-
3.	**Germen chamomile** *(Matricario cammomilla)*	Vallery	German Bulk	1994	Gamma rays (100 Gy)	High chamoza-lene and dry herb yield	High oil	-
4.	**Spearmint** *(Mentha spicata)*	Neera	-	1992		-	High alkaloid content	-
5.	**Opium poppy** *(Papaver so-maniferum L.)*	BC-28/9/4 (Vivek)	Shweta	1992	Gamma rays (50 Gy)	Big size capsule	Improved alkaloid content	Resistant to lodg-ing
XIII. CRRI, Cuttack								
1.	**Rice** *(Oryza sativa L.)*	Indira	Tainan-3	1980	EMS	Early maturity	-	-

TABLE 6.3 (Continued)

| S.No. | Crop | Mutant variety | Parent Variety | Year | Mutagenic source | Induced/improved characters | | Resistance |
						Quantitative	Qualitative	
		Sattari (CRM-13-3241)	Irradiation of hybrid seeds of cross (NSJ 200 x Padma)	1983	Gamma rays	Early maturity and high yield	-	-
		Padmini (IET-10561)	CR 1014	1988	Gamma rays	Early maturity	-	-
		Dharitri (IET-6272)	Pankaj x mutant variety Jagannath (BSS-873)	1989	X-rays (300 Gy)	Semi-dwarf, and high yield	Short grains	-
		Gayatri (IET-8020)	Pankaj x mutant variety Jagannath (BSS-873)	1989	X-rays (300 Gy)	Semi-dwarf, high yield	Short grains	Resistant to BLB, moderately resistant to blast and GM
		Radhi (CRM-40 and IET-12413)	Swarnaprab-ha	1997	-	-	Long bold grains	Tolerance to blast & BPH
		CRM 53	IR 50	1999	EMS (0.66%)	-	-	Resistant to blast disease
		Ramchandi (IET-13354)	IR-17494-32-2-2-1with mutant Jagannath (BSS-873)	1999	X-rays (300 Gy)	Semi-dwarfness, photo sensitive	Grains medium bold, white	-

S.No.	Crop	Mutant variety	Parent Variety	Year	Mutagenic source	Induced/improved characters		
						Quantitative	Qualitative	Resistance
		CRM 51	IR 50	1999	Sodium azide (NaN3) (0.001M)	-	-	Resistant to blast disease
		CRM 49	IR 50	1999	Sodium azide (NaN3) (0.001M)	-	-	Resistant to blast disease
		CRM 2007-1	Basmati 370	2005	Gamma rays (100 Gy)	Semi dwarf stature, shorter duration	-	-

XIV. CRIJAF, Kolkata, West Bengal

S.No.	Crop	Mutant variety	Parent Variety	Year	Mutagenic source	Quantitative	Qualitative	Resistance
1.	Tosa jute (Corchorus olitorius L.)	Bast fibreshy (bfs) mutant	JRO 632	2012	Thermal neutrons	Early maturing with significant reduction in bast fiber wood yield and poor fiber quality	-	-

XV. CSAU, Kanpur

S.No.	Crop	Mutant variety	Parent Variety	Year	Mutagenic source	Quantitative	Qualitative	Resistance
1.	(Hordeum vulgare L.)	K-257	Crossing with one mutant variety (RDB-1) x Vijaya	1980	Neutrons (4.5 x 1012 NP/cm2)	Medium tall plant, high yield	Altered ear morphology (long ears)	-

TABLE 6.3 (Continued)

S.No.	Crop	Mutant variety	Parent Variety	Year	Mutagenic source	Induced/improved characters		Resistance
						Quantitative	Qualitative	

XVI. DHFP, Chaubattia, Uttarakhand.

S.No.	Crop	Mutant variety	Parent Variety	Year	Mutagenic source	Quantitative	Qualitative	Resistance
1.	Apple *(Malus domestica)*	Chaubattia Agrim	Early Shanburry	1984	Gamma Rays	High yield	Fruits small, egg shaped, slightly conical, yellow colored with slight red tinge. Flesh is dirty yellowish white, Sour to sweet in taste	-
2.	Apricot (Prunus armeniaca)	Chaubattia Madhu	Turkey x Charmagz Budwood	1984	Gamma Rays	Early ripening, highly productive, regular in bearing.	-	-
		Chaubattia Kasri	St. Ambroise x Charmagz Budwood	1984	Gamma Rays	-	Mid season ripening with good quality fruits	-
		Chaubattia Alanker	Kaisha x Charmgaze Budwood	1984	Gamma Rays	Regular bearing, low chilling and very early maturity.	Good keeping quality	-

XVII. DMAPR (ICAR), Anand

S.No.	Crop	Mutant variety	Parent Variety	Year	Mutagenic source	Quantitative	Qualitative	Resistance
1.	(Isabgol *(Plantago ovata* L.)	DPO14	GI-2	2011	DES (0.4%)	Early maturing, high seed yield and high harvest index.	-	-

S.No.	Crop	Mutant variety	Parent Variety	Year	Mutagenic source	Induced/improved characters		
						Quantitative	Qualitative	Resistance
		DPO 296-4	GI-2	2014	DES (0.4%)	-	Golden yellow leaf color	-
XVIII. DOR, Hyderabad								
1.	**Castor (*Ricinus communis* L.)**	Yellow stem mutant castor (IC0587750)	DCS 99	2011	Spontaneous Mutant	High yield	Unique color and single bloom with synchronous maturity	-
XIX. GBPUA&T, Pantnagar								
1	**Blackgram (*Vigna mungo* L. Hepper)**	Manikya	-	1988	-	High yield, early maturity.	Large seed size	-
2.	**Mungbean (*Vigna radiata* L. Wiczeck)**	Pant Moong-2	ML-26	1982	Gamma rays (100 Gy)	More pods and high yield	-	Resistant to virus diseases (MYMV)
3.	**Poplar (*Populus deltoids* Marshall)**	PP-5	L 89-4	2011	-	Higher height and diameter	-	-
4.	**Tosa jute (*Corchorus olitorius* L.)**	JRO 3690 (Savitri)	Crossing with 2 mutants (long internode mutant x tobacco leaf mutant)	1985	X-rays (500 Gy)	Higher yield	-	Resistant to apion and yellow mite

TABLE 6.3 (Continued)

S.No.	Crop	Mutant variety	Parent Variety	Year	Mutagenic source	Induced/improved characters			Resistance
						Quantitative	Qualitative		

XX. GKVK Campus, Bangalore

S.No.	Crop	Mutant variety	Parent Variety	Year	Mutagenic source	Quantitative	Qualitative	Resistance
1.	Finger millet (*Eleusine coracana* (L.) Gaertn.)	PS-1	Mutant of GPU 28	2014	-	-	Partial sterility, useful in hybridization, easy maintenance	-

XXI. HAU, Hisar

S.No.	Crop	Mutant variety	Parent Variety	Year	Mutagenic source	Quantitative	Qualitative	Resistance
1.	Groundnut (*Arachis hypogaea* L.)	MH-2	Selection from Gujarat dwarf mutant	1978	Spontaneous mutant	Higher yield, very dwarf, early maturity and easy digging	-	Resistant to Cercospora personata
2.	Desi cotton (*Gossypium arborium* L.)	DS-1	G-27	1985	Gamma rays (200 Gy)	Semi-dwarf and high lint	-	-
3.	Barley (*Hordeum vulgare* L.)	BH-75	RD-150 (=RDB-1 x EB-795) x Ahor-131/68.	1983	Neutrons (4.5 x 1012 NP/cm2)	Semi-dwarf, good tillering ability, early maturity	-	Resistant to yellow rust and CCN
4.	Isabgol (*Plantago ovata* L.)	Haryana Isabgol-5	Local race	1994	Mutation	Profuse tillering,	Long and compact spikes	Moderately resistant to downy mildew
5.	Foxtail millet (*Setaria italica* L.)	PS 4	SIA 2616	1999	EMS (0.2%)	High yield and adopted to rainfed conditions	-	-

S.No.	Crop	Mutant variety	Parent Variety	Year	Mutagenic source	Induced/improved characters		
						Quantitative	Qualitative	Resistance
XXII. HPAU, Palampur								
1.	Rice (*Oryza sativa* L.)	HPU 8020	Bala	1984	Gamma rays (200 Gy)	Late maturity, synchronous til-lering and high yield	-	-
XXIII. IARI, New Delhi								
1.	Papaya (*Carica papaya* L.)	Pusa nanha	Ranchi	1987	Gamma rays (150 Gy)	Reduced plant height from 218 cm to 106 cm, 50% higher, girth of trunk and length of leaf from 193 to 86 cm	-	-
2.	Chrysanthe-mum (*Chrysan-themum* sp.)	Pusa Anmol	Ajay	2000	Gamma rays (10 Gy)	-	Yellowish pink flowers, photo and thermo- insensitive	-
		Pusa Cente-nary	Thai Chen Queen	2000	Gamma rays	-	Yellow flowers	-
3.	Chickpea (*Ci-cer arietinum* L.)	Pusa 408 (Ajay)	G-130	1985	Gamma rays (600 Gy)	High yield, semi-erect	Improved plant architecture	Blight resistance

TABLE 6.3 (Continued)

S.No.	Crop	Mutant variety	Parent Variety	Year	Mutagenic source	Induced/improved characters		Resistance
						Quantitative	Qualitative	
		Pusa 413 (Atul)	G-130	1985	Gamma rays (600 Gy)	High yield, semi-erect	-	Resistant to wilt, stunt virus, foot root, root rot & moderately resistant to blight
		Pusa 417 (Girnar)	BG 203	1985	Gamma rays (600 Gy)	High yield, short, semi-erect, profused branched, high pod number	-	Wilt resistance, moderate resistance to stunt virus, collar rot, foot rot, root rot, low pod borer and nematode damage
		BGM 547	BG 256	2005	Gamma rays	High yield, bold grain size	Attractive golden brown color	Moderate resistance to wilt, root rot, stunt and Helicoverpa armigera
		Pusa 547	BG-256	2006	Gamma rays (600 Gy)	High yield, good cooking quality	-	Tolerance to Fusarium wilt, stunt virus and root rot
4.	American cotton (*Gossypium hirsutum* L.)	Rasmi	MCU 5	1976	Gamma rays (300 Gy)	Photoperiod insensitive, wider adaptability, superior quality and improved yield	-	-

S.No.	Crop	Mutant variety	Parent Variety	Year	Mutagenic source	Induced/improved characters		
						Quantitative	Qualitative	Resistance
		Pusa Ageti	Stoneville 213	1978	Gamma rays (250 Gy)	Improved ginning capacity of fiber, early maturity	-	Tolerance to jassids
5.	Barley (*Hordeum vulgare L.*)	DL-253	Ratna	1981	Gamma rays (200 Gy) + EMS (0.30%)	Seed yield, good tillering	-	Resistant to smut and yellow rust
		Karan-15	Crossing with one mutant variety (RDB-1) x EB-20	1982	Neutrons (4.5 x 1012 NP/cm2)	Semi-dwarf, good tillering ability and high yield potential	-	-
		Karan-201	Crossing of two mutants RDB-1 x Riso mutant 1508	1982	Neutrons (4.5 x 1012 NP/cm2)	Semi-dwarf	High protein content (16%), hull less grain	-
		Karan-265	Crossing of two mutants (RDB-1 x EB-7725) x Riso mutant 1508.	1982	Neutrons (4.5 x 1012 NP/cm2)	Dwarf, high tillering ability. thick short narrow leaves	Hull less grain	Resistant to lodging

TABLE 6.3 (Continued)

S.No.	Crop	Mutant variety	Parent Variety	Year	Mutagenic source	Induced/improved characters		
						Quantitative	Qualitative	Resistance
		Karan-3	Crossing of two mutants (RDB-1 x EB-7576) x Riso mutant 1508.	1982	Neutrons (4.5 x 1012 NP/cm2)	Semi-dwarf, erect leaves	Hull less amber color grains	Resistant to lodging
		Karan-4	Crossing of EB-7576 with mutant variety RDB-1 amd	1983	Neutrons (4.5 x 1012 NP/cm2)	Semi-dwarf	Erect and narrow leaves, hull less and amber colored leaves	Resistant to lodging
6.	Tomato (Lycopersicon esculentum M.)	Pusa Lal Meeruti	Meeruti	1972	Gamma rays (300 Gy)	High yield	Uniform fruit ripening, and improved fruit color	-
7.	Rice (Oryza sativa L.)	Pusa-NR-162 (Renu)	Basmati-370 x mutant Jaya	1988	Gamma rays	Early maturity	Fine grains	Tolerance to low temperature
		Pusa-NR-166	Basmati-370 x mutant IR-8	1989	Gamma rays	-	Synchronous flowering	-
		Pusa-NR-381 (IET-9208)	Basmati-370 x mutant Tainan-3	1989	Gamma rays	-	-	Resistant to blast

S.No.	Crop	Mutant variety	Parent Variety	Year	Mutagenic source	Induced/improved characters		Resistance
						Quantitative	Qualitative	
		Pusa-NR-519	[Tainan-3 mut. x Basmati-370] x NR-417-3)	1990	Gamma rays	-	-	Resistant to pests and diseases
		Pusa-NR-555-5	[(Dular mut. x N-22 mut.) x (Tainan-3 mut. x Basmati-370)] x MTU-17 mut	1990	Gamma rays	Early maturity	-	Resistant to pests and diseases
		Pusa-NR-570-17	[(Gora mut. x MW-10 mut.) x N-22 mut.]	1990	Gamma rays	Early maturity	-	Resistant to pests and diseases
		Pusa-NR-571	MW 10 mutant x PMR 351 (Tainan 3mut./Basmati 370)	1990	Gamma rays	Semi-dwarfness and early maturity	-	-
		Pusa-NR-550-1-2 (JD-8)	Dular mut. x N-22 mut.	1997	Gamma rays	Semi-dwarf and high yield	-	-
		Pusa-NR-551-4-20 (JD-6)	Dular mut. x N-22 mut.	1997	Gamma rays	Semi-dwarf and high yield	-	-

TABLE 6.3 (Continued)

S.No.	Crop	Mutant variety	Parent Variety	Year	Mutagenic source	Induced/improved characters		
						Quantitative	Qualitative	Resistance
		Pusa-NR-555-28 (JD-10)	Dular mutant /N-22/PNR 351/mutant 17	1997	Gamma rays	Semi-dwarf and high yield	-	-
		Pusa-NR-546 (IET-11347)	F$_2$ seeds of cross (PNR 125-2 x PNR 130-2)	1998	Gamma rays	Semi dwarf, and high yield	Super fine aromatic grain	Tolerance to brown spot, leaf blast, sheath blight, BLB, SB, WBPH, GM
		Pusa-NR-555-5 (JD-3)	Dular mutant /N 22/PNR 351/mutant 17	1998	Gamma rays	High yield	-	-
		PNR-519	Basmati-370 PNR-417-3 x Tainang-3 mutant	2000	Gamma rays	Semi-dwarf	Grains large size, brown	Resistant to blast, Sh.B, BS, SB, GM, LF and WM,
8.	Pearl millet (*Pennisetium typhoides* L.)	New Hybrid Bajra 5	Mutagenic treatment of Hybrid F$_1$ (Male sterile inbred line Tift 23A)	1974	Gamma rays (350 Gy)	-	-	Resistant to mildew
		NHB 3 (Hybrid)	MS 5071A x Tift 23 B	1975	Gamma rays (350 Gy)	-	-	Resistant to Sclerospora graminicola

S.No.	Crop	Mutant variety	Parent Variety	Year	Mutagenic source	Induced/improved characters		
						Quantitative	Qualitative	Resistance
		NHB 4 (Hybrid)	K560-230 x Tift 23 B	1975	Gamma rays (350 Gy)	-	-	Resistant to Sclerospora graminicola
		Pusa 46	Mutagenic treatment of Hybrid F$_1$ (J104xK559)	1982	-	-	-	Resistant to fungal diseases
9.	French bean (*Phaseolus vulgaris* L.)	Pusa Parvathi	Yellow "wax pod" EC 1906 from USA	1994	-	Plants bushy	Flowers pink. pods are round green, straight flattish round, stringless and green in color	Highly susceptible to rust and MYMV
10.	Pea(*Pisum sativum* L.)	Hans	P 1163	1979	EI	Early maturity, high yield and better seed quality	-	-
11.	Castor (*Ricinus communis* L.)	Aruna	-	1969	Thermal neutrons (14 Gy)	Very early maturity. slightly higher yield	-	-
		Sowbhagya (157-B)	(Aruna mut. x dwarf mut. of HC-6) x (dwarf mut. of HC-6 x Mauthners dwarf)	1976	Thermal neutrons (14 Gy)	Late maturity. dwarf	Suitable for intercropping	Resistant to shattering

TABLE 6.3 (Continued)

S.No.	Crop	Mutant variety	Parent Variety	Year	Mutagenic source	Induced/improved characters		
						Quantitative	Qualitative	Resistance
13.	Rose (Rosa sp.)	Abhisarika	A mutant of the popular H.T. rose Kiss of Fire.	1963	Gamma rays	-	Apricot color changes to deep pink on aging with silvery reverse of petals and deep pink to light purplish streaks.	-
		Angara	Montezuma	1975	Gamma rays	-	Improved plant architecture (very vigorous, compact and profuse blooming), flower color (dark reddish orange) and more fragrant than the parent	-
		Madhosh H.T.	Gulzar	1975	EMS (0.25%)	-	Petals have mauve colored stripes contrasting with deep red base and vigorous growth	-
		Striped Christian Dior	Christian Dior	1975	Gamma rays (75 Gy)	-	White and pink streaked petals on deep pink to red base	-

S.No.	Crop	Mutant variety	Parent Variety	Year	Mutagenic source	Induced/improved characters		
						Quantitative	Qualitative	Resistance
		Pusa Christina	Christian Dior	1975	Gamma rays (100 Gy)	-	Pink flower color, globular bud and vigorous growth	-
		Saroda	Queen Elizabeth	1983	Gamma rays (30 Gy)	-	Very light pink color instead of carmine rose flowers	-
		Sharada	Queen Elizabeth	1983	Gamma rays	-	Flower color	-
		Striped Contempo	Contempo	1983	Gamma rays (30 Gy)	-	Yellow stripes on orange background instead of orange flowers with yellow eye	-
		Sukumari	America's Junior Miss	1983	Gamma rays (30 Gy)	-	White flower color instead of coral-pink flowers	-
		Tangerine Contempo	Contempo	1983	Gamma rays (30 Gy)	-	Tangerine orange flower color with yellow eye instead of orange flowers with yellow eye	-

TABLE 6.3 (Continued)

S.No.	Crop	Mutant variety	Parent Variety	Year	Mutagenic source	Induced/improved characters		
						Quantitative	Qualitative	Resistance
		Curio	Imperator	1986	Gamma rays (30 Gy)	-	Cherry red flower color and different flower morphology	-
		Pink Contempo	Contempo	1986	Gamma rays (30 Gy)	-	Pink flower color instead of orange flowers with yellow eye	-
		Twinkle	Imperator	1986	Gamma rays (30 Gy)	-	Striped flowers (Pink stripes on cherry red background)	-
		Yellow Contempo	Contempo	1986	Gamma rays (30 Gy)	-	Empire yellow flower color instead of orange flowers with yellow eye	-
		Light Pink Prize	First Prize	1989	Gamma rays (30 Gy)	-	Light pink flower color instead of blend of light red and deep pink	-
		Minalini Stripe	Minalini	1991	Gamma rays	-	Flower color	-

S.No.	Crop	Mutant variety	Parent Variety	Year	Mutagenic source	Induced/improved characters		Resistance
						Quantitative	Qualitative	
13.	Wheat (*Triticum aestivum* L.)	NP 836	NP 799	1961	x-rays (160 Gy)	Awned spike, high yield (10%), medium early maturity (140 days)	-	Resistant to leaf rust
		Sharbati Sonora	Sonora 64	1967	Gamma rays (200 Gy)	early maturity,	Amber grain color, high protein content (16.5%), good in bread and chapati making	-
		Pusa Lerma	Lerma rojo 64-A	1971	Gamma rays	-	Higher extensibility and elasticity of flour and higher Pelshenke value, stronger gluten, amber grain color	-
14.	Cowpea (*Vigna unguiculata* L. Walp.)	V16 (Amba)	Pusa Phalguni	1981	DMS	High yield	-	Resistant to fungal and bacterial diseases
		V37 (Shreshtha)	Pusa Phalguni	1981	DMS	High yield, high vegetative growth, suitable as fodder	-	-

TABLE 6.3 (Continued)

S.No.	Crop	Mutant variety	Parent Variety	Year	Mutagenic source	Induced/improved characters			Resistance
						Quantitative	Qualitative		
		V38 (Swarna)	Pusa Phal-guni	1981	DMS	High yield, early maturity, synchronous flowering, better quality pods and grains	-		Resistant to diseases
		V240	Pusa Phal-guni	1984	DMS (0.8%)	High yield	-		Resistant to fungal, viral and bacterial diseases
15.	Papaya (*Carica papaya* L.)	Pusa nanha	Ranchi	1987	Gamma rays (150 Gy)	Reduced plant height from 218 cm to 106 cm, 50% higher, girth of trunk and length of leaf from 193 to 86 cm	-		-

XXIV. ICAR Research, Complex for Eastern Region, Research Centre for Makhana, Darbhanga, Bihar

S.No.	Crop	Mutant variety	Parent Variety	Year	Mutagenic source	Quantitative	Qualitative	Resistance
1.	Makhana (*Euryale ferox* Salisb.)	Sel-5	A mutant was selected from a local landrace	2014	Diethyl sulphonate (0.4%)	-	White flower mutant	-

S.No.	Crop	Mutant variety	Parent Variety	Year	Mutagenic source	Induced/improved characters		Resistance
						Quantitative	Qualitative	
XXV. ICRISAT, India								
1.	**Pearl millet (*Pennisetium typhoides* L.)**	ICMH 451	Mutagenic treatment of Hybrid F₁ (Mildew resistant mutant 81 A/B x Tift 23 DB)	1986	Gamma rays (300 Gy)	High yield	-	Resistant to mildew
XXVI. IIHR, Bangalore								
1.	**Chrysanthe-mum (*Chrysan-themum* sp.)**	Yellow Gold	Flirt	1998	Gamma rays	-	Small flowered, double Korean type, red cultivar, good for both loose and cut flower.	-
		Usha Kiran	Kirti	2000	-	-	Yellow, Garland floral decoration	-
2.	**Carnation (*Dianthus caryophyllus L.)**	IIHRP-1	Accession CG 109	2006	-	-	Dark red colored flowers with blunt flower margin.	-
3.	**Gladiolus (*Gladiolus* sp.)**	Shobha	Wild Rose	1988	Gamma rays (10 Gy)	-	Shell pink flower color instead of roseine purple	-

TABLE 6.3 (Continued)

S.No.	Crop	Mutant variety	Parent Variety	Year	Mutagenic source	Induced/improved characters		
						Quantitative	Qualitative	Resistance
1.	**Bougainvillea (*Bougainvillea spectabilis* Wild)**	Jayalakshmi Variegata	Jayalakshmi	1977	Gamma rays (65 Gy)	-	Variegated leaves	-
XXVII. IPR, Kanpur, Uttar Pradesh								
1.	**Blackgram (*Vigna mungo* L. Hepper)**	IPU 99-12005 (IC0594172)	IPU 99-167	2012	Spontaneous Mutant	Male sterile mutant	Flower with a protruded stigma and crumpled petals	-
XXVIII. IPSI, Indore								
1.	**American cotton (*Gossypium hirsutum* L.)**	Indore-2	MU-4 (Dhar Kambodia)	1950	X-rays	High yield	-	-
		Badnawar-1	Hybridization MU-4 with mutant variety Indore-2	1961	X-rays	High Yield	-	-
		Khandwa-2	Hybridization MU-4 with mutant variety Indore-2	1971	X-rays	High Yield	-	-

S.No.	Crop	Mutant variety	Parent Variety	Year	Mutagenic source	Induced/improved characters		
						Quantitative	Qualitative	Resistance
XXIX. IIT, Kharagpur								
1.	Rice (*Oryza sativa* L.)	IIT 48	IR 8	1972	Ethyleneox-ide (0.3% for 8h)	Early maturity	Improved grain size	-
		IIT 60	IR 8	1972	EMS (0.5% for 8h)	Early maturity and high yield	-	-
XXX. Jammu & Kasmir								
1.	Rice (*Oryza sativa* L.)	K-84	T-65	1967	Gamma rays	Early maturity	-	-
XXXI. KAU, Kerala								
1.	Okra (*Abelmoschus esculentus* L. Moench)	Anjitha	F_1 and F_2 seeds of cross Kiran X *A. manihot*	2006	Gamma rays (300Gy)	High yield	-	Resistant to virus
2.	Rice (*Oryza sativa* L.)	Vellayani	PTB-10	1968	Thermal Neutrons	Early Maturity	-	-
3.	Coleus (*Solenostemon rotundifolius* poir.)	Suphala	A tissue culture mutant derived from Paipra local cultivar	2006	Gamma rays (50 Gy)	High yielding and photo-insensitive	-	-
XXXII. Mysore								
1.	American cotton (*Gossypium hirsutum* L.)	M.A.9	Mutant from Co-2	1948	X-rays	-	Drought tolerance	

TABLE 6.3 (Continued)

S.No.	Crop	Mutant variety	Parent Variety	Year	Mutagenic source	Induced/improved characters		
						Quantitative	Qualitative	Resistance
XXXIII. NBRI, Lucknow								
1.	**Bougainvillea (*Bougainvillea spectabilis* Wild)**	Arjuna	Partha	1976	Gamma rays (5Gy)	–	Variegated leaves, central part of the leaves has white, cream, pale and dark green variegations	–
		Pallavi	Roseville's Delight	1986	Gamma rays (10 Gy)	–	Chlorophyll variegated leaves	–
		Los Banos Variegata	Los Banos beauty	1990	Gamma rays (10 Gy)	–	Changed leaf color from green to variegated	–
		Mahara variegata	Mahara	1994	Gamma rays (10 Gy)	–	Chlorophyll variegated leaves	–
		Los Banos Variegata silver margin	Los Banos beauty	2002	Gamma rays (10 Gy)	–	Variegated leaves with combination of green and silver margin	–
		Mahara Variegata abnormal leaves	Mahara	2002	Gamma rays (10 Gy)	–	Variegated leaves and margin is unevenly undulated and leaf lamina is asymmetrical	–

S.No.	Crop	Mutant variety	Parent Variety	Year	Mutagenic source	Induced/improved characters		
						Quantitative	Qualitative	Resistance
		Los Banos Variegata-Jayanthi	Los Banos Beauty	2008	EMS (0.02%)	-	Two types of variegated leaves. One is mosaic with four visible colors and other has green central portion with white margin	-
		Pixie varie-gata	Los Banos Variegata-Jayanthi	2009	EMS (0.02%)	-	Leaf margins are creamish and their central region is green, some leaves have creamish margin and deco-rated with green dots	-
2.	Chrysanthe-mum (*Chrysan-themum* sp.)	Basant	Paul	1974	Gamma rays (10 Gy)	-	Yellow flower color	-
		Ashankit	Undaunted	1974	Gamma rays (15 Gy)	-	Large flower heads with bright mauve, semi-quilled and fringed ray-florets with silver reverse	-
		Aruna	Undaunted	1974	Gamma rays (15 Gy)	-	Dark reddish flower color	-

TABLE 6.3 (Continued)

S.No.	Crop	Mutant variety	Parent Variety	Year	Mutagenic source	Induced/improved characters		
						Quantitative	Qualitative	Resistance
		Kapish	E-13	1974	Gamma rays (20 Gy)	-	Brown flower color and small flowers about 4 cm in diameter	-
		Lohita	E-13	1974	Gamma rays (20 Gy)	-	Dark reddish flower color	-
		Kansya	Rose Day	1974	Gamma rays (15 Gy)	-	Extra large, bronze flower color with broad reflexing ray-florets	-
		Gairik	Belur Math	1974	Gamma rays (10 Gy)	-	Salmon light flower color and almost incurved flower heads	-
		Himani	E-13	1974	Gamma rays (20 Gy)	-	White flower color, small flowers, pompon type	-
		Pingal	Pink Casket	1974	Gamma rays (10 Gy)	-	Terracotta flower color and larger spoon type ray-florets	-
		Shveta	Fish tail	1974	Gamma rays (20 Gy)	-	Almost white flower color	-

S.No.	Crop	Mutant variety	Parent Variety	Year	Mutagenic source	Induced/improved characters		
						Quantitative	Qualitative	Resistance
		Tamra	Goldie	1974	Gamma rays (15 Gy)	-	Changed flower color (coppery red color on yellow background)	-
		Anamika	E-13	1975	Gamma rays (20 Gy)	-	Light reddish flower color, pom-pon type	-
		Asha	Hope	1975	Gamma rays (15 Gy)	-	Creamish white flower color	-
		Jhalar	Undaunted	1975	Gamma rays (15 Gy)	-	Flat and fringed ray florets	-
		Kanak	Undaunted	1975	Gamma rays (15 Gy)	-	Dark brown flower color	-
		Nirbhaya	Undaunted	1975	Gamma rays (15 Gy)	-	Changed flower morphology (large exhibition type flower heads with lighter mauve, semi-quilled, fringed ray-florets)	-
		Kunchita	Undaunted	1975	Gamma rays (15 Gy)	-	Incurved flower heads	-

TABLE 6.3   (Continued)

S.No.	Crop	Mutant variety	Parent Variety	Year	Mutagenic source	Induced/improved characters		
						Quantitative	Qualitative	Resistance
		Shafali	Undaunted	1975	Gamma rays (15 Gy)	-	Light reddish flower color and altered flower morphology (in-curved ray-florets)	-
		Nirbhik	Undaunted	1975	Gamma rays (10 Gy)	-	Changed flower morphology (large exhibition type flower heads with lighter mauve, almost flat, fringed ray-florets)	-
		Pitaka	Kansya	1978	Gamma rays	-	Extra large type, yellow flower color and reflecting ray-florets	-
		Pitambar	Otome-Zakura	1978	Gamma rays	-	Yellow flower color	-
		Purnima	Otome-Zakura	1978	Gamma rays	-	White flower color and pompon type flower-heads	-
		Basanti	E-13	1979	X-rays (15 Gy + 15 Gy after one day)	-	Yellow flower color	-

S.No.	Crop	Mutant variety	Parent Variety	Year	Mutagenic source	Induced/improved characters		
						Quantitative	Qualitative	Resistance
		Hemanti	Megami	1979	Gamma rays (15 Gy)	-	Dwarf, decorative, small early flowering, chinese yellow flower color	-
		Taruni	Kingsford Smith	1979	Gamma rays (20 Gy)	-	Azalea pink flower color and large flowers, parent variety has fuchsia purple color	-
		Rohit	Kingsford Smith	1979	Gamma rays (20 Gy)	-	Rodonit red flower color	-
		Alankar	D-5	1982	Gamma rays (15 Gy)	-	Spanish orange flower color instead of magnolia purple	-
		Man Bhawan	Flirt	1982	Gamma rays (15 Gy)	-	Bicolored red and yellow flower color instead of red flowers, at the fading stage flowers are completely yellow	-

TABLE 6.3 (Continued)

| S.No. | Crop | Mutant variety | Parent Variety | Year | Mutagenic source | Induced/improved characters | | Resistance |
						Quantitative	Qualitative	
		Cosmonaut	Nimrod	1984	Gamma rays (20 Gy)	-	Anemone double type flowers instead of single Korean type with white flowers	-
		Colchi Bahar	Sharad Bahar	1985	Colchicine (0.0625%)	-	Terracotta red flower color instead of purple color	-
		Sheela	Himani	1985	Gamma rays (20 Gy)	-	Canary yellow flower color instead of white	-
		Tulika	M-24	1985	Gamma rays (15 Gy)	-	Purple flower color and length of tubular portion of flower reduces to give a brush like appearance of the florets	-
		Agnisikha	D-5	1987	Gamma rays (15 Gy)	-	Changed flower color to erythrite red from magnolia purple	-

S.No.	Crop	Mutant variety	Parent Variety	Year	Mutagenic source	Induced/improved characters		
						Quantitative	Qualitative	Resistance
		Kumkum	M-71	1987	Gamma rays (20 Gy)	-	Lighter terracotta or garnet brown flower color instead of pink-purple	-
		Navneet	Kalyani Mauve	1987	Gamma rays (15 Gy)	-	Changed flower color to creamish white from pansy violet	-
		Shabnam	D-5	1987	Gamma rays (15 Gy)	-	Magnolia purple flower color and small appendage-like structure develops at the tip of each floret	-
		Subarna	Flirt	1990	Gamma rays (20 Gy)	-	Yellow flower color	-
		Jugnu	Lalima	1991	Gamma rays (15 Gy)	-	Altered flower color	-
		Surekha Yellow	Surekha	1992	Gamma rays (15 Gy)	-	Yellow flower color	-
		Sharad Har	Sharad Mala	1992	Gamma rays (15 Gy)	-	Yellow flower color	-
		Navneet Yellow	Navneet	1993	Gamma rays (15 Gy)	-	Yellow flower color	-

TABLE 6.3 (Continued)

S.No.	Crop	Mutant variety	Parent Variety	Year	Mutagenic source	Induced/improved characters		Resistance
						Quantitative	Qualitative	
3.	**Dahlia (*Dahlia* sp.)**	Bichitra	Kenya	1978	Gamma rays (20 Gy)	-	Altered plant architecture and mimosa yellow flower color	-
		Black Beauty	Black Out	1978	Gamma rays (20 Gy)	-	Altered plant architecture and changed flower color (dark crimson color with neyron rose stripes)	-
		Happiness	Croydon Monarch	1978	Gamma rays (20 Gy)	-	Altered plant architecture and deep ruby red flower color	-
		Jayaprakash	Croydon Apricot	1978	Gamma rays (20 Gy)	-	Altered plant architecture and phlox pink flower color	-
		Jubilee	Kenya	1978	Gamma rays (20 Gy)	-	Altered plant architecture and flower color (orange-yellow with occasional pink stripes)	-

S.No.	Crop	Mutant variety	Parent Variety	Year	Mutagenic source	Induced/improved characters		Resistance
						Quantitative	Qualitative	
		Jyoti	Kenya	1978	Gamma rays (20 Gy)	-	Altered plant architecture and mallow purple flower color	-
		Netaji	Eagle Stone	1978	Gamma rays (20 Gy)	-	Altered plant architecture and yellow flower color	-
		Pearl	Eagle Stone	1978	Gamma rays (20 Gy)	-	Improved plant architecture and white flower color	-
		Pride of Sindri	Kenya	1978	Gamma rays (20 Gy)	-	Improved plant architecture, flower color and morphology	-
		Twilight	Kenya	1978	Gamma rays (20 Gy)	-	Improved plant architecture, purplish red flower color and improved flower morphology	-
		Vivekananda	Croydon Master	1978	Gamma rays (20 Gy)	-	Improved plant architecture, flower color (spirea red), ray florets are divided at the tips	-

TABLE 6.3 (Continued)

S.No.	Crop	Mutant variety	Parent Variety	Year	Mutagenic source	Induced/improved characters		
						Quantitative	Qualitative	Resistance
4.	Gladiolus (*Gladiolus* sp.)	Tambari	Oscar	1991	Gamma rays	-	Altered flower color	-
5.	Hibiscus (*Hibiscus sinensis* L.)	Purnima	Alipore Beauty	1979	Gamma rays (200 Gy)	-	Variegated smaller leaves	-
		Anjali	Alipore Beauty	1987	Gamma rays (40 Gy)	-	Light carmine red color like Alipore beauty, five petal single type instead of double flower type	-
6.	Tuberose (*Polyanthus tuberosa* L.)	Rajat Rekha	Single flowered cv.	1974	Gamma rays (20 Gy)	-	Altered leaf color (leaves with silvery white streaks)	-
		Swarna Rekha	Double flowered cv.	1974	Gamma rays (20 Gy)	-	Leaf color (leaves with golden yellow streaks)	-
7.	Portulaca (*Portulaca grandiflora* L.)	Karna Phul	*P. grandiflora*	1970	Gamma rays (10 Gy)	-	Altered flower morphology (gerbera type flower)	-

S.No.	Crop	Mutant variety	Parent Variety	Year	Mutagenic source	Induced/improved characters		
						Quantitative	Qualitative	Resistance
		Karna Pali	-	1973	Gamma rays (10 Gy)	-	Altered flower morphology and flowers are about 3 cm in diameter with nearly 47 dissected acutely-tipped petals per flower	-
		Lalita	Vibhuti	1973	Gamma rays (10 Gy)	-	Altered flower morphology, flowers incurving, about 2.5 cm in diameter with 55 petals	-
		Mukta	Portulaca double	1973	Gamma rays (10 Gy)	-	Altered flower morphology and flowers are about 2 cm in diameter with nearly 43 curved petals with white tips per flower	-
		Vibhuti	Perennial Portulaca Double	1973	Gamma rays (40 Gy)	-	Altered flower morphology	-
		Jhumka	Karna Pali	1974	Gamma rays	-	Variegated leaves	-

TABLE 6.3 (Continued)

S.No.	Crop	Mutant variety	Parent Variety	Year	Mutagenic source	Induced/improved characters		
						Quantitative	Qualitative	Resistance
		Ratnam	Perennial Portulaca Double	1974	Gamma rays (40 Gy)	-	Altered flower morphology	-
		Five Petal	-	1974	Gamma rays (25 Gy)	-	Altered flower morphology with 5 petals per flower	-
		Pink color	Vegetative propagates	1974	Gamma rays (150 Gy)	-	Pink flower color	-
		Rosy Green	Vegetative propagates	1974	Gamma rays (100 Gy)	-	Altered flower morphology, non opening flowers, with rosy green periphery and light green centre	-
		Semi-double	Vegetative propagates	1974	Gamma rays (100 Gy)	-	Altered flower morphology with 5 outer petals per flower and petals in 4 rows	-
XXXIV. NDDB, Noida, Uttar Pradesh								
1.	Mustard (*Brassica juncea* L.)	NDUH-YJ-6	NU-6 x EH-1 (Mutant of Heera)	2013	-	-	Low glucosinolate content in seed (<10 moles/gram of seed), high oil content (>45%)	-

S.No.	Crop	Mutant variety	Parent Variety	Year	Mutagenic source	Induced/improved characters		
						Quantitative	Qualitative	Resistance
XXXV. NRCS, Indore								
1.	Soybean (*Glycine max* L.)	NRC 12 (Ahilya-2)	Bragg (M-95-10)	-	-	High seed longevity	-	-
		NRC 2 (Ahilya-1)	Bragg	-	Gamma rays (250 Gy) + UV (260 nm 2h)	High seed longevity	-	-
XXXVI. OUAT, Bhubaneswar								
1.	Turmeric (*Curcuma domestica* Val.)	Suroma	Clonal selection from T. Sunder	-	X-rays	Round and plumpy rhizome, duration 253 days	Curcumin 6.1%, oleoresin 13.1%, essential oil 4.4% and dry recovery 26.0%	Tolerance to leaf blotch, leaf spot and rhizome scale
2.	Citronella (*Cymbopogon winterianus* Jowitt.)	Bhanumati (OJC-11)	Subirrsourav (CKS-CW-S-1)	1987	X-rays (60 Gy)	High oil content under marginal condition	-	-
		Bibhuti (OJC-5)	Subirrsourav (CKS-CW-S-1)	1987	X-rays (90 Gy)	High oil content, adaptability to marginal condition	-	-
		Niranjan (OJC-6)	Subirrsourav (CKS-CW-S-1)	1987	X-rays (90 Gy)	High oil content under marginal condition	-	-

TABLE 6.3 (Continued)

S.No.	Crop	Mutant variety	Parent Variety	Year	Mutagenic source	Induced/improved characters		Resistance
						Quantitative	Qualitative	
		Phullara (OJC-22)	Subirrsourav (CKS-CW-S-1)	1987	X -rays (90 Gy)	High oil content under marginal condition	-	-
		Sourav (OJC-3)	Subirrsourav (CKS-CW-S-1)	1987	X -rays (60 Gy)	High oil content under marginal condition	-	-
		Subir (OJC-31)	Subirrsourav (CKS-CW-S-1)	1987	X -rays (90 Gy)	High oil content under marginal condition	-	-
3.	Rice (*Oryza sativa* L.)	Jagannath (BSS-873)	T-141	1969	X -rays (300 Gy)	Semi dwarf, improved grain size, wider adaptability	Good cooking quality	Resistant to lodging, blast & SB
		Keshari (IET-6215)	Kumar x Jagannath mutant (BSS-873)	1980	X -rays	Short stem, high yield, early maturity	-	Moderately resistant to GLH, blast and BLB
		Savitri (Ponmani) (IET-5897)	Pankaj x mutant Jagannath	1983	Gamma rays	High yield,semidwarf, day length sensitivity, late maturity, wide adaptability,	Short grains, good milling quality	Tolerance to blast & Sh.B

S.No.	Crop	Mutant variety	Parent Variety	Year	Mutagenic source	Induced/improved characters		
						Quantitative	Qualitative	Resistance
4.	Sesame *(Sesamum indicum* L.)	Kalika (BM 3-7)	Binayak	1980	EMS (1% for 6 hr)	Semi-dwarf, compact growth, higher number of seeds per capsule and increased yield (15%)	-	-
5.	Ginger *(Zingiber officinale)*	Surabhi	Induced mutant of Rudrapur local	2000	-	10.2% Oleoresin, 2.1% essential oil, 4.0% crude fiber, 23.6% dry recovery.	Plumpy rhizome, dark skinned yellow fleshed	-
1.	Mungbean *(Vigna radiata* L. Wiczeck)	Dhauli (TT9E)	Local type x T-51	1979	-	High yield, early type	-	Resistant to YMV
		OUM 11-2	Mutant of Dhauli	2002	-	High yielding	-	Moderately resistant to YMV and CLS
XXXVII. Onattukara RRS, Kayamkulam								
1.	Rice *(Oryza sativa* L.)	Dhanu	Mutant of Ptb. 20	2000	Gamma rays (220 Gy)	Semi tall	Short, bold grains	Tolerance to blast, Sh. B, BS and SB

TABLE 6.3 (Continued)

S.No.	Crop	Mutant variety	Parent Variety	Year	Mutagenic source	Induced/improved characters			Resistance
						Quantitative	Qualitative		
XXXVIII. PAU, Ludhiana									
1.	Mustard (*Brassica juncea* L.)	RLM 198	RL 18	1975	Gamma rays	Early maturity (5-6 days) and high yield	Improved oil content		-
		RLM-514	RL-18	1980	Gamma rays (2000 Gy)	Early maturity, high grain yield (22 % more), large grain size	High oil content, less erucic acid (11%)		Resistant to shattering
		RLM-619	RL-18	1985	Gamma rays (2000Gy)	Early maturity	-		Tolerant to white rust, downey mildew and aphids
		RLM-1359	Crossing with one mutant variety (RLM-514 X Varuna)	1987	Gamma rays (2000 Gy)	Short duration, high yield, high TGW, erect plant type	Oil content (43%)		Tolerance to aphids
2.	Barley (*Hordeum vulgare* L.)	PL-56	RS-17	1978	EMS (0.2%)	Semi-tall, high tillering and high yield, suited for rainfed condition	Bold grain		-
3.	Tomato (*Lycopersicon esculentum* M.)	S.12	Sioux	1969	Gamma rays	Dwarf and high yield (30%)	-		-

S.No.	Crop	Mutant variety	Parent Variety	Year	Mutagenic source	Induced/improved characters		
						Quantitative	Qualitative	Resistance
4.	Rice (*Oryza sativa* L.)	Hybrid Mutant 95	Irradiation of hybrid seeds of cross Jhona 349 x Taichung Native-1	1973	Gamma rays (500 Gy)	Semi-dwarf habit, high yield. wide adaptability.	High protein and high lysine content	-
5.	Egyptian clover (*Trifolium alexandrium* L.)	BL-22	Mescavi	1984	Gamma rays (400 Gy)	Late maturity, quick growing and increased dry matter yield	-	-
6.	Cowpea (*Vigna unguiculata* L. Walp.)	Cowpea-88 (Forage cowpea)	Irradiation of F₁ (Cowpea-74 x virus resistant strain H-2)	1990	-	High grain yield and green fodder yield	-	Resistant to yellow mosaic virus

XXXIX. RARS, Palem, Mahabubnagar

S.No.	Crop	Mutant variety	Parent Variety	Year	Mutagenic source	Quantitative	Qualitative	Resistance
1.	Sorghum (*Sorghum bicolor*)	SSG 226	SSG 59-3	2013	-	High leaf :stem ratio	High digestibility and hydrocyanic acid (HCN).	-

XL. RARS, Pattambi

S.No.	Crop	Mutant variety	Parent Variety	Year	Mutagenic source	Quantitative	Qualitative	Resistance
1.	Rice (*Oryza sativa* L.)	Rasmi	Oorpandy	1976	Gamma rays (220 Gy)	High yield, tall plant type	Awn less	Tolerance to salinity

TABLE 6.3 (Continued)

S.No.	Crop	Mutant variety	Parent Variety	Year	Mutagenic source	Induced/improved characters		Resistance
						Quantitative	Qualitative	
XLI. RAU, Dholi								
1.	Lentil (*Lens culinaris* L. Medik.)	PL 77-12 (Arun)	BR 25	1986	Irradiation	-	Medium bold seeded	Tolerance to rust
XLII. RRL, Jammu								
1.	Khasi kateri (Solanum viarum dunal syn. S. Khasianum clarke.)	RRL-20-2	Dehradun local	1975	Gamma rays (75 Gy)	Solasodine content	-	Resistant to diseases and spineless plants
XLIII. PORS, Berhampore								
1.	Lentil (*Lens culinaris* L. Medik.)	S-256 (Ranjan)	B 77 (Asha)	1981	X-rays	High yield and spreading type	-	-
XLIV. SBI, Coimbatore								
1.	Sugarcane (*Sachharum officinarum* L.)	Co 449	-	-	Gamma rays	High cane yield	-	Resistant to red rot
		Co 527	-	-	Gamma rays	High cane yield	-	-
		Co 312	-	1932	Gamma rays	High cane yield and suited for drought conditions	-	-

S.No.	Crop	Mutant variety	Parent Variety	Year	Mutagenic source	Induced/improved characters			Resistance
						Quantitative	Qualitative		
		Co 6608 mutant	Co 449	1966	Gamma rays (30 Gy)	High cane yield	-		Resistant to red rot
		Co 997 mutant	Co 997	1967	Gamma rays (30 Gy)	-	-		Resistant to red rot
		Co 775	-	1971	Gamma rays	High cane yield	-		-
		Co 8153	Irradiation of hybrid seeds form cross Co 6304 x Co 6806	1981	Gamma rays (150 Gy)	High cane yield	Improved juice quality		-
		Co 85017	Co 740	1985	Gamma rays (150 Gy)	High cane yield and Improved adaptability	High sucrose percentage		Resistant to Ustilago scitaminea.
		Co 85035	Co 740	1985	Gamma rays (150 Gy)	High cane yield and early planting in the east coast zone	High sucrose percentage		Resistant to Ustilago scitaminea

XLV. SBIRC, KARNAL

S.No.	Crop	Mutant variety	Parent Variety	Year	Mutagenic source	Quantitative	Qualitative	Resistance
1.	Sugarcane (*Sachharum officinarum* L.)	Co 89003	Co 7314 x Co 775	2001	Crossing with mutant	Early maturity	-	-
		Co 0239 (Karan-4)	CoLk 8102 x Co 775	2009	Crossing with mutant	Early maturity	-	-

TABLE 6.3 (Continued)

S.No.	Crop	Mutant variety	Parent Variety	Year	Mutagenic source	Induced/improved characters		Resistance
						Quantitative	Qualitative	
		Co 05011 (Karan-9)	CoS 8436 x Co 89003 (Co 7314 x Co 775)	2012	Crossing with mutant	Medium maturity	-	-
XLVI. SKN COA, RAJAU, Jobner, Rajasthan								
1.	Fenugreek (*Trigonella foenum-grae-cum* Linn.)	RMt 303	Mutation breeding from variety RMt 1	1975	-	Medium maturity	Seeds bold with typical yellow color	less susceptible to powdery mildew
		RMt-305 (UM-305)	Mutation of RMt-1 (Pure line selection)	2004	-	Determinant type, early maturing, and wider adaptability	Multi pod	Resistant to powdery mildew and rootknot nematodes
2.	Cumin (*Cuminum cyminum* L.)	RZ-223	UC-216	1988	-	Superior in yield and seed quality Plants bushy, semi erect, medium duration (120-130 days) wider adaptability,	Long bold attractive seeds, with 3.0 to 3.5% volatile oil content,	Resistant to wilt

S.No.	Crop	Mutant variety	Parent Variety	Year	Mutagenic source	Induced/improved characters		Resistance
						Quantitative	Qualitative	
3.	**Coriander** *(Coriandrum sativum L.)*	RCr 684	Induced mutant of Rcr-20	1994	Gamma rays	High yield,adapted to medium, heavy textured soil and sandy loam soil under irrigated conditions	-	Resistant to stem gall and less susceptible the powdery mildew.

XLVII. TNAU, Coimbatore

S.No.	Crop	Mutant variety	Parent Variety	Year	Mutagenic source	Quantitative	Qualitative	Resistance
1.	**Okra** *(Able-moschus esculentus L. Moench)*	MDU 1 (MDU.2)	Pusa Sawani	1978	DES (0.04%)	High yield	Light-green long fruits, less crude fiber	Field tolerant to YMV
		CO-3	F$_1$ hybrid between Parbhani Kranti x MDU.1 (Mutant)	1991	-	High yield	-	Moderately resistance to YMV
2.	**Groundnut** *(Arachis hypogaea L.)*	Co-2	Pol-1	1984	EMS (0.2%)	High yield	Higher shelling percentage	-
		VRI-2	Crossing with one mutant (JL 24 x Co-2)	1989	EMS (0.2%)	Large seed size	-	-

TABLE 6.3 (Continued)

S.No.	Crop	Mutant variety	Parent Variety	Year	Mutagenic source	Induced/improved characters		
						Quantitative	Qualitative	Resistance
3.	Pigeon pea (Cajanus cajan Milsp.)	Co 3	Co 1	1977	EMS (0.6%)	High yield, bold seeded	Higher shelling percentage	-
		Co 5	Co 1	1984	Gamma rays (160 Gy)	Early maturity, photoperiod in-sensitivity	-	Drought tolerance
4.	Chilli (Capsicum annum L.)	MDU.1	K-1	1976	Gamma rays (30 KR)	Compact growth, high yield	High capsaicin content	-
5.	Turmeric (Curcuma domestica Val.)	Co-1	Erode local	1983	X-rays	-	Rhizome color	-
6.	Hyacinh bean) (Dolichos lablab L.)	Co 10	Co 6	1983	Gamma rays (240 Gy)	High yield bushy type	Greenish white tubular pods	-
		CO-2	CO.1 Pitchi	1996	Gamma rays (15 Gy)	Bold pink buds, early	-	-
7.	Bitter gourd (Momordica charantia L.)	MDU 1	Mutant of local cultivar (MC 103)	1984	Gamma rays	Early with higher sex ratio of 1:20 of female and male flowers	-	-

S.No.	Crop	Mutant variety	Parent Variety	Year	Mutagenic source	Induced/improved characters		Resistance
						Quantitative	Qualitative	
		COBgoH1	Crossing of MC.84 x MDU.1 mutant	2001	Gamma rays	High yield	-	-
8.	Tomato (Lycopersicon esculentum M.)	CO-3 (Marutham)	Mutant from CO-1	1980	EMS	Determinate and dwarf,	-	-
		PKM 1	Mutant from a local variety Annanji	1988	-	Determinate, fruits flat to round	Red color with prominent green shoulders even after ripening	-
		Paiyur 1	Crossing of Pusa Ruby with CO-3 (Marutham) mutant	1988	EMS	Early	-	Tolerance to fruit borer, leaf spot and leaf curl virus
9.	Brinjal (Solanum melongena)	PKM1	'Puzhuthikathiri'.	1984	Gamma rays	Small fruits with green stripes		Drought tolerant
10.	Sorghum (Sorghum bicolor)	Mothi (SPV-141)	IS 6928	1978	Gamma rays	Early maturity, suitable for late Kharif and	-	-

TABLE 6.3 (Continued)

| S.No. | Crop | Mutant variety | Parent Variety | Year | Mutagenic source | Induced/improved characters | | Resistance |
						Quantitative	Qualitative	
		SPV-126 (CSV-9)	CS-3541 x CSV-4	1983	Gamma rays	-	-	Resistant charcoal rot downey mildew, head moulds and stem borer
		Co-21 (SPV-80)	CSV-5	1986	Gamma rays (400 Gy)	Tall and high grain yield	Fodder type, sweet stem	Tolerance to major insects and pathogens
11.	Snake gourd (*Trichosanthus anguina* L.)	PKM 1	Mutant from H.375	1979	-	-	Dark green fruit color with white stripes on outer side and light green inside	-
12.	Blackgram (*Vigna mungo* L. Hepper)	Co 4	Co 1	1978	EMS (0.02%)	Early maturity, erect and determinate	-	-
		Vamban 2	Mutant of T 9	1997	Gamma rays	-	Pod glabrous	Tolerance to drought and resistant to MYMV
13.	Mungbean (*Vigna radiata* L. Wiczeck)	Co-4	Co 1	1981	Gamma rays (200 Gy)	High yield, early maturity	-	Resistant to drought

S.No.	Crop	Mutant variety	Parent Variety	Year	Mutagenic source	Induced/improved characters		
						Quantitative	Qualitative	Resistance
14.	Cowpea (*Vigna unguiculata* L. Walp.)	Co 5	Co 1	1986	Gamma rays (300 Gy)	High yield (16%), good for intercropping with fodder cereals	More nutritive forage	-
15.	Ridged gourd (*Luffa acutangula* Roxb.)	PKM-1	Mutant from the type H-160	1964	X-rays	-	Dark green fruits	Tolerance to pumpkin beetle, fruit fly and leaf spot
XLVIII. UAS, Bangalore								
1.	Rice (*Oryza sativa* L.)	Intan Mutant	Intan	1988	EI (0.2% for 4 hr)	Early maturity. Photoperiod insensitivity	-	-
2.	Castor (*Ricinus communis* L.)	RC8	Rc 1188-54	1978	Gamma rays (400 Gy)	Late maturity, higher TGW, higher plant height, more internodes and higher yield	-	-
3.	Sorghum (*Sorghum bicolor*)	SbABM	A1 variety	2010	Somaclonal Mutant	Productive ear heads are branched with grain size at par with main ear head	-	-

TABLE 6.3 (Continued)

S.No.	Crop	Mutant variety	Parent Variety	Year	Mutagenic source	Induced/improved characters			Resistance
						Quantitative	Qualitative		

XLIX. UAS, Dharwad

S.No.	Crop	Mutant variety	Parent Variety	Year	Mutagenic source	Quantitative	Qualitative	Resistance
1.	Groundnut (*Arachis hypogaea* L.)	Mutant 28-2	VL-1	1993	EMS (0.5%)	Large seed	-	Resistant to late leaf spot and Spodoptera litura and thrips
		GPBD 5	JL-24	2010	Gamma rays (300Gy)	Large seed	-	Fungal disease resistance

L. UAS, Raichur

1.	Groundnut (*Arachis hypogaea* L.)	R9251	Original variety TG-23 (Crossing with one mutant)	1997	Gamma rays	Early maturity	-	Tolerance to peanut bud necrosis disease

LI. Others

1.	Groundnut (*Arachis hypogaea* L.)	TMV-10	Natural mutant from Argentina	1975	-	Large pods and kernels	-	-
		Kaushal	Natural mutant of Type-28	1985	-	-	-	Resistant to leaf spot and rust diseases
2.	Bougainvillea (*Bougainvillea spectabilis*) Wild	Jaya	Jayalakshmi	1977	Gamma rays (25 Gy)	-	Ornamental novelty, erect plant habit, deep mellow purple bracts color	-

S.No.	Crop	Mutant variety	Parent Variety	Year	Mutagenic source	Induced/improved characters		
						Quantitative	Qualitative	Resistance
3.	Rapeseed (*Brassica napus* L.)	NUDB-38	Westar	2004	Gamma + EMS	Early maturity and high yield	-	-
		NUDB-26-11	Westar	2007	Gamma + EMS	High yielding, early maturity	Double low erucic acid	-
4.	Chrysanthemum (*Chrysanthemum* sp.)	Sonali	Ratna	1968	Gamma rays (20 Gy)	-	Yellow flower color	-
		Shukla	Mrs. H. Gubby	1974	Gamma rays (15 Gy)	-	White flower color and early flowering variety	-
		Svarnim	Undaunted	1975	Gamma rays (15 Gy)	-	Light brown in-curved ray-florets	-
		Batik	Flirt	1994	Gamma rays (20 Gy)	-	Changed flower color (combination of yellow stripes on red background)	-
		Raktima	Shyamal	1996	Gamma rays	-	Crimson flower color	-
5.	Chickpea (*Cicer arietinum* L.)	Kiran (RSG-2)	RS-10	1984	Neutrons, 4.5 x 10 12 n/cm2	High yield, early maturity, erect plant type and increase pod number	-	Salt tolerance

TABLE 6.3 (Continued)

S.No.	Crop	Mutant variety	Parent Variety	Year	Mutagenic source	Induced/improved characters			Resistance
						Quantitative	Qualitative		
6.	White jute (*Corchorus capsularies* L.)	JRC-7447 (Shyamalia)	JRC 212	1967	X-rays	High yield (10% more), recommended for white jute belt except low lying areas	-		-
		Hyb 'C' (Padma)	Crossing with one mutant (JRC 919 x JRC-6165)	1983	X-rays	High yield	-		Tolerance to water logging, resistant to pests and diseases
7.	Tosa jute (*Corchorus olitorius* L.)	IR-1	JRO 632	1978	Gamma rays (100 Gy)	Good plant vigour and fiber yield	-		-
8.	Cumin (*Cuminum cyminum* L.)	Guj Cumin 2	Pure line selection from M₂ irradiated seeds from MC-43	-	-	Bushy plant, good branching habit, suitable for late sowing season	Bold, medium sized, lustrous grain		Tolerance to wilt and blight
9.	Turmeric (*Curcuma domestica* Val.)	BSR-2	Erode local	-	-	High yielding, short duration variety	Bigger rhizomes		Resistant to scale insects
10.	Soybean (*Glycine max* L.)	MACS-450	Crossing with one mutant (Bragg x MACS 111)	2000	-	High yield, medium maturity and semi determinate growth habit	-		-

S.No.	Crop	Mutant variety	Parent Variety	Year	Mutagenic source	Induced/improved characters		
						Quantitative	Qualitative	Resistance
		MACS 111	Kalitur	2000	Gamma rays + Ethyleneimine	High yield	-	-
11.	Wild sage (*Lantana depressa* L.)	Lantana depressa bicolored	Lantana depressa	1986	Gamma rays (10-50 Gy)	-	Leaf color (green) and flower color (yellow and white instead of yellow)	-
		Lantana depressa variagata	Lantana depressa	1986	Gamma rays (10 Gy)	-	Yellow flowers like original variety, variegated leaves instead of green leaves	-
		Niharika	Lantana depressa	1936	Gamma rays (10-50 Gy)	-	Leaf color is green and flower color is light yellow instead of yellow	-
12.	Lentil (*Lens culinaris* L. Medik.)	PL 77-2	BR 25	1984	Irradiation	-	Small seeded	Tolerance to wilt and blight
		Rajendra Masoor 1 (NFL 92)	BR 25	1995	Gamma rays (100 Gy)	Early maturity and good for late sowing	-	Tolerance to low temperatures, resistant to rust and tolerant to wilt
13.	Mulberry (*Morus alba* L.)	S-54	Berhampore	1974	EMS	High leaf yield and leaf quality	-	-

TABLE 6.3 (Continued)

S.No.	Crop	Mutant variety	Parent Variety	Year	Mutagenic source	Induced/improved characters		Resistance
						Quantitative	Qualitative	
14.	**Tobacco (*Nicotiana tabacum* L.)**	GSH-3	Crossing with one mutant (CTR1 Special x LTH X M4)	1979	Neutrons	Improved leaf quality and high yield	-	-
15.	**Rice (*Oryza sativa* L.)**	PL-56	C-164	1975	EMS (0.2%)	High yield, high tillering and good adaptability	-	-
		Au-1	IR 8	1976	Gamma rays	Early maturity	-	-
		CNM 25	IR 8	1979	X-rays (300 Gy)	Dwarf, Early maturity, increased tillering, higher yield	Long bold grains	Resistant to trips, moderately resistant to blast & SB, susceptible to RTV and BLB
		CNM 31	IR 8	1979	X-rays (300 Gy)	Semi dwarf, Early maturity, increased tillering, 9% higher yield	Long grain size	Resistant to BLB, BLS, BPH and brown spot
		CNM 20	IR 8	1980	X-rays (300 Gy)	Early maturity, increased tillering,	Long grain size	Resistant to BLB, BLS and BPH

S.No.	Crop	Mutant variety	Parent Variety	Year	Mutagenic source	Induced/improved characters		Resistance
						Quantitative	Qualitative	
		CNM 6 (Lakshmi)	IR 8	1980	X-rays (300 Gy)	Dwarf, early maturity, increased tillering, 10% higher yield	Long grain size	Resistant to drought, moderately resistant to blast
		Biraj (CNM-539)	OC 1393	1982	X-rays	High yield, late maturity, daylight sensitivity, suitable for low lands where water depth varies between 50-70 cm	-	Tolerance to submergence and regeneration
		Mohan (CSR4)	IR 8	1983	Gamma rays	-	-	Salt tolerance
		Prabhavati (PBN-1)	Ambemohor local	1984	EMS (0.2% for 6 hr)	High yield	Short culm	Resistant to lodging, tolerance to iron chlorosis
		Lunisree (IET-10678)	Nonasail variety	1992	-	High yield, for coastal saline areas	Long slender grain	Resistant to major diseases and pests

TABLE 6.3 (Continued)

S.No.	Crop	Mutant variety	Parent Variety	Year	Mutagenic source	Induced/improved characters		Resistance
						Quantitative	Qualitative	
		MDU-4 (ACM-15)	AC-2836 (CR-194-523) x mutant Jagannath (BSS-873)	1993	X-rays (300 Gy)	Semi-dwarf	-	Tolerance to cold stress, highly resistant to LB, neck infection, moderately resistant to BPH, WBPH & BS
		ADT-41 (JJ-92)	Basmati-370	1994	Gamma rays	Mild aroma, high yield	Semi-dwarf, large grains	Susceptible to major pest and disease
		Birsa Dhan-107	Gora Mutant x IAC 125	1996	-	Dwarf	Short bold, white grains	Resistant to blast, BLB, SB,G M, drought, moderately resistant to helminthosporium
		Gautam (IET-13439)	Rasi	1996	-	Dwarf	White long bold grains	-
		Tapaswini (IET-12168)	Mashuri x mutant Jagannath (BSS-873)	1997	Gamma rays x-rays (300 Gy)	-	-	Tolerance to WBPH, BB, moderately tolerance to LF and GM
		Malviya Dhan-36	Mutant from variety Masuri	1997	-	-	White fine long cylinder grain with excellent cooking quality	Resistant to major diseases

S.No.	Crop	Mutant variety	Parent Variety	Year	Mutagenic source	Induced/improved characters		
						Quantitative	Qualitative	Resistance
		MO 15 (Remanika) Culture M 20-19-4 (IET 13980)	Moncompu (MO) 1	1998	Gamma rays	High yield, dwarf	-	Tolerance to pests and diseases, particularly to brown plant hopper, sheath blight and sheath rot
		Padmanth (IET-11876)	AC-2836 (CR-194-523) Pankaj Nagoba x mutant Jagannath (BSS-873)	1999	X-rays (300 Gy)	-	Long bold grains	-
		Early Samba (RNRM-7) (IET 15845)	BPT 5204	2000	-	Dwarf	White medium grains	Tolerance to SB
		Kaum-20-19-4 (IET-14260.)	Mutant of MO-1	2002	-	Dwarf (90-93 cm)	Red medium bold grains	Resistant to BPH, tolerance to Sh. B and Sh. R, early maturity
		Anashwara	Mutant of PTB 20	2006	Gamma rays	-	Photoperiod sensitive, semi-tall suitable for rabi (post-rainy season)	-

TABLE 6.3 (Continued)

S.No.	Crop	Mutant variety	Parent Variety	Year	Mutagenic source	Induced/improved characters		
						Quantitative	Qualitative	Resistance
		Chingam	Mutant of Ptb-9 X Mutant (IR-8 X Ptb-8)	2006	-	Semi tall	-	Tolerance to Sh.B & LS
		Shivam (IET-17868) and (RR-272-829)	Mutant CR 314-5-10 (open floret mutant)	2006	-	Semi dwarf	-	Resistant to blast, GM biotype 4, SB and LF, moderate resistance to BS
		Abhishek (IET-17868) (RR-272-829)	CR 314-5-10 (Open Florat mutant)	2007	-	Semi dwarf	Short bold grains	Highly resistant to blast, GM (biotype 4), moderately resistant to BS, Sh.R, leaf folder, SB & LF
		CR Boro Dhan-2 (IET 17612)	Mutant 01-China-45	2008	-	Plant height 90 cm, semi dwarf	Grains medium slender	Tolerance to submergence; resistance to blast, BLB, tolerant to Shoot Borer, SB, BPH
		Jaldi Dhan-6 (IET 14359)	Dular mutant/Nagina 22 mutant	2008	-	Dwarf	Grains medium bold	Moderately tolerance to BS, leaf blast, Shoot Borer, BLB, SB, WBPH & GM

S.No.	Crop	Mutant variety	Parent Variety	Year	Mutagenic source	Induced/improved characters		Resistance
						Quantitative	Qualitative	
		Malaviya Sugandh-105 (HUR-105)	Mutant of MPR 7-2	2008	-	Semi dwarf	Long cylindrical grains	Tolerance to leaf & neck blast, BS. SB
		Malaviya Sugandh-4-3 (HUR-4-3)	Mutant of Lanjhi	2008	-	Semi dwarf 90-100 cm,	Long cylindrical grains	Resistant to leaf roller & BPH, moderately resistant to BLB.
		RC Mani-phou-7	Mutant culture from Punshi	2008	-	Semi dwarf	Long medium grains	Mod. resistant to leaf blast, Sh. B & SB
		PTB-58 (IET-17608)	PTB-20	2009	Gamma rays	Semi dwarf	Grains medium bold	Moderately resistance to blast, Sh.B, LF, SB & gall fly
16.	Guinea grass (Panicum maximum)	Marathakam	FR-600	1993	(Mutation)	Fodder yield	-	-
17.	Opium poppy (Papaver somniferum L.)	Soma	Indira	1978	Spontaneous mutant	Low morphine content/High thebaine	-	-
18.	Castor (Ricinus communis L.)	SA-2	TMV-1	1982	Spontaneous Mutant	High yield	-	Resistant to drought. TKW
19.	Sesame (Sesamum indicum L.)	UMA	Kanak	1990	Arsenic-q (10%)	Early and uniform maturity.	High oil content	-

TABLE 6.3 (Continued)

S.No.	Crop	Mutant variety	Parent Variety	Year	Mutagenic source	Induced/improved characters			Resistance
						Quantitative	Qualitative	
		USHA	Kanak	1990	Arsenic-q (10%)	High yield, uniform maturity	-	-
20.	Blackgram (*Vigna mungo* L. Hepper)	DU-1	irradiation F₁ TAU-1 (mutant variety) x No. 169	2007	Gamma rays (200 Gy)	High grain yield	-	-
21.	Mungbean (*Vigna radiata* L. Wiczeck)	ML 26-10-3	ML-26	1983	Gamma rays	High yield	-	Resistant to YMV
22.	Cowpea (*Vigna unguiculata* L. Walp.)	Gujarat cowpea-1	Barsati mutant	-	EMS (0.25%)	Improved yield traits, early maturity	-	Root knot resistance

* Sporeless mutant of *Pleurotus florida* (Mushroom)—IIHR (2009).

TABLE 6.4 List of Number of Released/Approved/Registered Mutant Varieties in Different Crop Species in India

Scientific Name	Common Name	Number of mutants	Scientific Name	Common Name	Number of mutants
Ablemoschus esculentus L. Moench	Okra	3	*Matricario cammomilla*	Germen chamomile	1
Arachis hypogaea L.	Groundnut	32	*Mentha spicata*	Spearmint	1
Bougainvillea spectabilis Wild	Bougainvillea	14	*Momordica charantia* L.	Bitter gourd	2
Brassica juncea L.	Indian Mustard	8	*Morus alba* L.	Mulberry	1
Brassica napus L.	Rapeseed	2	*Nicotiana tabacum* L.	Tobacco	1
Cajanus cajan Milsp.	Pigeon pea	7	*Oryza sativa* L.	Rice	68
Capsicum annum L.	Chilli	1	*Panicum maximum*	Guinea grass	1
Carica papaya L.	Papaya	1	*Papaver somniferum* L.	Opium poppy	2
Chrysanthemum sp.	Chrysanthemum	50	*Pennisetum typhoides* L.	Pearl millet	5
Cicer arietinum L.	Chickpea	6	*Phaseolus vulgaris* L.	Common bean	1
Corchorus capsularies L.	White jute	2	*Pisum sativum* L.	Pea	1
Corchorus olitorius L.	Tosa jute	4	*Plantago ovata* L.	Isabgol	3
Coriandrum sativum L.	Coriander	1	*Polyanthes tuberosa* L.	Tuberose	2
Cuminum cyminum L.	Cumin	2	*Populus deltoids* Marshall	Poplar	1
Curcuma domestica Val.	Turmeric	4	*Portulaca grandiflora* L.	Portulaca	11
Cymbopogon winterianus Jowitt.	Citronella	8	*Prunus armeniaca*	Apricot	3
Cyamopsis tetragonoloba L.	Cluster bean	1	*Ricinus communis* L.	Castor	5
Dahlia sp.	Dahlia	11	*Rosa* sp.	Rose	16
Dianthus caryophyllus L.	Carnation	1	*Saccharum officinarum* L.	Sugarcane	12

TABLE 6.4 (Continued)

Scientific Name	Common Name	Number of mutants	Scientific Name	Common Name	Number of mutants
Dolichos lablab L	Hyacinh bean	1	*Sesamum indicum* L.	Sesame	3
Eleusine coracana (L.) Gaertn.	Finger millet	1	*Sesbaniar rostrata*	Sesbania	1
Euryale ferox Salisb.	Makhana	1	*Setaria italica* L.	Foxtail millet	1
Gladiolus sp.	Gladiolus	2	*Solanum khasianum*	Khasianum	2
Glycine max L.	Soybean	7	*Solanum melongena* L.	Brinjal	1
Gossypium arborium L.	Desi cotton	1	Solenostemon	Colecus	1
Gossypium hirsutum L.	American cotton	8	*Sorghum bicolor* L	Sorghum	5
Helianthus annuus L.	Sunflower	1	*Trichosanthus anguina* L.	Snake gourd	1
Hibiscus sinensis L.	Hibiscus	2	*Trifolium alexandrium* L.	Egyptian clover	1
Hordeum vulgare L.	Barley	14	*Trigonella foenum-graecum* Linn.	Fenugreek	2
Hyocyamus niger	Indian henbane	2	*Triticum aestivum* L.	Wheat	4
Jasminum grandiflorum	Jasmin	1	*Vigna aconitifolia* Jacq. Marechal	Moth bean	5
Lantana depressa L.	Wild sage	3	*Vigna mungo* L. Hepper	Blackgram	12
Lens culinaris Medik	Lentil	5	*Vigna radiata* L. Wil.	Mungbean	17
Luffa acutangula Roxb.	Ridged gourd	1	*Vigna unguiculata* Walp.	Cowpea	10
Lycopersicon esculentum M.	Tomato	5	*Zingiber officinale*	Ginger	1
Malus domestica	Apple	1	*Pleurotus florida*	Mushroom	1

Mutant varieties in different crop species released in India.

- Most varieties of durum wheat are grown in Italy and used in making pasta that is marketed
- worldwide are induced mutants; their Cultivation of induced mutant lines/ varieties in worldwide generates tens of mil¬lions of dollars as additional income to farmers and the seed industry per annum
- The Rio Star variety of grape fruit in the USA is a mutant that accounts for 75% of the US grapefruit industry;
- The cultivation of the mutant Japanese pear variety, "Gold Nijesseiki," contributes US$30m in additional income to farmers annually;
- A mutant cotton variety, NIAB78, grown widely in Pakistan accounts for over US$20m in additional income to farmers annually;
- Most of the rice grown in Asia (especially in China, Japan. Vietnam, and India) and Australia are mutants. The additional income to farmers for growing and marketing these rice varieties is estimated in billions of US$ annually;
- Mutant barley varieties with enhanced adaptation to extremely harsh environmental conditions in high altitudes are literally extending the frontiers of arable lands in Peru. Their cultivation provides much need-ed employment for resource-poor farmers.

More recently, Kharkwal and Shu (2009) reviewed the contributions of induced crop mutants to global food security by highlighting some of the important varieties developed through induced mutagenesis mediated-strat-egies, their annual acreages in the countries of cultivation, and some other indices that underscore their importance in food security and income gen-eration. The crops covered in this review included rice in China, Thailand, Vietnam, and the USA; barley in European countries and Peru; durum wheat in Bulgaria and Italy; wheat in China; soybean in China and Vietnam; and some other food legumes in India and Pakistan. They concluded that there was considerable evidence that mutant crop varieties would continue to con-tribute significantly to address the food and nutritional securities of many countries especially in view of the potentials for harnessing novel traits in enhancing the adaptabilities of crops to climate change and variations.

Country's first sporeless mutant of Pleurotus was produced through UV radiation mutation. This mutant was produced to counter the respira-tory allergy caused due to inhalation of spores in the commercial farms. The sporeless mutant does not produce any spore. The yield of this mutant had increased by 38%.

6.3 MUTATION BREEDING IN GENOMICS ERA

Major drawbacks to induced mutagenesis include the requirement for generating large mutant populations, the incidence of chimeras, and the heterozygosity of the mutated loci. By integrating cell and molecular biology strategies in the induction and detection of mutations, these constraints are mitigated without compromising the imperative of generating large mutant population sizes. The judicious integration of these modern techniques and novel biotechnologies permits the rapid generation of large mutant populations with desired genetic backgrounds; These techniques target the specific regions of the genome that control a trait of interest for induced alterationsThe following are some modern techniques and molecular biology techniques with demonstrated potentials for enhancing the efficiency of induced crop mutagenesis.

6.3.1 MODERN TECHNIQUES

6.3.1.1 High Hydrostatic Pressure (HHP)

High hydrostatic pressure (HHP) is defined as an extreme thermo-physical factor that affects the multiple cellular processes like synthesis of DNA, RNA, proteins, and cell survival (Ishii et al., 2004) and is very effective in inducing mutagenesis in microorganisms (Zhang et al., 2013). The usage of HHP in mutation breeding started few years ago, and one of the examples is the creation of mutant varieties of rice (Zhang et al., 2013). The most important aspect of HHP is cost-effectiveness and ease of application. This technique will be used in more number of crops in the future.

6.3.1.2 Ion Beam Technology (IBT)

Plant breeding with heavy ion beams is a unique technology wherein heavy ions beams are generated by accelerating atomic ions using a particle accelerator. Ion beam technology was found to show high relative biological effectiveness (RBE) as compared to low linear energy transfer (LET) radiations, such as gamma rays, x-rays, and electrons (Blakely, 1992). Although the basic research on plant mutation by ion beams began in 1991 (Goodhead, 1995), it was utilized in full zeal only since last decade, and promising results have been achieved in agriculture. Datta (2012) argues that ion beam technology has been further modified as ion beam implantation into organisms, and energy deposit, mass deposit, charge transfer of the implanted ions into target organisms, long distance systemic effects in intact organisms, generation of reactive oxygen species etc. are the main topics of research (Feng et al., 2009; Datta, 2012).

6.3.1.3   Space Breeding

The experiments in space showed that the growth cycle of the seeds is shortened and the effective components are strengthened; thus, space breeding is a proven way and can be applicable in modern mutation breeding strategies. The two parameters in combination i.e., presence of cosmic rays and microgravity, affect the genetic diversity of the crop, and thus, they are the main causes of the changes in breeding new crop varieties (Gu and Shen, 1989; Mei et al., 1998). Plant mutation induction as an effective breeding approach in China has experienced more than 50 years. According to incomplete statistics by 2007, the total number of mutant varieties and mutant-derived varieties officially registered in China accumulates to 741, including 45 crops and ornamental species (Liu et al., 2007a, 2007b). The popularization and utilization of all these good mutant varieties have made important contribution to China's food production and social and economic development. Mutation technique has become one of the most fruitful and widely used methods for crop improvement in China (Liu et al., 2004). In the last 20 years, exploration and development of new mutagens such as space mutagenesis have been active to upgrade mutagenic efficiency in crop breeding.

6.3.2   *MOLECULAR TECHNIQUES*

6.3.2.1   Forward Genetic Approaches

6.3.2.1.1   *Next Generation Mutagenesis (Next GEM)*

Unfortunately, many mutagenesis methods, particularly pollen mutagenesis of maize is highly inefficient due to a relatively low number of transmitted mutations. Due to low mutation numbers, many promising uses of mutagenesis are not cost effective, and thus, not highly used. Therefore, there is an unmet need for a cost-effective method for generating mutant alleles with high frequency.

Gurmukh Johal and Brian Dilkes (2014), researchers at Purdue University, developed a new technique for stacking mutations to increase the mutation density per individual. The method involves subjecting isolated pollen to a chemical mutagen to produce a first generation seed; planting the first generation seed in (n) plots to produce first generation plants; recovering mutagenized pollen from the first generation plants; pollinating a female parent with the recovered mutagenized pollen to produce a next generation seed; subjecting pollen from the next generation seed to mutagenesis to pollinate the first generation plant, resulting in a new next generation seed; planting the new next generation seed to produce a new next generation plant; and subjecting pollen from the new next generation plant to mutagenesis to pollinate the new next generation plant.

Advantages

- Increases the efficiency of independent mutant alleles, thus opening up multiple applications that were impossible until now.
- The novel process utilizes a multigenerational design to maximize independent changes in the genome, allows screening of subtle differences in mutant phenotypes, and maximizes the frequency of polymorphisms of interest for enhancing elite breeding material.
- Desired traits can be mapped within an inbred line and the variation responsible can be cloned.
- Increasing the number of mutations per plant has decreased the number of plants needed and time required to perform testing.
- This is especially advantageous in controlled environment screens or with non-model species, which can be quite expensive per plant.
- This method generates a vast library of immortalized mutant seed and data, which can be reutilized for multiple independent purposes and benefits, rather than being inhibited by scale.

6.3.2.1.2 *Mutant-Assisted Gene Identification and Characterization (MAGIC)*

Mutant-assisted gene identification and characterization (MAGIC) is a forward genetics screen that uses the readily scorable phenotype of a mutant gene affecting the trait of interest as a reporter to discover and analyze relevant, interacting genes present naturally in diverse germplasm. MAGIC involves crossing a mutant to diverse germplasm and then evaluating the mutant progeny for transgressive changes (both suppressed and severe) in the mutant phenotype(s). This concept is further expounded and explored elsewhere (Johal et al., 2008; Chintamanani et al., 2010). If the mutation is recessive, the population needs to be advanced to the F_2 generation to detect and analyze such variation. However, for a dominant or partially dominant mutant, evaluations can be made immediately in the F_1 to discover lines that contain suppressors or enhancers of the trait (mutation) under study. Mutant F_1 progenies from such crosses can then be propagated further to identify, map, and clone genes/QTL that affect the trait positively or negatively. In the case of maize and other species for which genetically characterized mapping populations are available, modifying loci can be rapidly mapped by crossing a mutant line to each member of a mapping population and evaluating the resulting F_1 families. Chintamanani et al. (2010) provide a proof-of-concept for the MAGIC technique by using it to identify novel genetic loci that modify the maize hypersensitive response (HR).

6.3.2.2   Reverse Genetic Approaches

With the recent advances in genomics, it has been documented that the use of high-throughput platforms such as TILLING and EMAIL in the quick evaluation of mutants for specific genomic sequence alterations can be very much helpful in studying the genetic variability at the molecular level. With the advancement of molecular techniques, the mutation breeding comes into the era that is called the "Molecular Mutation Breeding." Ranalli (2012) defines the plant molecular mutation breeding as "Mutation breeding, in which molecular or genomic information and tools are used in the development of breeding strategies, screening, selection and verification of induced mutants, and in the utilization of mutated genes in the breeding process."

6.3.2.2.1   TILLING

The reverse genetic technique TILLING (McCallum et al., 2000), which was aptly characterized "Traditional Mutagenesis Meets Functional Genomics" (Henikoff et al., 2004), permits the high-throughput querying of putative mutants for point mutation events in specific genomic regions. With TILLING, the mutation events are identified by enzymatic cleavage of mismatches between a mutated strand and a wild-type one. It therefore has huge potentials for removing the major bottleneck to the routine application of induced mutagenesis: the need to produce, handle, and query large putative mutant populations of sufficient mutation density for low frequency recessive events. Since the proof-of-concept about 12 years ago, TILLING has been successfully applied in the detection of mutations in several plant species. These include *Arabidopsis*, rice, maize, wheat, sugar beet, barley, soybean, pea, beans, tomato and vegetatively propagated banana. The related technique, Ecotilling, is also a robust method for identifying spontaneous mutants and hence useful for characterizing germplasm in general (Till et al., 2010). Initially optimized with the chemical mutagen EMS, TILLING has also been effective in identifying point mutations induced by the physical mutagen (Tadele et al., 2009). The TILLING innovators used denaturing HPLC for detecting the mutations. This methodology lacked the high-throughput functionality needed for handling large mutant populations. To redress this, a more robust and high throughput platform that ran on LI-COR gel analyzer system (Lincoln, NE, USA) with analytical software was described by Colbert et al. (2001). This method became widely applicable and was the platform for using TILLING to identify mutants in the above plant species. More recently, Tsai et al. (2011) have described an alternative, TILLING by sequencing. Rather than using endonuclease to cut DNA strands at heteroduplexes produced by the mismatch between wild type and mutant strands, the next generation sequencer Illumina is used to sequence target genes from pooled DNA samples. Till et al. (2009) surveyed some ongoing TILLING projects and their activities; a number of these facilities also provided mutation discovery services to requestors. Table 6.5 shows some of the facilities providing TILLING services.

6.3.2.2.2 *Endonucleolytic Mutation Analysis by Internal Labeling (EMAIL)*

EMAIL has been developed by Cross et al. (2008) for detecting rare mutations in specific genes in pooled samples by using capillary electrophoresis. The inventors, Cross et al. (2008), advocate that this technique is an alter-

TABLE 6.5 List of Tilling Projects/Database and Services for Different Plant Species/Organisms

S.No	Plants/Organisms	TILLING Projects/Database	Link
1	Wheat, castor, other crops	Arcadia Biosciences TILLING	http://www.arcadiabio.com/toolbox.php
		Arcadia Biosciences. USA	
2	Rice	Rice TILLING platform	http://www.nig.ac.jp/section/kurata/kurata-e.html
		Nat. Inst. Of Genetics. Japan	
3	Rice, Hexaploid and Tetraploid Wheat and Tomato.	The UC Davis TILLING Core, University of California, USA	http://tilling.ucdavis.edu/index.php/Main_Page
5	Rapeseed (*Brassica napus*).	CAN-TILL, University of British Columbia, RevGenUK, John Innes Centre, Norwich	http://www3.botany.ubc.ca/can-till/Projects.html
6	Arabidopsis, Brassica oleracea, Populus (Ecotilling)		
	Arabidopsis	Seattle TILLING Project, Fred Hutchinson Cancer Research Center, North Seattle. WA, USA	http://tilling.fhcrc.org/
7	Pea, Brachypodium, Tomato, Pepper, Melon, Cucumber, Watermelon and Arabidopsis	URGV Plant Genomics Research	http://www-urgv.versailles.inra.fr/tilling/index.htm
8	Peanut	Peanut TILLING	http://www.gapeanuts.com/growerinfo/research/
		Georgia Peanut Commission	research2006.asp
9	Soybean	Soybean Mutation Project	portal.nifa.usda.gov/.../crisprojectpages/207533.htm
		Southern Illinois University, USA	http://www.soybeantilling.org/
10	Barley (*Hordeum vulgare* cv Morex)	TILLmore	http://www.distagenomics.unibo.it/
		DiSTA – University of Bologna, Italy	TILLMore/

TABLE 6.5 (Continued)

S.No	Plants/Organisms	TILLING Projects/Database	Link
11	Barley	Barley TILLING	http://www.scri.ac.uk/research/genetics/platformtechnologies/barleytilling
		SCRI, Scotland	
12	Arabidopsis, Barley, Potato,	GABI-TILL Project	https://www.genomforschung.uni-bielefeld.de/en/projects/former-projects/gabi-till/gabi-till-project
	Rapeseed, Rye and Sugar Beet	GABI Consortia. Germany	
13	Brassica sp. (*B. rapa* R-o-18, *B. napus* Cabriolet, *B. oleracea* DH1012 and *Capsella rubella*)	RevGenUK	http://revgenuk.jic.ac.uk/
		John Innes Centre	
	Lotus japonicus Gifu, *L. japonicus* MG20 and *Medicago truncatula* A17		
14	Maize	The Maize TILLING Project	http://genome.purdue.edu/maizetilling
		Purdue University, United States	
15	Pea, Tomato, Brachypodium and the new collection for Linum	French National Institute for Agriculture Research (INRA), UTILLdb Database	http://urgv.evry.inra.fr/UTILLdb
16	Oat (*Avena sativa*)	NordGen, CropTailor AB	http://www.nordgen.org, http://www.croptailor.com
17	Tomato (Red Setter)	LycoTILL	http://www.agrobios.it/tilling/ http://zamir.sgn.cornell.edu/mutants/
		Tomato Mutant Database	
18	Beans	USDA Bean TILLING	http://www.ars.usda.gov/researchipublications/publications.htm?SEQNO_115= 197922
		Project, USDA	

S.No	Plants/Organisms	TILLING Projects/Database	Link
19	Sorghum	USDA Sorghum TILLING Project, USDA	http://www.ars.usda.gov/researchipublications/publications.htm?SEQ_NO 115= 180731
Organisms			
1	Zebrafish	Moens Zebrafish TILLING Project	https://webapps.fhcrc.org/science/tilling/index.php.
		Moens lab/Zebrafish International Resource Center (ZIRC).	http://www.zfishtilling.org/zfish/
2	*C. elegans*	CAN-TILL. University of British Columbia. RevGenUK, John Innes Centre, Norwich	http://www3.botany.ubc.ca/can-till/Projects.html

nate approach to mismatch detection, in which amplicon labeling is achieved by incorporating fluorescent labeled deoxynucleotides. The strength of the EMAIL assay was demonstrated in the reclassification of a rice line as being heterozygous for the starch gene, which in previously used sequence studies had been described as being homozygous. Thus, this technique offers increased sensitivity in gene-specific mutant detection in pooled samples, thereby enabling enlarged pool sizes and improving throughput and efficiency (Lee et al., 2008). Prior to these methodologies, the principle of capillary electrophoresis was used in ecotilling processes (Cordeiro et al., 2006). Some workers are of the opinion that this technique is highly improved over TILLING approach and offers the plant breeder a new tool for efficient screening of induced point mutation at an early stage for variants in genes of specific interest before taking plants to field trial (Lee et al., 2008: Datta, 2012).

6.4   CONCLUSION AND FUTURE PERSPECTIVES

With recent advances in genomics, induced mutagenesis is gaining importance in molecular biology as a tool to identify and isolate novel genes and also serves to study their structure and function. In this regard, mutation breeding has clearly entered a new era of molecular mutation breeding.

KEYWORDS

- **Endonucleolytic Mutation Analysis by Internal Labeling (EMAIL)**
- **EMS**
- **ion beam technology (IBT)**
- **mutagenesis**
- **space breeding**
- **TILLING**

REFERENCES

Ahloowalia, B., S., Maluszynski, M., & Nichterlein, K., (2004). Global impact of mutation-derived varieties. *Euphytica., 135,* 187–204.

Auerbach, C., (1941). The effect of sex on the spontaneous mutations rate in *Drosophila melanogaster*. *J. Genet.*, *11*, 255–265.

Blakely, E. A., (1992). Cell inactivation by heavy charged particles, *Radiat. Environ. Biophys.*, *31*, 181–196.

Carriére, E. A., (1865). *Production and Fixation of the Varieties in the Plants*. The Vine, Simon Racon and Comp. Rue D'Erfurth-1, Paris, pp. 72.

Chintamanani, S., Hulbert, S. H., Johal, G. S., & Balint-Kurti, P. J., (2010). Identification of a maize locus that modulates the hypersensitive defense response, using mutant-assisted gene identification and characterization. *Genetics*, *184*, 813–825.

Colbert, T., Till, B. J., Tompa, R., Reynolds, S., Steine, M. N., Yeung, A. T., McCallum, C. M., Comai, L., & Henikoff, S., (2001). High-throughput screening for induced point mutations. *Plant Physiol.*, *126*, 480–484.

Cordeiro, G., Eliott, F. G., & Henry, R. J., (2006). An optimised ecotilling protocol for polyploids or pooled samples using a capillary electrophoresis system. *Analytical Biochemistry*, *355*, 145–147.

Cramer, P. J. S., (1907). Critical overview of the known cases of bud variation. Nature. Verh. Holl. Mij Wet., Haarlem, 6, 3: pp 474.

Cross, M. J., Daniel, L. E., Waters, L., Slade Lee., & Robert, J. H., (2008). Endonucleolytic mutation analysis by internal labeling (EMAIL). *Electrophoresis*, *29*, 1291–1301.

Darwin, C., (1859). The origin of the species by means of natural selection. London, John Murray, 1-466.

Datta, S. K., (2012). Success story of induced mutagenesis for development of new ornamental varieties. *Bioremedation, Biodiversity and Bioavailability*. *Global Science Books*, 6 *(1)*, pp. 20.

Feng, H. Y., Yang, G., & Yu, Z. L., (2009). Mutagenic mechanisms of ionimplantation in plants. In: *Induced Plant Mutations in the Genomics Era. Proceedings of an International JointFAO/IAEA Symposium.* International atomic energy agency, Vienna, Austria, 220–222.

Freisleben, R. A., & Lein, A., (1942). On the finding of a mildew-resistant mutant after X-irradiation of a susceptible pure line of spring barley. *Natural Sciences, 30*, pp. 608.

Freisleben, R. A., & Lein, A., (1944). Möglichkeiten und praktische Durchführung der Mutationszüchtung. *Kühn-Arhiv.*, *60*, 211–222.

Gager, C. S., & Blakeslee, A. F., (1927). Chromosome and gene mutations in Datura following exposure to radium rays. *Proceedings of the National Academy of Sciences of the USA, 13*, 75–79.

Goodhead, D. T., (1995). Molecular and cell models of biological effects of heavy ion radiation. *Radiat. Environ. Biophys.*, *34*, 67–72.

Gu Ruiqi., & Shen, H., (1989). Effects of space flight on the growth and some cytological characteristics of wheat seedlings. *Acta. Photophysiologica Sinica.*, *15*(4), 403–407.

Gurmukh, J., & Brian, D., (2014). *Next Generation Mutagenesis*, http://otc-prf.org/.

Gustafsson, A., (1936). The different stability of chromosomes and the nature of mitosis. *Hereditas.*, *22*, 281–335.

Gustafsson, A., (1938). Studies on the genetic basis of chlorophyll formation and the mechanism of induced mutating. *Hereditas.*, *24*, 33–93.

Henikoff, S., Till, B. J., & Comai, L., (2004). TILLING, traditional mutagenesis meets functional. genomics. *Plant Physiol.*, *135*, 630–636.

Huang, C., & Liang, J., (1980). Plant breeding achievements in ancient China. *Agronomic History Research, 1*, 1–10.

Hugo de Vries, (1901). Die Mutationstheorie, I. Leipzig: Veit & Co., 2, pp. 62.

IAEA/FAO mutant variety genetic stock database: http://mvgs.iaea.org.

Ishii, A., Sato, T., Wachi, M., Nagai, K., & Kato, C., (2004). Effects of high hydrostatic pressure on bacterial cytoskeleton FtsZ polymers *in vivo* and *in vitro*. *Microbiology, 6,* 1965–1972.

Johal, G. S., Balint-Kurti, P., & Weil, C. F., (2008). Mining and harnessing natural variation: a little MAGIC. *Crop Sci., 48,* 2066–2073.

Johannsen, W., (1909). *Elements of the hereditary genetics.* Jena: Gustav Fisher Verlag.

Kharkwal, M. C., Pandey R. N., & Pawar, S. E., (2012). Mutation breeding for crop improvement. In: *Plant Breeding-Mendelian to Molecular Approches*, Jain, H. K., & Kharkwal, M. C. Ed., Narosa publishing House, New Delhi., 620–637.

Kharkwal, M. C., & Shu, Q. Y., (2009). The role of induced mutations in world food security. In: *Induced Plant Mutations in the Genomics Era*, Shu, Q. Y., Ed., Food and Agriculture Organization of the United Nations: Rome, Italy, 33–38.

Kharkwal, M. C., (2011). A brief history of plant mutagenesis. In: *Plant Mutation Breeding and Biotechnology*, Shu, Q., Y., Forster, B. P., & Nakagawa, H., ed., Food and Agriculture Organization of the United Nations, Rome, Italy., 21–30.

Korschinsky, K., (1901). Heterogenesis and Evolution. *Flora., 89,* 240–363.

Lee, S. L., Cross, M. J., & Henry, R. J., (2008). 'EMAIL – a highly sensitive tool for specific mutation detection in plant improvement programs', paper presented to IAEA *International Symposium on Induced Mutation in Plants (ISIM),* Vienna, Austria, August. Southern Cross University ePublications@SCU, 12–15.

Liu, L., Van Zanten, L., Shu, Q. Y., & Maluszynski, M., (2004). Officially released mutant varieties in china. *Mutation Breeding Review., 14,* 1–62.

Liu, L., Guo, H., Zhao, L., & Zhao, S., (2007a). Advances in induced mutations for crop improvement in China. In: *Proceeding of China-Korea Joint Symposium on Nuclear Technique Application in Agriculture and Life Science*, 22–25, Hangzhou, China. 134–142.

Liu, L., Guo, H., Zhao, L., & Zhao, S., (2007b). Achievements in the past twenty years and perspective outlook of crop space breeding in China. *Journal of Nuclear Agricultural Sciences, 21*(6), 589–592.

McCallum, C. M., Comai, L., Greene, E. A., & Henikoff, S., (2000). Targeted screening for induced mutations. *Nat. Biotechnol., 18,* 455–457.

McClintock, B., (1950). The origin and behavior of mutable loci in maize. *Proc. Natl. Acad. Sci. USA, 36*(6), 344–355.

Mei, M., Qin, Y., & Sun, Y., (1998). Morphological and molecular changes of maize plants after seeds been flown on recoverable satellite. *Advances in Space Research, 22,* 1691–1697.

Morgan, T. H., (1910). Sex limited inheritance in Drosophila. *Science, 32,* 120–122.

Muller, H. J., (1927). Artificial transmutation of the gene. *Science, 66,* 84–87.

Oehlker, F., (1976). Chromosome mutation in meiosis by chemicals. In: *Mutation Research – Problems, Results and Prospects.* Auerbach, C. (ed.), Chapman and Hall, UK, 601–645.

Ranalli, P., (2012). The role of induced plant mutations in the present era. Bioremedation, Biodiversity and Bioavailability. *Global Science Books*, 1–5.

Rapoport, I. A., (1946). Carbonyl compounds and the chemical mechanism of mutation. *C. R. Doklady Acad. Sci. USSR., 54,* 65.

Rapoport, I. A., (1948). Alkylation of gene molecule. *C. R. Doklady Acad. Sci. USSR., 59,* 1183–86.

Sears, E. R., (1956). The transfer of leaf-rust resistance from *Aegilops umbellulata* to wheat. In: *Genetics in Plant Breeding,* Upton, New York, 1–22.

Shamel, A. D., & Pomeroy, C. S., (1936). Bud mutations in horticultural plants. *Journal of Heredity, 27,* 487–94.

Stadler, L. J., (1928). Genetic Effects of X-rays in maize. *Proc. Nat. Acad. Sci., 14,* 69–72.

Tadele, Z., Mba, C., & Till, B. J., (2009). TILLING for mutations in model plants and crops. In: *Molecular Techniques in Crop Improvement,* 2nd ed., Jain, S. M., Brar, D. S., Eds., Springer Publishing Inc.: Dordrecht, Holland, 307–322.

Till, B. J., Afza, R., Bado, S., Huynh, O. A., Jankowicz-Cieslak, J., Matijevic, M., & Mba, C., (2009). Global TILLING Projects. In: *Induced Plant Mutations in the Genomics Era, Proceedings of International Symposium on Induced Mutations in Plants,* Vienna, Austria, 11–15.

Till, B. J., Jankowicz-Cieslak, J., Sagi, L., Huynh, O. A., Utsushi, H., Swennen, R., Terauchi, R., & Mba C., (2010). Discovery of nucleotide polymorphisms in the Musa gene pool by Ecotilling. *Theor. Appl. Genet., 121,* 1381–1389.

Timofteff-Ressovsky, N. W., Zimmer, K. G., & Delbrock M., (1935). Genmutation und gen-struktur. *Nachr. Ges. Wiss.* (Gottingen), *1,* 189.

Tollenaar, D., (1938). Untersuchungen ueber Mutation bei Tabak. II. Einige kuenstlich er-zeugte Chromosom-Mutanten. *Genetica., 20*(3–4), 285–294.

Tsai, H., Howell, T., Nitcher, R., Missirian, V., Watson, B., Ngo, K. J., Lieberman, M., Fass, J., Uauy, C., Tran, R. K., Khan, A. I., Filkov, V., Tai, T. H., Dubcovsky, J., & Comai, L., (2011). Discovery of Rare Mutations in Populations: TILLING by Sequencing. *Plant Physiol., 156,* 1257–1268.

Van Harten, A. M., (1998). *Mutation Breeding: Theory and Practical Applications.* Cambridge University Press, Cambridge, U.K, 1–353

Vavilov, N. I., (1920). *The Law of Homologous Series in Variation.* Gubpoligrafotdel, Saratov, USSR.

Vavilov, N. I., (1926). Centres of origin of cultivated plants. *Bull. Appl. Bot. & Genet. Sel., 16*(2), 139–248.

Von Sengbusch., (1927). *Sweet lupins and oil lupins. Agricultural year book., 91,* 723–880.

Watson, J. D., & Crick, F. H. C., (1953). A structure for deoxyribose nucleic acid. *Nature, 171,* 737–738.

Zhang, Y., Zhang, F., Li, X., Baller, J. A., Qi, Y., & Starker, C. G., (2013). Transcription activator-like effectors nucleases enable efficient plant genome engineering. *Plant Physiol., 161* 20–27.

CHAPTER 7

APPLICATIONS OF MOLECULAR MARKER-ASSISTED BREEDING IN PLANTS

S. PANDEY[1] and V. C. PANDEY[2]

[1]Scientist, Department of Plant Breeding and Genetics, JNKVV, Jabalpur, Madhya Pradesh–482004, India, Tel.: 08305879759, E-mail: suneetagen@gmail.com

[2]Research and Development, Hester Biosciences Limited, 1st Floor, Pushpak Complex, Panchvati Circle, Ahmedabad, Gujarat – 380006, India

CONTENTS

7.1   INTRODUCTION

Conventional plant breeding is primarily based on phenotypic selection of superior individuals among segregating progenies resulting from hybridization. The period varies between 8 to 12 years, and even then, the release of improved variety is not guaranteed. Hence, breeders are extremely interested in new technologies that could make this procedure more efficient, often simply referred to as marker-assisted selection (MAS). MAS involve detectable genomic regions associated with the traits of interest called molecular markers that could make this more efficient than the traditional methods. MAS has become possible for traits governed by major genes as well as quantitative trait loci (QTLs) by the identification of an array of molecular markers and development of genetic maps in many crops. In particular, markers have provided a rapid method to screen parental germplasm for genetic variation, develop genetic linkage maps, and tag genes controlling important traits. Both high density maps and markers linked to traits can assist in selecting breeding progeny carrying desirable alleles. Thus, molecular markers brought a systematic basis in traditional breeding, thereby enhancing its precision and expediting the process (Kumar, 1999; Collard et al., 2005).

Plenty of information is already available on different kinds of marker systems and gene–marker associations. The practicalities of designing an MAS strategy, increasing their chances of success and efficiency, have not received adequate attention. A number of molecular investigations have revealed useful gene–marker associations in different crops, but routine implementation in ongoing plant breeding programs is still in its infancy. It is an opportune time. This chapter will cover the practical aspects of designing MAS strategy and high-throughput genotyping techniques that increase their efficiency significantly.

7.2   WHY USE MARKER-ASSISTED SELECTION IN CONVENTIONAL BREEDING

The development and use of marker-assisted selection in plant breeding fall into four broad areas that are pertinent to almost all target crops (Young and Tanksley, 1989; Ribaut and Hoisington, 1998; Xu, 2002, 2003; Koebner, 2004; Xu et al., 2005):

1. traits that are difficult to manage through conventional phenotypic selection, because they are expensive or time consuming to measure or have low penetrance or complex inheritance;
2. traits whose selection depends on specific environments or developmental stages that influence the expression of the target phenotype;
3. maintenance of recessive alleles during backcrossing or for speeding up backcross breeding in general; and
4. pyramiding multiple monogenic traits (such as pest and disease resistance or quality traits) or several QTLs for a single target trait with complex inheritance (such as drought tolerance or other adaptive traits).

There are many modeling and simulation studies regarding the power of markers to improve the pace and precision of backcross breeding. For most of the crops, over 90% of the recurrent parental genotype can be recovered within two generations when a suitable number of markers (e.g., one marker every 10 cM) and an adequate number of progeny is used for background selection (Tanksley et al., 1989). This represents a substantial saving in time compared to conventional backcross breeding.

Molecular markers intended for MAS can be selected based on their genome distribution; haplotype diversity and/or polymorphic information content indices; and their association with candidate genes and other agronomic traits excluding target introgression trait (Xu, 2003; Varshney et al., 2005). Introgression and pyramiding of multiple genes affecting the same trait is a great challenge to breeding programs. MAS has been shown to be valuable where there are many good varieties that need to be improved for just one simply inherited trait such as certain pest or disease resistances or a component trait for enhancing adaptation or stress tolerance (Johnson, 2004; Miklas et al., 2006; Dwivedi et al., 2007; Ragot and Lee, 2007). The target cropping environments of many breeding programs require a combination of diverse biotic stress resistance, agronomic and quality trait profiles, plus abiotic stress tolerances to improve performance and yield stability.

7.3 HISTORY OF MARKER-ASSISTED BREEDING (MAB)

Marker-assisted selection is a very young field. Sax (1923) and Thoday (1961) originally proposed the concept of MAS; subsequently, first true restriction fragment length polymorphism (RFLP) map in a crop plant was constructed for tomato in 1986 with only 57 loci (Chu et al., 1998). Selection of lines based on genotype rather than phenotype was enormously attractive to plant breeders. Analyzing plants at the seedling stage, screening multiple characteristics that would normally be epistatic with one another, minimizing linkage drag, and rapidly recovering a recurrent parent's genotype were the major advantages of MAS (Tanksley et al., 1989). DNA marker systems, viz., random-amplified polymorphic DNAs (RAPDs) (Williams et al., 1990), amplified fragment length polymorphisms (AFLPs) (Vos et al., 1995), and simple sequence repeats (SSRs) (Akkaya et al., 1992) were simple and boosted this area of research. Within a few years, high-density DNA marker maps were constructed for nearly every important crop species (O'Brien, 1993), each one offering the promise of strategic MAS to complement, even supplant, classical plant breeding techniques. Sophisticated computer algorithms were written to handle the mind-numbing amounts of genetic segregation data and extract every possible bit of useful information from complicated mapping results (Lander and Botstein, 1988; Knapp and Bridges, 1990; Zeng, 1993; Doerge et al., 1997).

It was only a matter of time, many believed, before most important traits would become "Mendelized" and subject to MAS-based breeding. At present, much of the plant breeding community has moved toward MAS with enthusiasm. Most of the agronomic traits were subjected to DNA marker mapping and QTL analysis. A (greatly) abbreviated list includes: drought tolerance (Martin et al., 1989), seed hardness (Keim et al., 1990), seed size (Fatokun et al., 1992), maturity and plant height (Lin et al., 1995), oil and protein content (Diers et al., 1992), soluble solids (Paterson et al., 1988), and of course, yield (Stuber et al., 1987). Before the advent of DNA marker technology, the idea of rapidly uncovering the loci controlling complex, multigenic traits seemed like a dream.

7.4   GENETIC MARKERS IN PLANT BREEDING: CONCEPTIONS, TYPES, AND APPLICATIONS

Genetic markers are the biological features that are determined by allelic forms of genes or genetic loci and can be transmitted from one generation to another; thus, they can be used as experimental probes or tags to keep track of an individual, a tissue, a cell, a nucleus, a chromosome, or a gene. Genetic markers used in conventional plant breeding can be classified into two categories: classical markers and DNA markers (Xu, 2010). Classical markers include morphological markers, cytological markers, and biochemical markers. DNA markers have developed into many systems based on different polymorphism-detecting techniques or methods (southern blotting – nuclear acid hybridization, PCR – polymerase chain reaction, and DNA sequencing) (Collard et al., 2005) such as RFLP, AFLP, RAPD, SSR, single nucleotide polymorphism (SNP), etc.

7.4.1   CLASSICAL MARKERS

7.4.1.1   Morphological Markers

During the ancient time, the morphological markers used primarily included visible traits such as leaf shape, flower color, pod color, seed color, seed shape, hilum color, awn type and length, fruit shape, rind (exocarp) color and stripe, flesh color, stem length, etc. These morphological markers generally represent genetic polymorphisms which are easily identified and manipulated. Therefore, they are usually used in construction of linkage maps by classical two- and/or three-point tests. However, they are available in limited number, and even have undesirable effects on the development and growth of plants.

7.4.1.2   Cytological Markers

In cytology, the structural features of chromosomes can be shown by chromosome karyotype and bands. The banding patterns, displayed in color, width, order, and position, reveal the differences in distribution of euchromatin and heterochromatin. For instance, Q bands are produced by quinacrine hydrochloride, G bands by Giemsa stain, and R bands are the reversed G bands. These chromosome landmarks are used not only for characterization of normal chromosomes and detection of chromosome mutation, but also widely used in physical mapping and linkage group identification. These types of markers have been very limited in genetic mapping and plant breeding.

7.4.1.3   Biochemical/Protein Markers

They are mainly the alternative forms or structural variants of an enzyme having same catalytic activity/function but different molecular weights and electrophoretic mobility. Isozyme, the products of different alleles having similar function but different molecular weight, results differences in electrophoretic mobility (Xu, 2010). These markers can be genetically mapped onto chromosomes and then used as genetic markers to map other genes. They are mostly used in seed purity test; besides this, there are only a small number of isozymes that are identified in most crop species. Therefore, their use is limited.

7.4.2   DNA MARKERS/MOLECULAR MARKERS

A molecular marker is a gene or DNA sequence with a known location on a chromosome that can be used to identify individuals or species. It can be described as a variation (which may arise due to mutation or alteration in the genomic loci) that can be observed. A genetic marker may be a short DNA fragment such as a sequence surrounding a single base-pair change or a long one. Such fragments are associated with a certain location within the genome and may be detected by certain molecular technology.

There are two basic methods to detect the polymorphism: Southern blotting, a nucleic acid hybridization technique (Southern 1975), and PCR, a polymerase chain reaction technique (Mullis, 1990). Using PCR and/or molecular hybridization, followed by electrophoresis, the variation in DNA samples or polymorphism for a specific region of DNA sequence can be identified based on the product features such as band size and mobility. Several new array chip techniques using DNA hybridization combined with labeled nucleotides have been developed. DNA markers are also called molecular markers in many cases and play a major role in molecular breeding. An ideal marker system should have these properties:

- High level of polymorphism
- Even distribution across the whole genome (not clustered in certain regions)
- Co-dominance in expression (so that heterozygotes can be distinguished from homozygotes)

- Clear distinct allelic features (so that the different alleles can be easily identified)
- Single copy and no pleiotropic effect
- Low cost to use (or cost-efficient marker development and genotyping)
- Easy assay/detection and automation
- High availability (unrestricted use) and suitability for duplicating/multiplexing so that the data can be accumulated and shared between laboratories
- Genome-specific in nature (especially with polyploids)
- No detrimental effect on phenotype

Substantial progress has been made in development and improvement of molecular techniques only after the use of DNA RFLP in human linkage mapping by Botstein et al. (1980). Most promising techniques for plant breeding are RFLP, AFLP, RAPD, microsatellites or SSR, and SNP. The marker techniques help in selection of multiple desired characteristics simultaneously using F_2 and backcross populations, near isogenic lines, doubled haploids, and recombinant inbred lines. The most commonly used marker systems are briefly addressed here (Gupta et al., 2001; Farooq and Azam, 2002a, 2002b; Xu, 2010).

7.4.2.1   RFLP Markers

These are the first generation of DNA markers and one of the important tools for plant genome mapping. In living organisms, mutation events (deletion and insertion) may occur at restriction sites or between adjacent restriction sites in the genome. RFLPs are detected by cutting genomic DNA with restriction enzymes. Each of these enzymes has a specific recognition sequence that is typically palindromic and that leads to restriction fragments of certain length when the DNA is digested. Changes within these sequences that can be caused by point mutations, insertions or deletions, result in DNA fragments of differing length and molecular weights. These fragments are size-separated with agarose gel electrophoresis and analyzed by Southern blots using either locus-specific or multilocus probes.

Most RFLP markers are co-dominant and locus-specific. RFLP markers are powerful tools for comparative and synteny mapping. By using an improved RFLP technique, i.e., cleaved amplified polymorphism sequence (CAPS), also known as PCR-RFLP, high-throughput markers can be developed from RFLP probe sequences. Very few CAPS are developed from probe sequences, which are complex to interpret. Most CAPS are developed from SNPs found in other sequences followed by PCR and detection of restriction sites. CAPS technique consists of digesting a PCR-amplified fragment and detecting the polymorphism by the presence/absence of restriction sites (Konieczny and Ausubel, 1993).

Disadvantages are the requirement of relatively large amounts of pure and intact DNA and the tedious experimental procedure. Radioactive autography involved in genotyping and physical maintenance of RFLP probes limit its use and share between laboratories. Since the last decade, fewer direct uses of RFLP markers in genetic research have been reported.

7.4.2.2   RAPD Markers

RAPD is a PCR-based marker system that detects nucleotide sequence polymorphism in DNA by using a single primer of arbitrary nucleotide sequence (oligonucleotide primer, mostly ten bases long). In this reaction, a single species of primer anneals to the genomic DNA at two different sites on complementary strands of DNA template. RAPD predominantly provides dominant markers. This system yields high levels of polymorphism and is simple and easy to be conducted. Advantages associated with RAPD analysis include: (i) nonrequirement of species specific probe, (ii) noninvolvement with radioactive assays, (iii) use of small amount of DNA, which makes it possible to work with population that is not accessible with RFLP, and (iv) the RAPD products of interest can be cloned, sequenced, and then converted into or used to develop other types of PCR-based markers such as sequence characterized amplified region (SCAR), SNP, etc.

Limitations are: (i) its polymorphisms are inherited as dominant or recessive characteristics, causing a loss of information relative to markers that show co-dominance, (ii) primers are relatively short, and a mismatch of even a single nucleotide can often prevent the primer from annealing, leading to a loss of band, and (iii) low reproducibility.

7.4.2.3   AFLP Markers

AFLPs are simply fragments of DNA that have been amplified using directed primers from restriction site of genomic DNA. It is a combination of RFLP and RADP methods. The DNA is enzymatically cut into small fragments (as with RFLP analysis), but only a fraction of fragments is studied following selective PCR amplification (Liu et al., 1994). AFLP is extremely sensitive technique and the added use of fluorescent primers for automated fragment analysis system and software packages to analyze the biallelic data makes it well suitable for high-throughput analysis. The major advantages of AFLP techniques are:

(i) no sequence information is required
(ii) generation of a large number of polymorphisms.

An ideal marker should have sufficient variation for the problem under study and be reliable and simple to generate and interpret. Unfortunately, neither AFLP nor other DNA markers exhibit these qualities. Thus, a specific technique or techniques selected on the basis of objectives should be utilized collectively to achieve the best results.

7.4.2.4 SSR Markers

SSRs, also called microsatellites, short tandem repeats (STRs), or sequence-tagged microsatellite sites (STMS), are PCR-based markers. They consist of tandemly repeated 2-7 base pair units arranged in repeats of mono-, di-, tri-, tetra, and penta-nucleotides (A, T, AT, GA, AGG, AAAG, etc.) with different lengths of repeat motifs. These repeats are widely distributed throughout the plants and animal genomes that display high level of genetic variation based on differences in the number of tandemly repeating units of a locus. The variation in the number of tandemly repeated units results in highly polymorphic banding patterns that are detected by PCR using locus-specific flanking region primers where they are known. One of the most important attributes of microsatellite loci is their high level of allelic variation, thus making them valuable genetic markers.

These types of markers are characterized by their hyper-variability, reproducibility, co-dominant nature, locus-specificity, and random genome-wide distribution in most cases. The advantages of SSR are:

(i) requires only very small DNA samples (~100 ng per individual) and low start-up costs for manual assay methods.
(ii) requires nucleotide information for primer design.

7.4.2.5   SNP Markers

Single nucleotide polymorphisms, frequently called SNPs (pronounced "snips"), are the most common type of genetic variation among people. Each SNP represents a difference in a single DNA building block, called a nucleotide. SNPs may be present within the coding sequences of genes, noncoding regions of genes, or in the intergenic regions between genes at different frequencies in different chromosome regions. SNPs provide the simplest form of molecular markers as a single nucleotide base is the smallest unit of inheritance, and thus, they can provide maximum markers.

SNPs can be categorized according to nucleotide substitutions either as transitions (C/T or G/A) or transversions (C/G, A/T, C/A or T/G). SNPs occur very commonly in animals and plants. Typically, SNP frequencies are in a range of one SNP every 100-300 bp in plants (Edwards et al., 2007; Xu, 2010).

A convenient method for detecting SNPs is RFLP (SNP-RFLP) or by using the CAPS marker technique. If one allele contains a recognition site for a restriction enzyme while the other does not, digestion of the two alleles will produce different fragments in length. A simple procedure is to analyze the sequence data stored in the major databases and identify SNPs. Four alleles can be identified when the complete base sequence of a segment of DNA is considered, and these are represented by A, T, G and C at each SNP locus in that segment. There are several SNP genotyping assays such as allele-specific hybridization, primer extension, oligonucleotide ligation, and invasive cleavage based on the molecular mechanisms (Sobrino et al., 2005), and different detection methods are used to analyze the products of each type of allelic discrimination reaction, such as gel electrophoresis, mass spectrophotometry, chromatography, fluorescence polarization, arrays or chips, etc.

The advantages of SNPs are:

(i) SNPs are co-dominant markers, and thus, they have become very attractive and potential genetic markers in genetic study and breeding.

(ii) Highly automated and quickly detected, with a high efficiency for detection of polymorphism.

High costs for start-up or marker development, requirement for high-quality DNA, and high technical/equipment demand limit to some extent the application of SNPs.

The features of the widely used DNA markers discussed above are compared in Table 7.1. The advantages or disadvantages of a marker system are relevant largely for the purpose of research, available genetic resources or databases, equipment and facilities, funding and personnel resources, etc. The choice and use of DNA markers in research and breeding is still a challenge for plant breeders. A number of factors need to be considered when a breeder chooses one or more molecular marker types (Semagn et al., 2006a). A breeder should make an appropriate choice that best meets the requirements according to the conditions and resources available for the breeding program.

7.5   MAS PROCEDURE

MAS is an indirect selection process where a trait of interest is selected based on a marker (morphological, biochemical, or DNA/RNA variation) linked to a trait of interest (e.g., productivity, disease resistance, abiotic stress tolerance, and quality) rather than on the trait itself. This process is used in plant and animal breeding.

The general procedure of MAS, taking a single cross as an example is described as follows:

a) Selection of parents, at least one (or both) possesses the DNA marker allele(s) for the desired trait of interest followed by intercross,

b) Plant F_1 population and conduct foreground selection to select heterozygote plants carrying desirable genes

c) Plant segregating F_2 population, screen individuals for the marker(s), and harvest the individuals carrying the desired marker allele(s).

d) Plant $F_{2:3}$ plant rows, and screen individual plants with the marker(s). Select and harvest the individuals with required marker alleles and other desirable traits.

e) In the subsequent generations (F_4 and F_5), conduct marker screening and make selection similarly as for previous generation, select superior individuals in each generation, but priorities should be given to superior individuals.

f) In $F_{5:6}$ or $F_{4:5}$ generations, bulk the best lines according to the phenotypic evaluation of target trait.

TABLE 7.1 Comparison of Commonly Used Molecular Markers

Feature and description	RFLP	RAPD	AFLP	SSR	SNP
Genomic abundance	High	High	High	Moderate to high	Very high
Genomic coverage	Low copy coding region	Whole genome	Whole genome	Whole genome	Whole genome
Expression/inheritance	Co-dominant	Dominant	Dominant/co-dominant	Co-dominant	Co-dominant
Number of loci	Small (<1,000)	Small (<1,000)	Moderate (1,000s)	High (1,000s – 10,000s)	Very high (>100,000)
Level of polymorphism	Moderate	High	High	High	High
Type of polymorphism	Single base changes, indels	Single base changes, indels	Single base changes, indels	Changes in length of repeats	Single base changes, indels
Type of probes/primers	Low copy DNA or cDNA clones	10 bp random nucleotides	Specific sequence	Specific sequence	Allele-specific PCR primers
Cloning and/or sequencing	Yes	No	No	Yes	Yes
PCR-based	Usually no	Yes	Yes	Yes	Yes
Radioactive detection	Usually yes	No	Yes or no	Usually no	No
Reproducibility/reliability	High	Low	High	High	High
Effective multiplex ratio	Low	Moderate	High	High	Moderate to high
Marker index	Low	Moderate	Moderate to high	High	Moderate
Genotyping throughput	Low	Low	High	High	High
Amount of DNA required	Large (5 – 50 µg)	Small (0.01-0.1 µg)	Moderate (0.5-1.0 µg)	Small (0.05-0.12 µg)	Small (≥ 0.05 µg)
Quality of DNA required	High	Moderate	High	Moderate to high	High

Technically demanding	Moderate	Low	Moderate	Low	High
Time demanding	High	Low	Moderate	Low	Low
Ease of use	Not easy	Easy	Moderate	Easy	Easy
Ease of automation	Low	Moderate	Moderate to high	High	High
Development/start-up cost	Moderate to high	Low	Moderate	Moderate to high	High
Cost per analysis	High	Low	Moderate	Low	Low
Number of polymorphic loci per analysis	1.0 – 3.0	1.5–5.0	20–100	1.0–3.0	1.0
Primary application	Genetics	Diversity	Diversity & genetics	All purposes	All purposes

(Reprinted from Jiang, G-L, Molecular Markers and Marker-Assisted Breeding in Plants, in Plant Breeding from Laboratories to Fields, 2in Plant Breeding from Laboratories to Fields, Sven Bode Andersen (ed.)], 2013, https://creativecommons.org/licenses/by/3.0/)

g) Plant yield trials and comprehensively evaluate the selected lines for yield, quality, resistance, and other characteristics of interest.

Selection of QTLs for MAS plays an important role. For a quantitatively inherited characteristic like yield, numerous QTLs or genes are usually involved. Typically, not more than three QTLs are regarded as an appropriate and feasible choice (Ribaut and Betran, 1999), although five QTLs were used in improvement of fruit quality traits in tomato via marker-assisted introgression (Konieczny and Ausubel, 1993). It is almost impossible to select all QTLs or genes simultaneously so that the selected individuals incorporate all the desired QTLs due to the limitation of resources and facilities. The number of individuals in the population increases exponentially with the increase in target loci involved. The relative efficiency of MAS decreases as the number of QTLs increases and their heritability decreases (Moreau et al., 1998). In other words, MAS will be less effective for a highly complex characteristic governed by many genes than for a simply inherited characteristic controlled by a few genes. Selection of more QTLs at the same time might be practicable with the help of SNP markers (Kumpatla et al., 2012).

Multi environmentally verified QTLs that possess medium to large effects are selected for MAS. The number of genes were limited to three to four if they are QTLs selected on the basis of linked markers or five to six if they are known loci selected directly (Hospital, 2003). The priority should be given to the major QTLs that can explain greatest proportion of phenotypic variation and/or can be consistently detected across a range of environments and different populations.

Flint-Garcia et al. (2003) developed an index for the selection of QTLs that weighs markers differently.

Another commonly asked question is that "how many markers should be used in MAS?" The more markers associated with a QTL are used, the greater is the opportunity of success in selecting the QTL of interest. However, efficiency is also important for a breeding program, especially when the resources and facilities are limited. From the point of both effectiveness and efficiency, for a single QTL, it is usually suggested to use two markers (i.e., flanking markers) that are tightly linked to the QTL of interest. The markers to be used should be close enough to the gene/QTL of interest (<5 cM) in order to ensure that only a minor proportion of the selected individuals will be recombinants. If a marker (e.g., the peak marker) is found to be located within the region of gene sequence of interest or in such a close proximity to the QTL/gene that no recombination occurs between the marker and the QTL/gene, such markers are only preferable.

However, if a marker is not tightly linked to a gene of interest, recombination between the marker and gene may reduce the efficiency of MAS because a single crossover may alternate the linkage association and lead to selection errors. The efficiency of MAS decreases as the recombination frequency (genetic distance) between the marker and gene increases. Use of two flanking markers rather than one may decrease the chance of such errors due to homologous recombination and increase the efficiency of MAS. In this case, only a double crossover (i.e., two single crossovers occurring simultaneously on both sides of the gene/QTL in the region) may result in selection errors, but the frequency of a double crossover is considerably rare. For instance, if two flanking markers with an interval of 20 cM or so between them are used, there will be higher probability (99%) for the recovery of the target gene rather than only one marker being used.

In practical MAS, a breeder is also concerned about how the markers should be detected, how many generations of MAS have to be conducted, and how large a size of the population is needed. In general, the detection of marker polymorphism is performed at early stages of plant growth. This is true especially for marker-assisted backcrossing and marker-assisted recurrent selection, because only the individuals that carry preferred marker alleles are expected to be used in backcrossing to the recurrent parent and/or inter-mating between selected individuals/progenies. The generations of MAS required vary with the number of markers used, the degree of association between the markers and the QTLs/genes of interest, and the status of marker alleles. In several cases, marker screening is performed for two to four consecutive generations in a segregating population. If fewer markers are used and the markers are in close proximity to the QTL or gene of interest, fewer generations are needed. Bonnett et al. (2005) discussed the strategies for efficient implementation of MAS involving several issues, e.g., breeding systems or schemes, population sizes, number of target loci, etc. Their strategies include F_2 enrichment, backcrossing, and inbreeding.

In MAS, phenotypic evaluation and selection are still very useful if the selected QTLs are not so stable across environments and the association between the selected markers and QTLs is not so close. The presence of a QTL or marker does not necessarily guarantee the expression of the desired trait. QTL data derived from multiple environments and different populations help in better understanding of the interactions of QTL × environment and QTL × QTL or QTL × genetic background, and thus helps in a better usage of MAS.

The situations favorable for MAS include:

- The selected characteristic is expressed late in plant development, like fruit and flower features or adult characteristics with a juvenile period
- The target gene is recessive
- Requirement of special conditions for breeding of disease and pest resistance, or the expression of target genes is highly variable with the environments.
- Two or more unlinked genes responsible for the phenotypic expression of a trait.

7.6   ACTIVITIES OF MAB

This involves the following activities:

a.   Planting the breeding populations selected according to priorities.
b.   Sampling of plant tissues at early stages of growth.
c.   DNA extraction from tissue sample of each individual.
d.   Running PCR for the molecular markers linked to the trait of interest.
e.   Separating and scoring the amplified products by using PAGE, AGE, etc.
f.   Identifying individuals carrying the desired marker alleles.
g.   Selecting the best individuals with both desired marker alleles for target traits and desirable phenotypes of other traits.
h.   Depending upon the association between the markers and the traits as well as the status of marker alleles, the above activities are repeated for several generations.

7.7   APPLICATION OF MARKER-ASSISTED SELECTION

7.7.1   IMPROVEMENT OF QUALITATIVE TRAITS

Many economically important characteristics, viz., resistance to diseases/pests, male sterility, self-incompatibility, etc. are controlled by major genes.

These traits are often of mono or oligogenic inheritance in nature. Even for some quality traits, one or a few major QTLs or genes can account for a very high proportion of the phenotypic variation of the trait (Bilyeu et al., 2006; Pham et al., 2012). Transfer of such gene to a specific line can lead to tremendous improvement of the trait in the cultivar under development. The marker loci that are tightly linked to major genes can be used for selection and are sometimes more efficient than direct selection for the target genes.

The most important example for improvement of qualitative traits is soybean cyst nematode (SCN). The SSR marker Satt 309 has been identified to be located only 1–2 cM away from the resistance gene *rhg1* (Cregan et al., 1999), with genotypic selection of 99% accuracy in predicting lines and 80% accuracy in a third population.

7.7.2 IMPROVEMENT OF QUANTITATIVE TRAITS

MAS for the improvement of quantitative traits are a complex and difficult task because it is related to many genes or QTLs involved, QTL × environment (E) interaction, and epistasis. Usually, each of these genes has a small effect on the phenotypic expression of the trait, and expression is affected by environmental conditions. To evaluate the stability across environments and characterization, repeated field tests are required. The QTL × E interaction reduces the efficiency of MAS, and epistasis can result in a skewed QTL effect on the trait. The application and utilization of the QTL mapping in plant breeding have been constrained by a number of factors (Collard and Mackill, 2008):

1. Lack of universally valid QTL–marker associations applicable across populations.
2. Strong QTL–environment interaction.
3. Overestimation or underestimation of the number of QTLs due to the deficiencies in QTL statistical analysis.
4. Due to the minor effect of most of the QTLs, a large number of QTLs have to be identified.

To improve the efficiency of MAS for quantitative traits, appropriate field experimental designs and approaches have to be employed. A saturated linkage map enables accurate identification of both targeted QTLs as well as linked QTLs in coupling and repulsion linkage phases.

Fusarium head blight (FHB) caused by *Fusarium* species is one of the most destructive diseases in wheat and barley worldwide. To combat this disease, great efforts from multiple fields have been dedicated since 1990s. Resistance to FHB in both wheat and barley is quantitatively inherited, and many QTLs have been identified from different resources of germplasm (Buerstmayr et al., 2009). In wheat, a major QTL designated as *Fhb1* was consistently detected across multiple environments and populations, and it explained 20–40% of phenotypic variation in most cases (Jiang et al., 2007a, 2007b; Buerstmayr et al., 2009). Pumphrey et al. (2007) compared 19 pairs of NIL for *Fhb1* derived from an ongoing breeding program and found that the average reduction in disease severity between NIL pairs was 23% for disease severity and 27% for kernel infection. Later investigation from the group also demonstrated successful implementation of MAS for this QTL (Miedaner et al., 2006; Anderson et al., 2007) demonstrated that MAS for three FHB resistance QTLs simultaneously was highly effective in enhancing FHB resistance in German spring wheat.

7.7.3 MAS IN CROP IMPROVEMENT

In this section, the main uses of DNA markers in plant breeding, emphasized with an important MAS schemes and classified into five broad areas: marker-assisted evaluation of breeding material; marker-assisted backcrossing; pyramiding; early generation selection; and combined MAS, although there may be overlap between these categories, generally, for line development.

7.7.3.1 Marker-Assisted Evaluation of Breeding Material

Prior to crossing (hybridization) and line development, there are several applications in which DNA marker data may be useful for breeding such as cultivar identity, assessment of genetic diversity and parent selection, and confirmation of hybrids. Traditionally, these tasks have been done based on visual selection and analyzing data based on morphological characteristics.

7.7.3.2 Cultivar Identity/Assessment of "Purity"

Markers can be used to confirm the true identity of individual plants. The maintenance of high levels of genetic purity is essential in cereal hybrid production in order to exploit heterosis. In hybrid rice, SSR and STS markers were used to confirm purity, which was considerably simpler than the standard "grow-out tests" that involves the growth of the plant to maturity and assessing morphological and floral characteristics (Yashitola et al., 2002).

7.7.3.3 Assessment of Genetic Diversity and Parental Selection

Breeding programs depends on a high level of genetic diversity for achieving progress from selection. Broadening the genetic base of core breeding material requires the identification of diverse strains for hybridization with elite cultivars (Xu et al., 2004; Reif et al., 2005). DNA markers have been an indispensable tool for characterizing genetic resources and providing breeders with more detailed information to assist in selecting parents. In some cases, information regarding a specific locus (e.g., a specific resistance gene or QTL) within breeding material is highly desirable. For example, the comparison of marker haplotypes has enabled different sources of resistance to Fusarium head blight, which is a major disease of wheat worldwide, to be predicted (Liu and Anderson, 2003; McCartney et al., 2004).

7.7.3.4 Study of Heterosis

For hybrid crop production, especially in maize and sorghum, DNA markers have been used to define heterotic groups that can be used to exploit heterosis (hybrid vigor). The development of inbred lines for use in producing superior hybrids is a very time-consuming and expensive procedure. Unfortunately, it is not yet possible to predict the exact level of heterosis based on DNA marker data, although there have been reports of assigning parental lines to the proper heterotic groups (Lee et al., 1989; Reif et al., 2003). The potential of using smaller subsets of DNA marker data in combination with phenotypic data to select heterotic hybrids has also been proposed (Jordan et al., 2003).

7.7.3.5   Identification of Genomic Regions under Selection

The identification of shifts in allele frequencies within the genome can be important information for breeders as it alerts them to monitor specific alleles or haplotypes and can be used to design appropriate breeding strategies (Steele et al., 2004). Other applications of the identification of genomic regions under selection are for QTL mapping: the regions under selection can be targeted for QTL analysis or used to validate previously detected marker–trait associations (Jordan et al., 2004). Ultimately, data on genomic regions under selection can be used for the development of new varieties with specific allele combinations using MAS schemes such as marker-assisted backcrossing or early generation selection (Ribaut et al., 2001; Steele et al., 2004).

7.7.4   MARKER-ASSISTED BACKCROSSING (MABC)

Backcrossing is used in plant breeding to transfer (introgress) favorable traits from a donor plant into an elite genotype (recurrent parent). In repeated crossings, the original cross is backcrossed with the recurrent parent until most of the genes stemming from the donor are eliminated. MABC is regarded as the simplest form of marker-assisted selection, and at present, it is most widely and successfully used method in practical molecular breeding.

The prime aim of backcross breeding is to transfer one or more genes of interest trait from donor parent into the background of the improved variety and recover the RP genome by eliminating the undesirable genes as quickly as possible. The general procedure of MABC is as follows:

a.   Select parents and make the cross between the RP (superior in performance) and donor parent (desired gene of interest) to produce an F1 hybrid.

b.   Plant F1 population and detect the presence of the marker allele at early stages of growth. Discard undesirable plants.

c.   Crossing the F1 plants back to RP to produce the first backcross generation (BC1F1). Screen individuals for the presence of phenotypic trait as well as marker allele.

d.   The next backcross generation is made by crossing selected BC1 plants (that have been screened for the target trait) with the RP to produce the BC2. Subsequent backcross populations are made by repeatedly crossing the selected backcross (BC) plants with the RP. Individuals are screened for the presence of marker allele in each subsequent generation.

e. After the final backcross (BC_4F_1) generation, selected individuals are self-pollinated so that selected lines will be homozygous for the target trait.

The expected recurrent parent (RP) genome recovery would be 99.2% by six backcrosses, which is most similar to improved variety. The proportion of the RP genome is recovered at a rate of $1 - (1/2)^{n+1}$ for each of the generations of backcrossing. However, for any specific backcross progeny (BC3 or BC2), they will be deviated during crossing over, resulting in a great chance to get the expected result that is not possible to detect phenotypically. For example, in BC1 population, theoretically the average percentage of the RP genome is 75% for the entire population. But some individuals possess more or less of the RP genome than others. Those individuals that contain the highest RP genome are selected.

Selection using marker-flanking QTLs and evenly spaced markers from other chromosomes of the recurrent parent accelerated the process of introgression of genes and recovery of the RP genome (Collard et al., 2005). There are two types of selection in MABC: foreground selection and background selection (Hospital, 2003).

Foreground selection is the selection of individuals carrying the donor genes with the help of the marker information to ensure that the genes are not lost during the backcross phase and are fixed rapidly during the intercross phase. The effectiveness of foreground selection depends on the number of genes involved in the selection, the marker-gene association or linkage distance, and the undesirable linkage to the target gene.

Background selection is the selection of genetic background of the recipient with the help of marker information to accelerate the recovery rate of recipient genetic back-ground and eliminate undesirable genes introduced from the DP. The recovery rate of RP genome depends on the number of markers used in background selection. The more markers evenly located on all the chromosomes are selected for the RP alleles.

Both foreground and background selection is usually conducted in the same program, either simultaneously or successively. Foreground selection and background selection are two respective aspects of MABC with different foci of selection (Figure 7.1). In many cases, they can be performed alternatively even in the same generation. The individuals that have the desired marker alleles for target trait are selected first (foreground selection). Then, the selected individuals are screened for other marker alleles again for the RP genome (background selection). It is understandable to do so because selection of the target gene/QTL is the essential and only critical point for backcrossing program, and the individuals that do not have the allele of target gene will be discarded; thus, it is not necessary to genotype them for other traits.

The success of MAB depends upon several factors, including the distance between the closest markers and the target gene, the number of target genes to be transferred, the genetic base of the trait, the number of individuals that can be analyzed and the genetic background in which the target gene has to be transferred, the type of molecular marker(s) used, and available technical facilities. Based on simulations of 1000 replicates, Hospital (2003)

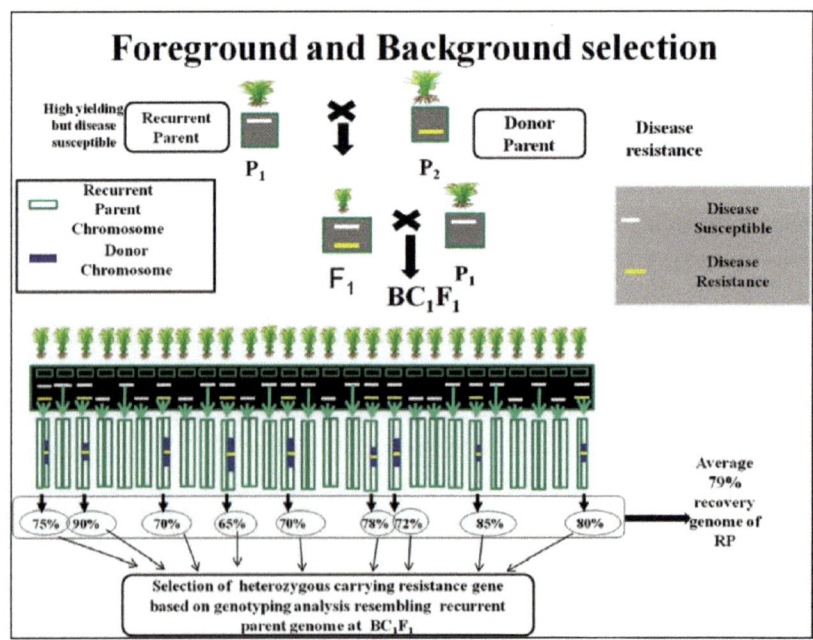

FIGURE 7.1 Schematic representation of selection of heterozygote carrying resistance gene based on genotyping analysis resembling recurrent parent genome at BC$_2$F$_1$. (Reprinted from Hasana, M. M.; Rafiiab, M.Y.; Ismaila, M.R.; Mahmoodc, M.; Rahimd, H.A.; Alama, M.A.; Ashkanibe, S.; Malekf, M.A.; & Latifg, M.A.; Marker-assisted backcrossing: a useful method for rice improvement. *Biotechnology & Biotechnological Equipment.* 2015, http://creativecommons.org/licenses/by/4.0/.

presented the expected results of a typical MABC program, in which hetero-zygotes were selected at the target locus in each generation, and RP alleles were selected for two flanking markers on target chromosome each located 2 cM apart from the target locus and for three markers on non-target chromosomes. As shown in Figure 7.2, a faster recovery of the RP genome could be achieved by MABC with combined foreground and background selection, compared to traditional backcrossing. Therefore, the use of markers can lead to considerable time-saving compared to that in conventional backcrossing (Frisch et al., 1999; Collard et al., 2005).

7.7.4.1 Application of MABC

MABC offers significant advantages in cases:

1. When phenotypic screening is expensive, difficult, or impossible.
2. When the trait is of low heritability.
3. When the selected trait is expressed late in plant development, like fruit and flower features or adult characteristics in species with a juvenile period.

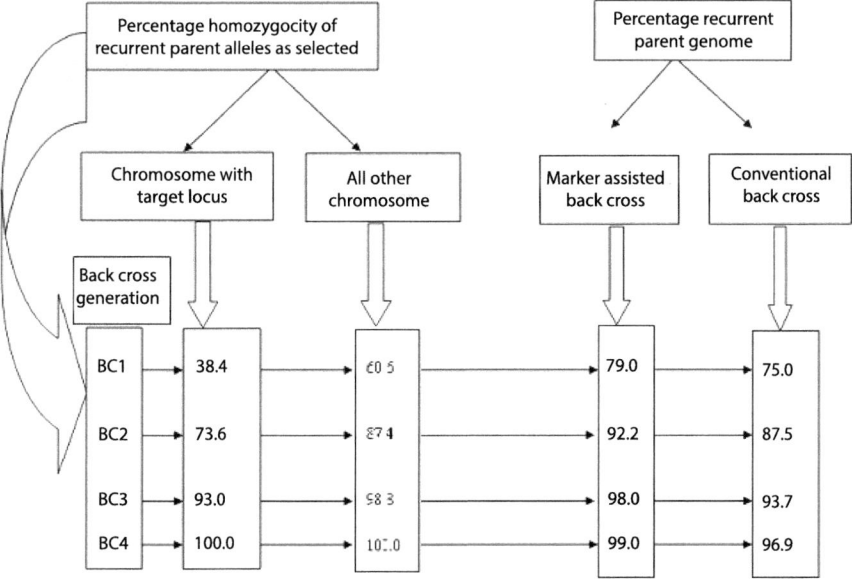

FIGURE 7.2 Average proportion of genes from the recurrent parent in different backcross generations. (Source: Adapted from Hospital, 2003)

4. For incorporating genes for resistance to diseases or pests that cannot be easily screened for due to special requirement for the gene to be expressed.

5. When the expression of the target gene is recessive.

6. To accumulate multiple genes for one or more traits within the same cultivar, a process called gene pyramiding.

MABC has been most widely and successfully used in plant breeding to date. It has been applied to different types of traits (e.g., disease/pest resistance, drought tolerance, and quality) in many species, e.g., rice, wheat, maize, barley, pearl millet, soybean, tomato, etc. (Collard et al., 2005; Dwivedi et al., 2007; Xu, 2010). In maize, for example, *Bacillus thuringiensis* is a bacterium that produces insecticidal toxins, which can kill corn borer larvae when they ingest the toxins in corn cells (Ragot et al., 1995). The integration of the *Bt* transgene into various corn genetic backgrounds has been achieved using MABC. Aroma in rice is controlled by a recessive gene, which is due to an eight base-pair deletion and three SNPs in a gene that codes for betaine aldehyde dehydrogenase 2 (Bradbury et al., 2005a). This discovery allows identification of the aromatic and non-aromatic rice varieties and discriminates homozygous recessive and dominant as well as heterozygous individuals in segregating population for the trait.

MABC has been used to select for aroma in rice (Bradbury et al., 2005b). High lysine *opaque 2* gene in corn was incorporated using MABC (Babu et al., 2005). However, the rate of success decreases when a large number of QTLs are targeted for introgression. Sebolt et al. (2000) used MABC for two QTL for seed protein content in soybeans. However, only one QTL was confirmed in $BC_3F_{4:5}$. When that QTL was introduced in three different genetic backgrounds, it had no effect on one background. In tomato, Tanksley and Nelson (1996) proposed an MABC strategy, called advanced backcross-QTL (AB-QTL), to transfer resistance genes from wild relative/unadapted genotype into elite germplasm. The strategy has proven effective for various agronomically important traits in tomato, including fruit quality and black mold resistance (Tanksley and Nelson, 1996; Bernacchi et al., 1998; Fulton et al., 2002). In addition, AB-QTL has been used in other crop species such as rice, barley, wheat, maize, cotton, and soybean, collectively demonstrating that this strategy is effective in transferring favorable alleles from the wild/ unadapted germplasm to elite germplasm (Concibido et al., 2003; Wang and Chee, 2010).

Currently, a cooperative marker-based backcrossing project for high oleic acid in soybean has been initiated among multiple US land-grant universities and USDA-ARS. Backcrossing and selection will be performed using the markers tightly linked to the high oleic genes/loci. Hopefully, the high oleic (80% or higher) traits will be successfully transferred from mutant lines or derived lines into other locally superior cultivars/lines or combined with other unique traits like low linolenic acid (Pham et al., 2012).

7.7.5 MARKER-ASSISTED GENE PYRAMIDING (MAGP)

Pyramiding is the simultaneous integration of multiple genes/QTLs into a single genotype. It is possible through conventional breeding, but it may be extremely difficult or impossible in early generations. In conventional phenotypic selection, individual plants should be screened for all phenotypic traits. Therefore, evaluation of certain types of plants, populations (e.g., F2), or traits (with destructive bioassays) are more difficult. Marker-assisted gene pyramiding (MAGP) is one of the most important applications of DNA markers to plant breeding. It is used for simultaneous incorporation of two or more resistance genes at a time into a plant for durable resistance. In rice, gene pyramids have been developed against bacterial blight and blast (Huang et al., 1997; Singh et al., 2001; Luo et al., 2012). Castro et al. (2003) reported a success in pyramiding qualitative gene and QTLs for resistance to stripe rust in barley.

Pyramiding of multiple genes or QTLs as a potential strategy for improvement of quantitatively inherited trait is recommended and proved the cumulative effects of multiple-QTL pyramiding in crop species like wheat, barley, and soybean by Richardson et al. (2006) Jiang et al. (2007a, 2007b), Li et al. (2010), and Wang et al. (2012). also proved the cumulative effects of multiple-QTL pyramiding in crop species like wheat, barley, and soybean.

Commonly used approaches for pyramiding are multiple-parent crossing or complex crossing, backcrossing, and recurrent selection. A suitable breeding scheme for MAGP depends on the number of genes/QTLs required for improvement of traits, the number of parents that contain the required genes/QTLs, the heritability of traits of interest, and other factors.

Three-way, four-way or double crossing is performed for pyramiding three or four desired genes/QTLs. Convergent backcrossing or stepwise backcrossing may also be used for pyramiding.

For pyramiding of more than four genes, complex or multiple crossing and/or recurrent selection may be preferred. Some of the crop plants in which the genes have been transferred through gene pyramiding are listed in Table 7.2.

In general, for MABC-based gene pyramiding, three strategies or breeding schemes are used: stepwise, simultaneous/synchronized, and convergent backcrossing or transfer.

7.7.5.1 Stepwise Backcrossing

In the stepwise backcrossing, only one gene/QTL is targeted or selected at one time, followed by the next step of backcrossing for another gene/QTL, until all target genes/QTLs have been introgressed into the recurrent parent. The advantage is that due to the incorporation of one gene at a time, the population size and genotyping amount will be small. The disadvantage is that it takes a longer time to complete.

TABLE 7.2 MAS in Gene Pyramiding

Crop	Trait (Combination of genes)	References
Rice	Bacterial blight resistance (xa4+xa5+xa13+Xa21; xa5+xa13+Xa21)	Huang et al., (1997), Sanchez et al., (2000), Singh et al., (2001)
	Blast resistance (Pi1+Piz-5+Pita) (Pi-tq5, Pi-tq1, Pi-tq6, Pi-lm2: pyramids of 2 to 4 genes)	Hittalmani et al., (2000)
	Multiple resistance: bacterial blight (Xa21)	Tabien et al., (2000)
	Sheath blight (RC7)	Datta et al., (2002)
Wheat	Powdery mildew resistance (Pm2+Pm4a; Pm2+Pm21; Pm4a+Pm21)	Liu et al., (2000)
Barley	Stripe rust resistance (3 QTL)	Castro et al., (2003b, c)
Broccoli	Diamondback moths resistance (cry1Ac+cry1c)	Cao et al., (2002)
Soybean	Lepidopteran resistance (cry1Ac+corn earworm QTL)	Walker et al., (2002)

7.7.5.2 Simultaneous/Synchronized Backcrossing

In this scheme, the recurrent parent is first crossed to each of the donor parents (suppose 4 parents) to produce four single-cross F_1s. Two of the four single-cross F_1s are crossed with each other to produce two double-cross F_1s, and these two double-cross F_1s are crossed again to produce a hybrid integrating all four target genes. The hybrid is subsequently backcrossed to recurrent parent with the objective to recover the recurrent parent, followed by one generation of selfing. The advantage of this method is that it takes the shortest time to complete. However, it requires a large population and more genotyping.

7.7.5.3 Convergent Backcrossing

This approach combines the advantages of stepwise and synchronized backcrossing. First, the four target gene/QTLs are transferred separately from the donors into the recurrent parent by single crossing, followed by backcrossing based on markers linked to the target genes/QTLs to produce four improved lines. Two of the improved lines are crossed with each other, and the two hybrids are then intercrossed to integrate all four genes/QTLs together and develop the final improved line having all four genes This scheme took less time as well as easily assured gene fixation in pyramiding.

Theoretically, computer simulations have been used for estimating efficiency of MABC for gene pyramiding (Ribaut et al., 2002; Servin et al., 2004; Ye and Smith, 2008). Practical application of MABC to gene pyramiding has been reported in crops, viz., rice, wheat, barley, cotton, soybean, common bean and pea, especially for developing durable resistance to stress in crops.

Very limited information is available about the release of commercial cultivars resulted from this strategy. Somers et al. (2005) implemented a molecular breeding strategy to introduce multiple pest resistance genes into Canadian wheat. They used high-throughput SSR genotyping and half-seed analysis to process backcrossing and selection for six FHB resistance QTLs, plus orange blossom wheat midge resistance gene *Sm1* and leaf rust resistance gene *Lr21*. They also used 45-76 SSR markers to perform background selection in backcrossing populations to accelerate the restoration of the RP genetic background. This strategy resulted in 87% fixation of the elite genetic background at the BC_2F_1 on average and successfully introduced all (upto 4) of the chromosome segments containing FHB, *Sm1*, and *Lr21* resistance genes in four separate crosses (Somers et al., 2005).

If all the parents are improved cultivars/lines and have different or complementary genes or favorable alleles for the traits of interest, marker-assisted complex, or convergent crossing, MACC is a proper option of breeding scheme. In MACC, the hybrid of convergent crossing is subsequently self-pollinated and marker-based selection for target traits is performed for several consecutive generations until genetically stable lines with desired marker alleles and traits have been developed.

Different markers may be used and selected in different generations, depending on their relative importance. The markers for the most important genes/QTLs can be detected and selected first in early generations and less important markers later. Once homozygous alleles of the markers for a gene/ locus are detected, they may not be necessarily detected again in the subsequent generations. Instead, phenotypic evaluation should be conducted if conditions permit.

7.7.6 MARKER-ASSISTED RECURRENT SELECTION

The improvement of complex traits via phenotypic recurrent selection is generally possible, but the long selection cycles impose restrictions on the practicability of this breeding method. For the improvement of polygenic traits, recurrent selection is widely regarded as an effective strategy. However, the effectiveness and efficiency of selection are not so satisfactory in some cases because phenotypic selection is highly dependent upon environment and genotypic selection takes a longer time.

Marker-assisted recurrent selection (MARS) is a scheme that allows performing genotypic selection and intercrossing in the same crop season for one cycle of selection (Figure 7.3). Therefore, it enhances the efficiency of recurrent selection and accelerate the progress of the procedure (Jiang et al., 2007a), particularly helping in integrating multiple favorable genes/QTLs from different sources.

Ribaut et al. (2010) defined MARS, a recurrent selection scheme using molecular markers for the identification and selection of multiple genomic regions involved in the expression of complex traits to assemble the best-performing genotype within a single or across related populations. For com-

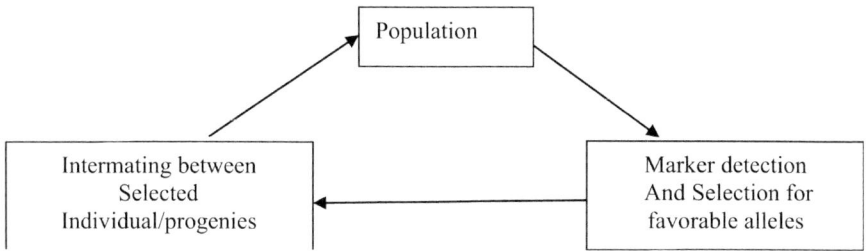

FIGURE 7.3   General procedure of marker-assisted recurrent selection (MARS).

plex traits, MARS has been proposed for "forward breeding" of native genes and pyramiding multiple QTLs (Ragot et al., 2000; Eathington, 2005; Crosbie et al., 2006; Ribaut et al., 2010).

Eathington (2005) and Crosbie et al. (2006) conducted studies on reference maize populations and found twice genetic gain than that of phenotypic selection (PS). Yi et al. (2004) reported significant effectiveness of MARS for resistance to *Helicoverpa armigera* in upland cotton. The mean levels of resistance in improved populations after recurrent selection were significantly higher than those of preceding populations.

7.7.7   GENOMIC SELECTION/GENOME-WIDE SELECTION (GWS)

It is a form of marker-based selection, which was defined by Meuwissen (2007) as the simultaneous selection for many (tens or hundreds of thousands) markers, which cover the entire genome in a dense manner so that all genes are expected to be in linkage disequilibrium with at least some of the markers. In GS genotyping, data (genetic markers) across the whole genome are used to predict complex traits with accuracy sufficient to allow selection on that prediction alone. Selection of desirable individuals is based on genomic estimated breeding value (GEBV) (Nakaya and Isobe, 2012), which is a predicted breeding value calculated using an innovative method based on genome-wide dense DNA markers (Meuwissen et al., 2001).

Significant testing and identifying a subset of markers associated with the trait (Meuwissen et al., 2001) are not required in GS. However, it does first need to develop GS models, i.e., the formulae for GEBV prediction

(Nakaya and Isobe, 2012). In this process (training phase), phenotypes and genome-wide genotypes are investigated in the training population (a subset of a population) to predict significant relationships between phenotypes and genotypes using statistical approaches. Subsequently, GEBVs are used for the selection of desirable individuals in the breeding phase, instead of the genotypes of markers used in traditional MAS.

GS can be possible only when high-throughput marker technologies, high-performance computing, and appropriate new statistical methods become available. This approach has become feasible due to the discovery and development of large number of SNPs by genome sequencing and new methods to efficiently genotype large number of SNP markers.

Goddard and Hayes (2007) suggested the ideal method to estimate the breeding value from genomic data is to calculate the conditional mean of the breeding value given the genotype at each QTL. This conditional mean can only be calculated by using a prior distribution of QTL effects, and thus this should be part of the research to implement GS. In practice, this method of estimating breeding values is approximated by using the marker genotypes instead of the QTL genotypes, but the ideal method is likely to be approached more closely as more sequence and SNP data are obtained.

The application of GS was proposed by Meuwissen et al. (2001) to breeding populations. Since then, theoretical, simulation, and empirical studies have been conducted, mostly in animals (Goddard and Hayes, 2007; Jannink et al., 2010). In oil palm, GS was superior to MARS and PS in terms of gain per unit cost and time (Wong and Bernardo, 2008).

GS has been highlighted as a new approach for MAS in recent years and is regarded as a powerful, attractive and valuable tool for plant breeding. However, GS has not become a popular methodology in plant breeding, and there might be a long way to go before the extensive use of GS in plant breeding programs. The major reason might be the unavailability of sufficient knowledge of GS for practical use. Statistics and simulation discussed in terms of formulae in GS studies are most likely too specific and hard for plant breeders to understand and to use in practical breeding programs. Development of easily understandable formulae for GEBVs and user-friendly software packages for GS analysis is helpful in facilitating and enhancing the application of GS in plant breeding.

7.8 CHALLENGES AND FUTURE PERSPECTIVES

Various types of molecular markers in crop plants were developed during the 1980s and 1990s.

Molecular markers may complement plant breeding in three general ways: (i) Provide a reliable genetic-diversity measure that can be used for determining relationships among inbred lines and cultivars, assessing changes in genetic diversity over time, protecting intellectual property rights, registering germplasm, evaluating new germplasm, and selecting parents to hybridize in a breeding program. (ii) Their linkage with alleles with large effects/small effects may improve screens for many traits. (iii) Provide the first understanding of the biology and the architecture of many quantitative traits. Molecular MAB have following significant advantages over traditional breeding methods:

a) Selection for all kinds of traits at the seedling stage.
b) Selective neutral behaviors: The DNA sequences of any organism are neutral to environmental conditions or management practices.
c) Due to identification of multiple/individual genes at the same time and in the same individuals, used for gene pyramiding.
d) Co-dominant inheritance: determination of homozygous and heterozygous states of diploid organisms.
e) Higher effectiveness and efficiency in terms of time, resources, and efforts.
f) Easy exchange of data between laboratories.

MAB is extensively used in crop improvement programs particularly since 2000s. However, it is limited to simply inherited traits such as diseases/pests, although quantitative traits were also involved (Collard and Mackill, 2008; Wang and Chee, 2010). Currently, MAS does not play a major role in genetic improvement programs in any of the agricultural sectors. The application of molecular technologies to plant breeding is still facing the following drawbacks:

a) Higher startup expenses and labor costs.
b) Unavailability of breeder-friendly markers.
c) Due to lack of marker polymorphism, all markers cannot be applicable across populations.

d) Due to recombination between the markers and the genes of interest, false selection may occur sometimes.

e) Requirement of a large number of breeding programs for a large-scale adoption of MAB in practice.

f) Complicated methods/schemes and terminology hinder large-scale adoption of MAB.

With newer technologies and through advances in the field of genomics, the challenge for plant breeders is to judiciously utilize these novel tools in molecular MAB for developing commercially viable improved cultivars that address specific problems in different crops. High costs and technical or equipment demands of MAB will continue to be a major obstacle for its large-scale use in the near future, especially in the developing countries (Collard and Mackill, 2008; Ribaut et al., 2010). Therefore, integration of MAB into traditional breeding programs rather than substitute will be an optimistic strategy for crop improvement in the future.

We expect that the cost of MABC can be decreased, resources pooled and shared, and a capacity to be developed if we make good partnership between developing and developed countries, including public- private sector collaboration. Currently, the cost of utilizing markers is possibly the most important factor that limits the implementation of MABC. Therefore, new marker technology can potentially reduce the cost of MAS considerably. If the effectiveness of the new methods is validated and the equipment can be easily obtained, this should allow MABC to become more widely applicable for crop breeding programs.

7.9   CONCLUDING REMARKS

Molecular MAB or MAS is gaining considerable importance as it would improve the efficiency of plant breeding through precise transfer of genomic regions of interest (foreground selection) and accelerate the recovery of the recurrent parent genome (background selection). It has been widely used for simple inherited traits than polygenic traits. They can be used to monitor DNA sequence variation in and among the species and create new sources of genetic variation by introducing new and favorable traits from land races, wild relatives, and related species and to fasten the time taken in conventional breeding, germplasm characterization, genetic mapping, gene tagging, and gene introgression from exotic and wild species. The success of molecular MAB depends on many critical factors such as the number of target genes to be transferred, the distance between the target gene and the flanking markers, number of genotypes selected in each breeding generation, the nature of germplasm, and the technical options available at the marker level. The power and efficiency of genotyping are expected to improve with the advent of markers like SNPs. Achieving a substantial impact on crop improvement by molecular MAB represents the great challenge for agricultural scientists in the next few decades.

KEYWORDS

- **conventional breeding**
- **gene introgression**
- **marker-assisted selection**
- **molecular markers**
- **QTL mapping**

REFERENCES

Akkaya, M. S., Bhagwatt, A. A., & Cregan, P. B., (1992). Length polymorphisms of simple sequence repeat DNA in soybean. *Genetics, 132,* 1131–1139.

Anderson, J. A., Chao, S., & Liu, S., (2007). Molecular breeding using a major QTL for *Fusarium* head blight resistance in wheat. *Crop Sci., 47,* 112–119.

Babu, R., Nair, S. K., Kumar, A., Venkatesh, S., Sekhar, J. C., Singh, N. N., Srinivasan G., & Gupta, H. S., (2005). Two-generation marker-aided backcrossing for rapid conversion of normal maize lines to quality protein maize (QPM*). Theor. Appl. Genet., 111,* 888–897.

Bernacchi, D., Beck-Bunn, T., Emmatty, D., Eshed, Y., Inai, S., Lopez, J., Petiard, V., Sayama, H., Uhlig, J., Zamir, D., & Tanksley, S. D., (1998). Advanced backcross QTL analysis of tomato: II. Evaluation of near-isogenic lines carrying single-donor introgressions for desirable wild QTL-alleles derived from *Lycopersicon hirsutum* and *L. pimpinellifolium. Theor. Appl. Genet., 97,* 170–180.

Bilyeu, K., Palavalli, L., Sleper D. A., & Beuselinck, P., (2006). Molecular genetic resources for development of 1% linolenic acid soybeans. *Crop Sci., 46,* 1913–1918.

Bonnett, D. G., Rebetzke, G. J., & Spielmeyer, W., (2005). Strategies for efficient implementation of molecular markers in wheat breeding. *Mol. Breeding., 15,* 75–85.

Botstein, D., White, R. L., Skolnick, M., & Davis, R. W., (1980). Construction of a genetic linkage map in man using restriction fragment length polymorphisms. *Am. J. Hum. Genet.*, *32*, 314–331.

Bradbury, L. M. T., Henry, R. J., Jin, Q., Reinke R. F., & Waters, D. L. E., (2005b). A perfect marker for fragrance genotyping in rice. *Mol. Breeding*, *16*, 279–283.

Bradbury, L. M. T., Fitzgerald, T. L., Henry, R. J., Jin, Q., & Waters, D. L. E., (2005a). The gene for fragrance in rice. *Plant Biotech. J.*, *3*, 363–370.

Buerstmayr, H., Ban T., & Anderson, J. A., (2009). QTL mapping and marker-assisted selection for Fusarium head blight resistance in wheat: a review. *Plant Breeding*, *128*, 1–26.

Cao, J., Zhao, J. Z., Tang, J. D., Shelton A. M., & Earle, E. D., (2002). Broccoli plants with pyramided cry1Ac and cry1C Bt genes control diamondback moths resistant to Cry1A and Cry1C proteins. *Theor. Appl. Genet.*, *105*, 258–264.

Castro, A. J., Capettini, F., Corey, A. E., Filichkina, T., Hayes, P. M., Kleinhofs, A., Kudrna, D., Richardson, K., Sandoval-ISLAS, S., Rossi, C., & Vivar, H., (2003). Mapping and pyramiding of qualitative and quantitative resistance to stripe rust in barley. *Theoretical and Applied Genetics.*, *107*, 922–930.

Castro, A. J., Chen, X., Corey, A. E., Filichkina, T., Hayes, P. M., Mundt, C., Richardson, K., Sandoval-Islas, S., & Vivar, H., (2003b). Pyramiding and validation of quantitative trait locus (QTL) alleles determining resistance to barley stripe rust: effects on adult plant resistance. *Crop Sci.*, *43*, 2234–2239.

Castro, A. J., Chen, X., Hayes, P. M., & Johnston, M., (2003c). Pyramiding quantitative trait locus (QTL) alleles determining resistance to barley stripe rust: effects on resistance at the seedling stage. *Crop Sci.*, *43*, 651–659.

Chu, S., De Risi, J., Eisen, M., Mulholland, J., Botstein, D., Brown, P. O., & Herskowitz, I., (1998). The transcriptional program of sporulation in budding yeast. *Science*, *282*, 699–705.

Collard, B. C. Y., & Mackill, D. J., (2008). Marker-assisted selection: an approach for precision plant breeding in the twenty-first century. *Phil. Trans. R. Soc.*, *363*, 557–572.

Collard, B. C. Y., Jahufer, M. Z. Z., Brouwer, J. B., & Pang, E. C. K., (2005). An introduction to markers, quantitative trait loci (QTL) mapping and marker-assisted selection for crop improvement: the basic concepts. *Euphytica.*, *142*, 169–196.

Concibido, V. C., La Vallee, B., Mclaird, P., Pineda, N., Meyer, J., Hummel, L., Yang, J., Wu, K., & Delannay, X., (2003). Introgression of a quantitative trait locus for yield from *Glycine soja* into commercial soybean cultivars. *Theor. Appl. Genet.*, *106*, 575–582.

Cregan, P. B., Mudge, J., Fickus, E. W., Danesh, D., Denny, R., & Young, N. D., (1999). Two simple sequence repeat markers to select for soybean cyst nematode resistance conditioned by the *rhg1* locus. *Theor. Appl. Genet.*, *99*, 811–818.

Crosbie, T. M., Eathington, S. R., Johnson, G. R., Edwards, M., Reiter, R., Stark, S., Mohanty, R. G., Oyervides, M., Buehler, R. E., Walker, A. K., Dobert, R., Delannay, X., Pershing, J. C., Hall, M. A., & Lamkey, K. R., (2006). Plant breeding: past, present and future. In: Lamkey, K. R., & Lee, M., (eds.), *Plant Breeding*: the Arnel, R., Hallauer International Symposium, Blackwell Publishing, Oxford, UK, pp. 3–50.

Datta, K., Baisakh, N., Maung Thet, K., Tu, J., & Datta, S. K., (2002). Pyramiding transgenes for multiple resistance in rice against bacterial blight, yellow stem borer and sheath blight, *Theor. Appl. Genet.*, *106*, 1–8.

Diers, B. W., Keim, P., Fehr, W. R., & Shoemaker, R. C., (1992). RFLP analysis of soybean seed protein and oil content. *Theor. Appl. Genet.*, *83*, 608–612.

Doerge, R. W., Zeng, Z. B., & Weir, B. S., (1997). Statistical issues in the search for genes affecting quantitative traits in experimental populations. *Stat. Sci., 12,* 195–219.

Dwivedi, S. L., Crouch, J. H., Mackill, D. J., Xu, Y., Blair, M. W., Ragot, M., Upadhyaya H. D., & Oritiz, R., (2007). The molecularization of public sector crop breeding: progress, problems and prospects. *Advances in Agronomy., 95,* 163–318.

Eathington, S. R., (2005). Practical applications of molecular technology in the development of commercial maize hybrids. In: P*roceedings of the 60th Annual Corn and Sorghum Seed Research Conferences,* American Seed Trade Association, Washington, DC.

Edwards, D., Forster, J. W., Chagne, D., & Batley, J., (2007). What is SNPs? In: Oraguzie, N. C., Rikkerink, E. H. A., Gardiner, S. E., & de Silva, H. N., (eds.), *Association Mapping in Plants,* Springer, 41–52.

Farooq, S., & Azam, F., (2002a). Molecular markers in plant breeding-I: Concepts and characterization. *Pakistan J. Biol. Sci., 5,* 1135–1140.

Farooq, S., & Azam, F., (2002b). Molecular markers in plant breeding-II: Some prerequisites for use. *Pakistan J. Biol. Sci., 5,* 1141–1147.

Fatokun, C. A., Menancio-Hautea, D., Danesh, D., & Young, N. D., (1992). Evidence for orthologous seed weight genes in cowpea and mungbean based on RFLPs. *Genetics, 132,* 841–846.

Flint-Garcia, S. A., Darrah, L. L., McMullen, M. D., & Hibbard, B. E., (2003). Phenotypic versus marker-assisted selection for stalk strength and second-generation European corn borer resistance in maize. *Theor. Appl. Genet., 107,* 1331–1336.

Frisch, M., Bohn, M., & Melchinger, A. E., (1999). Comparison of selection strategies for marker-assisted backcrossing of a gene. *Crop Sci., 39,* 1295–1301.

Fulton, T. M., Bucheli, P., Voirol, E., Lopez, J., Peetiard, V., & Tanksley, S. D., (2002). Quantitative trait loci (QTL) affecting sugars, organic acids and other biochemical properties possibly contributing to flavor, identified in four advanced backcross populations of tomato. *Euphytica., 127,* 163–177.

Goddard, M. E., & Hayes, B. J., (2007). Genomic selection. *J. Anim. Breed. Genet., 124,* 323–330.

Gupta, P. K., Roy, J. K., & Prasad, M., (2001). Single nucleotide polymorphisms: a new paradigm for molecular marker technology and DNA polymorphism detection with emphasis on their use in plants. *Current Science, 80,* 524–535.

Hasana, M. M.; Rafiiab, M.Y.; Ismaila, M.R.; Mahmoodc, M.; Rahimd, H.A.; Alama, M.A.; Ashkanibe, S.; Malekf, M.A.; & Latifg, M.A.; Marker-assisted backcrossing: a useful method for rice improvement. *Biotechnology & Biotechnological Equipment.* 2015.

Hittalmani, S., Parco, A., Mew, T. V., Zeigler, R. S., & Huang, N., (2000). Fine mapping and DNA marker-assisted pyramiding of the three major genes for blast resistance in rice, *Theor. Appl. Genet., 100,* 1121–1128.

Hospital, F.; Marker-assisted breeding. In: H.J. Newbury (ed.), Plant molecular breeding. Blackwell Publishing and CRC Press, Oxford and Boca Raton, **2003,** 30–59.

Hospital, F., & Decoux, G., (2002). Popmin: a program for the numerical optimization of population sizes in marker-assisted backcross breeding programs. *Journal of Heredity, 93,* 383–384.

Huang, N., Angeles, E. R., Domingo, J., Magpantay, G., Singh, S., Zhang, G., Kumaravadivel, N., Bennett, J., &. Khush, G. S., (1997). Pyramiding of bacterial blight resistance genes in rice: marker-assisted selection using RFLP and PCR, *Theor. Appl. Genet., 95,* 313–320.

Jannink, J. L., Lorenz, A. J., & Iwata, H., (2010). Genomic selection in plant breeding: from theory to practice. *Briefings in Functional Genomics, 9,* 166–177.

Jiang, G. L., Shi, J., & Ward, R. W., (2007a). QTL analysis of resistance to *Fusarium* head blight in the novel wheat germplasm CJ 9306. I. Resistance to fungal spread. *Theor. Appl. Genet., 116,* 3–13.

Jiang, G. L., Dong, Y., Shi, J., & Ward, R. W., (2007b). QTL analysis of resistance to *Fusarium* head blight in the novel wheat germplasm CJ 9306. II. Resistance to deoxynivalenol accumulation and grain yield loss. *Theor. Appl. Genet., 115,* 1043–1052.

Johnson, R., (2004). Marker assisted selection. In: Jannick, J., (ed.), *Plant Breed. Rev., 24*(1), 293–309.

Jordan, D. R., Tao, Y., Godwin, I. D., Henzell, R. G., Cooper, M., & McIntyre, C. L., (2003). Prediction of hybrid performance in grain sorghum using RFLP markers. *Theor. Appl. Genet., 106,* 559–567.

Jordan, D. R., Tao, Y., Godwin, I. D., Henzell, R. G., Cooper, M., & McIntyre, C. L., (2004). Comparison of identity by descent and identity by state for detecting genetic regions under selection in a sorghum pedigree breeding program. *Mol. Breed., 14,* 441–454.

Keim, P., Diers, B. W., & Shoemaker, R. C., (1990). Genetic analysis of soybean hard seededness with molecular markers. *Theor. Appl. Genet., 79,* 465–469.

Knapp, S. J., & Bridges, W. C., (1990). Using molecular markers to estimate quantitative trait locus parameters: Power and genetic variances for unreplicated and replicated progeny. *Genetics, 126,* 769–777.

Koebner, R. M., (2004). Marker assisted selection in the cereals: The dream and the reality. In: Gupta, P. K., & Varshney, R. K., (ed.) *Cereal Genomics.* Kluwer Academic, Dordrecht, the Netherlands., 317–329.

Konieczny, A., & Ausubel, F., (1993). A procedure for mapping Arabidopsis mutations using co-dominant ecotype-specific PCR based markers. *The Plant Journal, 4,* 403–410.

Kumar, L. S., (1999). DNA markers in plant improvement: An overview. *Biotechnol. Adv., 17,* 143–182.

Kumpatla, S. P., Buyyarapu, R., Abdurakhmonov, I. Y., & Mammadov. J. A., (2012). Genomics-assisted plant breeding in the 21st century: technological advances and progress. In: Abdurakhmonov, I. Y., (ed.), *Plant Breeding,* 131–184.

Lander, E. S., & Botstein, D., (1988). Mapping Mendelian factors underlying quantitative traits using RFLP linkage maps. *Genetics, 121,* 185–99.

Lee, M., Godshalk, E. B., Lamkey, K. R., & Woodman, W. W., (1989). Association of restriction fragment length polymorphisms among maize inbreds with agronomic performance of their crosses. *Crop Sci., 29,* 1067–1071.

Li, X., Han. Y., Teng, W., Zhang, S., Yu, K., Poysa, V., Anderson, T., Ding, J., & Li, W., (2010). Pyramided QTL underlying tolerance to Phytophthora root rot in mega-environment from soybean cultivar 'Conrad' and 'Hefeng 25'. *Theor. Appl. Genet., 121,* 651–658.

Lin, Y. R., Schertz, K. F., & Paterson, A. H., (1995). Comparative analysis of QTLs affecting plant height and maturity across the Poaceae in reference to an interspecific sorghum population. *Genetics, 141,* 391–411.

Liu, J., Liu, D., Tao, W., Li, W., Wang, S., Chen, P., Cheng, S., & Gao, D., (2000). Molecular marker-facilitated pyramiding of different genes for powdery mildew resistance in wheat. *Plant Breed, 119,* 21–24.

Liu, S. X., & Anderson, J. A., (2003). Marker assisted evaluation of *Fusarium* head blight resistant wheat germplasm. *Crop Sci., 43,* 760–766.

Luo, Y., Sangha, J. S., Wang, S., Li, Z., Yang, J., & Yin, Z., (2012). Marker-assisted breeding of *Xa4*, *Xa21* and *Xa27* in the restorer lines of hybrid rice for broad-spectrum and enhanced disease resistance to bacterial blight. *Mol. Breeding*, DOI 10. 1007/ s11032–012–9742–7.

Martin, B., Nienhuis, J., King, G., & Schaefer, A., (1989). Restriction fragment length polymorphisms associated with water use efficiency in tomato. *Science*, *243*, 1725–1728.

McCartney, C. A., Somers, D. J., Fedak, G., & Cao, W., (2004). Haplotype diversity at *fusarium* head blight resistance QTLs in wheat. *Theor. Appl. Genet.*, *109*, 261–271.

Meuwissen, T., (2007). Genomic selection: marker assisted selection on a genome wide scale. *J. Anim. Breed. Genet.*, *124*, 321–322.

Meuwissen, T. H. E., Hayes B. J., & Goddard, M. E., (2001). Prediction of total genetic value using genome wide dense marker maps. *Genetics*, *157*, 1819–1829.

Miedaner, T., Wilde, F., Steiner, B., Buerstmayr, H., Korzun, V., & Ebmeyer, E., (2006). Stacking quantitative trait loci (QTL) for Fusarium head blight resistance from non-adapted sources in an European elite spring wheat background and assessing their effects on deoxynivalenol (DON) content and disease severity. *Theor. Appl. Genet.*, *112*, 562–569.

Miklas, P. N., Kelly, J. D., Beebe, S. E., & Blair, M. W., (2006). Common bean breeding for resistance against biotic and abiotic stresses: From classical to MAS breeding. *Euphytica.*, *147*, 105–131.

Moreau, L., Charcosset, A., Hospital, F., & Gallais, A., (1998). Marker-assisted selection efficiency in populations of finite size. *Genetics.*, *148*, 1353–1365.

Mullis, K., (1990). The unusual origin of the polymerase chain reaction. *Scientific American*, *262*(4), 56–61, 64–65.

O'Brien, S. J., (1993). *Genetic Maps: Locus Maps of Complex Genomes*. Cold Spring Harbor Laboratory Press, Cold Spring Harbor, New York.

Paterson, A. H., Lander, E. S., Hewitt, J. D., Peterson, S., Lincoln, S. E., & Tanksley, S. D., (1988). Resolution of quantitative traits into Mendelian factors using a complete linkage map of restriction fragment length polymorphisms. *Nature*, *335*, 721–726.

Pham, A. T., Shannon, J. G., & Bilyeu, K. D., (2012). Combinations of mutant FAD2 and FAD3 genes to produce high oleic acid and low linolenic acid soybean oil. *Theor. Appl. Genet.*, *125*, 503–515.

Pumphrey, M. O., Bernardo, R., & Anderson, J. A., (2007). Validating the *Fhb1* QTL for *Fusarium* head blight resistance in near-isogenic wheat lines developed from breeding populations. *Crop Sci.*, *47*, 200–206.

Ragot, M., & Lee, M., (2007). Marker-assisted selection in maize: Current status, potential, limitations and perspectives from the private and public sectors. In: Guimaraes, E. P., et al., (ed.) *Marker-Assisted Selection, Current Status and Future Perspectives in Crops, livestock, forestry, and fish*. FAO, Rome, pp. 117–150.

Ragot, M., Gay, G., Muller, J. P., & Durovray, J., (2000). Efficient selection for the adaptation to the environment through QTL mapping and manipulation in maize. In: Ribaut, J. M., & Oland, D., (eds.), *Molecular Approaches for the Genetic Improvement of Cereals for Stable Production in Water-limited Environments*. CIMMYT, Mexico, 128–130.

Ragot, M., Biasiolli, M., Dekbut, M. F., Dell'orco, A., Margarini, L., Thevenin, P., Vernoy, J., Vivant, J., Zimmermann, R., & Gay, G., (1995). Marker-assisted backcrossing: a practical example. In: Berville, A., & Tersac, M., (eds.), Les Colloques, No. *72*, INRA, Paris, 45–46.

Reif, J. C., Melchinger, A. E., Xia, X. C., Warburton, M. L., Hoisington, D. A., Vasal, S. K., Beck, D., Bohn, M., & Frisch, M., (2003). Use of SSRs for establishing heterotic groups in subtropical maize. *Theor. Appl. Genet., 107,* 947–957.

Reif, J. C., Hamrit, S., Heckenberger, M., Schipprack, W., Maurer, H. P., Bohn, M., & Melchinger, A. E., (2005). Trends in genetic diversity among European maize cultivars and their parental components during the past 50 years. *Theor. Appl. Genet., 111,* 838–845.

Ribaut, J. M., & Hoisington, D., (1998). Marker-assisted selection: New tools and strategies. *Trends Plant Sci., 3,* 236–239.

Ribaut, J. M., & Betran, J., (1999). Single large-scale marker-assisted selection (SLSMAS). *Mol. Breeding, 5,* 531–541.

Ribaut, J. M., William, H. M., Khairallah, M., Worland, A. J., & Hoisington, D., (2001). Genetic basis of physiological traits. In: *Application of Physiology in Wheat Breeding* (Reynolds, M. P., Ortiz-Monasterio, J. I., & McNab, A., eds.). Mexico, DF: CIMMYT.

Ribaut, J. M., De Vicente, M. C., & Delannay, X., (2010). Molecular breeding in developing countries: challenges and perspectives. *Current Opinion in Plant Biology, 13,* 1–6.

Richardson, K. L., Vales, M. I., Kling, J. G., Mundt, C. C., & Hayes, P. M., (2006). Pyramiding and dissecting disease resistance QTL to barley stripe rust. *Theor. Appl. Genet., 113,* 485–495.

Sanchez, A. C., Brar, D. S., Huang, N., Li, Z., & Khush, G. S., (2000). Sequence tagged site marker-assisted selection for three bacterial blight resistance genes in rice, *Crop Sci., 40,* 792–797.

Sax, K., (1923). The association of size differences with seed-coat pattern and pigmentation in *Phaseolus vulgaris. Genetics, 8,* 552–560.

Sebolt, A. M., Shoemaker, R. C., & Diers, B. W., (2000). Analysis of a quantitative trait locus allele from wild soybean that increases seed protein concentration in soybean. *Crop Sci., 40,* 1438–1444.

Semagn, K., Bjornstad, A., & Ndjiondjop, M. N., (2006). Progress and prospects of marker assisted backcrossing as a tool in crop breeding programs. *Afr. J. Biotechnol., 5,* 2588–2603.

Semagn, K., Bjornstad, A., & Ndjiondjop, M. N., (2006a). An overview of molecular marker methods for plants. *Afr. J. Biotechnol., 5,* 2540–2568.

Servin, B., Martin, O. C., Mezard, M., & Hospital, F., (2004). Toward a theory of marker-assisted gene pyramiding. *Genetics, 168,* 513–523.

Singh, S., Sidhu, J. S., Huang, N., Vikal, Y., Li, Z., Brar, D. S., Dhaliwal, H. S., & Khush, G. S., (2001). Pyramiding three bacterial blight resistance genes (xa5, xa13 and Xa21) using marker-assisted selection into indica rice cultivar PR106, *Theor. Appl. Genet., 102,* 1011–1015.

Sobrino, B., Briona, M., & Carracedoa, A., (2005). SNPs in forensic genetics: a review on SNP typing methodologies. *Forensic Science International, 154,* 181–194.

Somers, D. J., Thomas, J., DePauw, R., Fox, S., Humphreys, G., & Fedak, G., (2005). Assembling complex genotypes to resist *Fusarium* in wheat (*Triticum aestivum* L.). *Theor. Appl. Genet., 111,* 1623–1631.

Song, Q., Jia, G., Zhu, Y., Grant, D., Nelson, R. T., Hwang, E. Y., Hyten, D. L., & Cregan, P. B., (2010). Abundance of SSR motifs and development of candidate polymorphic SSR markers in soybean. *Crop Sci., 50,* 1950–1960.

Southern, E. M., (1975). Detection of specific sequences among DNA fragments separated by gel electrophoresis. *Journal of Molecular Biology, 98,* 503–517.

Steele, K. A., Edwards, G., Zhu, J., & Witcombe, J. R., (2004). Marker-evaluated selection in rice: shifts in allele frequency among bulks selected in contrasting agricultural environments identify genomic regions of importance to rice adaptation and breeding. *Theor. Appl. Genet.*, *109,* 1247–1260.

Stuber, C. W., Edwards, M. D., & Wendel, J. F., (1987). Molecular marker facilitated investigations of quantitative trait loci in maize. II. Factors influencing yield and its component traits. *Crop Sci.*, *27,* 639–648.

Tabien, R. E., Li, Z., Patterson, A. H., Marchetti, M. A., Stansel, J. W., & Pinson, S. R. M., (2000). Mapping of four major rice blast resistance genes from Lemont and Teqint and evaluation of their combinatorial effect for field resistance, *Theor. Appl. Genet.*, *101,* 1215–1225.

Tanksley, S. D., & Nelson, J. C., (1996). Advanced backcross QTL analysis: a method for the simultaneous discovery and transfer of valuable QTLs from unadapted germplasm into elite breeding lines. *Theor. Appl. Genet.*, *92,* 191–203.

Thoday, J. M., (1961). Location of polygenes. *Nature*, *191,* 368–370.

Varshney, R. K., Graner, A., & Sorrells, M. E., (2005). Genic microsatellite markers in plants: Features and applications. *Trends Biotechnol.*, *23,* 48–55.

Vos, P., Hogers, R., Bleeker, M., Reijans, M., Van de Lee, T., Hornes, M., Frijters, A., Pot, J., Peleman, J., Kuiper, M., & Zabeau, M., (1995). AFLP: a new technique for DNA fingerprinting. *Nucl. Acids Res.*, *23,* 4407–4414.

Walker, D., Boerma, H. R., All, J., & Parrott, W., (2002). Combining cry1Ac with QTL alleles from PI 229358 to improve soybean resistance to lepidopteran pests. *Mol. Breed.*, *9,* 43–51.

Wang, B., & Chee, P. W., (2010). Application of advanced backcross quantitative trait locus (QTL) analysis in crop improvement. *Journal of Plant Breeding and Crop Science*, *2,* 221–232.

Williams, J., Kubelik, A., Livak, K., Rafalski, J., & Tingey, S., (1990). DNA polymorphisms amplified by arbitrary primers are useful as genetic markers. *Nucl. Acids Res.*, *18,* 6531–6535.

Wong, C. K., & Bernardo, R., (2008). Genome wide selection in oil palm: increasing selection gain per unit time and cost with small populations. *Theor. Appl. Genet.*, *116,* 815–824.

Xu, Y., (2010). *Molecular Plant Breeding*, CAB International. ISBN 1845933923, 9781845933920, 21–43.

Xu, Y. B., Beachell, H., & McCouch, S. R., (2004). A marker based approach to broadening the genetic base of rice in the USA. *Crop Sci.*, *44,* 1947–1959.

Xu, Y., (2002). Global view of QTL: Rice as a model. In: Kang, M. S., (ed.) *Quantitative Genetics, Genomics, and Plant Breeding*. CABI Publishing, Wallingford, UK, 109–134.

Xu, Y., (2003). Developing marker-assisted selection strategies for breeding hybrid rice. *Plant Breed. Rev.*, *23,* 73–174.

Xu, Y., McCouch, S. R., & Zhang, Q., (2005). How can we use genomics to improve cereals with rice as a reference genome? *Plant Mol. Biol.*, *59,* 7–26.

Yashitola, J., Thirumurugan, T., Sundaram, R. M., Naseerullah, M. K., Ramesha, M. S., Sarma, N. P., & Sonti, R. V., (2002). Assessment of purity of rice hybrids using microsatellite and STS markers. *Crop Sci.*, *42,* 1369–1373.

Ye, G., & Smith, K. F., (2008). Marker-assisted gene pyramiding for inbred line development: Basic principles and practical guidelines. *Intern. J. Plant Breed.*, *2,* 1–10.

Yi, C., Guo, W., Zhu, X., Min, L., & Zhang, T., (2004). Pyramiding breeding by marker as-
    sisted recurrent selection in upland cotton II. Selection effects on resistance to *Helicov-
    erpa armigera. Scientia Agricultura Sinica., 37,* 801–807.
Zeng, Z. B., (1993). Theoretical basis of precision mapping of quantitative trait loci. *Proc.
    Natl. Acad. Sci. USA, 90,* 10972–10976.

CHAPTER 8

IMPORTANCE OF GENOMIC SELECTION IN CROP IMPROVEMENT AND FUTURE PROSPECTS

SATISH KUMAR, SENTHILKUMAR K. MUTHUSAMY,
CHANDRA N. MISHRA, VIKAS GUPTA, and KARNAM VENKATESH

*Indian Institute of Wheat and Barley Research, Karnal,
Haryana – 132001, India, E-mail: kumarsatish227@gmail.com*

CONTENTS

8.1   INTRODUCTION

The use of marker-assisted selection (MAS) in plant breeding has continued to increase in the public and private sectors. However, the use of MAS has stagnated for the improvement of most of the quantitative traits. Biparental mating designs for the detection of loci affecting these traits impede their application, and the statistical methods used are ill-suited to these traits, which are polygenic in nature. Most applications of MAS have been constrained to simple, monogenic traits. MAS has been used as a strategic method in backcrossing of major genes into elite varieties. Backcrossing, on the other hand, is most conservative breeding method because it focuses on pyramiding of only a few target genes. Gene pyramiding for quantitative traits controlled by many small additive effects at a large number of loci is rendered inefficient. Current MAS methods are better suited for traits governed by a few major genes. Most of the economically important traits such as yield are governed by polygenes and are crucial for the success of a crop varieties. It may also be understood that the statistical methods used to identify target loci and implement MAS have been inadequate for improving polygenic traits controlled by many loci having small additive effects.

It has been predicted for more than two decades that molecular marker technology would reshape breeding programs and facilitate rapid gains from selection. Genomic selection (GS) proposed by Meuwissen in 2001 does have a potential to emerge as a solution to the limitations of traditional MAS. GS uses high marker densities on a large breeding population for improving the traits of importance. GS provides us with an exciting opportunity that could revolutionize crop improvement strategies in future. Genomic selection predicts the breeding values of lines in a population by analyzing their phenotypes and high-density marker scores. It looks to fulfill the drawbacks of MAS by using genome-wide marker coverage to accurately estimate breeding values, accelerate the breeding cycle, and introduce greater flexibility in the relationship between phenotypic evaluation and selection. A key to the success of GS is that it incorporates all marker information in the prediction model, thereby avoiding biased marker effect estimates and capturing more of the variation due to small-effect QTL. The correlation between true breeding value and the genomic estimated breeding value (GEBV) has been reported in some studies to be as high as 0.85 even for polygenic low heritability traits. This level of accuracy is sufficient to consider selecting for agronomic performance using marker information alone. Such selection would substantially accelerate the breeding cycle, enhancing gains per unit time. It would dramatically change the role of phenotyping, which would then serve to update prediction models and no longer to select lines.

For GS to be successful in crop improvement, it must shift from theory to practice. GS has almost exclusively been tested through simulation, and therefore, its potential value should be assessed with caution. The accuracy of GS and its cost effectiveness must now be evaluated in breeding programs to provide the empirical evidence needed to warrant the addition of GS to the plant breeding methods. While research to date shows the exceptional promise of GS, work remains to be done to validate it empirically and to incorporate it into breeding schemes.

In the next 30 years, the production of staple food grains like rice, wheat, maize, etc., must be doubled to cope up the demands of the ever-growing population (Godfray et al., 2010). The human-induced climate change resulted in increased instances of biotic and abiotic stresses that negatively influence the crop production (Jaggard et al., 2010; Powell et al., 2012). It is bit more challenging for breeders to develop a new line(s) to adopt for the unpredictable challenging environment in future while, at the same time, maintaining and improving yield and ensuring quality improvements over existing varieties. Traditionally, breeders relay mostly on pedigree methods and visual observations to screen/evaluate the breeding lines, and this process is extremely time consuming. On an average, it takes up to 10–12 years for elite lines to be developed and identified (Collard and Mackill, 2008). Knowing the breeding value of a breeding population is a prerequisite for a breeder to develop an elite variety. Introduction of molecular marker technology in breeding programs revolutionize the selection scheme in plant breeding programs. MAS has successfully utilized in breeding programs for the evaluation of breeding lines and/or introgression of few major genes in breeding programs (Francia et al., 2005; Collard and Mackill, 2008). However, the use of MAS has stagnated for the improvement of most of the quantitative traits (Xu and Crouch, 2008). Biparental mating designs for the detection of loci affecting these traits impede their application, and the statistical methods used are ill-suited to these traits, which are polygenic in nature (Xu and Crouch, 2008). Most applications of MAS have been constrained to simple, monogenic traits (Francia et al., 2005). MAS been used as a strategic method in backcrossing of major genes into elite varieties. Backcrossing, on the other hand, is the most conservative breeding method because it focuses on pyramiding of only a few target genes. Gene pyramiding for quantitative traits controlled by many small additive effects at a large number of loci is rendered inefficient (Collard and Mackill, 2008; Xu and Crouch, 2008). Current MAS methods are better suited for traits governed by a few major genes. Most of the economically important traits such as yield are governed by polygenes and are crucial for the success of a crop varieties. It may also be understood that the statistical methods used to identify target loci and implement MAS have been inadequate for improving polygenic traits controlled by many loci having small additive effects (Spindel et al., 2015). It has been predicted for more than two decades that molecular marker technology would reshape breeding programs and facilitate rapid gains from selection (Xu and Crouch, 2008). The genomic selection (GS) model proposed by Meuwissen et al. (2001) have a potential to emerge as a solution to the limitations of traditional MAS (Windhausen et al., 2012; Lado et al., 2013; Crossa et al., 2014; Spindel et al., 2015).

Genomic selection (GS) is emerging as a promising molecular breeding tool for crop plants and has potential to minimize the negative effect of complex traits in estimating the breeding value. GS is a new molecular breeding method in which the genome-wide markers are utilized to estimate the whole genome genetic merit of individuals of a breeding population based on the statistical prediction model already developed using training population (TP). Thus, by using GS, the GEBV of the breeding line(s) can be estimated using genome-wide markers before the phenotype is measured. Several studies have shown that the genetic gain obtained from the GS approach is far better than the conventional breeding schemes, and GS will be potentially exploited in crop improvement programs (Poland et al., 2012; Lopez-Cruz et al., 2015; Spindel et al., 2015; He et al., 2016). Introduction of GS in hybrid breeding increases the efficiency of selection of right parental crosses combinations for genetic gain enhancement. GS differs from MAS and association breeding in that the underlying genetic control and biological function is not known. The most important advantages are reductions in the length of the selection cycle, resulting in greater genetic gain per year.

Three major steps are involved in GS (i) presumption of the genotype-phenotype relationships in a training population, (ii) prediction of GEBV in a target population (breeding population) based on inference of TP (Training Population), and (iii) selection of breeding line(s) based on their GEBV. Unlike MAS, no prior knowledge of an association between the markers and the trait of interest is required for GS. In GS, genome-wide markers are utilized, so that all QTLs are likely to be in LD (Linkage Disequilibrium) with at least one SNP. The GEBV of a breeding line is predicted using statistical models with enough accuracy to allow the selection of individuals based on that prediction alone. Figure 8.1 depicts the methodology for the utilization of GS in breeding programs:

TRAINING POPULATION   BREEDING POPULATION

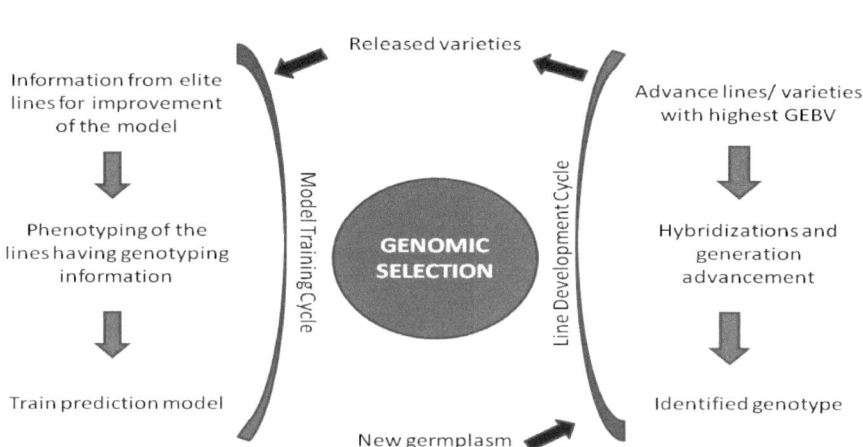

FIGURE 8.1   Genomic selection model for a breeding program (Adapted from Heffner et al., 2009).

Several studies showed that GS can greatly accelerate the breeding cycle while using marker(s) information to maintain genetic diversity and potentially prolong gain beyond that possible with phenotypic selection (Heffner et al., 2011b). Thus, GS is known to perfectly suit for breeding highly polygenic traits such as resource use efficiency, yield, and stress tolerance. Here, we discuss the various statistical models utilized in GS programs, the status of GS in crop plants, and its effective utilization in breeding programs.

8.2   GENOMICS TOOLS FOR GS STUDIES

Recent breakthrough in high-throughput genomic technologies enabled the process of sequencing of the genomes and genotyping of the breeding lines to be faster and highly cost-effective (Varshney et al., 2014). The genomic sequence information of the major food crops have been deciphered and populated in the publicly available genomics database (Table 8.1). In addition to these, custom-made gene chips are available for genotyping. These genomic resources can be effectively utilized for high resolution-genotyping and identification of marker-trait associations for GS studies (Figure 8.2).

TABLE 8.1 Examples of Genetic and Genomic Resources Available for Crop Plants

Crop	Genome Size	Database name	Genetic and genomic resources	Url	Reference
Rice	389 Mb	Gramene	Rice genetic and genomic resources	http://archive.gramene.org/Oryza_sativa/Info/Index	International Rice Genome Sequencing Project, 2005; Kawahara et al., 2013
		Ensembl Plants		http://plants.ensembl.org/Oryza_sativa/Info/Index	
		MSU Rice Genome Annotation Project		http://rice.plantbiology.msu.edu/	
		Phytozome		https://phytozome.jgi.doe.gov/pz/portal.html	
Wheat	17 Gb	Ensembl Plants	Wheat genetic and genomic resources	http://plants.ensembl.org/Triticum_aestivum/Info/Index	Mayer et al., 2014
		Gramene		http://archive.gramene.org/Triticum_aestivum/Info/Index	
		Phytozome		https://phytozome.jgi.doe.gov/pz/portal.html	
Barley	5.1 Gb	Gramene	Barley genetic and genomic resources	http://archive.gramene.org/Hordeum_vulgare/Info/Index	Mayer et al., 2012
		Ensembl Plants		http://plants.ensembl.org/Hordeum_vulgare/Info/Index	
		Phytozome		https://phytozome.jgi.doe.gov/pz/portal.html	
Maize	2.5 Gb	Gramene	Maize genetic and genomic resources	http://archive.gramene.org/Zea_mays/Info/Index	Schnable et al., 2009

		Ensembl Plants		http://plants.ensembl.org/Zea_mays/Info/Index	
		MaizeGDB		http://www.maizegdb.org/	
		Phytozome		https://phytozome.jgi.doe.gov/pz/portal.html	
Sorghum	730 Mb	Gramene	Sorghum genetic and genomic resources	http://archive.gramene.org/Sorghum_bicolor/Info/Index	Paterson et al., 2009
		Ensembl Plants		http://plants.ensembl.org/Sorghum_bicolor/Info/Index	
		Phytozome		https://phytozome.jgi.doe.gov/pz/portal.html	

Genetic resources:

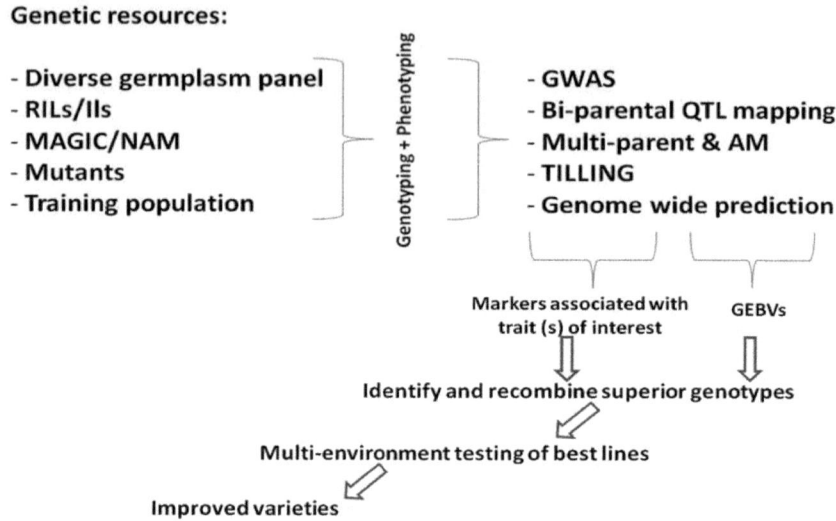

FIGURE 8.2   Role of new genetic/genomic tolls for utilization on genomic selection (modified from Varshney et al., 2014).

8.3   STATISTICAL METHODS

Statistical methods have been utilized in plant breeding programs to ensure the validity and reproducibility of the breeding research. The prediction accuracy of several statistical models depends upon the treating of marker effects and assumption(s) of the variations of complex traits. Several models have been proposed for GS (Desta and Ortiz, 2014). However, to date, no single model uniformly outperforms the other GS models across all traits. The selection of models for GS studies should depend on the objectives of our breeding programs (Desta and Ortiz, 2014). For GS studies, validation needs to be done with combination of models. The salient features of the various models are discussed below.

8.3.1   BEST LINEAR UNBIASED PREDICTION

Random regression-best linear unbiased prediction (RR-BLUP) was first proposed for biparental mapping population (Whittaker et al., 2000). This model assumes equal variances for all the markers with small but zero-effect (Heffner et al., 2011b). The realized-relation matrixes were computed based on the marker data. This model allows markers to have uneven effects, relates homogeneous shrinkage of predictors toward zero, and only few marker loci of QTLs are in LD (Whittaker et al., 2000; Heffner et al., 2011b; Desta and Ortiz, 2014). The RR-BLUP model has been successfully exploited in many crop improvement GS studies (Table 8.2). The genomic-best linear unbiased prediction (G-BLUP) developed by modifying the BLUP method. In G-BLUP, the genomic information is included in the form of a genomic relationship matrix that outlines the additive genetic covariance among breeding lines (Clark and van der Werf, 2013). Further, combining fixed variables of selected markers having fixed effects using genome-wide-association study (GWAS) on the RR-BLUP model increases the prediction accuracy (Spindel et al., 2016).

8.3.2 LEAST ABSOLUTE SHRINKAGE AND SELECTION OPERATOR (LASSO) AND ELASTIC NET (EN)

Unlike RR-BLUP, least absolute shrinkage and selection operator (LASSO) includes both variable selection and shrinkage methods (Meier et al., 2008; Friedman et al., 2010). RR-BLUP outperforms LASSO when two or more predictor variables are in multicollinearity condition (Li and Sillanpää, 2012). Bayesian version of LASSO (Bayesian LASSO) has an exponential prior on marker variance, which produces a double exponential (DE) distribution, and it does not incorporate the regression on covariates other than the markers for which a different shrinkage approach might be desired (Park and Casella, 2008; Li and Sillanpää, 2012). Thus, Bayesian LASSO, not LASSO, is desirable in dense marker-based regression studies (De Los Campos et al., 2009). Elastic net (EN) is highly useful when the number of observations (n) is much lesser than the number of predictors (p), and it outperforms LASSO when strongly correlated predictors present in the study (Zou and Hastie, 2005; Zou and Zhang, 2009; Friedman et al., 2010).

8.3.3 BAYESIAN METHODS

Bayes A model employs an inverted chi-square distribution on marker variances that results in a scaled t-distribution pattern for marker values. It

TABLE 8.2 Examples of Genomic Selection in Crop Plants

S. No.	Crop	Breeding material	Parameters studied	Prediction Model (s) utilized	Reference
1	Rice	278 hybrids	Yield, number of tillers per plant, number of grains per panicle, 1000 grain weight	Genome-BLUP LASSO SSVS	Xu et al., 2014
2	Rice	363 elite breeding lines	Trait's genetic architecture, Training population size, Marker number	RR-BLUP Bayesian LASSO RKHS RF	Spindel et al., 2015
3	Rice	Four inter-related synthetic populations (343 lines)	Training population, Marker density in relation with LD and Minor allele frequency (MAF), Relatedness between the training and breeding population	G-BLUP RR-BLUP LASSO Bayesian LASSO Bayesian ridge regression (BRR)	Grenier et al., 2015
4	Wheat	322 lines from 15 public and 3 private breeding programs	Fusarium head blight (FHB) resistance	RR Bayesian LASSO RKHS RF MLR	Rutkoski et al., 2012
5	Wheat	376 elite wheat varieties	Seven key traits	Mixed-linear modeling Elastic Net	Bentley et al., 2014
6	Wheat	2325 winter wheat lines	grain yield	RRBLUP BayesCπ RKHS EGBLUP	He et al., 2016
7	Wheat	254 advanced breeding lines	Days to heading Thousand-kernel weight Yield	RF Multivariate Normal Expectation Maximization Algorithm	Poland et al., 2012
8	Wheat	~800 advanced breeding lines	Grain yield	GBLUP	Lopez-Cruz et al., 2015

S. No.	Crop	Breeding material	Parameters studied	Prediction Model (s) utilized	Reference
9	Wheat	Six parents and the five wheat populations	Rust resistance	Bayesian LASSO RRBLUP SVR (Support Vector Regression – Linear and Nonlinear Kernels)	Ornella et al., 2012
10	Wheat	384 wheat (*Triticum aestivum*) genotypes	yield, thousand-kernel weight, number of kernels per spike heading date	GBLUP	Lado et al., 2013
11	Wheat	306 wheat lines and 599 wheat lines	Stem Rust	Bayesian LASSO GBLUP RKHS ayesian Regularized NN	Crossa et al., 2014
12	Wheat		Yield	GBLUP RRBLUP Bayesian Ridge Regression LASSO	Charmet and Storlie, 2012
13	Maize	255 diverse maize hybrids	grain yield, anthesis date, and anthesis-silking interval	RR-BLUP	Windhausen .. 2012
14	Maize	19 tropical maize biparental populations (3273 lines)	Anthesis day, grain yield and plant height	Random Effects Model	Zhang et al., 2015
15	Maize	two bi-parental populations comprising 249 and 250 F_2 test-cross individuals, respectively	Drought stress	Bayesian LASSO GBLUP RKHS NN	Crossa et al., 2014
16	Maize	788 testcross progenies	Grain moisture Grain yield	RR-BLUP	Zhao et al., 2012
17	Maize	930 testcross progenies	Grain moisture Grain yield	RR-BLUP	Zhao et al., 2013

TABLE 8.2 (Continued)

S. No.	Crop	Breeding material	Parameters studied	Prediction Model (s) utilized	Reference
18	Maize	479 inbreds	Grain yield Grain moisture Stalk lodging Root lodging	BLUP RR-BLUP	Massman et al., 2013
14	Barley	148 genotypes	12 malting quality traits	RR-BLUP	Schmidt et al., 2016
15	Barley	863 breeding lines	100 quantitative trait loci	Bayesian methods, multivariate regre sion methods RF SVM	Iwata and Jannink, 2011
16	Barley	647 lines	DON concentration, FHB resistance, Yield Plant height	RR–BLUP Gaussian kernel model Exponential kernel model Bayes Cπ	Sallam et al., 2015
17	Barley	42 two-row inbred lines of spring barley	20 or 80 additive QTLs	RR–BLUP Bayes-B Bayesian shrinkage regression method BLUP from a mixed model analysis	Zhong et al., 2009
18	Sor-ghum	300 sorghum accessions	Biomass yield, plant height, stem diam-eter, stalk number, stalk lodging, and root lodging		Yu et al., 2014

shrinks tiny marker effects toward zero and retains the larger marker effects (Meuwissen et al., 2001; De Los Campos et al., 2009). The Bayesian models Bayes B and Bayes C are similar to the Bayes A model, but these models apply both variable and shrinkage selection methods and consist of point of mass at zero (0) in their slab priors (Meuwissen et al., 2001; De Los Campos et al., 2009). Bayes C consists of gaussian distribution (De Los Campos et al., 2009; Kizilkaya et al., 2010; de los Campos et al., 2013). The new statistical method Bayes Cπ designed to alleviate the deficiency of Bayes A and Bayes B (Kizilkaya et al., 2010; Habier et al., 2011).

8.3.4 REPRODUCING KERNEL HILBERT SPACES (RKHS)

Reproducing Kernel Hilbert Spaces (RKHS) model was proposed by Gianola et al. (2006) and Gianola and van Kaam (2008) for estimating the GEBV by capturing non-additive effects. In this model, the distribution of QTL effects was regulated based on the kernel function and genetic distance with a smoothing ANOVA function (Gianola and van Kaam, 2008; De los Campos et al., 2010). Sun et al. (2012) proposed pRKHS model, which combines the features of RKHS and supervised principal component analysis (SPCA).

8.3.5 RANDOM FOREST (RF)

The random forest (RF) model uses regression model, and a number of trees are constructed by introducing two layers of randomness to improve the predictive accuracy (Breiman, 2001; Cordell, 2009; Holliday et al., 2012). The main advantage of the RF model is that it takes the averages of all the tree nodes to discover the best prediction model and records the interactions between marker loci (Breiman, 2001; Cordell, 2009).

8.4 STATUS OF GS IN CROP PLANTS

Most of the studies on genomic selection derived from different prediction models have been reported from animal systems. The breeding objectives and the programs are different as far as plant systems are considered. Availability of large population size and the efficiency of selection are quite high in plant systems. Hence, the applicability of prediction-based genomic selection seems low. However, recent advances in genomic resources and understanding of the statistics-based prediction models have led to the utilization of GS in plants, especially major crops. We review here the recent progress on GS in major cereal crops.

8.4.1 RICE

The genome size of rice is 389 Mb, and the availability of the genome sequence and other genetic and genomic resources has made rice a suitable crop for GS breeding programs (Table 8.2) (Yang et al., 2013). Spindel et al. (2015) studied the effect of various components influencing the prediction accuracy in rice genomic selection. Parameters like trait's genetic architecture, training population composition, number of markers, and prediction models were studied, and it was found that the genomic prediction models outperformed the prediction based on pedigree records alone; the usage of one marker every 0.2 cM is sufficient for genomic selection in rice, and combining GWAS and interpretation increases the prediction efficiency of genomic selection (Spindel et al., 2015). In another study, Grenier et al. (2015) investigated the effect of different parameters (trait genetic architecture, population structure, characteristics of the incidence matrix, and regression models) of GS in estimating the GEBV of four inter-related synthetic populations and found that the prediction accuracy of GEBV across all 540 cross-validation experiments with 100 replicates ranged from 0.12 to 0.54 with an average of 0.30. The breeding value of the hybrid and its performance were also successfully predicted using the GS approach in rice (Xu et al., 2014). By using the phenotypic data such as yield, number of tillers per plant, number of grains per panicle, and 1000 grain weight of 278 existing hybrids derived from 210 RIL parents, the genomic values of all 21,945 potential hybrids were successfully predicted (Xu et al., 2014). The average yield of the hybrids derived from top 100 crosses displayed 16% increase in average yield compared with all the potential hybrids (Xu et al., 2014). Combining fixed variables of selected markers with fixed effects using de novo GWAS with the GS model RR-BLUP increases the prediction accuracy and outperformed the models that made use of historical GWAS data (Spindel et al., 2016).

8.4.2   WHEAT

Wheat is a polyploid crop that consists of three homologous chromosomes and has a very large genome size of 17 Gb. Because of the complex nature of the genome, the availability of genomic resources was delayed in wheat (Brenchley et al., 2012; Mayer et al., 2014; Muthusamy et al., 2016). The heterozygous nature of three sub-genomes makes it more challenging for marker development and genetic association studies. With the advance in technologies and the combined efforts of International Wheat Genome Sequencing Consortium (IWGSC), the first chromosome-based draft of the genome was released on 2014 (Mayer et al., 2014). Studies have shown that GS can be done for complex traits, e.g., grain yield with the breeding lines having a population size of 250 (Poland et al., 2012) and 800 (Lopez-Cruz et al., 2015). The genetic gain obtained for a bi-parental population through GS was higher than that for both phenotypic selection and conventional MAS program (Heffner et al., 2011b). Using GS, Rutkoski et al. (2012) studied a quantitative trait, Fusarium head blight (FHB) resistance, which is difficult to screen and evaluate in the field. GS could accelerate FHB resistance breeding. They found that the use of genome-wide markers as well as information of correlated traits, QTLs, and the phenotypic data from large trails to train training population increase the prediction accuracy of GEBV of the breeding lines. The use of statistical models that can predict additive and epistatic effects increases the prediction accuracy up to 5 % (He et al., 2016). GS can be combined with association mapping to dissect the quantitative traits, and this approach is highly useful for polyploid crops like wheat (Bentley et al., 2014).

8.4.3   MAIZE

Maize is utilized as a staple food and animal feed and as a fuel. Global demand for maize is increasing, and huge efforts have been made for yield and quality improvement. The genome size of maize is 2.5 Gb (Schnable et al., 2009). Several attempts have been made by utilizing GS for yield and quality improvement in maize (Table 8.2). Zhao et al. (2012) analyzed the prediction accuracy of GS in 788 elite breeding lines; the prediction accuracy is higher for grain moisture content of 0.9, whereas the accuracy for grain yield is 0.58. The prediction accuracy of GS for grain yield was realized in the training population model having precession phenotyping. Studies have shown that GS can be effectively utilized in pre-breeding programs for introgression of agronomically important traits from wild relatives to breeding lines (Gorjanc et al., 2016) and also in hybrid development programs of maize (Windhausen et al., 2012).

8.4.4 BARLEY

Barley has a genome size of 5.1 Gb. Availability of whole genome sequence of barley and other genetic and genomic resources facilitate GS studies (Sreenivasulu et al., 2008; Mayer et al., 2012). Iwata and Jannink (2011) evaluated the accuracy of GS using 1325 SNPs among 863 breeding lines for 100 QTLs. Statistical models like Bayesian methods, multivariate regression methods, and machine learning methods were used for building the prediction models. Ridge regression and Bayesian methods were known to have higher predicting accuracies, and GS can be efficiently used for barley breeding programs (Iwata and Jannink, 2011). Zhong et al. (2009) evaluated the effect of marker density, level of LD, number of QTLs, sample size, and level of replication on the accuracy of GS by using four simulation models. Presence of trade-off among methods in capturing marker-QTL LD vs. marker-based relatedness of individuals was shown.

8.4.5 SORGHUM

Sorghum is an important food and fodder crop, now emerging as a potential biofuel crop (Paterson et al., 2009). The sorghum genome has been sequenced, and it has a genome size of 730 Mb; the genomic sequences and other genetic and genomic resources are accessible in public databases (Paterson et al., 2009; Katiyar et al., 2015). However, only fewer efforts have mean made to exploit GS for crop improvement in sorghum. Yu et al. (2014) analyzed 300 breeding lines using GS and estimated GEBV of about 700 accessions using GS with six phenotypic traits (biomass yield, plant height, stem diameter, stalk number, stalk lodging, and root lodging).

8.5 FACTORS AFFECTING GS PREDICTION ACCURACIES

The prediction accuracy of GEBV of the several statistical models depends upon the assumption(s) of the variations in complex traits. For GS, the choice of statistical model, population size, marker density, etc., should depend on the objective of study. Bayesian least absolute shrinkage, Bayesian LASSO, and RR models outperform the support vector regression for predicting GEBVs for traits controlled by additive gene effects (Ornella et al., 2012). Several studies have shown that the increase in training population and increase in marker density of the biparental crosses increase the prediction accuracy of GS (Zhong et al., 2009; Heffner et al., 2011a, 2011b; Zhao et al., 2012). In complex genomes like wheat, the genome-wide markers used for GS must be able to tag all the loci and the genotyping platforms must cover the whole genome, implying LD (Heffner et al., 2011a; Jonas and De Koning, 2013). The prediction accuracy of GS increases when the training populations have been phenotyped in multiple environments to record GXE (Heslot et al., 2013). However, the main factors affecting the prediction accuracy of GEBV are: (i) level and distribution of LD between markers and QTL, (ii) size of training population and relationship to the breeding population, and (iii) distribution of QTL effects, and iv) heritability of the trait.

8.6   FUTURE RESEARCH

In recent times, GS has emerged as a strong genomic tool for enhancing the efficiency and accuracy of the breeding programs. GS is a new technique that helps accelerate the rate of genetic gain in breeding by using whole-genome data to predict the breeding value of offspring. GS can be utilized in breeding new genotypes through effective utilization of phenotypic data/historical data by training the statistical models. The major advantage of using GS is that we can handle a large set of data and population size. Breeding crops for complex traits such as yield, nutrient use efficiency, abiotic and biotic stresses, etc., can be easily done using GS approaches (Rutkoski et al., 2010). More recently, a new model GS + de novo GWAS (Spindel et al., 2016) has been proposed; it is a two-part breeding design that can be used to efficiently integrate novel variation into elite breeding populations, thus expanding genetic diversity and enhancing the potential for sustainable productivity gains. Both marker density and the rates of LD decay across the genome strongly influence the prediction accuracy of GEBV. Genetic variance of the elite breeding lines can be effectively improved using GS (Gorjanc et al., 2016). Thus, GS can be effectively utilized in hybrid development programs for parental selection and in pre-breeding programs for introgression of agronomic traits from wild relatives to breeding lines (Windhausen et al., 2012; Xu et al., 2014; Gorjanc et al., 2016). Thus, GS offers an alternative to MAS and conventional phenotypic selection. GS has the capacity to use full-genome data to increase breeding efficiency.

KEYWORDS

- **disequilibrium**
- **GEBV**
- **genomic selection**
- **linkage**
- **MAS**
- **prediction models**
- **training population**

REFERENCES

Bentley, A. R., et al., (2014). Applying association mapping and genomic selection to the dissection of key traits in elite European wheat. *Theor. Appl. Genet., 127*(12), 2619–2633.

Breiman, L., (2001). Random forests. *Mach. Learn., 45*(1), 5–32.

Brenchley, R., et al., (2012). Analysis of the bread wheat genome using whole-genome shotgun sequencing. *Nature, 491*(7426), 705–710.

Charmet, G., & Storlie, E., (2012). Implementation of genome-wide selection in wheat. *Russ. J. Genet. Appl. Res., 2*(4), 298–303.

Clark, S. A., & Van der Werf, J., (2013). Genomic best linear unbiased prediction (gBLUP) for the estimation of genomic breeding values. *Methods Mol. Biol., 1019*, 321–330.

Collard, B. C. Y., & Mackill, D. J., (2008). Marker-assisted selection: an approach for precision plant breeding in the twenty-first century. *Philos. Trans. R. Soc. Lond. B. Biol. Sci., 363*(1491), 557–572.

Cordell, H. J., (2009). Detecting gene-gene interactions that underlie human diseases. *Nat. Rev. Genet., 10*(6), 392–404.

Crossa, J., et al., (2014). Genomic prediction in CIMMYT maize and wheat breeding programs. *Heredity (Edinb), 112*(1), 48–60.

De Los Campos, G., et al., (2009). Predicting quantitative traits with regression models for dense molecular markers and pedigree. *Genetics, 182*(1), 375–385.

De los Campos, G., et al., (2010). Semi-parametric genomic-enabled prediction of genetic values using reproducing kernel Hilbert spaces methods. *Genet. Res. (Camb), 92*(4), 295–308.

De los Campos, G., et al., (2013). Whole-genome regression and prediction methods applied to plant and animal breeding. *Genetics, 193*(2), 327–345.   .

Desta, Z. A., & Ortiz, R., (2014). Genomic selection: Genome-wide prediction in plant improvement. *Trends in Plant Sci., 19*(9), 592–601.

Francia, E., et al., (2005). Marker assisted selection in crop plants. *Plant Cell. Tissue Organ Cult., 82*(3), 317–342.

Friedman, J., Hastie, T., & Tibshirani, R., (2010). Regularization paths for generalized linear models via coordinate descent. *J. Stat. Softw., 33*(1), 1–22.

Gianola, D., & Van Kaam, J. B. C. H. M., (2008). Reproducing kernel hilbert spaces regression methods for genomic assisted prediction of quantitative traits. *Genetics, 178*(4), 2289–2303.

Gianola, D., Fernando, R. L., & Stella, A., (2006). Genomic-assisted prediction of genetic value with semiparametric procedures, *Genetics, 173*(3), 1761–1776.

Godfray, H. C. J., et al., (2010). Food security: the challenge of feeding 9 billion people. *Science, 327,* 812–818.

Gorjanc, G., et al., (2016). Initiating maize pre-breeding programs using genomic selection to harness polygenic variation from landrace populations. *BMC genomics, 17*(1), 30.

Grenier, C., et al., (2015). Accuracy of genomic selection in a rice synthetic population developed for recurrent selection breeding. *PLoS One, 10*(8), e0136594.

Habier, D., et al., (2011). Extension of the bayesian alphabet for genomic selection. *BMC Bioinformatics, 12,* 186.

Heffner, E.L., M.E. Sorrells, and J.L. Jannink. 2009. Genomic selection for crop improvement. Crop Sci. 49:1–12.

He, S., et al., (2016). Genomic selection in a commercial winter wheat population. *Theor. Appl. Genet., 129*(3), 641–651.

Heffner, E. L., et al., (2011). Genomic selection accuracy for grain quality traits in biparental wheat populations. *Crop Sci., 51*(6), 2597–2606.

Heffner, E. L., Jannink, J., & Sorrells, M. E., (2011). Genomic selection accuracy using multifamily prediction models in a wheat breeding program. *Plant Genome, 4*(1), 65–75.

Heslot, N., Jannink, J. L., & Sorrells, M. E., (2013). Using genomic prediction to characterize environments and optimize prediction accuracy in applied breeding data. *Crop Sci., 53*(3), 921–933.

Holliday, J. A., Wang, T., & Aitken, S., (2012). Predicting adaptive phenotypes from multilocus genotypes in Sitka spruce *Picea sitchensis*, using random forest. *G3 (Bethesda), 2*(9), 1085–1093.

International Rice Genome Sequencing Project, (2005). The map-based sequence of the rice genome. *Nature, 436*(7052), 793–800.

Iwata, H., & Jannink, J. L., (2011). Accuracy of genomic selection prediction in barley breeding programs: A simulation study based on the real single nucleotide polymorphism data of barley breeding lines. *Crop Sci., 51*(5), 1915–1927.

Jaggard, K. W., Qi, A., & Ober, E. S., (2010). Possible changes to arable crop yields by 2050. *Philos. Trans. R. Soc. Lond. B. Biol. Sci., 365*(1554), 2835–2851.

Jonas, E., & De Koning, D. J., (2013). Does genomic selection have a future in plant breeding? *Trends in Biotechnology, 31*(9), 497–504.

Katiyar, A., et al., (2015). Identification of novel drought-responsive microRNAs and trans-acting siRNAs from *Sorghum bicolor* L., Moench by high-throughput sequencing analysis. *Front. Plant Sci., 6*, 506.

Kawahara, Y., et al., (2013). Improvement of the *Oryza sativa* Nipponbare reference genome using next generation sequence and optical map data. *Rice, 6*(4), 1–10.

Kizilkaya, K., Fernando, R. L., & Garrick, D. J., (2010). Genomic prediction of simulated multibreed and purebred performance using observed fifty thousand single nucleotide polymorphism genotypes. *J. Anim. Sci., 88*(2), 544–551.

Lado, B., et al., (2013). Increased genomic prediction accuracy in wheat breeding through spatial adjustment of field trial data. *G3 (Bethesda), 3*(12), 2105–2114.

Li, Z., & Sillanpaa, M. J., (2012). Overview of LASSO-related penalized regression methods for quantitative trait mapping and genomic selection. *Theor. Appl. Genet. 125*(3), 419–435.

Lopez-Cruz, M., et al., (2015). Increased prediction accuracy in wheat breeding trials using a marker x environment interaction genomic selection model. *G3, 5*(4), 569–582.

Massman, J. M., et al., (2013). Genomewide predictions from maize single-cross data. *Theor. Appl. Genet., 126*(1), 13–22.

Mayer, K. F. X., et al., (2012). A physical, genetic and functional sequence assembly of the barley genome. *Nature, 491*(7426), 711–716.

Mayer, K. F. X., et al., (2014). A chromosome-based draft sequence of the hexaploid bread wheat *Triticum aestivum*, genome. *Science, 345*(6194), 1251788.

Meier, L., Van De Geer, S., & Buhlmann, P., (2008). The group lasso for logistic regression. *J. R. Stat. Soc. Ser. B. Statistical Methodol., 70*(1), 53–71.

Meuwissen, T. H. E., Hayes, B. J., & Goddard, M. E., (2001). Prediction of total genetic value using genome-wide dense marker maps. *Genetics, 157*(4), 1819–1829.

Muthusamy, S. K., et al., (2016). Differential regulation of genes coding for organelle and cytosolic *ClpATPases* under biotic and abiotic stresses in wheat. *Front. Plant Sci., 7*, 929.

Ornella, L., et al., (2012). Genomic prediction of genetic values for resistance to wheat rusts. *Plant Genome J., 5*(3), 136.

Park, T., & Casella, G., (2008). The Bayesian Lasso. *J. Am. Stat. Assoc., 103*(482), 681–686.

Paterson, A. H., et al., (2009). The Sorghum bicolor genome and the diversification of grasses. *Nature, 457*(7229), 551–556.

Poland, J., et al., (2012). Genomic selection in wheat breeding using genotyping-by-sequencing. *Plant Genome, 5*(3), 103–113.

Powell, N. A., et al., (2012). Yield stability for cereals in a changing climate. *Functional Plant Biol., 39*, 539–552.

Rutkoski, J. E., Heffner, E. L., & Sorrells, M. E., (2010). Genomic selection for durable stem rust resistance in wheat. *Euphytica, 179*(1), 161–173.

Rutkoski, J., et al., (2012). Evaluation of genomic prediction methods for fusarium head blight resistance in wheat. *Plant Genome J., 5*(2), 51.

Sallam, A. H., et al., (2015). Assessing genomic selection prediction accuracy in a dynamic barley breeding population. *Plant Genome, 8*(1), 1–15.

Schmidt, M., et al., (2016). Prediction of malting quality traits in barley based on genome-wide marker data to assess the potential of genomic selection. *Theor. Appl. Genet., 129*(2), 203–213.

Schnable, P. S., et al., (2009). The B73 maize genome: complexity, diversity, and dynamics. *Science, 326*(5956), 1112–1115.

Spindel, J. E., et al., (2016). Genome-wide prediction models that incorporate de novo GWAS are a powerful new tool for tropical rice improvement. *Heredity (Edinb), 116*(4), 395–408.

Spindel, J., et al., (2015). Genomic selection and association mapping in rice (*Oryza sativa*): effect of trait genetic architecture, training population composition, marker number and statistical model on accuracy of rice genomic selection in elite, tropical rice breeding lines. *PLoS Genet., 11*(2), e1004982.

Sreenivasulu, N., Graner, A., & Wobus, U., (2008). Barley genomics: An overview. *Int. J. Plant Genomics, 486258.*

Sun, X., Ma, P., & Mumm, R. H., (2012). Nonparametric method for genomics-based prediction of performance of quantitative traits involving epistasis in plant breeding. *PLoS One, 7*(11), e50604.

Varshney, R. K., Terauchi, R., & McCouch, S. R., (2014). Harvesting the promising fruits of genomics: Applying genome sequencing technologies to crop breeding. *PLoS Biol., 12*(6), 1–8.

Whittaker, J. C., Thompson, R., & Denham, M. C., (2000). Marker-assisted selection using ridge regression. *Genet. Res., 75*(2), 249–252.

Windhausen, V. S., et al., (2012). Effectiveness of genomic prediction of maize hybrid performance in different breeding populations and environments. *G3 (Bethesda), 2*(11), 1427–1436.

Xu, S., Zhu, D., & Zhang, Q., (2014). Predicting hybrid performance in rice using genomic best linear unbiased prediction. *Proc. Natl. Acad. Sci., 111*(34), 12456–12461.

Xu, Y., & Crouch, J. H., (2008). Marker-assisted selection in plant breeding: From publications to practice. *Crop Sci., 48*(2), 391–407.

Yang, Y., Li, Y., & Wu, C., (2013). Genomic resources for functional analyses of the rice genome. *Curr. Opin. Plant Biol., 16*(2), 157–163.

Yu, X., et al., (2014). Genomic selection of biomass traits in a global collection of 976 sorghum accessions. In: *Plant and Animal Genome XXII Conference Plant and Animal Genome*, San Diego, California, Poster No. 011.

Zhang, X., et al., (2015). Genomic prediction in biparental tropical maize populations in water-stressed and well-watered environments using low-density and GBS SNPs. *Heredity (Edinb), 114*(3), 291–299.

Zhao, Y., et al., (2012). Accuracy of genomic selection in European maize elite breeding populations. *Theor. Appl. Genet., 124*(4), 769–776.

Zhao, Y., et al., (2013). Choice of shrinkage parameter and prediction of genomic breeding values in elite maize breeding populations. *Plant Breed, 132*(1), 99–106.

Zhong, S., et al., (2009). Factors affecting accuracy from genomic selection in populations derived from multiple inbred lines: a Barley case study. *Genetics, 182*(1), 355–64.

Zou, H., & Hastie, T., (2005). Regularization and variable selection via the elastic net. *J. R. Stat. Soc. Ser. B Statistical Methodol., 67*(2), 301–320.

Zou, H., & Zhang, H. H., (2009). On the adaptive elastic-net with a diverging number of parameters. *Ann. Stat., 37*(4), 1733–1751.

PART III

MOLECULAR SELECTION TOOLS IN PLANT BREEDING

CHAPTER 9

APPLICATION AND ACHIEVEMENTS OF RECOMBINANT DNA TECHNOLOGY IN CROP IMPROVEMENT

RAKESH KUMAR, H. M. MAMRUTHA, KARNAM VENKATESH, KASHI NATH TIWARI, and SATISH KUMAR

ICAR–Indian Institute of Wheat and Barley Research (IIWBR), Karnal – 132001, India, E-mail: kumarsatish227@gmail.com

CONTENTS

9.1   INTRODUCTION

The world food production has increased substantially over the last few decades. However, a substantial increase in the world population constantly increases the demand for food. The number of undernourished people in the world has increased in spite of lower proportion of such population. By 2050, the world population growth is estimated to increase by more than 3 billion people, hence increasing the food demand. The intensive nature of world agriculture has put increased pressure of the inputs, which are ever diminishing. The natural resources are limited, and their exploitation has led to adverse climatic phenomenon, leading to losses in crop yield worldwide.

In the last 3-4 decades, the field of genetic engineering has developed rapidly and has emerged as a potential tool for manipulating crop plants for betterment of mankind. The term recombinant DNA technology or *genetic engineering* is used to describe the process by which the genetic makeup of an organism can be altered using *recombinant DNA technology*. This involves the use of laboratory tools to manipulate the genetic architecture of the plant by inserting pieces of DNA from another organism that contain one or more genes of interest. By using conventional breeding methods, there is little or no guarantee of obtaining any particular gene combination, and there are also chances of linkage drag associated with the introgression. Genetic engineering allows the direct transfer of one or only a few genes of interest, between either closely or distantly related organisms.

The global area under genetically modified or transgenic crops has been increasing consistently from 1996 to date. Herbicide-tolerant soybean, maize, cotton, and canola; insect-resistant maize, cotton, potato, and rice; and virus-resistant squash and papaya are successful examples of transgenic crops grown around the world. With genetic engineering, more than one trait can be incorporated into a plant without any undesirable trait. Herbicide-tolerant and insect-resistant maize and cotton are such examples.

9.2 TRANSFORMATION TECHNIQUES IN CEREALS

The gene transfer methods can be categorized into two types: indirect method, namely *Agrobacterium* mediated, and direct gene transfer methods like biolistic, electroporation, and microinjection. All these available methods have their own merits and demerits.

9.2.1 AGROBACTERIUM-MEDIATED TRANSFORMATION

About three decades ago, *Agrobacterium* emerged as a new engineering tool in the field of agricultural biotechnology to improve and study the plant genome. Currently, a number of agronomically important plant species are being modified using *Agrobacterium*-mediated transformation. *Agrobacterium* is a gram-negative soil bacterium that causes tumor in plants. Generally, *Agrobacterium* sp. causes tumor at the site of injured tissue or at the junction between the root and the shoot. *Agrobacterium* has the ability to transfer its DNA segment (T-DNA) from itself to plant genome, and *Agrobacterium tumefaciens* is the most commonly studied species in this genus. Tumors are incited by the conjugative transfer of a DNA segment from the bacterial tumor inducing (Ti) plasmid and causes crown gall disease. The second most studied species *A. rhizogenes* induces root tumors and carries a distinct root-inducing (Ri) plasmid. Other closely related species includes *A. radiobacter*, which is an "avirulent" species; *A. rubi*, which causes cane gall disease; and *A. vitis*, which is restricted to grapevines (Gelvin et al., 2003).

9.2.2   CALLUS-MEDIATED METHOD

In most of the *Agrobacterium*-mediated transformation protocols, callus was used as the target material for *Agrobacterium* infection. But for successful transformation, a robust regeneration protocol is required. Callus is a cluster of undifferentiated cells that is capable to regenerate into any part of plant or into whole plant. The whole process consists of a number of steps discussed in the following subsections.

9.2.2.1   Genotype and Explant Selection

A number of different explants are commonly used in different crops such as mature embryos, immature embryos, leaf, stem etc. The selection of genotype and explant highly depends upon its regeneration capacity and crop species. In cereal crops, *Agrobacterium*-mediated transformation has been achieved mainly in genotypes selected for their good response in tissue culture, for example, in rice japonica (Hiei et al., 1994), javanica (Dong et al., 1996), and indica (Rashid et al., 1996); in wheat *Triticum aestivum* cv. bobwhite (Cheng et al., 1997; Han et al., 2012), Chinese spring (Langridge et al., 1992; Agarwal et al., 2009), HD2329 (Patnaik et al., 2006; Chugh et al., 2012), and CPAN1676 (Chugh and Khurana, 2003; Patnaik et al., 2006; Chauhan and Khurana, 2011), in maize cv. inbred line A188 (Ishida et al., 1996), Hi-II (Zhao et al., 2001; Frame et al., 2002; Vega et al., 2008) and inbred B104 (Frame et al., 2006); in barley cv. Golden Promise (Tingay et al., 1997), Dissa (Wu et al., 1998), and Schooner (Wang et al., 2001); in sorghum cv inbred P898012 (Casas et al., 1993; Zhao et al., 2000; Gurel et al., 2009), and PHI391 (Zhao et al., 2000), Tx430 (Gao et al., 2005a; Liu and Godwin, 2012), C401 (Gao et al., 2005a, 2005b), and Pioneer 8505 (Gao et al., 2005a, 2005b).

9.2.2.2 Sterilization and Preculture

The selected explants should be sterilized using different surface sterilizing agents such as sodium hypochlorite, hydrogen peroxide, mercuric chloride, or ethanol for different duration (Li et al., 2011; Kumar et al., 2017). The explants should be precultured for callus induction. The preculturing treatment can vary from few hours to 20 days depending on plant species and explant response toward callus induction media (Ding et al., 2009).

9.2.2.3 Inoculation

Agrobacterium strain containing plasmid with gene of interest will be grown overnight in liquid culture. The agro culture is generally prepared in half MS media containing 200–500 µM acetosyringone (phenolic derivative) for inoculation (Hensel et al., 2009; Chugh et al., 2012). The precultured explants are submerged with agro culture in dark for 15 minutes to 12 hours (Jones et al., 2005; Ding et al., 2009) with intermittent shaking. It is found that the dark environment during the inoculation step enhances transformation efficiency (Sahoo et al., 2011).

9.2.2.4 Co-Cultivation

After removing the excess agro culture, the explants are placed on co-cultivation media or on filter papers wet with liquid containing acetosyringone. These explants are generally placed in dark at 22–26°C for 1 to 3 days (Cheng et al., 1997; Supartana et al., 2006) depending on the crop.

9.2.2.5 Selection

The explants are sterilized with distilled water followed by washing with antibiotic solution. Generally, antibiotics like cefotaxime, timintin, carbenicillin, etc. are used to suppress/inhibit excess *Agrobacterium* growth. The use of antibiotics is highly dependent on the virulence of the strain. The selective agents are used to screen positive putative transformants from untransformed ones, which depends on the selection marker gene present in T-DNA. A number of selection marker genes are available to select transformants. The antibiotics and selective agents will remain present in subsequent media to avoid *Agrobacterium* regrowth and gene escaping problem, respectively. A number of scorable marker genes are available to cross-check the transformants by using specific substrate/conditions for the respective marker gene with distinct observable phenotype.

9.2.2.6   Regeneration

The positive transformants are further placed in regeneration media to obtain shoots and roots.

9.2.3   INPLANTA METHOD

In this method, seeds are allowed to germinate for 2 days at 20–24°C. Germinated seeds are punctured about 1–2 mm deep with the help of sterilized needle and incubated with agro culture for 15 minutes to 1 hour in dark. After removing excess culture, these seeds are further incubated on a wet filter paper in dark at 20–24°C for the next 1 to 3 days. The infected seeds are washed with distilled water, followed by the antibiotic solution to suppress/inhibit excess *Agrobacterium*. The washed seeds are placed in pots containing soil mix and allowed to grow. The T_1 seeds can be screened for transformants according to genes present in T-DNA.

A highly efficient semi-*in-planta* method has been described (Risacher et al., 2009). In this system, 16–18 days after anthesis, wheat tillers were harvested, and the immature seeds were inoculated with *Agrobacterium* strain EHA105 by a syringe injection. After 2–3 days of inoculation, the seeds were isolated from the ear and surface-sterilized. Immature embryos were dissected from the infected seeds and processed for callus induction and selection. An average of 5% transformation efficiency was achieved with this method in spring wheat NB1 (Risacher et al., 2009).

9.2.4   FLORAL-DIP METHOD

The inflorescence of plants is carefully submerged in *Agrobacterium* culture and co-cultured for 1 to 3 days. The floral-dip transformation has no requirement of tissue culture; hence, it is less labor intensive, but it has low transformation frequency (Rod-in et al., 2014). A number of cereal crops have been transformed using the floral-dip method including *Triticum aestivum* (Agarwal et al., 2009; Zale et al., 2009) and *Oryza sativa* (Dong et al., 2001; Rod-in et al., 2014).

9.2.5   AGROINFECTION METHOD

Introduction of a viral genome into the plant cells by placing it within the T-DNA of a Ti plasmid, and using the *Agrobacterium* cell containing this recombinant plasmid for co-culture with the plant cells is called agroinfection. The induction of virus resistance is done by utilizing virus-encoded genes-virus coat proteins, movement proteins, transmission proteins, satellite RNA, antisense RNAs, and ribozymes. The virus coat protein-mediated approach is the most successful one to provide virus resistance to plants. This strategy can be used with viruses that have to be transmitted by an insect vector for successful infection, e.g., geminiviruses. Agroinfection has been shown for at least two geminiviruses, viz., maize streak virus (MSV) (Grimsley et al., 1986) and wheat dwarf virus (WDV) (Hayes et al., 1988) and in rice to study rice tungro bacilliform virus (Dasgupta et al., 1991).

9.2.6   BIOLISTIC OR GENE GUN OR MICROPROJECTILE METHOD

In recent years, it has been shown that DNA delivery to plant cells is also possible, when heavy microparticles (tungsten or gold) coated with the DNA of interest are accelerated to a very high initial velocity (1400 ft per sec) (Li et al., 2011). This method is called as "Biolistic method." The biolistic (gene gun or microprojectile method) is widely used to transform cereal crops. The concept of biolistic method was first introduced by Sanford et al. (1987). Klein et al. (1987) evolved a method by which the delivery of DNA into cells of intact plant organs or cultured cells is done by a process called projectile bombardment. The micro-projectiles (small high-density particles) are accelerated to high velocity by a particle gun apparatus. These particles with high kinetic energy penetrate the cells and membranes and carry foreign DNA inside the bombarded cells.

The microprojectiles are of normally 1–3 μm in diameter and carried by a "bullet" and are accelerated into living plant cells (target cells can be pollen, cultured cells, cells in differentiated tissues, and meristems) so that they can penetrate cell walls of intact tissue. The acceleration is achieved either by an explosive charge (cordite explosion) or by using shock waves initiated by a high voltage electric discharge or by exerting high pressure using gas, e.g., helium. A number of influencing factors were listed by Li et al. (2011) that affect the transformation efficiency using the gene gun method: (a) biolistic transformation parameters such as rupture pressure, bombarding distance, DNA purity and concentration, $CaCl_2$ concentration, and spermidine concentration and (b) biological parameters such as explant types and physiological status, culture conditions before and after bombardment, screening procedures, and regeneration rates of putative transgenic plants. Transformed plants using the above technique have been obtained in all major cereal crops such as maize, rice, wheat, sorghum, barley, etc. In most cases, immature embryos derived from calli were used as target explants for this technique. After bombardment, the treated explants will be tissue cultured to regenerate transformants. But multiple gene insertion is the major drawback with this technique.

9.2.7 ELECTROPORATION

Electroporation is another efficient method for the incorporation of foreign DNA into protoplasts and thus for direct gene transfer into plants. This method was introduced by Fromm and his coworkers in 1986. This method is a technique that uses high-intensity short electrical impulses to create pores in the cell membrane, which increase the permeability of the protoplast membrane and facilitate the entry of DNA molecules into the cells, if the DNA is in direct contact with the membrane (Patnaik and Khurana, 2001). The electroporation pulse is generated by discharging a capacitor across the electrodes in a specially designed electroporation chamber. Either a high voltage (1.5 kV) rectangular wave pulse of short duration or a low voltage (350 V) pulse of long duration is used. Protoplasts in an ionic solution with vector DNA are suspended between the electrodes, electroporated, and then cultured for development of transformants.

Using the electroporation method, successful transfer of genes was achieved with the protoplasts of maize, rice, wheat, and sorghum. In most of the cases, the *cat* gene associated with a suitable promoter sequence was transferred. Transformation frequencies can be further improved by (i) using field strength of 1.25 kV/cm, (ii) adding PEG after adding DNA, (iii) heat shock treatment of protoplasts at 45°C for 5 minutes before adding DNA, and (iv) by using linear instead of circular DNA. A number of other factors also affect efficient electroporation such as electroporation voltage, pulse length, electroporation buffer volume, medium, pre-electroporation medium osmoticum, and incubation time and temperature (Patnaik and Khurana, 2001).

In addition to the above-described methods, a number of other methods are also available with very low transformation efficiency and time-consuming. These methods include polyethylene glycol (PEG)-mediated DNA uptake (Uchimiya et al., 1986; Datta et al., 1990), silicon carbide fibers (Kaeppler et al., 1992; Frame et al., 1994) and micro-targeting (Sautter et al., 1993).

9.3 CURRENT STATUS OF TRANSGENICS IN CEREAL CROPS

Cereals are one of the most versatile foods in the world, but their quality and yield have been decreased due to diseases, insect-pests, and suboptimal environmental conditions. Genetic modification has been used via different transformation methods in cereal crops to study gene function and improving agricultural traits to gain resistance to both biotic and abiotic stresses. The success stories in gene transformation in important cereals like rice, maize, wheat, sorghum, and barley are discussed below.

9.3.1 *RICE*

The first *Agrobacterium*-mediated transformation in *Oryza sativa* L. cv. Fujisaka 5 was reported by Raineri et al. (1990). They demonstrated kanamycin resistance and *gus* gene activity in transformed calli using supervirulent strain A281 (pTiBo542). Chan et al. (1993) first reported successful inheritance of chimeric genes of β-glucuronidase (*uidA*) and neomycin phosphotransferase (*nptII*) in a japonica type of rice (*Oryza sativa* L. cv. Tainung 62) using the *Agrobacterium*-mediated gene transfer system. They expressed the *uidA* reporter gene driven by the α-amylase promoter. A number of researchers successfully transformed different rice varieties such as japonica (Hiei et al., 1994), javanica (Dong et al., 1996), and indica (Rashid et al., 1996). A number of studies were carried out to improve the existing transformation method in rice (Aldemita and Hodges, 1996; Toki, 1997; Cheng et al., 1998; Khanna and Raina, 1999; Mohanty et al., 1999, 2000; Jiang et al., 2000; Pons et al., 2000; Upadhyaya et al., 2000; Lin and Zhang, 2004; Rachmawati et al., 2004; Hiei and Komari, 2006; Nishimura et al., 2006; Shri et al., 2013).

The first insect-resistant transgenic rice plant with *Bt* (*Bacillus thuringiensis*) toxin was reported by Fujimoto et al. (1993). Further investigation proved that Bt rice was resistant to rice leaf folder (*Cnaphalocrocis medinalis*, RLF), striped stemborer (*Chilo suppressalis*), and yellow stemborer (*Scirpophaga incertulas*) (Nayak et al., 1997; Cheng et al., 1998; Shu et al., 2000; Tu et al., 2000; Ye et al., 2001, 2003; Ramesh et al., 2004; Bashir et al., 2005). In a different study, rice was transformed with two synthetic *cryIA(b)* and *cryIA(c)* coding sequences from *B. thuringiensis* to generate striped stem borer- and yellow stem borer-resistant plant using *Agrobacterium*-mediated transformation (Cheng et al., 1998). Similarly, *BT* (*cryIA(b)* and *cryIA(c)*) along with hygromycin resistance (*hptII*) and β-glucuronidase (*uidA*) genes were introduced in indica rice Basmati 370 variety via the *Agrobacterium*-mediated system (Ahmad et al., 2002). A rice cultivar IR64 was genetically engineered with a fully modified (plant codon optimized) synthetic *cryIC* coding sequences that confer resistance to the RLF *C. medinalis*, which is an important pest of rice that causes severe damage in many areas of the world, along with the marker genes *hptII* and *uidA* (Ignacimuthu and Raveendar, 2011). In a recent study, two *B. thuringiensis* insecticidal genes, *CryIA(c)* and *CryII(g)*, and a modified glyphosate-tolerant 5-enolpyruvylshikimate-3-phosphate synthase (*EPSPS*) gene (*G10*) were combined into a single transferred DNA (T-DNA) fragment and introduced into rice through *Agrobacterium*-mediated transformation. A transgenic line named GAI-14 with single-copy T-DNA insertion was found to be highly resistant to striped stem borer and RLF and tolerant to glyphosate (Zhao et al., 2015).

A high-yielding *indica* variety IR72 was transformed with the bacterial blight resistance gene *Xa21* through transformation (Tu et al., 1998), which showed excellent performance in field experiment (Tu et al., 2000). To provide protection against yellow stem borer and stripped stem borer, the cowpea trypsin inhibitor gene was integrated in *japonica* rice (Xu et al., 1996). The viral capsid protein genes were reported to be effective against virus in transgenic rice (Hayakawa et al., 1992; Sivamani et al., 1999). In transgenic rice, the constitutive expression of RsAFP2 enhanced inhibition to *Magnaporthe oryzae* and *Rhizoctonia solani* (Jha and Chattoo, 2010). This study suggested that the transgene expression of *Rs-AFP2* was not accompanied by an induction of pathogenesis-related (PR) gene expression, the expression of *Rs-AFP2* directly inhibits the pathogens. Other genes such as spider venom, agglutinin, and snow drop lectin (GNA) also showed resistance to RLF, striped stem borer, aphids, and rice brown planthopper, respectively (Qiu et al., 2001; Xiaofen et al., 2001; Khan et al., 2006; Ye et al., 2009a, 2009b).

The hairpin RNA (*hpRNA*) T-DNA constructs containing gene-specific tags of the phytoene desaturase (*OsPDS*) and the SLENDER 1 (*OsSLR1*) genes were introduced in rice (*O. sativa* L.) cv. japonica and *indica* to trigger RNAi of target genes using the leaf agroinfection method (Andrieu et al., 2012). Folate is produced in a multistep process from a pterin ring, p-aminobenzoate (pABA) and glutamate residues in plants. Rice plants transgenic for wheat 6-hydroxymethyl-7,8-dihydropterin pyrophosphokinase/7,8-dihydropteroate synthase (*HPPK/DHPS*), which operates at a central point in the production pathway, gives elevated folate levels (Gillies et al., 2008). A wheat *Glu-Dy10* gene encoding the high-molecular-weight glutenin subunits (*HMG-GS*) from the Korean wheat cultivar "Jokyeong" was transferred in rice using *Agrobacterium*-mediated co-transformation for increasing quality processing of bread and noodles (Park et al., 2014).

Signaling pathways induced in response to various abiotic stresses are very complex and involve the coordinated action of many genes (Sikdar et al., 2015). In most of the eukaryotic cells, intracellular calcium ion (Ca^{2+}) concentration and a second messenger were produced in response to most abiotic stresses (Hofer and Brown, 2003; Kolukisaoglu et al., 2004). This indicates that there are some proteins that sense changes in Ca^{2+} levels and may play an important role in stress signaling. In plants, many calcium-based signaling molecules have been identified, which include calmodulin (CaM) and calmodulin-related proteins (Luan et al., 2002), Ca^{2+}-dependent protein kinases (CDPKs) (Pandey et al., 2000; Sanders et al., 2002), calcineurin B-like (CBL) protein, and CBL-interacting protein kinases (CIPK) (Cheong et al., 2003; Mahajan et al., 2006). Different isoforms of CBL were shown to be unregulated upon exposure to stress conditions. CBL specifically interacts with CIPK and regulates its kinase activity (Batistic and Kudla, 2009). The N- and C-terminal domains of the CIPK participate in salt-tolerant calcium signaling mechanism. Kinase activity of CIPK remains inactive under normal conditions due to the activity of the FISL motif, which functions as an auto-inhibitory domain (Xiang et al., 2007). The transgenic rice overexpressing *OsCBL8* found tolerant at 150 mM NaCl stress treatment (Gu et al., 2008). In another study, it was found that the *PsCBL* and *PsCIPK* genes are upregulated in pea during stress response to cold, salinity, or wounding (Mahajan et al., 2006). *Agrobacterium*-mediated transformation was applied to transfer the *Pisum sativum CBL* (*PsCBL*) and *P. sativum CIPK* (*PsCIPK*) genes for the development of salt-tolerant rice (*O. sativa* L.) (Sikdar et al., 2015). In a comparative study, the leaf Na+ content between wildtype rice and transgenic rice expressing a vacuolar Na^+/H^+ antiporter from *Atriplex gmelini* showed that the transgenic line accumulates a similar level of Na^+ to control plants after 3 days of salt stress (Ohta et al., 2002). A similar approach was implemented to create salt-tolerant rice by using *OsNHX1* (Fukuda et al., 2004). A rice strain was genetically modified with the barley (*Hordeum vulgare*) C-repeat binding factor 4 (*HvCBF4*) gene, resulting in improved tolerance to drought, high salinity, and low temperature stresses without affecting plant growth (Oh et al., 2007). Similarly, the overexpression of the late embryogenesis abundant 3-1 (*OsLEA3-1*) gene in *O. sativa* improved grain yield under drought conditions (Xiao et al., 2007). The transgenic rice transformed with the *Escherichia coli catalase* (*katE*) gene was found to increase salt tolerance trait (Prodhan et al., 2008; Motohashi et al., 2010). In the same way, the overexpression of *O. sativa* trehalose 6- phosphate synthase 1 (*OsTPS1*) gene enhanced abiotic stress tolerance in rice (Li et al., 2011a).

There are many abiotic stress upregulated transcription factors that have been acknowledged so far from various plant species including rice, wheat, and barley (Robertson, 2004; Xue et al., 2006; Fang et al., 2008; Gao et al., 2010; Xia et al., 2010). Among these, the NAC transcription factor family have been widely studied in rice, which participates in the regulation of a number of abiotic stresses especially drought, cold, and salt stress (Hu et al., 2006; Hu et al., 2008; Nuruzzaman et al., 2010). It has been recognized that some abiotic stress that upregulated members of the NAC family have an adverse effect on plant growth when they are constitutively overexpressed. The overexpression of an abiotic stress-upregulated rice NAC gene, OsNAC6, showed an increase in the transcript levels of a rice ZIM family gene (Os03g0180900) (Nakashima et al., 2007). In other study, the grain yield was improved in transgenic rice overexpressing a drought-upregulated NAC gene in water-limited environments (Hu et al., 2006; Jeong et al., 2010). Similarly, overexpression of a rice homolog (OsNAC10 or OsNAC5) of TaNAC69 in transgenic rice resulted in upregulation of chitinase (Jeong et al., 2010; Takasaki et al., 2010).

Overexpression of a rice nitrate transporter1 (NRT1)/peptide transporter (PTR) gene, OsPTR9, increases the recycling of nitrogen from leaves to panicles and improves both plant growth and yield in rice; knockdown of the expression of this transporter has the opposite effect (Fang et al., 2013). It has been demonstrated that the rate of ammonium uptake enhanced in transgenic rice overexpressing a high-affinity AMT1 transporter, which promotes higher expression levels of genes in the nitrogen assimilation pathway. It also improves grain yield under both suboptimal and optimal nitrogen conditions in transgenic lines as compared with wild-type plants (Ranathunge et al., 2014).

A construct, rolC::Hd3a-GFP, was introduced in black rice cv. Cempo Ireng to improve the rice productivity (Purwestri et al., 2015). Hd3a (Heading date 3a) was identified by quantitative trait locus (QTL), which showed flowering in rice under short-day conditions. An earlier study showed that Hd3a expressed under the control of rolC promoter in rice cv. japonica exhibited early flowering phenotype (Tamaki et al., 2007). A study demonstrated the regulation of plant height and yield potential under optimal growth conditions for transgenic rice plants overexpressing the rice gene encoding the HAP3H subunit (nuclear factor YB), DTH8 [QTL]) for days to heading on chromosome 8], which appears to play an important role as a suppressor in the signal network of photoperiodic flowering. DTH8 delayed flowering of transgenic rice by negatively influencing the expression of the Ehd1 and Hd3a genes under long-day conditions (Wei et al., 2010). A number of reports have been published regarding the correlation of plant growth rate, cell wall modification, and phenotype changes with the regulation of expansin gene expression or heterologous overexpression in rice (Lee and Kende, 1997; Choi et al., 2003; Ma et al., 2013). These reports support the role of expansin as a major protein to induce cell wall extension, larger cells, taller plants, and longer roots. A correlation was found in rice expansin upregulation with root system and growth improvement (Ma et al., 2013), and antisense expression was found to cause stunting and early flowering (Choi et al., 2003).

Other reports are also available for agronomical important trait-specific transformation in rice such as drought tolerance (Xiao et al., 2009; Yokotani et al., 2009), salt stress tolerance (Rohila et al., 2002; Zheng et al., 2009; Zhang et al., 2010), insect-resistant rice (Ye et al., 2009), resistance to sheath blight disease (Molla et al., 2013), and development of rice having enhanced cytokinin level (Lin et al., 2002).

9.3.2   WHEAT

After rice, wheat is the most transformed cereal crop worldwide. Cheng et al. (1997) first reported wheat transformation with the marker gene *uidA* using *A. tumefaciens*. Gopalakrishnan et al. (2000) has established a highly efficient microprojectile bombardment method of wheat transformation in CPAN004, Sonalika, and UP2338 varieties of Indian wheat, which were successfully transformed with the *bar* gene. The mature embryos of *Triticum aestivum* and *T. durum* were transformed with LBA4404 agro strain and putative transformants were obtained for the *bar* gene (herbicide resistance) and proteinase inhibitor (*pin 2*) gene for insect resistance in wheat (Patnaik et al., 2006).

The particle bombardment method of gene transfer in *T. aestivum* (CPAN1676) and *T. durum* (PDW215) was done using mature embryos. The transgenics were developed with the *HVA1* gene, and transformation was confirmed under controlled conditions in T_0 plants (Patnaik and Khurana, 2003). The *AtCBF3* gene was transformed in HDR77-A wheat variety using the particle bombardment method and validated the putative T_1 plants under moisture stress conditions, which showed 8% yield advantage over wild-type plants under stress (Kasirajan et al., 2013). The microprojectile method was used to deliver the vector containing *AtMYB12* fused with GFP and GUS. It was evident that *AtMYB12* can be used as invaluable visible marker (Gao et al., 2011).

Most studies were centered on the overexpression of single sequences that encode antimicrobial proteins to develop fungal pathogen-resistant wheat (Patnaik and Khurana, 2000). Plants express a number of defense genes in response to microbial infection, and most of these genes belong to the so-called pathogenesis related (PR) group, such as glucanases and chitinases. Other antimicrobial peptides (AMPs) like defensins and thionins have shown important antibacterial and antifungal activities (Silverstein et al., 2007; Tavares et al., 2008). A transgenic wheat line harboring a chitinase and a β-1,3-glucanase delayed susceptibility to *Fusarium graminearum* compared with nontransgenic controls under controlled conditions, but was unsuccessful under field conditions (Anand et al., 2003). Similarly, the expression of defense response genes wheat α-1-purothionin, barley thaumatin-like protein 1, and barley β-1,3-glucanase in transgenic wheat showed improved resistance against *F. graminearum* in both greenhouse and field trials (Mackintosh et al., 2007). A barley chitinase gene was transferred in Bobwhite, which resulted in enhanced FHB resistance as compared to the non-transgenic control in the greenhouse (Shin et al., 2008). Another small cysteine-rich antifungal protein defensin *RsAFP2* from radish (*Raphanus sativus*) was transformed into Chinese wheat variety Yangmai 12 via biolistic bombardment and showed resistance against agronomically important fungal pathogens such as *F. graminearum* and *Rhizoctonia cerealis* (Li et al., 2011). Transgenic wheat lines constitutively expressing the *S. chacoense SN1* gene were challenged with *Blumeria graminis* f.sp. *tritici*, and enhanced resistance to the pathogen was observed. It delayed the development of the disease and reduced the infection compared with the wild-type variety ProINTA Federal. A correlation was observed between high resistance to the pathogen and a high level of snakin-1 transcripts in the plant (Faccio et al., 2011). In a study, a rice thaumatin-like protein-1 (*tlp-1*) was transferred in wheat, which resulted in enhanced FHB resistance during the early stages of disease progression in the greenhouse (Chen et al., 1999). A construct containing a bovine *lactoferrin* cDNA was used to transform wheat using an *Agrobacterium*-mediated DNA transfer system to express this antimicrobial protein in wheat (Han et al., 2012). The resulted transformants showed resistance against head blight caused by *F. graminearum* Schwabe, a devastating disease of wheat (*T. aestivum* L.) and barley (*Hordeum vulgare* L.) that reduces both grain yield and quality.

A number of reports evidenced that the subunits of NF-Y TFs were shown to be involved in development and stress responses in different cereal crops (Miyoshi et al., 2003; Nelson et al., 2007; Wei et al., 2010; Ito et al., 2011; Stephenson et al., 2011; Qu et al., 2015). Heterotrimeric nuclear factors Y (*NF-YA, NF-YB,* and *NF-YC*) are involved in the regulation of various vital functions in all eukaryotic organisms. It was anticipated that a large number of heterotrimeric combinations in plants provides significant flexibility for the association of *NF-Y* complexes in regulation of multiple plant developmental processes (Edwards et al., 1998). Recently, the total number of identified *NF-Y* genes in wheat reached 80. Transgenic wheat overexpressing TaNFYA-B1, a low nitrogen and low phosphorus inducible NFYA transcript factor (gene locus present on chromosome 6B), showed significant enhanced uptake of both nitrogen and phosphorus and higher grain yield under different nitrogen and phosphorus supplies in a field experiment. The increased nitrogen and phosphorus uptake may be due to overexpression of *TaNFYA-B1* that stimulate root development and upregulate the expression of both nitrate and phosphate transporters in roots (Qu et al., 2015). In an experiment, it was revealed that the activation of the *NF-Y* genes by drought is ABA independent in wheat, because no significant changes in expression of all three genes (*NF-YA, NF-YB,* and *NF-YC*) were induced by treatment with 200 µM ABA. Tillering in monocots is strictly associated with grain yield. However, knowledge of the physiology and genetics of tillering is not clear to date (Yadav et al., 2015). Excessive tillering in cereal crops can, however, lead to yield reduction because tillers compete for nutritional resources and most of secondary tillers are not fertile (Kebrom et al., 2012). In a recent study, the *TaNF-YB4* gene was overexpressed under a constitutive promoter to enhance grain yield in wheat. Overexpression of *TaNF-YB4* provided an enhancement of nutrient uptake and/or more efficiency of nutrient utilization, which resulted in increased tiller number in transgenic *TaNF-YB4* lines; most of these were fertile and reached maturity, although the underlying mechanism is as yet unknown. The transgenic wheat lines overexpressing *TaNF-YB4* maintained uniformity in yield performance under water-limited conditions, but their yield did not increase under drought conditions. The reasons behind this may be due to higher demand of wheat plants for water under drought, leading to a decline in the number of tillers (Yadav et al., 2015).

Transgenic wheat expressing a vacuolar Na⁺/H⁺ antiporter gene *AtNHX1* from *Arabidopsis thaliana* showed better yield under saline soil. An improved biomass production at the vegetative growth stage and germination rates in *AtNHX1* expressing transgenic wheat lines under severe saline conditions (Xue et al., 2004). The *Vigna aconitifolia*_1-pyrroline-5-carboxylate synthetase *"P5CS"* cDNA that encodes enzymes required for the biosynthesis of proline, plasmid DNA (pBI-P5CS) along with the selectable neomycin phosphotransferase-II *"npt II"* gene for kanamycin resistance and the reporter β-glucuronidase *"gus"* gene were delivered into wheat using *Agrobacterium*-mediated gene transfer via the indirect pollen system. The resulted transgenics showed improved salinity tolerance due to higher level of proline concentration (Sawahel and Hassan, 2002).

As stated in the rice section, *NAC* transcription factors were found to play major roles in the regulation of root and leaf growth and development, hormone responses, and tolerance to multiple biotic and abiotic stresses, including drought, salt, and low temperature stress (Nakashima et al., 2012; Nuruzzaman et al., 2013). The analysis of *TaNAC69* in transgenic wheat revealed that overexpression of *TaNAC69* driven under the barley drought-inducible *HvDhn4s* promoter enhanced in the leaves and roots of transgenic lines under drought conditions. More shoot biomass was observed in *HvDhn4s:TaNAC69* transgenic lines under combined mild salt stress and water-limitation conditions. Longer root and more root biomass were also observed under polyethylene glycol-induced dehydration. Hence, it revealed the constitutive overexpression of *TaNAC69* in wheat enhanced the expression levels of several stress-upregulated genes (Xue et al., 2011). Similarly, transgenic wheat plants overexpressing *TaNAC-S* showed delayed leaf senescence (stay-green phenotype). Different parameters such as grain yield, aboveground biomass, harvest index, and total grain nitrogen content were unaffected, except grain nitrogen content. *NAC*-overexpressing lines had higher grain nitrogen concentrations at similar grain yields compared to non-transgenic controls. This indicates that *TaNAC-S* is a negative regulator of leaf senescence and is responsible for delayed leaf senescence, which may lead not only to increased grain yields but also to increased grain protein concentrations. Overexpression of *TaNAC2-5A* in wheat also enhanced root growth and nitrate influx rate and, hence, increased the root's ability to acquire nitrogen (Zhao et al., 2014).

Wheat was transformed with the *ipt* gene under the control of senescence induce promoter pSEE1 from maize using the biolistic method, although no significant phenotypic difference was noted with respect to nontransformed ones (Pant et al., 2009). A combined strategy was used to enhance β-carotene content in wheat endosperm. In a "block strategy" (silencing *TaHYD*), endosperm-specific silencing of the carotenoid hydroxylase gene (*TaHYD*) increased 10.5-fold to 1.76 µg g−1. In another "push strategy" (overexpressing *CrtB*), the overexpression of *CrtB* introduced an additional flux into wheat, accompanied by increase in β-carotene by 14.6-fold to 2.45 µg g−1. But when both "push strategy" (overexpressing *CrtB*) and "block strategy" (silencing *TaHYD*) were combined in wheat metabolic engineering, a significant level of β-carotene accumulation was obtained, corresponding to an increase of up to 31-fold to 5.06 µg g^{-1} (Zeng et al., 2015).

9.3.3   MAIZE

An attempt of maize transformation by injected genomic DNA directly into apical meristems of developing maize seedlings was made, but expected morphological changes were not observed (Coe and Sarkar, 1966). The exogenous DNA could be introduced into maize using *Agrobacterium tumefaciens* (Grimsley et al., 1986). The successful transgenic maize was generated from protoplasts of embryogenic suspension cultures of the inbred maize line A188 (Rhodes et al., 1988), but these plants were unable to set seed. The first fertile transgenic maize was reported from protoplasts that had been isolated from embryogenic suspension cultures of genotype HE/89 (Golovkin et al., 1993; Omirulleh et al., 1993). The first maize transformation was performed with the biolistic gun method by Gordon-Kamm et al. (1990) and with the *Agrobacterium*-mediated method by Ishida et al. (1996). These are now routinely used to transform maize genotypes in many laboratories around the world. The stable transformation of maize cultivar A188 and its hybrids using freshly isolated immature embryos with *Agrobacterium* harboring a super binary vector was reported (Ishida et al., 1996). The first herbicide-resistant transgenic maize was obtained with phosphinothricin (PPT) (Gordon-Kamm et al., 1990). The PPT is considered as efficient and widely used as the selection system (Fromm et al., 1990; Gordon-Kamm et al., 1990; Ishida et al., 1996; Zhao et al., 2001; Frame et al., 2002, 2006). Another herbicide resistant maize was reported by Howe et al. (2002) by introducing the *5-enolpyruvylshikimate-3-phosphate synthase* (*EPSPS*) gene, which showed resistance to the glyphosate selection medium.

European corn borer-resistant transgenic maize plants have been developed that express *B. thuringiensis* genes (Snow and Palma, 1997; Pilcher and Rice, 2001). Considerably improved drought tolerance was observed in constitutive overexpressing *ZmNF-YB2* gene in transgenic maize subjected to mild drought in controlled conditions and during field trials. An improvement in different parameters such as chlorophyll content, stomatal conductance, leaf temperature, reduced wilting, and maintenance of photosynthesis was observed. These stress adaptations contributed to improved grain yield for transgenic maize lines under water-limited conditions (Nelson et al., 2007).

It has been reported that the expression of *O. sativa* Na^+/H^+ antiporter gene (*OsNHX1*) in maize enhanced salt tolerance. The transgenic maize plants overexpressing *OsNHX1* accumulated more biomass when grown in the presence of 200 mM NaCl in greenhouse conditions and higher Na^+/K^+ content was observed in transgenic leaves in comparison with wild-type leaves when treated with 100~200 mM NaCl. However, the osmotic potential and the proline content in transgenic leaves were lower than those in wild-type maize (Chen et al., 2007).

As described above, *expansin* plays an important role as a major protein to induce cell wall extension, larger cells, taller plants, and longer roots (Ma et al., 2013). Cucumber *expansin* has been overexpressed in maize kernels to produce enough protein to assess its potential to serve as an industrial enzyme for applications, particularly in biomass conversion (Yoon et al., 2015).

9.3.4   BARLEY

In the first report on barley transformation, different explants (immature embryos, immature embryos derived calli and microspore-derived embryos) of barley *H. vulgare* cv. Golden Promise were bombarded with DNA carrying the bialaphos selection system (Wan and Lemaux, 1994). The *Agrobacterium*-mediated transformation in barley cv. Golden Promise obtained 54 transgenic lines for the *bar* gene (Tingay et al., 1997). Another *Agrobacterium*-mediated transformation in immature embryos from Australian elite cultivars for the hygromycin selection system was reported (Murray et al., 2004). The *hpt* gene as a selection marker and the luciferase (*luc*) gene as a reporter gene were introduced in barley using immature embryos via *Agrobacterium* (Bartlett et al., 2008). A number of laboratories around the world reported successful production of transgenic barley plants with different genes (Shrawat and Lörz, 2006; Ji et al., 2013).

9.3.5 SORGHUM

Among cereal crops, sorghum is least successfully manipulated for tissue culture and *Agrobacterium*-mediated transformation. A number of investigations have been made to generate transgenic sorghum plants via biolistic (Casas et al., 1993, 1997; Zhu et al., 1998; Able et al., 2001; Emani et al., 2002; Jeoung et al., 2002; Devi and Sticklen, 2003; Tadesse et al., 2003; Girijashankar et al., 2005; Liu and Godwin, 2012) or *Agrobacterium*-mediated transformation (Zhao et al., 2000; Carvalho et al., 2004; Gao et al., 2005; Howe et al., 2006; Van Nguyen et al., 2007; Gurel et al., 2009; Lu et al., 2009; Mall et al., 2011). Among them, Liu and Godwin (2012) were able to achieve an average 20.7% transformation efficiency from three separate experiments. The efficiency of sorghum transformation mediated by *A. tumefaciens* is low as compared to the biolistic method ranging from 2.1–4.5% (Zhao et al., 2000; Gao et al., 2005; Howe et al., 2006).

A transgenic sorghum was developed by transferring the *chiII* gene, encoding rice chitinase under the control of the constitutive CaMV35S promoter, for resistance to stalk rot (*F. thapsinum*) (Zhu et al., 1998). Chitinases are pathogenesis-related proteins (PR-proteins) produced in plants as defense against pathogen attack. This anti-fungal strategy involves the direct targeting of the fungal cell wall component chitin, a 1, 4-linked homopolymer of N-acetyl-D-glucosamine, by the enzyme chitinase. Another chitinase gene *ECH2* and the *bar* gene were introduced in sorghum to produce disease- and herbicide-resistant transgenic plants, respectively (Devi and Sticklen, 2003). The *cry1Ac* gene from *B. thuringiensis* under the control of the wound-inducible promoter from the maize protease inhibitor gene (*mpiC1*) was introduced in the genome of sorghum. Transgenic sorghum plants showed up to 60% reduction in leaf damage by the spotted stem borer (*Chilo partellus*) (Girijashankar et al., 2005). In a different study, sorghum was genetically transformed to express the *TaALMT1* (Sasaki et al., 2004) gene from wheat to resist toxic aluminum concentrations present in acidic and contaminated soils (Brandão, 2007). The *ALMT1* gene codes a malate transporter Al+3 activated protein that is highly expressed in wheat root apices of aluminum-tolerant cultivars. Transgenic sorghum plants grown in hydroponic culture under stress of Al+3 showed a higher level of aluminum tolerance when compared with isogenic non-transgenic control plants.

9.4 FUTURE PROSPECTIVE

The optimization of parameters that are considered crucial for cereal transformation and screening of highly competent explants and genotypes should broaden the scope for genetic transformation with genes of interest. As foreign gene transfer is highly genotype sensitive, it is very much required to standardize the transformation method in each genotype or to develop genotype-insensitive transformation method. In near future, there is a tremendous need to develop transgenic cereals to meet future food demand. The information from genome sequencing of different cereal crops can be explored to develop transgenic plants.

KEYWORDS

- agrobacterium
- callus
- regeneration
- rice
- sorghum
- transgenic
- wheat

REFERENCES

Able, J. A., Rathus, C., & Godwin, I. D., (2001). The investigation of optimal bombardment parameters for transient and stable transgene expression in sorghum. *In Vitro Cell. Dev. Biol., 37,* 341–348.

Agarwal, S., et al., (2009). Floral transformation of wheat. *In: Jones H., Shewry P. (eds) Transgenic Wheat, Barley and Oats. Methods in Molecular Biology (Methods and Protocols), Humana Press, 478, 105-113.*

Ahmad, A., et al., (2002). Expression of synthetic *cry1ab* and *cry1ac* genes in basmati rice (*Oryza sativa* L.) variety 370 *via Agrobacterium*-mediated transformation for the control of the european corn borer (Ostrinia nubilalis). *In Vitro Cell. Dev. Biol. Plant, 38,* 213–220.

Aldemita, R. R., & Hodges, T. K., (1996). *Agrobacterium tumefaciens* mediated transformation of japonica and indica rice varieties. *Planta., 199,* 612–617.

Anand, A., et al., (2003). Greenhouse and field testing of transgenic wheat plants stably expressing genes for thaumatin-like protein, chitinase and glucanase against *Fusarium graminearum. J. Exp. Bot., 54,* 1101–1111.

Bartlett, J. G., et al., (2008). High-throughput *Agrobacterium*-mediated barley transformation. *Plant Methods, 4,* 22.

Bashir, K., et al., (2005). Novel Indica basmati line (B-370) expressing two unrelated genes of *Bacillus thuringiensis* is highly resistant to two lepidopteran insects in the field. *Crop Prot., 24,* 870–879.

Batistic, O., & Kudla, J., (2009). Plant calcineurin B-like proteins and their interacting protein kinases. *Biochim. Biophys. Acta., 1793,* 985–992.

Brandao, R. L., (2007). Genetic transformation of Sorghum bicolor (Moench, L.) targeting Al+3 tolerance. Doctoral thesis. Federal University of Lavras, UFLA, Lavras, Brazil, 127.

Carvalho, C. H. S., et al., (2004). *Agrobacterium*-mediated transformation of sorghum: factors that affect transformation efficiency. *Genet. Mol. Biol., 27,* 259–269.

Casas, A. M., et al., (1997). Transgenic sorghum plants obtained after microprojectile bombardment of immature inflorescences. *– In. Vitro Cell Dev. Biol., 33,* 92–100.

Casas, A. M., et al., (1993). Transgenic sorghum plants *via* microprojectile bombardment. *Proc. Natl. Acad. Sci. U. S. A., 90,* 11212–11216.

Chan, M. T., et al., (1993). *Agrobacterium*-mediated production of transgenic rice plants expressing a chimeric *alpha amylase* promoter/β-*glucuronidase* gene. *Plant Mol. Biol., 22,* 491–496.

Chauhan, H., & Khurana, P., (2011). Use of doubled haploid technology for development of stable drought tolerant bread wheat (*Triticum aestivum* L.) transgenics. *Plant Biotechnol. J., 9,* 408–417.

Chen, M., et al., (2007). Expression of *OsNHX1* gene in maize confers salt tolerance and promotes plant growth in the field. *Plant Soil Environ., 53*(11), 490–498.

Chen, W. P., et al., (1999). Development of wheat scab symptoms is delayed in transgenic wheat plants that constitutively express a rice thaumatin-like protein gene. *Theor. Appl. Genet., 99,* 755–760.

Cheng, M., et al., (1997). Genetic transformation of wheat mediated by *Agrobacterium tumefaciens. Plant Physiol., 115,* 971–980.

Cheng, X., et al., (1998). *Agrobacterium*-transformed rice plants expressing synthetic *cryIA(b)* and *cryIA(c)* genes are highly toxic to striped stem borer and yellow stem borer. *Appl. Biol. Sci., 95,* 2767–2772.

Cheong, Y. H., et al., (2003). CBL1, a calcium sensor that differentially regulates salt, drought, and cold responses in *Arabidopsis. Plant Cell, 15,* 1833–1845.

Choi, D., et al., (1998). Regulation of expansin gene expression affects growth and development in transgenic rice plants. *Society 2003, 15,* 1386–1398. doi:10.1105/tpc. 011965.

Chugh, A., & Khurana, P., (2003). Herbicide-resistant transgenics of bread wheat (*T. aestivum*) and emmer wheat (*T. dicoccum*) by particle bombardment and *Agrobacterium* mediated approaches. *Curr. Sci., 84,* 78–83.

Chugh, A.,et al., (2012). A Novel approach for *Agrobacterium*-mediated germ line transformation of Indian bread wheat (*Triticum aestivum*) and pasta wheat (*Triticum durum*). *J Phytol., 4*(2), 22–29.

Coe, E. H., & Sarkar, K. R., (1966). Preparation of nucleic acids and a genetic transformation attempt in maize. *Crop Sci., 6,* 432–435.

Dasgupta, I., et al., (1991). Rice *tungro bacilliform* virus DNA independently infects rice after *Agrobacterium*-mediated transfer. *J. Gen. Virol., 72*(6), 1215–1221.

Datta, S. K., et al., (1990). Genetically engineered fertile Indica-rice plants recovered from protoplasts. *Bio/Technol., 8,* 736–740.

Devi, P., & Sticklen, M., (2003). *In vitro* culture and genetic transformation of sorghum by microprojectile bombardment. *Plant Biosystems, 137,* 249–254.

Ding L., et al., (2009). Optimization of *Agrobacterium* mediated transformation conditions in mature embryos of elite wheat. *Mol. Biol. Rep., 36,* 29–36.

Dong, J., et al., (2001). Characterization of rice transformed *via* an *Agrobacterium*-mediated inflorescence approach. *Mol. Breed., 7,* 187–194.

Dong, J., et al., (1996). *Agrobacterium*-mediated transformation of Javanica rice. *Mol. Breed., 2,* 267–276.

Edwards, D., Murray, J. A., & Smith, A. G., (1998). Multiple genes encoding the conserved CCAAT-box transcription factor complex are expressed in *Arabidopsis. Plant Physiol., 117,* 1015–1022.

Emani, C., Sunilkumar, G., & Rathore, K. S., (2002). Transgene silencing and reactivation in sorghum. *Plant Sci., 162,* 181–192.

Faccio, P., et al., (2011). Increased tolerance to wheat powdery mildew by heterologous constitutive expression of the *Solanum chacoense snakin-1* gene. *Czech. J. Genet. Plant Breed, 47,* 135–141.

Fang, Y., et al., (2008). Systematic sequence analysis and identification of tissue-specific or stress responsive genes of NAC transcription factor family in rice. *Mol. Genet. Genomics, 280,* 547–563.

Fang, Z., et al., (2013). Altered expression of the *PTR/NRT1* homologue *OsPTR9* affects nitrogen utilization efficiency, growth and grain yield in rice. *Plant Biotechnol. J., 11,* 446–458.

Frame, B. R., et al., (1994). Production of fertile transgenic maize plants by silicon carbide whisker-mediated transformation. *Plant J., 6,* 941–948.

Frame, B. R., et al., (2002). *Agrobacterium tumefaciens* -mediated transformation of maize embryos using a standard binary vector system. *Plant Physiol., 129,* 13–22.

Frame, B. R., et al., (2006). Improved *Agrobacterium*-mediated transformation of three maize inbred lines using MS salts. *Plant Cell Rep., 25,* 1024–1034.

Fromm, M. E., et al., (1990). Inheritance and expression of chimeric genes in the progeny of transgenic maize plants. *Biotechnol., 8,* 833–839.

Fujimoto, H., et al., (1993). Insect resistant rice generated by introduction of a modified endotoxin gene *Bacillus thuringiensis. Bio/Technol., 11,* 1151–1155.

Fukuda, A., et al., (2004). Function, intracellular localization and the importance in salt tolerance of a vacuolar Na^+/H^+ antiporter from rice. *Plant Cell Physiol., 45,* 146–159.

Gao, X., et al., (2011). *AtMYB12* gene: a novel visible marker for wheat transformation. *Mol. Biol. Rep., 38,* 183–190.

Gao, F., et al., (2010). *OsNAC52*, a rice NAC transcription factor, potentially responds to ABA and confers drought tolerance in transgenic plants. *Plant Cell Tissue Organ Cult., 100*, 255–262.

Gao, Z., et al., (2005). *Agrobacterium tumefaciens*-mediated sorghum transformation using a mannose selection system. *Plant Biotechnol. J., 3*, 591–599.

Gelvin, S. B., et al., (2003). *Agrobacterium*-mediated plant transformation: the biology behind the "Gene-Jockeying" tool. *Microbiol. Mol. Biol. Rev., 67*(1), 16–37.

Girijashankar, V., et al., (2005). Development of transgenic sorghum for insect resistance against the spotted stem borer (*Chilo partellus*). *Plant Cell Rep., 24*, 513–522.

Golovkin, M. V., et al., (1993). Production of transgenic maize plants by direct DNA uptake into embryogenic protoplasts. *Plant Sci., 90*, 41–52.

Gopalakrishnan, S., et al., (2000). Herbicide-tolerant transgenic plants in high yielding commercial wheat cultivars obtained by microprojectile bombardment and selection on Basta. *Curr. Sci., 79*, 1094–1100.

Gordon-Kamm, W. J., et al., (1990). Transformation of maize cells and regeneration of fertile transgenic plants. *Plant Cell, 2*, 603–618.

Grimsley, N., et al., (1986). "Agroinfection," an alternative route for viral infection of plants by using the Ti plasmid. *Proc. Natl. Acad. Sci. USA, 83*, 3282–3286.

Gu, Z., et al., (2008). Expression analysis of the calcineurin B-like gene family in rice (*Oryza sativa* L.) under environmental stresses. *Gene, 415*, 1–12.

Gurel, S., et al., (2009). Efficient, reproducible *Agrobacterium*-mediated transformation of sorghum using heat treatment of immature embryos. *Plant Cell Rep., 28*, 429–444.

Han, J., et al., (2012). Transgenic expression of lactoferrin imparts enhanced resistance to head blight of wheat caused by *Fusarium graminearum*. *BMC Plant Biol., 12*, 33. http://www.biomedcentral.com/1471–2229/12/33.

Hayakawa, T., et al., (1992). Genetically engineered rice resistant to stem stripe virus an insect transmitted virus. *Proc. Natl. Acad. Sci. USA, 89*, 9865–9869.

Hayes, R. J., et al., (1988). Agroinfection of *Triticum aestivum* with cloned DNA of wheat dwarf virus. *J. Gen. Virol., 69*, 891–896.

Hensel, G., et al., (2009). *Agrobacterium*-mediated gene transfer to cereal crop plants: Current protocols for barley, wheat, triticale, and maize. *Int. J. Plant Genomics*, 1–9, doi: 10.1155/2009/835608.

Hiei, Y., & Komari, T., (2006). Improved protocols for transformation of Indica rice mediated by *Agrobacterium tumefaciens* . *Plant Cell Tissue Organ Cult., 85*, 271–283.

Hiei, Y., et al., (1994). Efficient transformation of rice (*Oryza sativa* L.) mediated by *Agrobacterium* and sequence analysis of the boundaries of the T-DNA. *Plant J., 6*, 271–282.

Hofer, A. M., & Brown, E. M., (2003). Extracellular calcium sensing and signalling. *Nat. Rev. Mol. Cell Biol., 4*, 530–538.

Howe, A., et al., (2006). Rapid and reproducible *Agrobacterium*-mediated transformation of sorghum. *Plant Cell Rep., 25*, 784–791.

Hu, H., et al., (2008). Characterization of transcription factor gene *SNAC2* conferring cold and salt tolerance in rice. *Plant Mol. Biol., 67*, 169–181.

Hu, H., et al., (2006). Overexpressing a NAM, ATAF, and CUC (NAC) transcription factor enhances drought resistance and salt tolerance in rice. *Proc. Natl Acad. Sci. USA, 103*, 12987–12992.

Ignacimuthu, I., & Raveendar, S., (2011). *Agrobacterium* mediated transformation of indica rice (*Oryza sativa* L.) for insect resistance. *Euphytica, 179*, 277–286. Doi: 10.1007/s10681–010–0308–7.

Ishida, Y., et al., (1996). High efficiency transformation of maize (*Zea mays*, L.) mediated by *Agrobacterium tumefaciens* . *Nat. Biotechnol., 14*, 745–750.

Ito, Y., et al., (2011). Aberrant vegetative and reproductive development by over expression and lethality by silencing of *OsHAP3E* in rice. *Plant Sci., 181*, 105–110.

Jeong, J. S., et al., (2010). Root-specific expression of *OsNAC10* improves drought tolerance and grain yield in rice under field drought conditions. *Plant Physiol., 153*, 185–197.

Jeoung, J. M., et al., (2002). Optimization of sorghum transformation parameters using genes for green fluorescent protein and glucuronidase as visual markers. *Hereditas, 137*, 20–28.

Jha, S., & Chattoo, B. B., (2010). Expression of a plant defensin in rice confers resistance to fungal phytopathogens. *Trans. Res., 19*, 373–384.

Ji, Q., Xu, X., & Wang, K., (2013). Genetic transformation of major cereal crops. *Int. J. Dev. Biol., 57*, 495–508.

Jiang, J. D., et al., (2000). High efficiency transformation of US rice lines from mature seed derived calli and segregation of glufosinate resistance under field conditions. *Crop Sci., 40*, 1729–1741.

Jones, H. D., Doherty, A., & Wu, H., (2005). Review of methodologies and a protocol for the *Agrobacterium*-mediated transformation of wheat. *Plant Methods, 1*, 5.

Kaeppler, H. F., et al., (1992). Silicon carbide fiber-mediated stable transformation of plant cells. *Theor. Appl. Genet., 84*, 560–566.

Kasirajan, L., Boomiraj, K., & Bansal, K. C., (2013). Optimization of genetic transformation protocol mediated by biolistic method in some elite genotypes of wheat (*Triticum aestivum* L.). *Afr. J. Biotechnol., 12*(6), 531–538.

Kebrom, T. H., et al., (2012). Inhibition of tiller bud outgrowth in the tin mutant of wheat is associated with precocious internode development. *Plant Physiol., 160*, 308–318.

Khan, S. A., et al., (2006). Spider venom toxin protects plants from insect attack. *Trans. Res., 15*, 349–357.

Khanna, H. K., & Raina, S. K., (1999). *Agrobacterium*-mediated transformation of indica rice cultivars using binary and superbinary vectors. *Aust. J. Plant Physiol., 26*, 311–324.

Klein, T. M., et al., (1987). High velocity microprojectiles for delivering nucleic acids into living cells. *Nature, 327*, 70–73.

Kolukisaoglu, U., et al., (2004). Calcium sensors and their interacting protein kinases: genomics of the *Arabidopsis* and rice CBL-CIPK signaling networks. *Plant Physiol., 134*, 43–58.

Kumar, R., et al., (2017). Development of an efficient and reproducible regeneration system in wheat (*Triticum aestivum* L.). *Physiol. Mol. Biol. Plants, 23*, 945-954. https://doi.org/10.1007/s12298-017-0463-6

Langridge, P., et al., Transformation of cereals *via Agrobacterium* and the pollen pathway: a critical assessment. *Plant J., 2*, 631–638.

Lee, Y., & Kende, H., (1997). Expression of β-expansins is correlated with internodal elongation in deepwater rice. *Plant Cell, 9*, 1661–1671. doi:10.1105/tpc. 9. 9. 1661.

Li, H., W., et al., (2011a). Overexpression of the trehalose-6-phosphate synthase gene *OsTPS1* enhances abiotic stress tolerance in rice. *Planta.* doi: 10.1007/s00425–011–1458–0123.

Li, Z., et al., (2011). Expression of a radish defensin in transgenic wheat confers increased resistance to Fusarium graminearum and Rhizoctonia cerealis. *Funct. Integr. Genomics, 11*, 63–70. doi: 10.1007/s10142–011–0211-x.

Lin, Y. J., & Zhang, Q., (2004). Optimising the tissue culture conditions for high efficiency transformation of indica rice. *Plant Cell Rep., 23*, 540–547.

Lin, Y. J., et al., (2002). Cultivating rice with delaying leaf senescence by *PSAG12-IPT* gene transformation. *Acta Bot. Sin., 44*, 1333–1338.

Liu, G., & Godwin, I. D., (2012). Highly efficient sorghum transformation. *Plant Cell Rep., 31*, 999–1007.

Lu, L., et al., (2009). Development of markerfree transgenic sorghum [*Sorghum bicolor* (L.) Moench] using standard binary vectors with bar as a selectable marker. *Plant Cell Tissue Organ Cult., 99*, 97–108.

Luan, S., et al., (2002). Calmodulins and calcineurin B-like proteins: calcium sensors for specific signal response coupling in plants. *Plant Cell, 14*, 389–400.

Ma, N., et al., (2013). Overexpression of *OsEXPA8*, a root-specific gene, improves rice growth and root system architecture by facilitating cell extension. *PLoS One, 8*, e75997. doi: 10. 1371/journal.pone. 0075997.

Mackintosh, C. A., (2007). Overexpression of defense response genes in transgenic wheat enhances resistance to *Fusarium* head blight. *Plant Cell Rep., 26*, 479–488.

Mahajan, S., Sopory, K., & Tuteja, N., (2006). Cloning and characterization of CBL-CIPK signalling components from a legume (*Pisum sativum*). *FEBS J., 273*, 907–925.

Mall, T. K., et al., (2011). Expression of the rice CDPK-7 in sorghum: molecular and phenotypic analyses. *Plant Mol. Biol., 75*, 467–479.

Miyoshi, K., et al., (2003). *OsHAP3* genes regulate chloroplast biogenesis in rice. *Plant J., 36*, 532–540.

Mohanty, A., et al., (2000). Analysis of the activity of promoters from two photosynthesis-related genes, *psaF* and *petH*, of spinach in a monocot plant, rice. *Indian J. Biochem. Biophy., 37*, 447–452.

Mohanty, A., Sarma, N. P., & Tyagi, A. K., (1999). *Agrobacterium*-mediated high frequency transformation of an elite indica rice variety Pusa Basmati 1 and transmission of the transgenes to R2 progeny. *Plant Sci., 147*, 127–137.

Molla, K. A., et al., (2013). Rice oxalate oxidase gene driven by green tissue-specific promoter increases tolerance to sheath blight pathogen (*Rhizoctonia solani*) in transgenic rice. *Mol. Plant Pathol., 14*(9), 910–922.

Motohashi, T., et al., (2010). Production of salt stress tolerant rice by overexpression of the catalase gene, *katE*, derived from *Escherichia coli*. *Asia Pac. J. Mol. Biol. Biotechnol., 18*, 37–41.

Murray, F., et al., (2004). Comparison of *Agrobacterium*-mediated transformation of four barley cultivars using the *GFP* and *GUS* reporter genes. *Plant Cell Rep., 22*, 397–402.

Nakashima, K., et al., (2007). Functional analysis of a NAC-type transcription factor *OsNAC6* involved in abiotic and biotic stress-responsive gene expression in rice. *Plant J., 51*, 617–630.

Nakashima, K., et al., (2012). NAC transcription factors in plant abiotic stress responses. *Biochim. Biophys. Acta., 1819*, 97–103.

Nelson, D. E., et al., (2007). Plant nuclear factor Y (NF-Y) B subunits confer drought tolerance and lead to improved corn yields on water-limited acres. *Proc. Natl. Acad. Sci. USA, 104*, 16450–16455.

Nishimura, A., Aichi, I., & Matsuoka, M., (2006). A protocol for *Agrobacterium*-mediated transformation in rice. *Nat. Prot., 1*(6), 2796–2802.

Nuruzzaman, M., et al., (2010). Genome-wide analysis of NAC transcription factor family in rice. *Gene, 465*, 30–44.

Nuruzzaman, M., et al., (2013). Roles of NAC transcription factors in the regulation of biotic and abiotic stress responses in plants. *Front. Microbiol., 4*, 248.

Oh, S. J., et al., (2007). Expression of barley *HvCBF4* enhances tolerance to abiotic stress in transgenic rice. *Plant Biotechnol. J., 5,* 646–656.

Ohta, M., et al., (2002). Introduction of a Na+/H+ antiporter gene from *Atriplex gmelini* confers salt tolerance to rice. *FEBS Lett., 532,* 279–282.

Omirulleh, S., et al., (1993). Activity of a chimeric promoter with the doubled CaMV 35S enhancer element in protoplast-derived cells and transgenic plants in maize. *Plant Mol. Biol., 21,* 415–428.

Pandey, S., et al., (2000). Calcium signaling: linking environmental signals to cellular functions. *Crit. Rev. Plant Sci., 19,* 219–318.

Pant, D. R., et al., (2009). Genetic transformation of Nepalese spring wheat (*Triticum aestivum* L.) cultivars with *ipt* gene under the regulation of a senescence enhanced promoter from maize. *Pak. J. Biol. Sci., 12*(2), 101–109.

Park, S. K., et al., (2014). Development of marker-free transgenic rice expressing the wheat storage protein, Glu-1Dy10, for increasing quality processing of bread and noodles. *J. Life Sci., 24*(6), 618–625.

Patnaik, D., & Khurana, P., (2001). Wheat Biotechnology: A mini review. *Electron. J. Biotechnol., 4,* 74–102.

Patnaik, D., & Khurana, P., (2003). Genetic transformation of Indian bread (*T. aestivum*) and pasta (*T. durum*) wheat by particle bombardment of mature embryo-derived calli. *BMC Plant Biol., 3,* 5.

Patnaik, D., Vishnudasan, D., & Khurana, P., (2006). *Agrobacterium*-mediated transformation of mature embryos of *Triticum aestivum* and *Triticum durum*. *Curr. Sci., 91,* 307–317.

Pilcher, C. D., & Rice, M. E., (2001). Effect of planting dates and *Bacillus thuringiensis* corn on the population dynamics of European corn borer (*Lepidoptera*: Crambidae). *J. Econ. Entomol., 94,* 730–742.

Pons, M. J., et al., (2000). Regeneration and genetic transformation of Spanish rice cultivars using mature embryos. *Euphytica, 114,* 117–122.

Prodhan, S. H., et al., (2008). Improved salt tolerance and morphological variation in indica rice (*Oryza sativa* L.) transformed with a catalase gene from E. *coli. Plant Tissue Cult. Biotechnol., 18,* 57–63.

Purwestri, Y. A., et al., (2015). *Agrobacterium tumefaciens* mediated transformation of rolC::Hd3a-GFP in black rice (*Oryza sativa* L. cv. Cempo Ireng) to promote early flowering. *Procedia Chem., 14,* 469–473.

Qiu, H. J., et al., (2001). *Agrobacterium tumefaciens*-mediated transformation of rice with the spider insecticidal gene conferring resistance to leaf folder and striped stem borer. *Cell Res., 11,* 149–155.

Qu, B., et al., (2015). A wheat CCAAT box-binding transcription factor increases the grain yield of wheat with less fertilizer input. *Plant Physiol., 167,* 411–423.

Rachmawati, D., et al., (2004). *Agrobacterium*-mediated transformation of javanica rice cv. Rojolele. *Biosci. Biotechnol. Biochem., 68,* 1193–1200.

Raineri, D. M., et al., (1990). *Agrobacterium*-mediated transformation of rice (*Oryza sativa* L.). *Bio/Technol., 8,* 33–38.

Ramesh, S., et al., (2004). Production of transgenic indica rice resistant to yellow stem borer and sap-sucking insects, using super-binary vectors of *Agrobacterium tumefaciens*. *Plant Sci., 166,* 1077–1085.

Ranathunge, K., et al., (2014). *AMT1*, 1 transgenic rice plants with enhanced NH^{4+} permeability show superior growth and higher yield under optimal and suboptimal NH^{4+}conditions. *J. Exp. Bot., 65,* 965–979.

Rashid, H., et al., (1996). Transgenic plant production mediated by *Agrobacterium* in Indica rice. *Plant Cell Rep., 15,* 727–730.

Rhodes, C. A., et al., (1988). Genetically transformed maize plants from protoplasts. *Sci., 240,* 204–207.

Risacher, T., et al., (2009). Highly efficient *Agrobacterium*-mediated transformation of wheat *via* in planta inoculation. In: *Transgenic Wheat, Barley and Oats: Production and Characterization Protocols* (Jones, H. D., Shewry, P. R., Eds.). *Methods in Molecular Biology,* vol. *478.* Humana Press, Totowa, N. J., pp. 115–124.

Robertson, M., (2004). Two transcription factors are negative regulators of gibberellin response in the HvSPY-signaling pathway in barley aleurone. *Plant Physiol., 136,* 2747–2761.

Rod-In, W., Sujipuli, K., & Ratanasut, K., (2014). The floral-dip method for rice (*Oryza sativa*) transformation. *J. Agri. Tech., 10*(2), 467–474.

Rohila, J. S., Jain, R. K., & Wu, R., (2002). Genetic improvement of Basmati rice for salt and drought tolerance by regulated expression of a barley *Hva1* cDNA. *Plant Sci., 163,* 525–532.

Sahoo, K. K., et al., (2011). An improved protocol for efficient transformation and regeneration of diverse indica rice cultivars. *Plant Methods, 7,* 49.

Sanders, D., et al., (2002). Calcium at the crossroads of signaling. *Plant Cell, 14,* 401–417.

Sanford, J. C., et al., (1987). Delivery of substances into cells and tissues using particle bombardment process. *J. Particle Sci. Tech., 5,* 27–37.

Sasaki, T., et al., (2004). A wheat gene encoding an aluminum-activated malate transporter. *Plant J., 37,* 645–653.

Sautter, C., et al., (1993). Development of a microtargeting device for particle bombardment of plant meristems. *Plant Cell Tissue Organ Cult., 33,* 251–257.

Sawahel, W. A., & Hassan, A. H., (2002). Generation of transgenic wheat plants producing high levels of the osmoprotectant proline. *Biotechnol. Lett., 24,* 721–725.

Shin, S., et al., (2008). Transgenic wheat expressing a barley class II *chitinase* gene has enhanced resistance against *Fusarium graminearum. J. Exp. Bot., 59,* 2371–2378.

Shrawat, A. K., & Lorz, H., (2006). *Agrobacterium*-mediated transformation of cereals: a promising approach crossing barriers. *Plant Biotechnol. J., 4,* 575–603.

Shri, M., et al., (2013). An improved *Agrobacterium*-mediated transformation of recalcitrant indica rice (*Oryza sativa,* L.) cultivars. *Protoplasma, 250,* 631–636.

Shu, Q. Y., et al., (2000). Transgenic rice plants with a synthetic *cry1Ab* gene from *Bacillus thuringiensis* were highly resistant to eight *lepidopteran* rice pest species. *Mol. Breed., 6,* 433–439.

Sikdar, S. U., et al., (2015). *Agrobacterium*-mediated *PsCBL* and *PsCIPK* gene transformation to enhance salt tolerance in indica rice (*Oryza sativa*). *In. Vitro Cell. Dev. Biol. Plant.,* doi: 10. 1007/s11627–014–9654–9.

Silverstein, K. A., et al., (2007). Small cysteine-rich peptides resembling antimicrobial peptides have been under-predicted in plants. *Plant J., 51,* 262–280.

Sivamani, E., et al., (1999). Rice plant (*Oryza sativa* L.) containing rice tungro spherical virus (RTSV) coat protein transgenes are resistance to virus infection. *Mol. Breed., 5,* 177–185.

Snow, A. A., & Palma, P. M., (1997). Commercialization of transgenic plants: potential ecological risks. *Biosci., 47,* 86–96.

Stephenson, T. J., et al., (2011). *TaNF-YB3* is involved in the regulation of photosynthesis genes in *Triticum aestivum. Funct. Integr. Genomics, 11,* 327–340.

Supartana, P., et al., (2006). Development of simple and efficient in planta transformation method for wheat (*Triticum aestivum*, L.) using *Agrobacterium tumefaciens* . *J. Biosci. Bioengg.*, *102*, 162–170.

Tadesse, Y., et al., (2003). Optimisation of transformation conditions and production of transgenic sorghum (*Sorghum bicolor*) *via* microparticle bombardment–Rev. *Plant Biotechnol. Appl. Geneti.*, *75*, 1–18.

Takasaki, H., et al., (2010). The abiotic stress-responsive NAC-type transcription factor *OsNAC5* regulates stress-inducible genes and stress tolerance in rice. *Mol. Genet. Genomics*, *284*, 173–183.

Tamaki, S., et al., (2007). Hd3a protein is a mobile flowering signal in rice. *Sci.*, *316*, 1033–1036.

Tavares, L. S., et al., (2008). Biotechnological potential of antimicrobial peptides from flowers. *Peptides*, *29*, 1842–1851.

Tingay, S., et al., (1997). *Agrobacterium tumefaciens*-mediated barley transformation. *Plant J.*, *11*, 1369–1376.

Toki, S., (1997). Rapid and efficient *Agrobacterium*-mediated transformation in rice. *Plant Mol. Biol. Rep.*, *15*, 16–21.

Tu, J., et al., (2000). Field performance of *Xa21* transgenic rice (*Oryza sativa* L.) IR72. *Theor. Appl. Genet.*, *101*, 15–20.

Tu, J., et al., (1998). Transgenic rice variety IR72 with *Xa21* resistant to bacterial blight. *Theor. Appl. Genet.*, *97*, 31–36.

Uchimiya, H., et al., (1986). Expression of a foreign gene in callus derived from DNA-treated protoplasts of rice (*Oryza sativa* L.). *Mol. Gen. Genet.*, *204*, 204–207.

Upadhyaya, N. M., et al., (2000). *Agrobacterium*-mediated transformation of Australian rice cultivars Jarrah and Amaroo using modified promoters and selectable markers. *Aust. J. Plant Physiol.*, *27*, 201–210.

Van-Nguyen, T., et al., (2007). *Agrobacterium*-mediated transformation of sorghum (*Sorghum bicolor*, L.) (Moench) using an improved *in vitro* regeneration system. *Plant Cell Tissue Organ Cult.*, *91*, 155–164.

Vega, J. M., et al., (2008). Improvement of *Agrobacterium*-mediated transformation in Hi-II maize (*Zea mays*) using standard binary vectors. *Plant Cell Rep.*, *27*, 297–305.

Wan, Y., & Lemaux, P. G., (1994). Generation of large numbers of independently transformed fertile barley plants. *Plant Physiol.*, *104*, 37–48.

Wang, M. B., et al., (2001). *Agrobacterium tumefaciens*-mediated transformation of an elite Australian barley cultivar with virus resistance and reporter genes. *Aust. J. Plant Physiol.*, *28*, 149–156.

Wei, X., et al., (2010). DTH8 suppresses flowering in rice, influencing plant height and yield potential simultaneously. *Plant Physiol.*, *153*, 1747–1758.

Wu, H., et al., (2008). *Agrobacterium*-mediated stable transformation of suspension cultures of barley (*Hordeum vulgare*, L.). *Plant Cell Tissue Organ Cult.*, *54*, 161–167.

Xia, N., et al., (2010). Characterization of a novel wheat NAC transcription factor gene involved in defence response against stripe rust pathogen infection and abiotic stresses. *Mol. Biol. Rep.*, *37*, 3703–3712.

Xiang, Y., Huang, Y., & Xiong, L., (2007). Characterization of stress-responsive *CIPK* genes in rice for stress tolerance improvement. *Plant Physiol.*, *144*, 1416–1428.

Xiao, B., et al., (2007). Over-expression of a *LEA* gene in rice improves drought resistance under the field conditions. *Theor. Appl. Genet.*, *115*, 35–46.

Xiao, B. Z., et al., (2009). Evaluation of seven function-known candidate genes for their effects on improving drought resistance of transgenic rice under field conditions. *Mol. Plant, 2*, 73–83.

Xiaofen, S., et al., (2001). Transgenic rice homozygous lines expressing GNA showed enhanced resistance to rice brown plant hopper. *Chin. Sci. Bull., 46*, 1698–1703.

Xu, D., et al., (1996). Constitutive expression of a cowpea trypsin inhibitor gene, *CpTi*, in transgenic rice plants confers resistance to two major rice insect pests. *Mol. Breed., 2*(2), 167–173.

Xue G. P., et al., (2011). Overexpression of *TaNAC69* leads to enhanced transcript levels of stress up-regulated genes and dehydration tolerance in bread wheat. *Mol. Plant, 4*(4), 697–712.

Xue, G. P., et al., (2006). *TaNAC69* from the NAC superfamily of transcription factors is up-regulated by abiotic stresses in wheat and recognises two consensus DNA-binding sequences. *Func. Plant Biol., 33*, 43–57.

Xue, Z. Y., et al., (2004). Enhanced salt tolerance of transgenic wheat (*Triticum aestivum*, L.) expressing a vacuolar Na^+/H^+ antiporter gene with improved grain yields in saline soils in the field and a reduced level of leaf Na^+. *Plant Sci., 167*, 849–859.

Yadav, D., et al., (2015). Constitutive overexpression of the *TaNF-YB4* gene in transgenic wheat significantly improves grain yield. *J. Exp. Bot.*, doi: 10.1093/jxb/erv370.

Ye, G. Y., et al., (2003). High levels of stable resistance in transgenic rice with a *cry1Ab* gene from *Bacillus thuringiensis* Berliner to rice leaf folder, *Cnaphalocrocis medinalis* (Guenee) under field conditions. *Crop Prot., 22*, 171–178.

Ye, G. Y., et al., (2001). Transgenic IR72 with fused Bt gene *cry1Ab/cry1Ac* from *Bacillus thuringiensis* is resistant against four lepidopteran species under field conditions. *Plant Biotechnol., 18*, 125–133.

Ye, R., et al., (2009). Development of insect-resistant transgenic rice with *Cry1C*-free endosperm. *Pest Manage. Sci., 65*, 1015–1020.

Ye, R., et al., (2009a). Development of insect-resistant transgenic rice with *Cry1C* free endosperm. *Pest Manag. Sci., 65*, 1015–1020.

Ye, S. H., et al., (2009b). Transgenic tobacco expressing *Zephyranthes grandiflora* agglutinin confers enhanced resistance to aphids. *Appl. Biochem. Biotechnol., 158*, 615–630.

Yokotani, N., et al., (2009). Tolerance to various environmental stresses conferred by the salt-responsive rice gene *OsNAC063* in transgenic *Arabidopsis*. *Planta, 229*, 1065–1075.

Yoon, S., et al., (2015). Over-expression of the cucumber expansin gene (*Cs-EXPA1*) in transgenic maize seed for cellulose deconstruction. *Trans. Res.*, doi: 10.1007/s11248–015–9925–1.

Zale, J. M., et al., (2009). Evidence for stable transformation of wheat by floral dip in *Agrobacterium tumefaciens* . *Plant Cell Rep., 28*, 903–913.

Zeng, J., et al., (2015). Metabolic engineering of wheat provitamin a by simultaneously over-expressing *CrtB* and silencing carotenoid hydroxylase (*TaHYD*). *J. Agri. Food Chem., 63*, 9083–9092.

Zhang, H., et al., (2010). Functional analyses of ethylene response factor JERF3 with the aim of improving tolerance to drought and osmotic stress in transgenic rice. *Trans. Res., 19*, 809–818.

Zhao, D., et al., (2014). Overexpression of a NAC transcription factor delays leaf senescence and increases grain nitrogen concentration in wheat. *Plant Biol.*, doi: 10.1111/plb.12296.

Zhao, Q. C., et al., (2015). Generation of insect-resistant and glyphosate-tolerant rice by introduction of a T-DNA containing two Bt insecticidal genes and an *EPSPS* gene. J. *Zhejiang Univ-Sci. B.* (*Biomed & Biotechnol*), *16*(10), 824–831.

Zhao, Z, Y., et al., (2000). *Agrobacterium*-mediated sorghum transformation. *Plant Mol. Biol., 44,* 789–798.

Zhao, Z. Y., et al., (2001). High throughput genetic transformation mediated by *Agrobacterium tumefaciens* in maize. *Mol. Breed., 8,* 323–333.

Zheng, X., et al., (2009). Overexpression of a NAC transcription factor enhances rice drought and salt tolerance. *Biochem. Biophys. Res. Commun., 379,* 985–989.

Zhu, H., et al., (1998). Biolistic transformation of sorghum using a rice *chitinse* gene. *J. Geneti. Breed., 52,* 243–252.

CHAPTER 10

PLANT TISSUE CULTURE TECHNIQUES AND ITS ACHIEVEMENTS

S. PANDEY[1] and V. C. PANDEY[2]

[1]Scientist, Department of Plant Breeding and Genetics, JNKVV, Jabalpur, Madhya Pradesh–482004, India, Tel.: 08305879759, E-mail: suneetagen@gmail.com

[2]Executive- Research & Development, Hester Biosciences Limited, 1st Floor, Pushpak Complex, Panchavati Circle, Ahmedabad, Gujarat 380006, India

CONTENTS

10.1 INTRODUCTION

Tissue culture is the *in vitro* aseptic culture of cells, tissues, organs, or whole plant under controlled nutritional and environmental conditions to produce an identical copy similar to the selected genotype (Thorpe, 2007). It is an integral part of molecular approaches to plant improvement. Some of the simpler techniques that are more approachable and have been found to be applied directly in plant propagation and genetic improvement of plants are (i) micropropagation, (ii) meristem culture, (iii) somatic embryogenesis, (iv) somaclonal variation, (v) embryo culture, (vi) *in vitro* selection, (vii) anther culture, and (viii) protoplast culture (Smith and Drew, 1990).

Plant tissue culture technology is being widely used for large-scale plant multiplication. Apart from their use as a tool of research, plant tissue culture techniques have, in recent years, gained major industrial importance in the area of plant propagation, disease elimination, plant improvement, and production of secondary metabolites. Small pieces of tissue (named explants) can be multiplied into several thousand plants in a relatively short time period and space under controlled conditions, irrespective of the environmental conditions (Akin-Idowu et al., 2009). Endangered, threatened, and rare species have successfully been grown and conserved by micropropagation because of high coefficient of multiplication and small demand on number of initial plants and space. Certain type of callus cultures give rise to clones that have inheritable characteristics different from those of parent plants due to the possibility of occurrence of somaclonal variability (George, 1993), which leads to the development of commercially important improved varieties. Micropropagation is a rapid propagation process that leads to the production of virus-free plants (Garcia-Gonzales et al., 2010). *Coryodalis yanhusuo*, an important medicinal plant, was propagated by somatic embryogenesis from tuber-derived callus to produce disease-free tubers (Sagare et al., 2000). Meristem tip culture of banana plants devoid from banana bunchy top virus (BBTV) and brome mosaic virus (BMV) was produced (El-Dougdoug and El-Shamy, 2011).

10.2 BASIC REQUIREMENTS OF PLANT TISSUE CULTURE

In tissue culture techniques, there are some basic requirements, viz., (1) aseptic conditions, (2) control of temperature, (3) proper culture media, and (4) sub-culturing. These are briefly described as follows:

10.2.1 ASEPTIC CONDITIONS

The tissue culture laboratory should have aseptic conditions. A pathogen-free environment helps in maintaining good health of the callus, cell, or protoplast culture, resulting in recovery of healthy plants from such culture. The explant and glassware's should be properly sterilized before their entry into the tissue culture laboratory.

10.2.2 CONTROL OF TEMPERATURE

Air conditioning of the tissue culture laboratory is essential. Generally, temperature between 18–25°C is used. However, this varies from species to species. High temperature adversely affects the growth of the callus.

10.2.3 PROPER CULTURE MEDIUM

Culture media have been developed by various workers for different crop species. The medium should be modified according to the requirement of species. The culture media developed by Murashige and Skoog (1962) and Gamberg et al. (1968) are used with some modification in various crop species.

10.2.4 SUB-CULTURING

Transfer of tissue or callus from old culture media to fresh culture media is called sub-culturing. This is essential to maintain good health of the callus or tissues, because after some period, some nutrients are depleted in the culture media and change of media becomes essential.

10.3 NUTRITIONAL COMPONENTS REQUIRED

All the required chemical (Table 10.1) supplemented through growth medium in which water act as the principal biological solvent including physical environment (light, temperature, etc.) are required by the plant cells for *in vitro* culture. The growth medium should supply all the essential mineral ions required

TABLE 10.1   Some of the Elements Important for Plant Nutrition and Their Physiological Function

Element	Function
Nitrogen	Component of proteins, nucleic acids and some coenzymes. Element required in greatest amount
Potassium	Regulates osmotic potential, principal inorganic cataion
Calcium	Cell wall synthesis, membrane function, cell signaling
Magnesium	Enzyme cofactor, component of chlorophyll
Phosphorus	Component of nucleic acids, energy transfer, component of intermediates in respiration and photosynthesis
Sulphur	Component of some amino acids (methionine, cysteine) and some cofactors
Chlorine	Required for photosynthesis
Iron	Electron transfer as a component of cytochromes
Manganese	Enzyme cofactor
Cobalt	Component of some vitamins
Copper	Enzyme cofactor, electron-transfer reactions
Zinc	Enzyme cofactor, chlorophyll biosynthesis
Molybdenum	Enzyme cofactor, component of nitrate reductase

(Source: Slater et al. 2003)

and sometimes organic supplements such as amino acids and vitamins are additionally required as the cells do not have the biosynthetic capability similar to that of the parent plant. Many nonphotosynthetic plants also require the addition of a fixed carbon source in the form of a sugar (most often sucrose) in cell cultures. Physical factors such as temperature, pH, gaseous environment, light (quality and duration), and osmotic pressure have to be maintained within the acceptable limits.

10.4   PLANT CELL CULTURE MEDIA

Culture media used for the *in vitro* cultivation of plant cells are composed of three basic components: (1) essential elements, or mineral ions, supplied as a complex mixture of salts; (2) an organic supplement supplying vitamins and/or amino acids; and (3) a source of fixed carbon, usually supplied as the sugar sucrose. These elements have to be supplied by the culture medium in order to support the growth of healthy cultures *in vitro*. The essential elements are further divided into the following categories:

1. macroelements (or macronutrients);
2. microelements (or micronutrients); and
3. an iron source.

Complete plant cell culture medium is usually made by combining several different components, as outlined in Table 10.2, as given by Murashige

TABLE 10.2 Composition of a Typical Plant Culture Medium Given by Murashige and Skoog

Essential element	Concentration in stock solution (mg/L)	Concentration in medium (mg/L)
Macroelements[a]		
NH_4NO_3	33000	1650
KNO_3	38000	1900
$CaCl_2.2H_2O$	8800	440
$MgSO_4.7H_2O$	7400	370
KH_2PO_4	3400	170
Microelements[b]		
KI	166	0.83
H_3BO_3	1240	6.2
$MnSO_4.4H_2O$	4460	22.3
$ZnSO_4.7H_2O$	1720	8.6
$Na_2MoO_4.2H_2O$	50	0.25
$CuSO_4.5H_2O$	5	0.025
$CoCl_2.6H_2O$	5	0.025
Iron source[c]		
$FeSO_4.7H_2O$	5560	27.8
$Na_2EDTA.2H_2O$	7460	37.3
Organic supplement[c]		
Myoinositol	20000	100
Nicotinic acid	100	0.5
Pyridoxine-HCl	100	0.5
Thiamine-HCl	100	0.5
Glycine	400	2
Carbon source[d]		
Sucrose	Added as solid	30 000

a: 50 mL of stock solution used per liter of medium; b: 5 mL of stock solution used per liter of medium; c: Added as solid.

(Source: Slater et al. 2003)

and Skoog (MS). MS is an extremely widely used medium and forms the basis for many other media formulations. Many other commonly used plant culture media such as Gamborg's B5 and Schenk and Hildebrandt (SH) medium are similar in composition to MS medium and can be thought of as "high-salt" media.

10.5   COMPONENTS OF PLANT TISSUE CULTURE MEDIA

10.5.1   MACROELEMENTS

The stock solution supplies those elements that are required in large amounts for plant growth and development, including nitrogen, phosphorus, potassium, magnesium, calcium, and sulfur (and carbon, which is added separately); these are usually regarded as macroelements. These elements usually comprise at least 0.1% of the dry weight of plants. Nitrogen is most commonly supplied as a mixture of nitrate ions (from KNO_3) and ammonium ions (from NH_4NO_3). Theoretically, there is an advantage in supplying nitrogen in the form of ammonium ions, as nitrogen must be in the reduced form to be incorporated into macromolecules. Nitrate ions therefore need to be reduced before incorporation.

However, at high concentrations, ammonium ions can be toxic to plant cell cultures. and uptake of ammonium ions from the medium causes acidification of the medium. In order to use ammonium ions as the sole nitrogen source, the medium needs to be buffered. High concentrations of ammonium ions can also cause culture problems by increasing the frequency of vitrification (the culture appears pale and "glassy" and is usually unsuitable for further culture). Using a mixture of nitrate and ammonium ions has the advantage of weakly buffering the medium as the uptake of nitrate ions causes OH^- ions to be excreted. Phosphorus is usually supplied as the phosphate ion of ammonium, sodium, or potassium salts. High concentrations of phosphate can lead to the precipitation of medium elements as insoluble phosphates.

10.5.2   MICROELEMENTS

These elements are required in trace amounts for plant growth and development and have diverse roles. Manganese, iodine, copper, cobalt, boron, molybdenum, iron, and zinc usually comprise the microelements, although other elements such as nickel and aluminum are frequently used in some formulations. Iron can precipitate in free form and is therefore supplemented as iron sulfate or iron citrate usually with ethylene diamine tetraacetic acid (EDTA) to conjunct the iron complex, which allows the slow and continuous release of iron into the medium.

10.5.3   ORGANIC SUPPLEMENTS

Thiamine and myoinositol are only two essential vitamins; other vitamins are often added to plant cell culture media. Amino acids provide a source of reduced nitrogen (ammonium ions, uptake causes acidification of the medium); therefore, glycine is frequently used with the organic supplement. Arginine, asparagine, aspartic acid, alanine, glutamic acid, glutamine, and proline are also used, but inclusion of amino acids is not essential sometimes as casein hydrolysate can be used as a relatively cheap source of mixture of amino acids.

10.5.4   CARBON SOURCE

Sucrose is commonly used as a carbon source that is readily assimilated, relatively stable, cheap, and easily available. Other carbohydrates such as glucose, maltose, galactose, and sorbitol can also be used as a carbon source.

10.5.5   GELLING AGENTS

Depending on the type of culture being grown, media for plant cell culture can be used either liquid (suspension culture) or on the solid surface. For preparation solid surface to grow the plant cell or tissue, agar added into the nutrient media while for suspension culture without or low percentage of agar used in nutrient media. As agar is a natural product produced from seaweed, the agar quality depends on the purity, which is available in range of purer gelling agents.

10.5.6   PLANT GROWTH REGULATORS

Plant hormones are the chemical substances that profoundly influence the growth and differentiation of plant cells, tissues, and organs through intercellular communication, called the plant growth regulators. There are currently five recognized groups of plant hormones: auxins, gibberellins, cytokinins, abscisic acid (ABA), and ethylene. They work together coordinating the growth and development of cells. The specialized functions of different kinds of plant growth regulators are mentioned herewith:

10.5.6.1   Auxins

It is used to induce callus from explants and cause root and shoot morphogenesis. Auxins are often most effective in eliciting their effects when combined with cytokinins. Indole-3-acetic acid (IAA) is the principal auxin used in plant tissue culture. Several other indole derivatives, all as precursors to IAA, are known to express auxin activity, probably by converting to IAA in the tissue.

10.5.6.2   Cytokinins

Cytokinins are able to stimulate cell division and induce shoot bud formation in tissue culture. They usually act as antagonists to auxins. (Cytokinins are the derivatives of nitrogenous purine base adenine.). The most commonly used cytokinins in tissue culture include zeatin, N6- (2- isopentyl) adenine (2ip), kinetin, and 6-benzylaminopurine (BAP). They often inhibit embryogenesis and root induction.

10.5.6.3   Gibberellins

There are over 80 different gibberellin compounds in plants, but only giberrellic acid (GA3) is often used in plant tissue culture. They are involved in regulating cell elongation and are agronomically important in determining plant height and fruit-set.

10.5.6.4   Abscisic Acid

Abscisic acid (ABA) inhibits cell division. It is most commonly used in plant tissue culture to promote distinct developmental pathways such as somatic embryogenesis.

10.5.6.5   Ethylene

Ethylene is apparently not required for normal vegetative growth. However, it can have a significant impact on the development of root and shoots.

10.6   IMPORTANT STEPS IN PLANT TISSUE CULTURE TECHNIQUE

Tissue culture technique generally consists of four important steps, viz., (a) isolation of tissues, (b) regeneration and callus formation of culture medium, (c) embryogenesis, and (d) organogenesis. These are briefly described in the following subsections.

10.6.1   ISOLATION OF TISSUES

Tissues for regeneration can be isolated with the help of a sterilized blade from any plant part, viz., leaf, stem, apical bud, axillary bud, etc. The isolated tissues are sterilized and then grown on culture medium. Tissues should be isolated from disease-free portion.

10.6.2   REGENERATION AND CALLUS FORMATION

Tissues proliferate on the culture medium and give rise to a mass of cells called callus. The callus is generally of two types, viz., friable callus, which can be easily manipulated for subsequent culture, and compact callus, which is not suitable for suspension culture.

10.6.3   EMBRYOGENESIS

The process of formation of somatic embryos from the callus is called embryogenesis. Sometimes, somatic embryos are not formed; rather, somatic buds are formed, which after germination give rise to a plant. Sub-culturing leads to healthy growth of callus and rapid embryogenesis.

10.6.4   ORGANOGENESIS

The process of differentiation to shoot and root from the somatic embryos is called organogenesis. Sometimes, somatic embryos are not formed directly from the somatic bud. Usually, a plant develops from somatic embryos after germination. The plants thus obtained are transferred after some time to pot culture from the culture medium. The soil of pots should be sterilized to make it pathogen-free before transplantation of regenerated plants from the culture medium to pots.

10.7 TECHNIQUES OF PLANT TISSUE CULTURE

Some of the important tissue culture techniques used in crop improvement is discussed herewith:

10.7.1 MICROPROPAGATION

Micropropagation starts with the selection of plant tissues (explant) from a healthy, vigorous mother plant (Murashige, 1974). Any part of the plant (leaf, apical meristem, bud, and root) can be used as explant. Micropropagation involves the following stages:

- **Stage 0: Preparation of donor plant:** Any plant tissue can be introduced *in vitro*. To enhance the probability of success, the mother plant should be ex vitro cultivated under optimal conditions to minimize contamination in the *in vitro* culture (Cassells and Doyle, 2005).
- **Stage I: Initiation stage:** In this stage, an explant is surface sterilized and transferred into nutrient medium. The selection of products depends on the type of explant to be introduced. The surface sterilization of explant in chemical solutions is an important step to remove contaminants with minimal damage to plant cells (Husain and Anis, 2009). The most commonly used disinfectants are sodium hypochlorite (Marana et al., 2009; Tilkat et al., 2009), calcium hypochlorite (Garcia et al., 1999), ethanol (Singh and Gurung, 2009), and mercuric chloride ($HgCl_2$) (Husain and Anis, 2009). The cultures are incubated in growth chamber either under light or dark conditions according to the method of propagation.
- **Stage II: Multiplication stage:** The aim of this phase is to increase the number of propagules (Saini and Jaiwal, 2002). The number of propagules is multiplied by repeated subcultures until the desired (or planned) number of plants is attained.

- **Stage III: Rooting stage:** The rooting stage may occur simultaneously in the same culture media used for multiplication of the explants. However, in some cases, it is necessary to change media, including nutritional modification and growth regulator composition to induce rooting and the development of strong root growth.
- **Stage IV: Acclimatization Stage:** At this stage, the *in vitro* plants are weaned and hardened. Hardening is done gradually from high to low humidity and from low light intensity to high light intensity. The plants are then transferred to an appropriate substrate (sand, peat, compost etc.) and gradually hardened under greenhouse.

10.7.2 SOMATIC EMBRYOGENESIS AND ORGANOGENESIS

10.7.2.1 Somatic Embryogenesis

An *in vitro* method of plant regeneration widely used as an important bio-technological tool for sustained clonal propagation (Park et al., 1998). It is a process by which somatic cells or tissues develop into differentiated embryos. These somatic embryos can develop into whole plants without undergoing the process of sexual fertilization as done by zygotic embryos. The somatic embryogenesis can be initiated directly from the explants or indirectly by the establishment of mass of unorganized cells named callus. Plant regeneration via somatic embryogenesis occurs by the induction of embryogenic cultures from zygotic seed, leaf or stem segment and further multiplication of embryos. Mature embryos are then cultured for germination and plantlet development, and finally transferred to soil.

Somatic embryogenesis has been reported in many plants including trees and ornamental plants of different families. The phenomenon has been observed in some cactus species (Torres-Munoz and Rodriguez-Garay, 1996). There are various factors that affect the induction and development of somatic embryos in cultured cells. A highly efficient protocol has been reported for somatic embryogenesis on grapevine (Jayasankar et al., 1999) that showed higher plant regeneration sufficiently when the tissues were cultured in liquid medium.

Plant growth regulators play an important role in the regeneration and proliferation of somatic embryos. Highest efficiency of embryonic callus was induced by culturing nodal stem segments of rose hybrids on medium supplemented with various plant growth regulators (PGRs) alone or in combination (Xiangqian et al., 2002). This embryonic callus showed high germination rate of somatic embryos when grown on abscisic acid (ABA) alone. Somatic embryogenesis is not only a process of regenerating the plants for mass propagation, but it is also regarded as a valuable tool for genetic manipulation. The process can also be used to develop the plants that are resistant to various kinds of stresses (Bouquet and Terregrosa, 2003) and to introduce the genes by genetic transformation (Maynard et al., 1998). A successful protocol has been developed for the regeneration of cotton cultivars with resistance to *Fusarium* and *Verticillium* wilts (Han et al., 2009).

10.7.2.2 Organogenesis

Refers to the production of plant organs, i.e., roots, shoots, and leaves, which may arise directly from the meristem or indirectly from the undifferentiated cell masses (callus). Plant regeneration via organogenesis involves callus production and differentiation of adventitious meristems into organs by altering the concentration of plant growth hormones in nutrient medium. Skoog and Miller (1957) were the first who demonstrated that a high ratio of cytokinin to auxin stimulated the formation of shoots in tobacco callus, while a high auxin to cytokinin ratio induced root regeneration.

10.7.3 CELL SUSPENSION CULTURE

Cell suspension culture systems are used presently for large-scale culturing of plant cells from which secondary metabolites could be extracted. This culture is developed by transferring the relatively friable portion of the callus into liquid medium and is maintained under suitable conditions of aeration, agitation, light, temperature, and other physical parameters (Chattopadhyay et al., 2002). Cell cultures cannot only yield defined standard phytochemicals in large volumes but also eliminate the presence of interfering compounds that occur in the field-grown plants (Lila, 2005). The advantage of this method is that it can ultimately provide a continuous, reliable source of natural products (Rao and Ravishankar, 2002). The major advantage of the cell cultures include synthesis of bioactive secondary metabolites, running in controlled environment, independently from climate and soil conditions (Karuppusamy, 2009). A number of different types of bioreactors have been used for mass cultivation of plant cells. The first commercial application of large-scale cultivation of plant cells was carried out in stirred tank reactors of 200 liter and 750 liter capacities to produce shikonin by cell culture of *Lithospermum erythrorhizon* (Payne et al., 1987).

Cells of *Catharanthus roseus, Dioscorea deltoidea, Digitalis lanata, Panax notoginseng, Taxus wallichiana*, and *Podophyllum hexandrum* have been cultured in various bioreactors for the production of secondary plant products. A number of medicinally important alkaloids, anticancer drugs, recombinant proteins, and food additives are produced in various cultures of plant cells and tissues. Advances in the area of cell cultures for the production of medicinal compounds has made possible the production of a wide variety of pharmaceuticals like alkaloids, terpenoids, steroids, saponins, phenolics, flavonoids, and amino acids (Vijayasree et al., 2010, Yesil-Celiktas et al., 2010). Some of these are now available commercially in the market, for example, shikonin and paclitaxel (Taxol). Until now, 20 different recombinant proteins have been produced in plant cell culture, including antibodies, enzymes, edible vaccines, growth factors, and cytokines (Hellwig et al., 2004). Advances in scale-up approaches and immobilization techniques contribute to a considerable increase in the number of applications of plant cell cultures for the production of compounds with a high added value. Some of the secondary plant products obtained from cell suspension culture of various plants are given in Table 10.3.

10.7.4 ANTHER/POLLEN CULTURE

Regeneration of whole plant from anther/pollen in the culture medium is called anther or pollen culture. Anthers of appropriate stage are taken from the plant. The optimum stage may differ from species to species. This technique is used to obtain haploid plants. This technique has been used in different crops such as Brassica, tobacco, petunia, rice, barley, wheat, triticale, etc. The main uses of this technique in crop improvement are given below:

1. It is useful in development of haploids.
2. Production of homozygous diploids by doubling the chromosome number of haploids.

TABLE 10.3   List of Some Secondary Plant Product Produced in Suspension Culture

Secondary Metabolite	Plant name	Reference
Vasine	*Adhatodavasica*	Shalaka and Sandhya, 2009
Artemisinin	*Artemisia annua*	Baldi and Dixit, 2008
Azadirachtin	*Azadirachtaindica*	Sujanya et al., 2008
Cathin	*Bruceajavanica*	Wagiah et al., 2008
Capsiacin	*Capsicum annum*	Umamaheswai and Lalitha, 2007
Sennosides	*Cassia senna*	Shrivastava et al., 2006
Ajmalicine Secologanin Indole alkaloids Vincristine	*Catharanthusroseus*	Zhao et al., 2001 Contin et al., 1999 Moreno et al., 1993 Lee-Parsons and Rogce, 2006
Stilbenes	*Cayratiatrifoliata*	Roat and Ramawat, 2009
Berberin	*Cosciniumfenustratum*	Khan et al., 2008
Sterols	*Hyssopusofficinalis*	Skrzypek and Wysokinsku, 2003

(Adapted from Hussain et al., 2012)

10.7.5 OVULE CULTURE

Regeneration of whole plant from the ovule in the nutrient medium is called ovule culture. This is also known as ovary culture. Two types of ovules, viz., 1. Unfertilized, and 2. Fertilized may be available. This technique is used for embryo rescue in distant crosses. Haploid plants are obtained from culture of unfertilized ovules and diploid from the fertilized ovules. This technique has been used to a limited extent.

10.7.6 MERISTEM CULTURE

Regeneration of whole plant from tissues of an actively dividing plant part such as stem tip, root tip, or auxillary bud is known as meristem culture. Generally, shoot apical meristem is used for regeneration. In other words, tissues of shoot apex are used for culturing in the nutrient medium. Regeneration is obtained in a suitable culture medium. The medium differs from species to species. This technique has been widely used in vegetatively propagated crops such as sugarcane, potato, banana, and several fruit trees. The main applications of meristem culture in crop improvement are given below:

1. It is used for micropropagation
2. Production of virus-free plants

3. The germplasm can be conserved in the form of meristems at -196°C for long-term storage in liquid nitrogen. In other words, meristems are suitable for cryopreservation.
4. Useful for the exchange of germplasm of asexually propagated plant species.

10.7.7 HAIRY ROOT CULTURES

The hairy root system based on inoculation with *Agrobacterium rhizogenes* has become popular in the last two decades as a method of producing secondary metabolites synthesized in plant roots (Palazon et al., 1997). Organized cultures, and especially root cultures, can make a significant contribution in the production of secondary metabolites. Most of the research efforts that use differentiated cultures instead of cell suspension cultures have focused on transformed (hairy) roots. *Agrobacterium rhizogenes* causes hairy root disease in plants. The neoplastic (cancerous) roots produced by *A. rhizogenes* infection are characterized by high growth rate, genetic stability, and growth in hormone-free media (Hu and Du, 2006). High stability (Giri and Narasu, 2000) and productivity features allow the exploitation of hairy roots as a valuable biotechnological tool for the production of plant secondary metabolites (Pistelli et al., 2010).

These genetically transformed root cultures can produce levels of secondary metabolites comparable to that of intact plants (Shrivastava and Shrivastava, 2007). Hairy root technology has been strongly improved by increased knowledge of molecular mechanisms underlying their development. Optimizing the composition of nutrients for hairy root cultures is critical to gain a high production of secondary metabolites (Hu and Du, 2006). Some of the secondary plant products obtained from hairy root culture of various plants are shown in Table 10.4.

10.8 TECHNIQUES OF TISSUE CULTURE AND THEIR VARIOUS APPLICATIONS

The various tissue culture technique discussed above have many applications and limitations that are discussed briefly in Table 10.5.

TABLE 10.4   List of Some Secondary Plant Product Produced in Hairy Root Culture

Secondary Metabolite	Plant name	Reference
Rosmarinic acid	*Agastache rugosa*	Lee et al., (2007)
Deoursin	*Angelica gigas*	Xu et al., (2008)
Resveratol	*Arachys hypogaea*	Kim et al., (2008)
Tropane	*Brugmansia candida*	Marconi et al., (2008)
Asiaticoside	*Centella asiatica*	Kim et al., (2007)
Rutin	*Fagopyrum esculentum*	Lee et al., (2007)
Glucoside	*Gentiana macrophylla*	Tiwari et al., (2007)
Glycyrrhizin	*Glycyrrhiza glabra*	Mehrotra et al., (2008)
Shikonin	*Lithospermum erythrorhizon*	Fukui et al., (1998)
Glycoside	*Panax ginseng*	Jeong and Park, (2007)

(Adapted from Hussain et al., 2012)

10.9   SCOPE AND IMPORTANCE OF TISSUE CULTURE IN CROP IMPROVEMENT

Tissue-culture techniques are part of a large group of strategies and technologies, ranging through molecular genetics, recombinant DNA studies, genome characterization, gene-transfer techniques, aseptic growth of cells, tissues, and organs, and *in vitro* regeneration of plants that are considered to be plant biotechnologies. The use of the term biotechnology has become widespread recently but, in its most restricted sense, it refers to the molecular techniques used to modify the genetic composition of a host plant, i.e., genetic engineering. The applications of various tissue-culture approaches to crop improvement through breeding, wide hybridization, haploidy, somaclonal variation and micropropagation are discussed:

10.9.1   TISSUE CULTURE IN AGRICULTURE

As an emerging technology, the plant tissue culture has a great impact on both agriculture and industry, through providing plants needed to meet the ever-increasing world demand. It has made significant contributions to the advancement of agricultural sciences in recent times, and today, they constitute an indispensable tool in modern agriculture (Garcia-Gonzales et al., 2010). Biotechnology has been introduced into agricultural practice at a rate

TABLE 10.5 Techniques of Tissue Culture and Their Various Applications

Technique	Application	Comments and References
Micro-propa-gation	• Rapid propagation of a superior plant while maintaining the genetic make up under controlled conditions • To maintain stock • Germplasm storage	1. Immediate application; however, the ease with which plants can be micro-propagated varies from species to species and even certain genotypes within a species can prove to be more recalcitrant than others. In general, herba-ceous species are more amenable to tissue culture techniques than woody perennials. 2. Somaclonal variation can be a problem with some micropropagation tech-niques. Field evaluation of plants is required to verify trueness-to-type and to check for genetic variants (Hussey, 1978; Hussey, 1983; Withers, 1989)
Meristem culture	• Elimination of diseases (particularly viral diseases) from plant propagative material	1. Immediate application if plants amenable to tissue culture. 2. Heat treatment/meristem culture does not ensure that the material has been freed of virus. Quarantine and virus indexing are still recommended to verify that the material is in fact disease-free (Quak, 1977)
Somatic em-bryo-genesis	• Rapidly increase desirable plants while maintaining the genotype of the original plant	1. Involves regeneration from callus and cell suspensions and as such is more difficult to achieve than micropropagation. More research and develop-ment is needed to successfully develop the technique and a time frame that may involve years of trial and error with a recalcitrant species. 2. Orbital shakers, centrifuge and microscopes are needed when working with cell suspensions. 3. Regeneration is from undifferentiated cells the chances of somaclonal variations increase (Narayanaswamy, 1977)

TABLE 10.5   (Continued)

Technique	Application	Comments and References
Somaclonal variation	• To induce desirable, heritable changes in regenerated plants	1. Involves regeneration from callus and cell suspensions, therefore the constraints and limitations are the same as those above. 2. Not all changes are desirable; in fact most are deleterious or of no agronomic use. 3. Not recommended if suitable genetic diversity is already present in the species; better application in vegetatively propagated material with a limited gene base. 4. Screening of thousands plants for those with useful characters is expensive and time-consuming. If a selection pressure can be applied at the cellular level then better use of somaclonal variation and *in vitro* selection can be made (Larkin and Scowcroft, 1981; Evans, 1989)
Embryo culture	• To 'rescue' embryos during attempts at wide hybridization by sexual crosses between distantly related plants and culture them to maturity	1. Relatively easy to culture. 2. Immediate application. 3. Chances of success are good; but difficulty increases with more immature embryos. Hormone and growth factor requirements are more specific with early-stage embryos (Raghavan, 1976; Williams et al., 1982; Collins and Grosser, 1984)
In vitro selection	• To induce desirable, heritable changes in regenerated plants by subjecting a population of cells to a selection pressure	1. Involves regeneration from callus and cell suspension. Constraints and limitation as mentioned in embryo culture. 2. Important to have a reproducible system for the regeneration of large numbers of plants from stressed cells as the selecting agent may lower the ability to regenerate plants. 3. Important that tolerance to the stress operates at both the cellular and whole plant levels so that there is a greater chance of recovering desirable plant. Unfortunately, many of the agriculturally important traits are multigenic and depend on the structural and physiological integrity of the whole plant (Tomes and Swanson, 1982; Chaleff, 1983)

Technique	Application	Comments and References
Anther culture	• To produce homozygous, pure-breeding lines of plants for hybrid production and genetic studies • To improve the efficiency of *in vitro* selection	1. Involves regeneration from callus, cell suspension and pollen. Constraints and limitation as mentioned in embryo culture. 2. Important to have the frequency of regeneration and to be able to distinguish between plants regenerated from haploid somatic tissue found in the anther. 3. The use of colchicines may be needed to double the chromosome number of haploid plants (Collins and Genovesi, 1982; Hu and Zeng, 1984)
Protoplast culture	• To incorporate potentially useful genes from one plant species to another by fusion of protoplast and regeneration from the hybrid cell line • Somatic hybridization • To transfer specific genes into protoplasts and regenerate transgenic plants	1. Protoplast are cells from which the cell wall has been removed either by mechanical and/or enzymatic methods 2. Orbital shakers, centrifuge and microscopes are needed. 3. Regeneration of plants from protoplast is generally very difficult to accomplish and a long lead time is often needed to develop the techniques with a particular species (Bhojwani et al., 1977; Binding, 1986)

(Source: Chawla, 2002)

without precedent. Tissue culture allows the production and propagation of genetically homogeneous, disease-free plant material (Chatenet et al., 2001). Cell and tissue *in vitro* culture is a useful tool for the induction of somaclonal variation (Marino and Battistini, 1990).

Genetic variability induced by tissue culture could be used as a source of variability to obtain new stable genotypes. Interventions of biotechnological approaches for *in vitro* regeneration, mass micropropagation techniques, and gene transfer studies in tree species have been encouraging. *In vitro* cultures of mature and/or immature zygotic embryos are applied to recover plants obtained from intergeneric crosses that do not produce fertile seeds (Ahmadi et al., 2010). Genetic engineering can make possible a number of improved crop varieties with high yield potential and resistance against pests. Genetic transformation technology relies on the technical aspects of plant tissue culture and molecular biology for:

- Production of improved crop varieties
- Production of disease-free plants (virus)
- Genetic transformation
- Production of secondary metabolites
- Production of varieties tolerant to salinity, drought, and heat stresses

10.9.2 PRESERVATION OF GERMPLASM

In vitro cell and organ culture offers an alternative source for the conservation of endangered genotypes (Sengar et al., 2010). Germplasm conservation worldwide is increasingly becoming an essential activity due to the high rate of disappearance of plant species and the increased need for safeguarding the floristic patrimony of the countries (Filho et al., 2005). Tissue culture protocols can be used for the preservation of vegetative tissues when the targets for conservation are clones instead of seeds, to keep the genetic background of a crop, and to avoid the loss of the conserved patrimony due to natural disasters, whether biotic or abiotic stress (Tyagi et al., 2007). The plant species that do not produce seeds (sterile plants) or that have "recalcitrant" seeds that cannot be stored for a long period of time can successfully be preserved via *in vitro* techniques for the maintenance of gene banks.

Cryopreservation plays a vital role in the long-term *in vitro* conservation of essential biological material and genetic resources. It involves the storage of *in vitro* cells or tissues in liquid nitrogen that results in cryo-injury on the exposure of tissues to physical and chemical stresses. Successful cryo-preservation is often ascertained by cell and tissue survival and the ability to re-grow or regenerate into complete plants or form new colonies (Harding, 2004). It is desirable to assess the genetic integrity of recovered germplasm to determine whether it is "true-to-type" following cryopreservation (Day, 2004). The fidelity of recovered plants can be assessed at phenotypic, histological, cytological, biochemical, and molecular levels, although there are advantages and limitations of the various approaches used to assess genetic stability (Harding et al., 2005). Cryobionomics is a new approach to study genetic stability in the cryopreserved plant materials. The embryonic tissues can be cryopreserved for future use or for germplasm conservation (Correddoira et al., 2004).

10.9.3 EMBRYO RESCUE

Embryo culture techniques helps in making the interspecific crosses successful when there is postfertilization disharmony between the embryo and endosperm. The embryo in interspecific cross is removed before abortion and cultured in a nutrient medium. The technique has been developed to break seed dormancy, test the vitality of seeds, and for production of rare species and haploid plants (Burun and Poyrazoglu, 2002; Holeman, 2009). It is an effective technique that is employed to shorten the breeding cycle of plants by growing excised embryos and results in the reduction of long dormancy period of seeds. Intravarietal hybrids of an economically important energy plant "Jatropha" have been produced successfully with the specific objective of mass multiplication (Mohan et al., 2011). Somatic embryogenesis and plant regeneration has been carried out in embryo cultures of Jucara Palm for rapid cloning and improvement of selected individuals (Guerra and Handro, 1988). In addition, conservation of endangered species can also be attained by practicing embryo culture technique. Recently, a successful protocol has been developed for the *in vitro* propagation of *Khaya grandifoliola* by excising embryos from mature seeds (Okere and Adegey, 2011). The plant has a high economic value for timber wood and for medicinal purposes as well. This technique has an important application in forestry by offering a mean of propagation of elite individuals where the selection and improvement of natural population are difficult.

10.9.4   GENETIC TRANSFORMATION

Genetic transformation is the most recent aspect of plant cell and tissue culture that provides the mean of transfer of genes with desirable trait into host plants and recovery of transgenic plants (Hinchee et al., 1994). The technique has a great potential of genetic improvement of various crop plants by integrating in plant biotechnology and breeding programs. It has a promising role for the introduction of agronomically important traits such as increased yield, better quality, and enhanced resistance to pests and diseases (Sinclair et al., 2004). Genetic transformation in plants can be achieved by either vector-mediated (indirect gene transfer) or vector less (direct gene transfer) method (Sasson, 1993). Among vector-dependent gene transfer methods, *Agrobacterium*-mediated genetic transformation is most widely used for the expression of foreign genes in plant cells.

Successful introduction of agronomic traits in plants was achieved using root explants for the genetic transformation (Franklin and Lakshmi, 2003). Virus-based vectors offers an alternative way of stable and rapid transient protein expression in plant cells, thus providing an efficient mean of recombinant protein production on a large scale (Chung et al., 2006). Recently, successful transgenic plants of Jatropha were obtained by direct DNA delivery to mature seed-derived shoot apices via the particle bombardment method (Purkayastha et al., 2010). This technology has an important impact on the reduction of toxic substances in seeds (Misra and Misra, 1993), thus overcoming the obstacle of seed utilization in various industrial sector. Regeneration of disease or viral-resistant plants is now achieved by employing the genetic transformation technique. Researchers succeeded in developing transgenic plants of potato resistant to potato virus Y (PVY), which is a major threat to potato crop worldwide (Bukovinszki et al., 2007). In addition, marker-free transgenic plants of *Petunia hybrid* were produced using the multi-auto-transformation (MAT) vector system. The plants exhibited high level of resistance to *Botrytis cinerea,* causal agent of gray mold (Khan et al., 2011).

10.9.5   PROTOPLAST FUSION

Somatic hybridization is an important tool of plant breeding and crop improvement by the production of interspecific and intergeneric hybrids. The technique involves the fusion of protoplasts of two different genomes followed by the selection of desired somatic hybrid cells and regeneration of hybrid plants (Evans and Bravo, 1988). Protoplast fusion provides an efficient mean of gene transfer with desired trait from one species to another and has an increasing impact on crop improvement (Brown and Thorpe, 1995). Somatic hybrids were produced by fusion of protoplasts from rice and ditch reed using electrofusion treatment for salt tolerance (Mostageer and Elshihy, 2003).

In vitro fusion of the protoplast opens a way of developing unique hybrid plants by overcoming the barriers of sexual incompatibility. The technique has been applicable in horticultural industry to create new hybrids with increased fruit yield and better resistance to diseases. Successful viable hybrid plants were obtained when protoplasts from citrus were fused with other related citrinae species (Motomura et al., 1997). The potential of somatic hybridization is best illustrated by the production of intergeneric hybrid plants among the members of Brassicaceae (Toriyama et al., 1987). To resolve the problem of loss of chromosomes and decreased regeneration capacity, successful protocol has been established for the production of somatic hybrid plants by using two types of wheat protoplast as a recipient and the protoplast of *Haynaldia villosa* as a fusion donor. It is also employed as an important gene source for wheat improvement (Liu et al., 1988).

10.9.6 HAPLOID PRODUCTION

Tissue culture techniques enable to produce homozygous plants in relatively short time period through the protoplast, anther, and microspore cultures instead of conventional breeding (Morrison and Evans, 1998). Haploids are sterile plants with a single set of chromosomes, which are converted into homozygous diploids by spontaneous or induced chromosome doubling. The doubling of chromosomes restores the fertility of plants, resulting in the production of double haploids with potential to become pure breeding new cultivars (Basu et al., 2011). The term androgenesis refers to the production of haploid plants from young pollen cells without undergoing fertilization. Sudherson et al. (2008) reported haploid plant production of sturt's desert pea by using pollen grains as primary explants. The haploidy technology has now become an integral part of plant breeding programs by speeding up the production of inbred lines (Bajaj, 1990) and overcoming the constraints of seed dormancy and embryo nonviability (Yeung et al., 1981).

The technique has a remarkable use in genetic transformation by the production of haploid plants with induced resistance to various biotic and abiotic stresses. Introduction of genes with desired trait at the haploid state, followed by chromosome doubling led to the production of double haploids inbred (Chauhan and Khurana, 2011). At present, 250 plant species have been used to produce haploid plants by pollen, microspore, and anther culture. These include cereals (barley, maize, rice, rye, triticale, and wheat), forage crops (alfalfa and clover), fruits (grape and strawberry), medicinal plants *(Digitalis* and *Hyoscyamus),* ornamentals *(Gerbera* and sunflower), oil seeds (canola and rape), trees (apple, litchi, poplar, and rubber), plantation crops (cotton, sugarcane, and tobacco), and vegetable crops (asparagus, brussels sprouts, cabbage, carrot, pepper, potato, sugar beet, sweet potato, tomato, and wing bean).

10.9.7   MICROPROPAGATION

During the last 30 years, it has become possible to regenerate plantlets from explants and/or callus from all types of plants. As a result, laboratory-scale micropropagation protocols are available for a wide range of species, and at present, micropropagation is the widest use of plant tissue culture technology. The cost of the labor needed to transfer tissue repeatedly between vessels and the need for asepsis can account for up to 70% of the production costs of micropropagation. There are three methods used for micropropagation:

1.   Enhancing axillary-bud breaking;
2.   Production of adventitious buds; and
3.   Somatic embryogenesis.

In the latter two methods, organized structures arise directly on the explant or indirectly from the callus. Axillary-bud breaking produces the least number of plantlets, as the number of shoots produced is controlled by the number of axillary buds cultured, but it remains the most widely used method in commercial micropropagation and produces the most true-to-type plantlets. Adventitious budding has a greater potential for producing plantlets, as bud primordia may be formed on any part of the inoculum. Unfortunately, somatic embryogenesis, which has the potential of producing the largest number of plantlets, can only presently be induced in a few species.

10.9.8 PRODUCTION OF SYNTHETIC SEED

A synthetic or artificial seed has been defined as a somatic embryo encapsulated inside a coating and is considered to be analogous to a zygotic seed. There are several different types of synthetic seed: somatic embryos encapsulated in a water gel; dried and coated somatic embryos; dried and uncoated somatic embryos; somatic embryos suspended in a fluid carrier; and shoot buds encapsulated in a water gel. These artificial seeds can be utilized for the rapid and mass propagation of desired plant. Applications of synthetic seeds include the maintenance of male sterile lines, maintenance of parental lines for hybrid crop production, and preservation and multiplication of elite genotypes of woody plants that have long juvenile developmental phases. However, before the widespread application of this technology, somaclonal variation will have to be minimized, large-scale production of high-quality embryos must be perfected in the species of interest, and the protocols will have to be made cost-effective compared with existing seed or micropropagation technologies.

10.9.9 PRODUCTION OF SECONDARY METABOLITES

The most important chemicals produced using cell culture is secondary metabolites, which are defined as those cell constituents that are not essential for survival'. These secondary metabolites include alkaloids, glycosides (steroids and phenolics), terpenoids, latex, tannins, etc. It has been observed that as the cells undergo morphological differentiation and maturation during plant growth, some of the cells specialize to produce secondary metabolites (Table 10.6). The *in vitro* production of secondary metabolites is much higher from differentiated tissues when compared to nondifferentiated tissues.

The cell cultures contribute in several ways to the production of natural products. These are:

- A new route of synthesis to establish products, e.g., codeine, quinine, pyrethroids;
- A route of synthesis to a novel product from plants difficult to grow or establish, e.g., thebain from *Papaver bracteatum*;
- A source of novel chemicals in their own right, e.g., rutacultin from culture of *Ruta*;

TABLE 10.6   Plant Species and Secondary Metabolites Obtained from Them Using Tissue Culture Techniques

Product	Plant source	Uses
Artemisin	*Artemisia spp.*	Antimalarial
Azadirachtin	*Azadirachta indica*	Insecticidal
Berberine	*Coptis japonica*	Antibacterial anti inflammatory
Capsaicin	*Capsicum annum*	Cures Rheumatic pain
Codeine	*Papaver spp.*	Analgesic
Camptothecin	*Campatotheca accuminata*	Anticancer
Cephalotaxine	*Cephalotaxus harringtonia*	Antitumour
Digoxin	*Digitalis lanata*	Cardiac tonic
Pyrethrin	*Chrysanthemum cinerariaefolium*	Insecticide (for grain storage)
Morphine	*Papaver somniferum*	Analgesic, sedative
Quinine	*Cinchona officinalis*	Antimalarial
Taxol	*Taxus spp.*	Anticarcinogenic
Vincristine	*Cathranthus roseus*	Anticarcinogenic
Scopolamine	*Datura stramonium*	Antihypertensive

Source: http://www.biotechnology4u.com/plant_biotechnology_applications_cell_tissue_culture.html

- As biotransformation systems either on their own or as part of a larger chemical process, e.g., digoxin synthesis.

10.9.9.1   The Advantages of *In Vitro* Production of Secondary Metabolites

- The cell cultures and cell growth are easily controlled in order to facilitate improved product formation.
- The recovery of the product is easy.
- As the cell culture systems are independent of environmental factors, seasonal variations, pest and microbial diseases, and geographical location constraints, it is easy to increase the production of the required metabolite.
- Mutant cell lines can be developed for the production of novel and commercially useful compounds.
- Compounds are produced under controlled conditions according to the market demands.
- The production time is less and cost effective due to minimal labor involved.

10.9.10 PRODUCTION OF SOMATIC HYBRIDS AND CYBRIDS

The somatic cell hybridization/parasexual hybridization or protoplast fusion offers an alternative method for obtaining distant hybrids with desirable traits significantly between species or genera, which cannot be made to cross by the conventional method of sexual hybridization. The applications of somatic hybridization are as follows:

10.9.10.1 Creation of Hybrids with Disease Resistance

Many disease-resistance genes (e.g., tobacco mosaic virus, potato virus X, club rot disease) could be successfully transferred from one species to another, for example, resistance has been introduced in tomato against diseases such as TMV, spotted wilt virus, and insect pests.

10.9.10.2 Environmental Tolerance

Using somatic hybridization, the genes conferring tolerance to cold, frost, and salt were introduced in tomato.

10.9.10.3 Cytoplasmic Male Sterility

Using cybridization method, it was possible to transfer cytoplasmic male sterility.

10.9.10.4 Quality Characteristics

Somatic hybrids with selective characteristics have been developed, e.g., production of high nicotine content.

10.9.11 DEVELOPMENT OF AMINO ACID ANALOG RESISTANT MUTANTS

Cereal grains are deficient in lysine, tryptophan, and threonine, while pulses are deficient in methionine and tryptophan. Amino acid analog-resistant cells may be expected to show a relatively higher concentration of that particular amino acid, for example, carrot (*D. carota*) and tobacco (*N. tabacum*) cell lines resistant to tryptophan analog 5-methyl tryptophan show a 10- to 27-fold increase in the level of tryptophan. Similarly, rice cells resistant to lysine analog 5-(*B-aminoethyl*)-cysteine, show much higher levels of lysine. This technique may prove useful in the development of crop varieties with better-balanced amino acid content.

10.9.12 EVOLUTION OF NEW VARIETIES

The use of plant tissue culture to produce somaclonal variation is a means of generating variation that may be needed in a breeding program. This is particularly true in species that are traditionally propagated asexually or for which only few cultivars are available. Deliberate attempts to induce variations in tissue culture have been in progress for the last 60 years and a large number of variants in ornamentals and horticultural crops have been reported (Rout et al., 2006). However, there are only a few instances where somaclonal variations have produced agriculturally desirable changes in the progeny (Table 10.7). These include sugarcane – increase in cane and sugar yield, and resistance to eye-spot disease (Larkin & Scowcroft, 1983); potato – improvement of tuber shape, color and uniformity, and late blight resistance (Shepherd et al., 1980); tomato – increased solids, resistance to *Fusarium* race 2 (Evans, 1989). Among the other agricultural crops, CIEN BTA-03, a variant of Williams variety of Banana resistant to yellow Sigatoka disease; AT626 & BT 627 of sugarcane variant resistant to sugarcane mosaic virus A and B are released for commercial usage. In addition, "Ono," a sugarcane variant from variety, Pindar resistant to Fiji disease; ATCC 40463, a tobacco variety with enhanced flavor; DK 671, corn variety with higher yield, with lasting green color and higher seedling vigor are the varieties from overseas inventions. Bio-13, a variety of citronella released by CIMAP, India, and Bio- 902, Bio-YSR variants of *Brassica* parent "Varuna" with enhanced seed yields are from India that are being multiplied and are cultivated successfully on the commercial scale.

10.10 CONCLUDING REMARKS

Plant tissue culture represents the most promising areas of application at present and give a positive outlook in the future. The areas range from

TABLE 10.7 Somaclonal Variations in Some Agronomically Important Plant Species

Species	Explant	Variant characters	Transmission
Avena sativa	Immature embryo, apical meristem	Plant height, heading date, leaf striping, awns	Sexual
Triticuam Aestivum	Immature embryo	Plant height, spike shape, awns, maturity, tillering, leaf wax, gliadine, amylase	Sexual
Oryza sativa	Seed embryo	Number of tillers, panicle size, seed fertility, flowering date, plant height	Sexual
Saccharum Officinarum	Various	Eyespot, Fiji virus, downy mildew, caulm color, spot disease, auricle length, sterase isozyme, sugar yield	Asexual
Zea mays	Immature embryo	Endosperm and seedling mutants, *D. maydis* race T toxin resistance mtDNA sequence rearrangement	Sexual
Solanum Tuberosum	Protoplast, leaf callus	Tuber shape, yield, maturity date, plant habit, stem, leaf and flower morphology, early and late blight resistance	Asexual
Lycopersicon Esculentum	Leaf	Male sterilty, jointless pedicel, fruit color, indeterminate type	Sexual
Nicotaina Species	Anthers, protoplasts, leaf callus	Plant height, leaf size, yield grade index, alkaloids, reducing sugars, specific leaf chlorophyll loci	Sexual
Medicago Sativa	Immature Ovaries	Multifoliate leaves, petiole length, plant habit, plant height, dry matter yield	Asexual
Brassica Species	Anthers, embryos, meristems	Flowering time, growth habit, waxiness, glucosinolates, *Phoma lingam* tolerance	Sexual

micropropagation of ornamental and forest trees, production of pharmaceutically interesting compounds, and plant breeding for improved nutritional value of staple crop plants, including trees to cryopreservation of valuable germplasm. All biotechnological approaches like genetic engineering, haploid induction, or somaclonal variation to improve traits strongly depend on an efficient *in vitro* plant regeneration system. The rapid production of high-quality, disease-free, and uniform planting stock is only possible through micropropagation. New opportunities have been created for producers, farmers, and nursery owners for high-quality planting materials of fruits, ornamentals, forest tree species, and vegetables.

Plant production can be carried out throughout the year irrespective of season and weather. However, micropropagation technology is expensive as compared to conventional methods of propagation by means of seed, cuttings, and grafting etc. Therefore, it is essential to adopt measures to reduce the cost of production. Low cost production of plants requires cost-effective practices and optimal use of equipment to reduce the unit cost of plant production. It can be achieved by improving the process efficiency and better utilization of resources. Bioreactor-based plant propagation can increase the speed of multiplication and growth of cultures and reduce space, energy, and labor requirements when commencing commercial propagation. The cost of production may also be reduced by selecting several plants that provide the option for around the year production and allow cost flow and optimal use of equipment and resources. It is also essential to have sufficient mother culture and reduce the number of subculture to avoid variation and plan the production of plants according to the demand. Quality control is also very essential to assure high-quality plant production and to obtain confidence of the consumers.

The selection of explant source, disease-free material, authenticity of variety, and elimination of somaclonal variants are some of the most critical parameters for ensuring the quality of the plants. The *in vitro* culture has a unique role in sustainable and competitive agriculture and forestry and has been successfully applied in plant breeding for rapid introduction of improved plants. Plant tissue culture has become an integral part of plant breeding. It can also be used for the production of plants as a source of edible vaccines. Plant tissue culture is a noble approach to obtain these substances in large scale. Perhaps the most significant role that plant cell culture has to play in the future will be in its association with transgenic plants. The ability to accelerate the conventional multiplication rate can be of great benefit to many countries where a disease or some climatic disaster wipes out crops. The loss of genetic resources is a common story when germplasm is held in field genebanks. Tissue culture has been one of the main technological tools and reasons that have contributed to the "Second Green Revolution and Gene Revolution." India is being looked upon by the world as the main technology base for production and supply of economically important plant varieties.

KEYWORDS

- **micro-propagation**
- **somaclonal variation**
- **somatic embryogenesis**
- **tissue culture**

REFERENCES

Ahmadi, A., Azadfar, D., & Mofidabadi, A. J., (2010). Study of inter-generic hybridization possibility between *Salix aegyptica* and *Populus caspica* to achieve new hybrids. *Int. J. Plant Prod.*, *4*(2), 143–147.

Akin-Idowu, P. E., Ibitoye, D. O., & Ademoyegun, O. T., (2009). Tissue culture as a plant production technique for horticultural crops. *Afr. J. Biotechnol.*, *8*(16), 3782–3788.

Bajaj, Y. P. S., (1990). *In vitro* production of haploids and their use in cell genetics and plant breeding. In: Bajaj, Y. P. S., editor. *Biotechnol. Agr. Forest. Berlin.*, 3–44.

Baldi, A., & Dixit, V. K., (2008). Enhanced artemisinin production by cell cultures of *Artemisia annua*. *Current Trends in Biotechnology and Pharmacology*, *2*, 341–348.

Basu, S. K., Datta, M., Sharma, M., & Kumar, A., (2011). Haploid plant production technology in wheat and some selected higher plants. *Aust. J. Crop Sci.*, *5*(9), 1087–1093.

Bhojwani, S. S., Evans, P. K., & Cocking, E. C., (1977). Protoplast technology in relation to crop plants: Progress and problems. *Euphytica.*, *26*, 343–360.

Binding, H., (1986). In: *Cell Culture and Somatic Cell Genetics*, vol. *3*, Vasil, I. K., (ed.), Academic Press, Orlando, 259–274.

Bouquet, A., & Terregrosa, L., (2003). Micropropagation of grapevine (*Vitus spp.*). In: Jain, S. M., Ishii, K., editors. Micropropagation of woody trees and fruits. *The Netherlands*, *75*, 319–352.

Brown, D. C. W., & Thorpe, T. A., (1995). Crop improvement through tissue culture. *World J. Microbiol.& Biotechnol.*, *11*, 409–415.

Bukovinszki, A., Diveki, Z., Csanyi, M., Palkovics, L., & Balazs, E., (2007). Engineering resistance to PVY in different potato cultivars in a marker-free transformation system using a 'Shooter mutant' *A. tumefaciens*. *Plant Cell Rep.*, *26*(4), 459–465.

Burun, B., & Poyrazoglu, E. C., (2002). Embyo culture in barley (*Hordeum vulgare* L.). *Turk. J. Biol.*, *26*, 175–180.

Cassells, A. C., & Doyle, B. M., (2005). Pathogen and biological contamination management: the road ahead. In: Loyola-Vargas, V. M., Vázquez-Flota, F., editors. *Plant Cell Culture Protocols*, Humana Press. New York, USA, 35–50.

Chaleff, R. S., (1983). Isolation of agronomically useful mutants from plant cell cultures. *Science, 219*(4585), 676–82.

Chawla, H.S.; Introduction to Plant Biotechnology. Science Publishers, 2002 - Technology & Engineering, 1-538.

Chatenet, M., Delage, C., Ripolles, M., Irey, M., Lockhart, B. L. E., & Rott, P., (2001). Detection of sugarcane yellow leaf virus in quarantine and production of virus-free sugarcane by apical meristem culture. *Plant Disease, 85*(11), 1177–1180.

Chattopadhyay, S., Farkya, S., Srivastava A. K., & Bisaria, V. S., (2002). Bioprocess considerations for production of secondary metabolites by plant cell suspension cultures. *Biotechnol. Bioprocess. Eng., 7,* 138–149.

Chauhan, H., & Khurana, P., (2011). Use of double haploid technology for development of stable drought tolerant bread wheat (*Triticum aestivum* L.) transgenics. *Plant Biotechnol. J., 9*(3), 408–417.

Chung, S. M., Manjusha, V., & Tzfira, T., (2006). Agrobacterium is not alone: gene transfer to plants by viruses and other bacteria. *Trends Plant Sci., 11*(1), 1–4.

Collins, G. B., & Genovesi, A. D., (1982). In: *Application of Plant Cell and Tissue Culture to Agriculture and Industry*. Tomes, D. T., Ellis, B. E., Harney, P. M., et al.,(eds.), The University of Guelph, Guelph, 1–24.

Collins, G. B., & Grosser, J. W., (1984). Culture of embryos. In: *Cell Culture and Somatic Cell Genetics of Plants*, Vasil, I. K., (ed.), *1,* 241–257.

Contin, A., Van der Heijden, R., & Verpoorte, R., (1999). Effects of alkaloid precursor feeding and elicitation on the accumulation of secologanin in a *Catharanthus roseus* cell suspension culture. *Plant Cell Tissue Org. Cult., 56,* 111–119.

Corredoira, E., San-Jose, M. C., Ballester, A., & Vieitez, A. M., (2004). Cryopreservation of zygotic embryo axes and somatic embryos of European chestnut. *Cryo Lett., 25,* 33–42.

Day, J. G., (2004). Cryopreservation: fundamentals, mechanisms of damage on freezing/thawing and application in culture collections. *Nova Hedwigia., 79,* 191–206.

El-Dougdoug, K. A., & El-Shamy, M. M., (2011). Management of viral diseases in banana using certified and virus tested plant material. *Afr. J. Microbiol. Res., 5*(32), 5923–5932.

Evans, D. A., & Bravo, J. E., (1988). Agricultural applications of protoplast fusion. In: Marby, T. I., editor. *Plant Biotechnol. Austin.,* 51–91.

Evans, D. A., (1989). Somaclonal variation-Genetic basis and breeding applications. *Trends Genet., 5,* 46–50.

Evans, D. A., (1989). Somaclonal variation-genetic basis and breeding applications. *Trends Genet., 5,* 46–50.

Filho, A. R., Dal Vesco, L. L., Nodari, R. O., Lischka, R. W., Muller, C. V., & Guerra, M. P., (2005). Tissue culture for the conservation and mass propagation of *Vriesea reitzii* Leme and Costa, abromelian threatened of extinction from the Brazilian Atlantic Forest. *Biodivers. Conserv., 14*(8), 1799–1808.

Franklin, G., & Lakshmi, S. G., (2003). *Agrobacterium tumefaciens*-mediated transformation of egg plant (*Solanum melongena* L.) using root explants. *Plant Cell Rep., 21,* 549–554.

Fukui, H., Feroj, H., Ueoka, T., & Kyo, M., (1998). Formation and secretion of a new benzoquinone by hairy root cultures of *Lithospermum erythrorhizon*. *Phytochem., 47,* 1037–1039.

Gamberg, O. L., Miller, R. A., & Ojima, O., (1968). Nutrient requirements of suspension cultures of soybean root cell. *Exp. Cell Res.*, *50*, 151–158.

Garcia, R., Moran, R., Somonte, D., Zaldua, Z., Lopez, A., & Mena, C. J., (1999). Sweet potato (*Ipomoea batatas* L.) biotechnology: perspectives and progress. In: Altman, A., Ziv, M., Shamay, I., editors. Plant biotechnology and *in vitro* biology in 21st century. *The Netherlands*, 143–146.

Garcia-Gonzales, R., Quiroz, K., Carrasco, B., & Caligari, P., (2010). Plant tissue culture: Current status, opportunities and challenges. *Cien. Inv. Agr.*, *37*(3), 5–30.

George, E. F., (1993). *Plant Propagation by Tissue Culture*. 2nd edition. Part 1. *The technology*. Exegetics Ltd, pp. 574.

Giri, A., & Narasu, M., (2000). Transgenic hairy roots: recent Trends and application. *Biotechnol. Adv.*, *18*, 1–22.

Guerra, M. P., & Handro, W., (1988). Somatic embryogenesis and plant regeneration in embryo cultures of *Euterpe edulis* Mart. (Palmae). *Plant Cell Rep.*, *7*, 550–552.

Han, G. Y., Wang, X. F., Zhang, G. Y., & Ma, Z. Y., (2009). Somatic embryogenesis and plant regeneration of recalcitrant cottons (*Gossypium hirsutum*). *Afr. J. Biotechnol.*, *8*(3), 432–437.

Harding, K., (2004). Genetic integrity of cryopreserved plant cells: a review. *Cryo. Lett.*, *25*, 3–22.

Harding, K., Johnston, J., & Benson, E. E., (2005). Plant and algal cell cryopreservation: Issues in genetic integrity, concepts. In: Benett, I. J., Bunn, E., Clarke, H., McComb, J. A., editors. Cryobionomics and current European applications. In: *Contributing to a Sustainable Future.* Western Australia, 112–119.

Hellwig, S., Drossard, J., Twyman, R. M., & Fischer, R., (2004). Plant cell cultures for the production of recombinant proteins. *Nat. Biotechnol.*, *22*, 1415–1422.

Hinchee, M. A. W., Corbin, D. R., Armstrong, C. L., Fry, J. E., Sato, S. S., Deboer, D. L., Petersen, W. L., Armstrong, T. A., Connor-Wand, D. Y., Layton, J. G., & Horsch, R. B., (1994). *Plant Transformation in Plant Cell and Tissue Culture*. In: Vasil, L. K., Thorpe, T. A., editors. Dordrecht, Kluwer Academic., 231–270.

Holeman, D. J., (2009). Simple embryo culture for plant breeders: a manual of technique for the extracyion and *in vitro* germination of mature plant embryos with emphasis on the rose. First edition. *Rose Hybridizers Association*, 10.

Hu, H., & Zeng, J. Z., (1984). Development of new varieties of anther culture, in *Handbook of Plant Cell Culture* (Ammirato, P. V., Evans, D. A., Sharp, W. R., & Yamada, Y., eds.), vol. 3, Macmillan, New York, 65–90.

Hu, Z. B., & Du, M., (2006). Hairy Root and Its Application in Plant Genetic Engineering. *J. Integr. Plant Biol.*, *48*(2), 121–127.

Husain, M. K., & Anis, M., (2009). Rapid *in vitro* multiplication of *Melia azedarach* L. (a multipurpose woody tree). *Acta Physiologiae Plantarum.*, *31*(4), 765–772.

Hussain, A., Qarshi, I.A., Nazir, H. & Ullah, I. (2012). Plant Tissue Culture: Current Status and Opportunities, Recent Advances in Plant in vitro Culture. In: Laura Rinaldi, IntechOpen, DOI: 10.5772/50568. Available from: https://www.intechopen.com/books/recent-advances-in-plant-in-vitro-culture/plant-tissue-culture-current-status-and-opportunities. https://creativecommons.org/licenses/by/3.0/)

Hussey, G., (1978). The application of tissue culture to the vegetative propagation of plants. *Sci. Prog.* (Oxford), *65*, 185–208.

Hussey, G., (1983). *In vitro* propagation of horticultural and agricultural crops, pp. 111–138. In: *Plant Biotechnology.* Mantell, S. H., & Smith, H., (eds.). Cambridge Univ. Press, Cambridge.

Jayasankar, S., Gray, D. J., & Litz, R. E., (1999). High-efficiency somatic embryogenesis and plant regeneration from suspension cultures of grapevine. *Plant Cell Rep., 18,* 533–537.

Jeong, G. A., & Park, D. H., (2007). Enhanced secondary metabolite biosynthesis by elicitation in transformed plant root system. *Appl. Biochem. Biotechnol., 130,* 436–446.

Karuppusamy, S., (2009). A review on trends in production of secondary metabolites from higher plants by *in vitro* tissue, organ and cell cultures. *J. Med. Plant Res., 3*(13), 1222–1239.

Khan, R. S., Alam, S. S., Munir, I., Azadi, P., Nakamura, I., & Mii, M., (2011). *Botrytis cinerea* resistant marker-free *Petunia hybrida* produced using the MAT vector system. *Plant Cell Tissue Organ Cult., 106,* 11–20.

Khan, T., Krupadanam, D., & Anwar, Y., (2008). The role of phytohormone on the production of berberine in the calli culture of an endangered medicinal plant, turmeric (*Coscinium fenustratum* L.). *Afr. J. Biotechnol., 7,* 3244–3246.

Kim, J. S., Lee, S. Y., & Park, S. U., (2008). Resveratol production in hairy root culture of peanut, *Arachys hypogaea* L. transformed with different *Agrobacterium rhizogenes* strains. *Afr. J. Biotechnol., 7,* 3788–3790.

Kim, O. T., Bang, K. H., Shin, Y. S., Lee, M. J., Jang, S. J., Hyun, D. Y., Kim, Y. C., Senong, N. S., Cha, S. W., & Hwang, B., (2007). Enhanced production of asiaticoside from hairy root cultures of *Centella asitica* (L.) Urban elicited by methyl jasmonate. *Plant Cell Rep., 26,* 1914–1949.

Larkin, P. J., & Scowcroft, W. R., (1981). Somaclonal variation–a novel source of variability from cell cultures for plant improvement. *Theor. Appl. Genet., 60*(4), 197–214.

Larkin, P. J., & Scowcroft, W. R., (1983). Somaclonal variation and eyespot toxin tolerance in sugarcane. *Plant Cell Tiss. Org. Cult., 2,* 111–121.

Lee, S. Y., Cho, S. J., Park, M. H., Kim, Y. K., Choi, J. I., & Park, S. U., (2007). Growth and rutin production in hairy root culture of buck weed (*Fagopyruum esculentum*). *Prep. Biochem. Biotechnol., 37,* 239–246.

Lee-Parsons, C. W. T., & Rogce, A. J., (2006). Precursor limitations in methyl jasmonate-induced *Catharanthus roseus* cell cultures. *Plant Cell Rep., 25,* 607–612.

Lila, K. M., (2005). Valuable secondary products from *in vitro* culture, *Plant Development and Biotechnology* edited by Robert N. T., & Dennis J. G., CRC Press LLC, Chapter 24, 285–289.

Liu, Z. Y., Chen, P. D., Pei, G. Z., Wang, Y. N., Qin, B. X., & Wang, S. L., (1988). Transfer of *Haynaldia villosa* chromosomes into *Triticum aestivum*. *Proceeding of the 7th International Wheat Genetics Sumposium*, Cambridge, UK, 355–361.

Marana, J. P., Miglioranza, E., & De Faria, R. T., (2009). *In vitro* establishment of *Jacaratia spinosa* (Aubl.) ADC. *Semina-Ciencias Agrarias, 30*(2), 271–274.

Marconi, P. L., Selten, L. M., Cslcena, E. N., Alvarez, M. A., & Pitta-Alvarez, S. I., (2008). Changes in growth and tropane alkaloid production in long term culture of hairy roots of *Brugmansia candida*. *Elect. J. Integrative Biosci., 3,* 38–44.

Marino, G., & Battistini, S., (1990). Leaf-callus growth, shoot regeneration and somaclonal variation in *Actinidia deliciosa*: effect of media pH. *Acta Horticulturae, 280,* 37–44.

Maynard, C., Xiang, Z., Bickel, S., & Powell, W., (1998). Using genetic engineering to help save American chestnut: a progress report. *J. Am. Chestnut Found, 12,* 40–56.

Mehrotra, S., Kukreja, A. K., Kumar, A., Khanuja, S. P. S., & Mishra, B. N., (2008). Genetic transformation studies and scale up of hairy root culture of *Glycyrrhiza glabra* in bioreactor. *Electronic Journal of Biotechnology*, *11*(2), 15.

Misra, M., & Misra, A. N., (1993). Genetic transformation of grass pea. In: DAE *Symposium on Photosynth. Plant Molecular Biology*, BRNS/DAE, Govt. of India, 246–251.

Mohan, N., Nikdad, S., & Singh, G., (2011). Studies on seed germination and embryo culture of *Jatropha curcas* L. under *in vitro* conditions. *Biotechnol. Bioinf. Bioeng.*, *1*(2), 187–194.

Moreno, P. R. H., Van der Heijden, R., & Verpoorte, R., (1993). Effect of terpenoid precursor feeding and elicitation on formation of indole alkaloids in cell suspension cultures of *Catharanthus roseus*. *Plant Cell Rep. J.*, *12*, 702–705.

Morrison, R. A., & Evans, D. A., (1998). Haploid plants from tissue culture: New plant varieties in a shortened time frame. *Nat. Biotechnol.*, *6*, 684–690.

Mostageer, A., & Elshihy, O. M., (2003). Establishment of salt tolerant somatic hybrid through protoplast fusion between rice and ditch reed. *Arab. J. Biotech.*, *6*(1), 01–12.

Motomura, T., Hidaka, T., Akihama, T., & Omura, M., (1997). Protoplast fusion for production of hybrid plants between citrus and its related genera. *J. Japan. Soc. Hort. Sci.*, *65*, 685–692.

Murashige, T., (1962). Plant propagation through tissue culture. *Ann. Rev. Plant Physiol.*, *25*, 135–166. Murashige, T., & Skoog, F., A revised medium for rapid growth and bio assays with tobacco tissue cultures. *Physiologia Plantarum.*, *15*, 473–497.

Narayanaswamy, S., (1977). In: *Applied and Fundamental Aspects of Plant Cell, Tissue and Organ Culture*. Reinert, J., Bajaj, Y. P. S., (eds.), Springer-Verlag, Berlin, 179–248.

Okere, A. U., & Adegey, A., (2011). *In vitro* propagation of an endangered medicinal timber species *Khaya grandifoliola* C. Dc. *Afr. J. Biotechnol.*, *10*(17), 3335–3339.

Palazon, J., Pinol, M. T., Cusido, R. M., Morales, C., & Bonfill, M., (1997). Application of transformed root technology to the production of bioactive metabolites. *Recent Res. Dev. Plant Physiol.*, *1*, 125–143.

Park, Y. S., Barrett, J. D., & Bonga, J. M., (1998). Application of somatic embryogenesis in high value clonal forestry: development, genetic control and stability of cryopreserved clones. *In vitro Cell. Dev. Biol. Plant.*, *34*, 231–239.

Payne, G. F., Shuler, M. L., & Brodelius, P., (1987). Plant cell culture. J. In: Lydensen, B. K. (ed.). *Large Scale Cell Culture Technology*. Hanser Publishers, New York, USA, 193–229.

Pistelli, L., Giovannini, A., Ruffoni, B., Bertoli, A., & Pistelli, L., (2010). Hairy root cultures for secondary metabolites production. *Adv. Exp. Med. Biol.*, *698*, 167–184.

Purkayastha, J., Sugla T., Paul, A., Maumdar, P., Basu, A., Solleti, S. K., Mohommad, A., Ahmed, Z., & Sahoo, L., (2010). Efficient *in vitro* plant regeneration from shoot apices and gene transfer by particle bombardment in *Jatropha Curcas*. *Biologia Plantarum.*, *54*, 13–20.

Quak, F., (1977). In: *Applied and Fundamental Aspects of Plant Cell, Tissue and Organ Culture*, Reinert, J., Bajaj, Y. P. S., (eds.), Springer-Verlag, Berlin, 598–615.

Raghavan, V., (1976). *Experimental Embryogenesis in Vascular Plants,* Academic Press, New York, 382–408.

Rao, R. S., & Ravishankar, G. A., (2002). Plant tissue cultures, chemical factories of secondary metabolites. *Biotechnol. Adv.*, *20*, 101–153.

Roat, C., & Ramawat, K. G., (2009). Elicitor induced accumulation of stilbenes in cell suspension cultures of *Cayratia trifoliata* (L.) Domin. *Plant Biotechnol. Rep. J.*, *3*, 135–138.

Rout, G. R., Mohapatra, A., & Jain, S. M., (2006). Tissue culture of ornamental pot plant: A critical review on present scenario and future prospects. *Biotechnology Advances, 24,* 531–560.

Sagare, A. P., Lee, Y. L., Lin, T. C., Chen, C. C., & Tsay, H. S., (2000). Cytokinin-induced somatic embryogenesis and plant regeneration in *Coryodalis yanhusuo* (Fumariaceae)- a medicinal plant. *Plant Sci., 160,* 139–147.

Saini, R., & Jaiwal, P. K., (2002). Age, position in mother seedling, orientation, and polarity of the epicotyl segments of blackgram (*Vigna mungo* L. Hepper) determines its morphogenic response. *Plant Sci., 163*(1), 101–109.

Sasson, A., (1993). *Biotechnologies in Developing Countries, Present and Future,* vol. *1.* Paris, United Nations Educational, Scientific and Cultural Organization.

Sengar, R. S., Chaudhary, R., & Tyagi, S. K., (2010). Present status and scope of floriculture developed through different biological tools. *Res J. of Agri. Sci., 1*(4), 306–314.

Shalaka, D. K., & Sandhya, P., (2009). Micropropagation and organogenesis in *Adhatoda vasica* for the estimation of vasine. *Pharm. Mag., 5,* 539–363.

Shepherd, J. F., Bidney, D., & Shahin, E., (1980). Potato protoplasts in crop improvement. *Science, 208,* 17–24.

Shrivastava, N., Patel, T., & Srivastava, A., (2006). Biosynthetic potential of *in vitro* grown callus cells of *Cassia senna* L. var. senna. *Curr. Sci. J., 90,* 1472–1473.

Shrivastava, S., & Srivastava, A. K., (2007). Hairy root culture for mass-production of high value secondary metabolites. *Crit. Rev. Biotechnol., 27*(1), 29–43.

Sinclair, T. R., Purcell, L. C., & Sneller, C. H., (2004). Crop transformation and the challenge to increase yield potential. *Trend Plant Sci., 9,* 70–75.

Singh, K. K., & Gurung, B., (2009). *In vitro* propagation of R. maddeni Hook. F. an endangere Rhododendron species of Sikkim Himalaya. *Notulae Botanicae Horti Agrobotanici Cluj-Napoca., 37*(1), 79–83.

Skoog, F., & Miller, C. O. L., (1957). Chemical regulation of growth and organ formation in plant tissue cultured *in vitro. Symposia of the Society for Experimental Biology, 11,* 118–131.

Slater, A.; Scott, N.; and Fowler, M.; Plant Biotechnology - The Genetic Manipulation of Plants, Oxford University Press, Oxford. **2003,** 35–53.

Skrzypek, Z., & Wysokinsku, H., (2003). Sterols and titerpenes in cell cultures of *Hyssopus officinalis* L. *Ver Lag der Zeitschrift fur Naturforschung,* 312.

Smith, M. K., & Drew, R. A., (1990). Current applications of tissue culture in plant propagation and improvement. *Aust. J. Plant Physiol., 17,* 267–289.

Sudherson, C. S., Manuel, J., & Al-Sabah., (2008). Haploid plant production from pollen grains of sturt's desert pea via somatic embryogenesis. *Am-Euras. Sci. Res., 3*(1), 44–47.

Sujanya, S., Poornasri, D. B., & Sai, I., (2008). *In vitro* production of azadirachtin from cell suspension cultures of *Azadirachta indica. Biosci. J., 33,* 113–120.

Thorpe, T., (2007). History of plant tissue culture. *J. Mol. Microbial Biotechnol., 37,* 169–180.

Tilkat, E., Onay, A., Yildirim, H., & Ayaz, E., (2009). Direct plant regeneration from mature leaf explants of pistachio, *Pistacia vera* L. *Scientia Hort., 121*(3), 361–365.

Tiwari, K. K., Trivedi, M., Guang, Z. C., Guo, G. Q., & Zheng, G. C., (2007). Genetic transformation of Gentiana macrophylla with *Agrobacterium rhizogenes*: growth and production of secoiridoid glucoside gentiopicroside in transformed hairy root cultures. *Plant Cell Rep. J., 26,* 199–210.

Tomes, D. T., & Swanson, E. B., (1982). In: Appli*cation of Plant Cell and Tissue Culture to Agriculture and Industry* (Tomes, D. T., et al., eds.). University of Guelph., 25–43.

Toriyama, K., Hinata, K., & Kameya, T., (1987). Production of somatic hybrid plants, "*Brassico moricandia*," through protoplast fusion between *Moricandia arvensis* and *Brassica oleracea*. *Plant Sci.*, *48*(2), 123–128.

Torres-Munoz, L., & Rodriguez-Garay, B., (1996). Somatic embryogenesis in the threatened cactus *Turbinicarpus psudomacrochele* (Buxbaum & Backerberg). *J. PACD*, *1,* 36–38.

Tyagi, R. K., Agrawal, A., Mahalakshmi, C., Hussain, Z., & Tyagi, H., (2007). Low-cost media for *in vitro* conservation of turmeric (*Curcuma longa* L.) and genetic stability assessment using RAPD markers. *In Vitro Cell. Develop. Biol. Plant.*, *43,* 51–58.

Umamaheswai, A., & Lalitha, V., (2007). *In vitro* effect of various growth hormones in *Capsicum annum* L. on the callus induction and production of Capsiacin. *Plant Sci. J.*, *2,* 545–551.

Vijayasree, N., Udayasri, P., Aswani, K. Y., Ravi, B. B., Phani, K. Y., & Vijay, V. M., (2010). Advancements in the production of secondary metabolites. *J. Nat. Prod.*, *3,* 112–123.

Wagiah, M. E., Alam, G., Wiryowidagdo, S., & Attia, K., (2008). Improved production of the indole alkaloid cathin-6-one from cell suspension cultures of *Brucea javanica* (L.) Merr. *Sci. Technol. J.*, *1,* 1–6.

Williams, E. G., Verry, I. M., & Williams, W. M., (1982). Use of embryo culture in interspecific hybridization. In: *Plant Improvement and Somatic Cell Genetics,* Vasil, I. K., Scowcroft, W. R., Frey, K. J., (eds.), Academic Press, New York, 119–128.

Withers, L. A., (1989). In: *The Use of Plant Genetic Resources,* Brown, A. H. D., Marshall, D. R., Frankell, O. H., Williams, J. T., (eds.), Cambridge University Press, Cambridge, 309–334.

Xiangqian, L. I., Krasnyanski, F. S., & Schuyler, S. K., (2002). Somatic embryogenesis, secondary somatic embryogenesis and shoot organogenesis in Rosa. *Plant Physiol.*, *159,* 313–319.

Xu, H., Kim, Y. K., Suh, S. Y., Udin, M. R., Lee, S. Y., & Park, S. U., (2008). Deoursin production from hairy root culture of *Angelica gigas*. *Korea Soc. Appl. Biol. Chem. J.*, *51,* 349–351.

Yesil-Celiktas, O., Gurel, A., & Vardar-Sukan, F., (2010). Large scale cultivation of plant cell and tissue culture in bioreactors. *Trans. World Res. Network*, 1–54.

Yeung, E. C., Thorpe, T. A., & Jensen, C. J., (1981). *In vitro* fertilization and embryo culture in plant tissue culture: *Methods and Applications in Agriculture*, (ed.). Thorpe, T. A. Academic Press, New York, 253–271.

Zhao, J., Zhu, W., & Hu, Q., (2001). Enhanced catharanthine production in *Catharanthus roseus* cell cultures by combined elicitor treatment in shake flasks and bioreactors. *Enzyme Microbiol Technol. J.*, *28,* 673–681.

PART IV

GENETIC ENGINEERING AND TISSUE CULTURE IN PLANT BREEDING

CHAPTER 11

BREEDING STRATEGIES TO CONVERT C$_3$ PLANTS INTO C$_4$ PLANTS

AKSHAYA KUMAR BISWAL,[1,3] ARUN KUMAR SINGH,[2] VIVEK THAKUR,[1,4] SATENDRA KUMAR MANGRAUTHIA,[2] and REVATHI PONNUSWAMY[2]

[1]C4 Rice Center, International Rice Research Institute, DAPO 7777, Metro Manila, Philippines

[2]ICAR-Indian Institute of Rice Research, Hyderabad, India, E-mail: revathi.ponnusamy@gmail.com

[3]Department of Biology, University of North Carolina at Chapel Hill, NC – 27599, USA

[4]Ramalingaswami Fellow / Visiting Scientist (Bioinformatics), ICRISAT, Patencheru, Hyderabad, India

CONTENTS

11.1   INTRODUCTION

The plants can be divided into distinct physiological groups according to their photosynthetic pathways. Three different types of photosynthetic mechanisms have evolved in nature, among which the plants exhibiting the C$_4$ photosynthesis have the highest efficiency to convert the solar energy into biological energy. The C$_4$ plants contribute to 25% of the total terrestrial photosynthesis even though they form only 3% of the total vascular plants.

Most of the terrestrial plants assimilate carbon through C$_3$ photosynthesis, which is also called photosynthetic carbon reduction cycle (PCR). In this process, ribulose-1,5-bisphosphate carboxylase/oxygenase (Rubisco) catalytically incorporates inorganic, atmospheric CO$_2$ to the five-carbon sugar phosphate, ribulose-1,5-bisphosphate (RuBP). The resulting product is an unstable 6-carbon compound, which in turn, is hydrated and cleaved to produce two molecules of three-carbon compound called 3-phosphoglycerate or 3-phosphoglyceric acid (3-PGA) as the first stable product (hence the name C$_3$). This process of carbon fixation and photosynthesis in the C$_3$ plants occurs only in the mesophyll cells located on the surface of the leaf (Figure 11.1). The carboxylation process is interfered by the deviant interaction of Rubisco with oxygen, at a lower affinity. Rubisco has a slow turn over for CO$_2$ with the higher abundance of atmospheric O$_2$ (21%) than CO$_2$ (0.03%); some portion of the Rubisco is always engaged with oxygen. Rubisco's oxygenase activity produces only half the number of 3-PGA in comparison to carboxylase activity. The phosphoglycolate produced in this process undergoes detoxification in the peroxisome and mitochondria in a process called photorespiration, resulting in loss of energy and some of the previously fixed CO$_2$.

Rubisco, which evolved long ago in an atmosphere rich in CO$_2$, has undergone slight adaptive changes without much alterations in its oxygenase activity (Young et al., 2012). How can plants avoid photorespiration? Either the plants need to use another enzyme to fix CO$_2$ that does not react

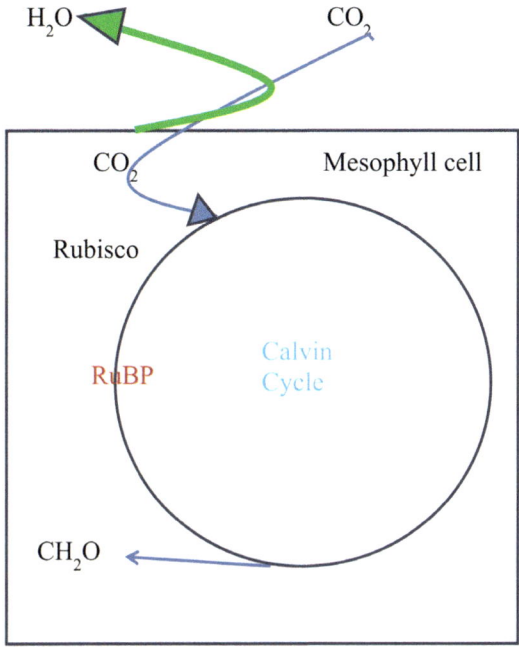

FIGURE 11.1 A schematic diagram of C_3 photosynthesis.

with O_2 or they need to fix CO_2 in an environment shielded from O_2. In fact, some of the plants have developed a CO_2 concentrating mechanism around Rubisco by sequestering it in bundle sheath (BS) cells. CO_2 is initially fixed by the cytosolic enzyme phosphoenolpyruvate carboxylase (PEPC) to form a four-carbon compound oxaloacetate (OAA), hence the name C_4 (Figure 11.2). The PEPC is insensitive to O_2 and specifically reacts with only CO_2. The OAA is then reduced to malate or aminated to aspartate, which are later diffused into BS cells where they are decarboxylated by NADP/NAD malic enzymes or phosphoenolpyruvate carboxykinase. Based upon the reduction of OAA, C4 plants can be classified to three subtypes according to decarboxylation modes: NADP-malicenzyme (NADP-ME), NAD-malic enzyme (NAD-ME), and PEP carboxykinase. The CO_2 produced increases its effective concentration around Rubisco up to 10 times over atmospheric concentration (Furbank and Taylor, 1995). Hence, the Rubisco works almost at its optimal rate, and C_4 plants also thrive well in a wide range of environmental conditions including arid conditions, high temperatures, and marginal environments. The C_4 photosynthesis works by compartmentalizing the activi-

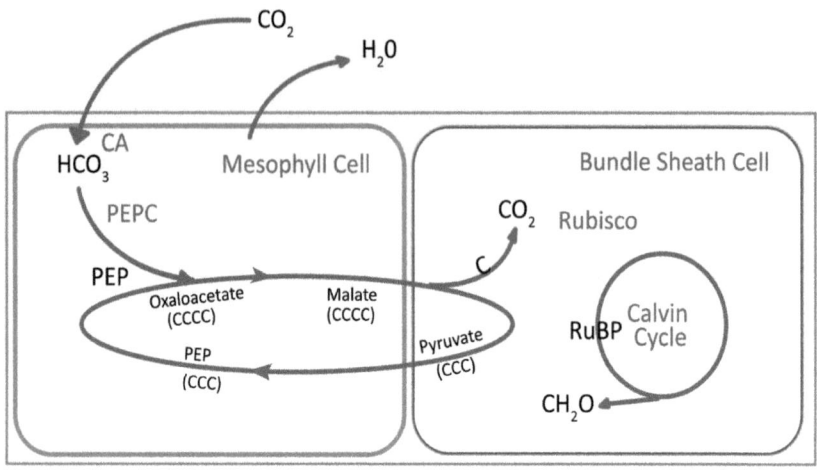

FIGURE 11.2   A schematic diagram of C_4 photosynthesis.

ties into two specialized cells and chloroplast types and by incorporating a biochemical CO_2 pump to increase the concentration of CO_2 around the Rubisco in the BS cells while shielding it from O_2. The C_4 photosynthesis system involves the combined function of differential expression of genes in mesophyll and BS cells and/or trafficking of enzymes and metabolite transporters of photosynthetic carbon assimilation pathway and electron transport complexes along with factors that direct the development of Kranz anatomy during early stages of leaf development.

The third group of plants is called CAM (Crassulacean Acid Metabolism) plants. It is suggested that CAM maybe the oldest alternative to conventional C_3 photosynthesis. This alternative way of photosynthesis is a variation on the C_3 pathway; it often exists in plants that are adapted to extreme environmental conditions. They can fix CO_2 just like the C_3 plants do by using the Rubisco enzyme to feed the Calvin cycle. However, instead of atmospheric CO_2 that the Rubisco fixes, in this case, CO_2 comes from an internal storage pool of malic acid that has been accumulated during the night. In the light, decarboxylation of this malic acid pool releases an internal CO_2 source. Thus, these plants are able to fix atmospheric CO_2 in complete darkness using a dark-regulated PEPC system and store this in the form of malate. This allows CAM plants to leave the stomata completely closed during the day avoiding water loss and photorespiration.

11.2   OPPORTUNITIES TO RAISE THE CEILING OF SOLAR ENERGY CONVERSION

The highest solar energy conversion efficiency across a full growing season based on solar radiation intercepted by the leaf canopy has been reported to be about 2.4% for C$_3$ crops and about 3.7% for C$_4$ crops, while the maximum theoretical conversion efficiency of solar energy to biomass can be as much as 4.6% for C$_3$ photosynthesis at 30°C and today's 380 ppm atmospheric (CO$_2$) and 6% for C$_4$ photosynthesis (Zhu et al., 2008). This major gap between the theoretical and practical values indicates that both C$_3$ and C$_4$ plants can be engineered to further boost the solar energy conversion. What are the possible options to close this gap? Although photosynthesis increases linearly at low light conditions, the rate of increment slowly decreases above about one quarter of full sunlight and eventually falls to zero above about half of full sunlight. Plant architecture with high leaf angle can increase the photosynthetic efficiency to some extent. Another potential option to substantially increase the photosynthetic capacity of individual leaves is by stoichiometric adjustment between Rubisco and proteins involved in the regeneration of RuBP, the cytochrome b6/f complex and sedoheptulose-1:7-bisphosphatase, etc., though its practical engineering is highly complex.

The maximal energy conversion efficiency of C$_3$ plants can be decreased by 49% at 30°C (Zhu et al., 2008). What are the realistic opportunities to achieve higher solar energy conversion efficiency? The oxygenation of RuBP by Rubisco results in the formation of one molecule of glycolate. Two molecules of glycolate are metabolized through the C$_2$ pathway that needs ATP and reductive energy to produce one molecule of 3-PGA while releasing one of the previously fixed CO$_2$. The oxygenation reaction is largely eliminated in C$_4$ plants, thereby reducing the energy loss by photorespiration. Clearly, the C$_3$ plants need to overcome the photorespiration, and this can be achieved by converting C$_3$ plants to C$_4$.

11.3   ECONOMIC POTENTIAL OF C$_4$ RICE

Rice serves as the primary source of energy for more than half of the world's population and 90% of the rice is grown and consumed in Asia. It is one of the oldest crops and is being cultivated over last 9,000 years (Molina et al., 2011). Worldwide, the total cultivated area of rice is 157.8 m.ha (www.statica.com). The International Food Policy Research Institute (IFPRI) reported that by 2050, rice prices will increase 32–37% and yield loss of rice would be 10–15% as a result of climate change. The demand for rice will be doubled by end of this century due to increasing world population, economic development, and diminishing land and water resources. Rice can have up to 50% increase in yield if the C_4 photosynthesis system can be incorporated into it (Hibberd et al., 2008). Some of the most important advantages of C_4 plants are outlined below:

1. C_4 plants have a higher potential efficiency of converting solar energy to biomass because of photorespiration suppressing modifications.
2. The C_4 mechanism is intrinsically linked to 1.3–4 times higher nitrogen use efficiency and water use efficiency (Sage and Zhu, 2011).
3. The higher water use efficiency is associated with a faster fixation of CO_2 by the O_2-insensitive PEP carboxylase. Therefore, stomatal opening for the uptake of CO_2 is shorter, leading to a reduction of leaf water evaporation (Byrt et al., 2011).
4. C_4 photosynthesis has advantageous characteristic for future climate scenarios (e.g., during drought, salinity areas, erratic rainfall, and cold conditions) in which the advantages of C_4 photosynthesis over the C_3 type are even more apparent.
5. C_4 species are the most productive plants on the Earth.

The broad differences between C_3 and C_4 plants are outlined in Table 11.1. However, a more systematic discussion about the biochemical and physiological difference between C_3 and C_4 plants will be made in the following section.

11.4   BIOCHEMICAL AND PHYSIOLOGICAL CHARACTERISTICS OF C_4 PHOTOSYNTHESIS

The biochemical and physiological characteristics of C_4 photosynthesis have been researched and elucidated by different research groups. It is known that the differentiation of two cell types, mesophyll (M) cells and BS cells, is required for efficient C_4 photosynthesis. When the rate of photorespiration approaches the rate of photosynthesis, it becomes a major limitation for the growth of many plants (Voet et al., 2013). This phenomenon is circumvented in C_4 plants by the localization of Rubisco in the BS cells and by the

TABLE 11.1   Difference between C$_3$ and C$_4$ Photosynthesis

S.No.	C$_3$ plants	C$_4$ plants
1	C$_3$ plants are called temperate or cool-season plants.	C$_4$ plants are often called tropical or warm season plants.
2	The carboxylase enzyme is Rubisco	The first carboxylase enzyme is PEP carboxylase and the second is Rubisco, which is sequestered in BS cells.
3	The primary CO$_2$ accepter is RuBP (a 5 carbon compound) and first stable product is PGA (a 3 carbon compound)	The primary CO$_2$ accepter is PEP (a 3 carbon compound) and first stable product is OAA (a 4 carbon compound).
3	Monomorphic chloroplast	Dimorphic chloroplast (The chloroplast of parenchymatous BS cells is different from that of mesophyll cells). The chloroplasts in BS cells are centripetally arranged and lack grana.
4	Each chloroplast has Photo-system I and II	The chloroplasts of BS cells lack the photo-system II. These are dependent on mesophyll chloroplasts for the supply of NADPH$^+$ and H$^+$.
5	BS cells are unspecialized.	The BS cells are highly developed with unusual construction of organelles.
6	The photosynthetic rate per unit of N$_2$ is less in C$_3$ plants.	C$_4$ plants are more efficient at gathering carbon dioxide and utilizing nitrogen from the atmosphere and soil. They utilize less water for the dry matter production.
7	High rate of photorespiration	Rate of photorespiration is minimal or absent.
8	Lower Radiation Use Efficiency (RUE)	RUE is up to 50% higher than that of C$_3$ due to reduced number of photons required to fix each molecule of CO$_2$

temporary fixation of CO$_2$ into PEP, forming oxaloacetate in the mesophyll cells. Unlike Rubisco, PEP carboxylase continues to be active at low CO$_2$ concentrations and is not inhibited by O$_2$, thereby allowing C$_4$ plants to adapt in hot and dry environments. The oxaloacetate thus formed is either reduced to malate at the expense of NADPH or converted to aspartate by transamination (Cox and Nelson, 2013). Aspartate and/or malate is shuttled into the BS cells where it is ultimately decarboxylated to yield CO$_2$ and proceeds downstream the Calvin-Benson cycle (Cox and Nelson, 2013; Voet et al., 2013). This "CO$_2$ pumping" mechanism in Kranz-type C$_4$ plants effectively minimizes photorespiration by providing high CO_2/O_2 ratio in the BS cells for the carboxylase activity of Rubisco to be more favored.

Thus, the leaves of C_4 plants have more complicated structural and functional features than those of C_3 plants (Hatch, 1999; Kanai and Edwards, 1999). Current studies are focused on elucidating the regulatory mechanisms of genetic and developmental events in C_4 photosynthesis, but most of them are yet poorly understood (Dengler and Nelson, 1999; Sheen, 1999). There are many indirect evidences that C_4 plants have evolved in parallel from C_3 plants among diverse taxonomic groups (Kellogg, 1999). Some C_3–C_4 intermediate plants such *as Flaveria spp.* provide a suitable system for studying the possible evolution of biochemical and physiological processes in C_4 plants (Ku et al., 1991; Westhoff et al., 1997). It is becoming clear that there is diversity not only in the structural and biochemical features, but also in the genetic and developmental aspects of C_4 photosynthesis (Dengler and Nelson, 1999; Sheen, 1999; Edwards et al., 2001).

A comparison of C_3 and C_4 metabolic networks has shown that C_3 network exhibit more dense topology structure than C_4, while the simulation of enzyme knockouts demonstrates that both C_3 and C_4 networks are very robust when optimizing CO_2 fixation (Wang et al., 2012). Nevertheless, C_4 plant has better robustness for both biomass synthesis and CO_2 fixation. All the essential reactions in C_3 network have been found to be essential for C_4 and as expected there are some other reactions specifically essential for C_4. The C_4 plants have better modularity with complex mechanism coordinates and pathways than that of C_3 plants. As a result, radiation use efficiency, biomass production, and CO_2 fixation in C_4 plants may be faster than that in C_3 plants. Wang et al. (2012) have predicted that the rate of CO_2 fixation and biomass production in PCK subtype of C_4 plants can be superior to NADP-ME and NAD-ME subtypes under enough supply of water and nitrogen.

Hibberd and Quick (2002) reported that tobacco, a typical C_3 plant, shows characteristics of C_4 photosynthesis in cells of stems and petioles that surround the xylem and phloem, and that these cells are supplied with carbon for photosynthesis from the vascular system and not from the stomata. These photosynthetic cells possess high activities of enzymes characteristic of C_4 photosynthesis, which allow the decarboxylation of four-carbon organic acids from the xylem. Ueno (2001) presented that the differentiation of photosynthetic characteristics in some amphibious species of *Eleocharis,* Cyperaceae, with particular reference to the sedge *Eleocharis vivipara. E.*

Vivipara has a unique nature that expresses C$_4$ characteristics under terrestrial conditions and C$_3$ characteristics under submerged aquatic conditions (Ueno et al., 1988). These plants may provide a scientific basis for converting C$_3$ plants to C$_4$ plants.

11.5   ANATOMICAL DIFFERENCES

Photosynthesis in C$_4$ plants involving two-cell system displays wreath-like cellular arrangement called Kranz Anatomy (Cox and Nelson, 2013). In Kranz anatomy, the vein is concentrically surrounded by the BS cells, which are also enclosed in a layer of mesophyll cells (Figure 11.3) (Gowik and Westhoff, 2011; Cox and Nelson, 2013). The C$_4$ plants have lower conductance to CO$_2$ diffusion across the interface of the BS and M cells. Though it has generally been presumed that BS cell walls of C$_4$ leaves are much thicker than their C$_3$ counterparts, von Caemmerer and Furbank (2003) observed that the thickness of BS cell walls in C$_4$ species is similar to cell wall thickness of C$_3$ mesophyll cells. C$_3$ leaves have thin cell walls and chloroplasts are pressed against the cell walls facing intercellular air-spaces of M cells, thus minimizing the liquid diffusion paths. Only C$_4$ plants have chloroplasts in BS cells. Most C$_4$ plants (with the exception of C$_4$ dicotelydons and the NAD-ME types) have one or several layers of suberin laid down in the BS cell wall adjacent to the mesophyll. The CO$_2$ diffusion resistance of the BS cells might be affected by the presence or absence of the suberin lamellae in the BS cell wall. NAD-ME type C$_4$ species, which do not have suberin

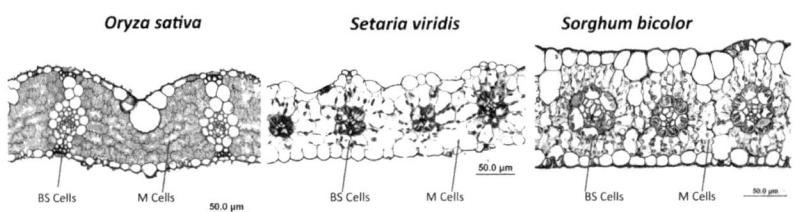

FIGURE 11.3   Comparison of rice (C3), Setaria (C4), and sorghum (C4) leaf anatomy. C4 plants show less interveinal distance and bigger BS cells with the presence of chloroplast (This figure was kindly provided by Ms. Jacqueline Dionora, C$_4$ Rice Center, IRRI.).

in their BS cell walls, do not appear to compensate for this with thicker BS cell walls. Though significance of suberin as a barrier is circumstantial, its potential role may be accounted for in modeling the diffusion path for CO_2 (von Caemmerer and Furbank, 2003). Though the CO_2 diffusion properties of membranes such as the plasma lemma, chloroplast, and mitochondrial membranes are difficult to estimate, the bundle sheath cytosol itself may account for more than half of the final calculated resistance value for CO_2 leakage. Hence, the location of the site of decarboxylation, its distance from the mesophyll interface, and the physical arrangement of chloroplasts and mitochondria in the BS cells are as important to the efficiency of the process as the properties of the BS cell wall (von Caemmerer and Furbank, 2003). In rice, 66% of protoplast volume is occupied by chloroplasts, and chloroplasts/ stromules cover >95% of the cell periphery. Mitochondria and peroxisomes occur in the cell interior and are intimately associated with chloroplasts/ stromules. The chlorenchyma architecture of rice has been hypothesized to enhance diffusive CO_2 conductance while maximizing the scavenging of photorespired CO_2 (Sage and Sage, 2009). The extensive chloroplast/stro-mule sheath forces photorespired CO_2 to exit cells via the stroma, where it can be refixed by Rubisco.

The cultivated rice *Oryza sativa* has 22 well-recognized wild relatives (2n = 24, 48) representing 10 genomes. The wild relatives of rice have long been recognized as a rich source of novel genes underlying traits of agricultural importance. The interveinal spacing (at the middle of the blade) for the wild relative of rice ranged from 113 to 322 μm (the value for IR72 was 170 μm). The range for the C_4 weeds was 93 to 136 μm. Wild rice types probably have some of the anatomical features peculiar to C_4 plants and the wild types may contain C_3–C_4 intermediates (Brar and Ramos, 2007; Hibberd et al., 2008).

11.6   EVOLUTION OF C_4 SPECIES AND C_3–C_4 INTERMEDIATES

Most of the terrestrial plant species are C_3 plants. These include the cereals (wheat, rice, barley and oats), peanuts, cottons, sugar beets, tobacco, spinach, soybeans, and most tree plants. Most lawn grasses such as rye and fescue are also C_3 plants. Sugarcane is a champion at photosynthesis under the right condition and is a prime example for C_4 plants. Cereal crops like maize, sorghum, and pearlmillet also use C_4 photosynthesis. Other annual

C$_4$ plants include sudan grass, big bluestem, Bermuda grass, switch grass, and old world bluestem. Moore et al. (1987) reported that weedy species *Partheniam hysterophorus* (Asteraceae) possesses C$_3$–C$_4$ intermediate. Various other research groups have also identified many naturally occurring plant species with photosynthetic characteristics intermediate to C$_3$ and C$_4$ plants (Holaday and Chollet, 1984; Monson et al., 1984; Edwards and Ku, 1987). Most C$_3$–C$_4$ species have at least a partially developed Kranz leaf anatomy and reduced levels of photorespiration relative to C$_3$ plants in common. Most of the intermediate species occur in general, which also have both C$_3$ and C$_4$ species [e.g. Poaceae (*Panicum* and *Neurachne*), Amaranthaceae (*Alternanthera*), Asteraceae (*Flaveria*), and Aizoaceae (*Mollugo*)]. Some of the species have one or more intermediates, but no known C$_4$ species, e.g., Moricandia (*Brassicaceae*) and Parthenium (*Asteraceae*). Such circumstantial evidence suggests that these are evolutionary intermediates of C$_3$ – C$_4$ species and not hybrid species. Different mechanisms for reducing photorespiration are likely utilized among the various C$_3$–C$_4$ intermediate species. In rice, some wild types have overlap with CO$_2$ compensation concentration and PEPC activity reported for C$_3$–C$_4$ intermediates. For example, some accessions of *O. rufipogon* were reported to have CO$_2$ compensation points of about 30 µmol mol^{-1} and PEP carboxylase activity of about 3 µmol min^{-1} mg^{-1} chlorophyll (Yeo et al., 1994). In addition, there is evidence for some C$_4$ characteristics in rice spikelets (Imaizumi et al., 1997).

How did the C$_4$ photosynthesis evolve in nature? What was the driving force for C$_4$ evolution? The C$_4$ photosynthesis has independently evolved in nature at least 66 times in multiple lineages (Sage et al., 2012). The polyphyletic appearance of C$_4$ photosynthesis has been associated with diverse and flexible evolutionary paths because no single sequence of acquisitions is capable of producing observed intermediate phenotypes. Similarly, several traits such as compartmentation of GDC into BS cells and the increased number of chloroplasts in the BS cells clearly display bimodal probability distributions for their acquisition (Williams et al., 2013). A commonly accepted hypothesis for the ecological drivers of C$_4$ evolution is that the declining CO$_2$ concentration in the oligocene decreased the rate of carboxylation by Rubisco, creating a strong pressure to evolve alternative photosynthetic strategies (Christin et al., 2008; Vicentini et al., 2008). Therefore, alterations to the localization and abundance of the primary carboxylases PEPC and Rubisco would be expected to occur early in the evolutionary trajectories. In contrary to this hypothesis, Williams et al. (2013) predicted that alterations to anatomy and cell biology preceded the majority of biochemical alterations, and that other enzymes of the C_4 pathway were recruited prior to PEPC and Rubisco. They reached this conclusion following a novel Bayesian approach based on a hidden Markov model. Further, they argue that enzymes such as PPDK and C_4 acid decarboxylases, function in processes not related to photosynthesis, and hence, the early changes to localization and abundance of these enzymes within C_4 lineages may have been driven by nonphotosynthetic drivers. It has been predicted that the changes to photorespiratory metabolism and BS cell specificity of GDC might have evolved prior to the development of the C_4 pathway (Heckmann et al., 2013). Evidence from physiological and ecological studies has identified a number of additional environmental pressures such as high evaporative demands and increased fire frequency that may have driven the evolution and radiation of C_4 lineages.

Leaves of almost all C_4 lineages separate the reactions of photosynthesis into the mesophyll (M) and bundle sheath (BS). A comparison of messenger RNA profiles of M and BS cells *Setaria viridis* with maize (*Zea mays*) revealed a high correlation (r = 0.89) between the relative abundance of transcripts encoding proteins of the core C_4 pathway, indicating significant convergence in transcript accumulation in these evolutionarily independent C_4 lineages (John et al., 2014). A vast majority of genes encoding proteins of the C_4 cycle in *S. viridis* are also syntenic to homologs used by maize. In both lineages, 122 and 212 homologous transcription factors were preferentially expressed in the M and BS, respectively. Fourteen of 16 shared regulators of chloroplast biogenesis were syntenic homologs in maize and *S. viridis*. The repeated recruitment of syntenic homologs from large gene families implies that parallel evolution of both structural genes and trans-factors underpins the polyphyletic evolution of C_4 photosynthesis in the monocotyledons.

11.7   FEASIBILITY OF ENGINEERING C_4 RICE

As a C_3 crop, rice (*Oryza sativa*) has less crop yield than its C_4 counterpart such as corn. With the justifications stated earlier, the engineering of C_4 rice may be a feasible solution to increase rice yield by a higher percentage.

There are certain reasons why C_3 plants like rice may be manipulated to become C_4 plants. The C_4 photosynthesis has independently evolved in nature in multiple lineages (Sage et al., 2012) and at least 20 times in grasses, making the later more plausible as the model of interest (Burnell, 2011). Being one of the most important examples of convergent evolution, it suggests that it is feasible to genetically convert C_3 to C_4 plants. All the genes and enzymes present in C_4 plants can also be found in C_3 plants (Sheehy and Mitchell, 2011). The transcriptomic analysis of mature leaf revealed only ~3% difference between closely related C_3 and C_4 plants (Bräutigam et al., 2011). Keeping the yield advantages and the need to boost rice yield in mind, it is promising to convert rice into a C_4 crop. In addition to numerous independent research groups, one of the largest consortia of plant biologists, the C_4 Rice Consortium, is pursuing to develop C_4 rice by genetic manipulation of biochemical pathway enzymes and transporters and more importantly incorporating the Kranz anatomy (von Caemmerer et al., 2012).

11.8   BREEDING EFFORTS FOR CONVERTING C_3 TO C_4 PLANTS

What is the genetic architecture of C_4 photosynthesis trait and how many genes are involved? Is this a quantitative trait? In fact, it has been proposed as a polygenic quantitative trait (Westhoff and Gowik, 2010) because of stepwise transition from C_3 to C_4 and presence of C_3–C_4 intermediates. Attempts have been made to map QTLs between closely related species from 1970s. In one of the crosses made by Malcolm Nobs and Olle Björkman between *Atriplexprostrata* (C_3) and *Atriplexrosea* (C_4), an F_2 population was generated that showed segregation of individual C_4 traits (Björkman et al., 1971). However, it could not proceed further due to aneuploidy. Similarly, crosses have been made between C_3, C_4 and C_3–C_4 intermediates of *Flaveria species* (Brown et al., 1993, 1986). Recently, a cross was made between *A. prostrata* with *A. rosea* at the University of Toronto by Oakley et al. (2014), which resembled C_3–C_4 intermediates in anatomy with disrupted C_4 metabolic cycle. Interestingly, they have successfully generated F_3 lines, and further analysis in this post-genomic era may lead to an insight into C_4 genetic framework. However, more intensive breeding efforts using modern biotechnological tools such as embryo rescue and wide hybridization may be helpful in getting a breakthrough.

The revolution in sequencing technology and availability of new and affordable methods for whole-genome sequencing has generated huge amount of data to understand the genetic difference between C_3 and C_4 plants. Yet, systematic analysis of these data and generating useful information have been a great challenge. Bioinformaticians and geneticists are trying to develop a catalog of genes associated with C_3 or C_4 photosynthesis through DNA and RNA sequencing and to understand how these genes act coordinately to bring about the convergent C_4 phenotype as discussed in the next section.

11.9   BIOINFORMATICS TOOLS AND RESOURCES

The bioinformatics resources for research on C_3/C_4 species are largely dispersed. The resources can be grouped into following categories: 1) mutant resources, 2) gene expression data, 3) gene expression visualization tools, 4) ortholog tools/databases, and 5) genome sequence resources.

Information on mutants of model C_3/C_4 plants is very useful resource for mutant screening work as well as for validating the role of candidate genes for both forward and reverse genetic studies. For C_3 rice, T-DNA insertion mutants for a large fraction of genome are available through a web-portal named Taiwan Rice Insertional Mutant database (TRIM; http://trim.sinica. edu.tw/). Availability of loss or gain of functional mutants for a gene of interest can be searched through its web interface. Even for *Arabidopsis*, information about a large collection of gene knockdown mutants is available through TAIR database (http://www.arabidopsis.org/portals/mutants/index. jsp). In case of C_4 maize, McCarty et al. (2009) generated transposon-based mutants that can be searched for availability at Maize Genetics and Genomics database (http://www.maizegdb.org/uniformmu).

In the last decade, C_4 biology research has received huge attention, and differential gene expression studies were frequently used for the purpose of gene discovery. A majority of these expression datasets are available at NCBI's Short Read Archive (SRA; www.ncbi.nlm.nih.gov/sra). There have been several studies comparing leaf transcriptome of: C_3 and C_4 species of Cleome (SRA accession: SRS002743.1 & SRS002744.2.); C_3, C_4, and intermediate species of Flaveria (accession: SRP006166); rice and maize (SRA accession: SRP035579); emerging C_4 model plant *Setaria viridis* (SRA accession: SRR784127); and around 100 samples derived from C_3/C_4 species, which have independently evolved (http://www.onekp.com/samples/list.php; not yet public). On the other hand, gene expression studies involving vein development have recently been carried out in maize (SRA accession: SRS394616–SRS394626) and rice (SRA accession: SRP062323).

Graphic representations of expression of genes across the plant tissues (expression atlas) or in tissues that are relevant for C$_3$/C$_4$ research are also available. One such resource is the Bio-Analytic Resource for plant biology (BAR) which hosts expression atlas for most of the model monocot and dicot species (http://bar.utoronto.ca). In addition, it also provides gene expression information for maize mesophyll and BS cells and maize and rice leaf developmental gradient samples. These resources have been found useful in prioritizing candidate gene for functional validation. Another equally useful web resource hosting expression atlas (based on microarrays) for a variety of plant species is Plead (http://www.plexdb.org/index.php). A similar resource only for rice is available at Riverport (http://ricexpro.dna.affrc.go.jp/).

Genomics research associated with C$_3$ to C$_4$ conversion often faces challenges when it comes to finding correct ortholog relationship for a gene of interest. Accuracy in orthology establishment is important as evolution of C$_4$ plants has been shown to have recruited genes, which are also present in their C$_3$ counterpart. Given this challenge, researchers developed improved methods to address this issue, which resulted into a dedicated database named Plant Homolog Database (http://phd.big.ac.cn) as well as a standalone tool named orthoFinder (http://www.stevekellylab.com/software/orthofinder).

The last but very important resource is the genome sequence resources. There are many species-specific web portals for genome resources such as TAIR, RAP-DB, and MaizeGDB (for *Arabidopsis*, rice, and maize respectively). However, a web portal named Phytozome (https://phytozome.jgi.doe.gov/pz/portal.html) serves as a better alternate where genomic resources for all sequenced genome projects are available.

While several unsuccessful attempts have already been made to generate C$_4$ rice by simply expressing some of the known genes that play important roles in C$_4$ biochemistry, transport, or anatomy, intensive efforts are being made to understand the fine genetic difference and regulation of these elements using model plants.

11.10   GENETIC MODELS TO STUDY C$_4$ PHOTOSYNTHESIS

Genetic model systems are highly useful for dissecting biological traits into component genes and understanding their functional mechanism. Lack of a small, rapid cycling genetic model system to study C$_4$ photosynthesis has limited progress in dissecting the regulatory networks underlying the C_4 syndrome. Maize (*Zea mays*) and sorghum (*Sorghum bicolor*) have been widely used for studying C_4 mechanism (Wang et al., 2013b, 2014; Rizal et al., 2015). The large plant size and longer growth period makes it quite difficult for using these plants for forward or reverse genetic models. Even though the sequencing information is available for both the plants, the quasitetraploid nature of maize makes the analysis relatively difficult. A mutant of maize with smaller plant size is being explored for this purpose. The Asteracean genus *Flaveria sp.* contains large numbers of C_3, C_3–C_4, and C_4 species and is suitable for exploring, but the lack of genome sequence makes it challenging for genetic analysis. Similarly, the genus cleome was also proposed mainly because its phylogenetic relationship with *Arabidopsis*. *Setaria viridis*, a member of the Panicoideae clade, is a true diploid with a relatively small genome of ~510 Mb. It has short stature, simple growth requirements, of and rapid life cycle. However, it has been challenging to genetically manipulate *Setaria* so far. Recently, there is some success in *Setaria* transformation, and it is also being used as a genetic model (Li and Brutnell, 2011; Martins et al., 2015; Saha and Blumwald, 2016). Clearly, more progress is needed to generate more efficient genetic transformation and genetic models for testing the effect of individual genes C_4 genes and their combination.

11.11   GENETIC TOOLS AVAILABLE

Though the C_4 photosynthetic system has naturally evolved several times during evolution, it is a complex multigenic trait that must modulate the anatomy, biochemistry, and physiology of the plant. Hence, the C_4 rice project needs cutting-edge technology for the discovery and precise systematic introduction of genes that will supercharge photosynthesis. Though a lot of effort is being made to find the master regulator(s), currently more than hundred genes have been predicted to affect several important C_4 phenotypes. Hence, the availability of modern technologies for precise introduction of these genes into rice and their expression is very important. The combination-based Gateway™ cloning technology circumvents traditional restriction enzyme-based cloning limitations, thereby enabling one to use several promoter-gene combinations in just a few simple steps. This system has been widely adopted for its time-saving nature and open access policy by which academic researchers can modify the backbone vectors according to their need and freely distribute among collaborators. For example, one can clone a BS-cell specific promoter into a Gateway vector and share it with other researcher to validate their genes. Keeping the large number of genes in mind, researchers have been looking for a more efficient methodology. Recent availability of the Golden gate cloning technology has given a boost to plant genetic research, which can substantially reduce the time required for cloning and development of multigene constructs using Type IIS restriction enzymes (Engler et al., 2009). The site of integration in the host genome also greatly affects the transgene expression. The recently developed transcription activator-like effect or nuclease (TALEN) technology is a versatile tool that has been described as the customizable molecular scissors for genome engineering of plants. In the past few years, there has been an explosion in the number and diversity of applications of this technology. Similarly, zinc-finger nucleases (ZFNs) have also been used widely for similar flexibility. The DNA-binding domain of ZFNs and TALENs can be customized to recognize virtually any sequence and combined with numerous effect or domains to affect genomic structure and function including nucleases, transcriptional activators and repressors, recombinases, transposases, DNA and histone methyltransferases, and histone acetyltransferases (Gaj et al., 2013). The single base recognition of TALE–DNA binding repeats gives greater design flexibility than triplet-confined zinc-finger proteins. However, cloning of TALE repeat arrays has been a major technical challenge because of the long and highly repetitive nature of the DNA binding domains of TALEs. The genome editing, using programmable RNA-guided DNA endonucleases, is the most recent addition to the toolbox of genetic engineers that is redefining the boundaries of biological research. In contrast to ZFNs and TALENS, the design of CRISPR construct is much simpler. In type II CRISPR–Cas system, a short segment of foreign DNA of 20 nucleotides, termed "spacer" is integrated within the CRISPR genomic loci and transcribed and processed into short CRISPR RNA (crRNA). The crRNA guides the Cas9 endonuclease to the target site where the later makes a targeted double strand break in the genomic DNA. The host cell DNA repair mechanism based on error-prone non-homologous end joining (NHEJ) adds mutation to the target sequences. The CRISPR–Cas system can thereby be retargeted to cleave virtually any DNA sequence by redesigning the crRNA. All applications of ZFNs/TALENs are being explored by the CRISPR–Cas9 system with relative ease.

11.12   DEVELOPMENT OF A C₄ PROTOTYPE

Though a functional C_4 prototype is yet to be developed, numerous transgenic lines have been developed with either single genes related to biochemical pathway, Kranz anatomy, transporters, or plastid biogenesis. Recently, a rudimentary version of the supercharged photosynthesis has been shown in rice by introducing key C_4 photosynthesis genes into a rice plant by the C_4 consortium led at the International Rice Research Institute (IRRI), though the altered rice plants still rely primarily on their usual form of photosynthesis (Bullis, 2015). The development of a fully functional rice plant may need integrated engineering of its biochemical pathway, anatomy, transporters, and plastid biogenesis. Some of the developments in each case have been discussed below.

11.12.1   C₄ BIOCHEMICAL PATHWAY

It is interesting to note that most of the enzymes needed for the C_4 photosynthesis are present in C_3 plants, but they have different substrate specificity, efficiency, or activity. For example, the isoform of PEPC involved in C_4 photosynthesis has 15- to 30-fold higher activity than C_3 leaves (Furbank et al., 2004). *In vivo*, it is under constant "restraint" via allosteric metabolite and covalent modification regulatory mechanisms. Overexpression of the intact maize C_4-specific PEPC gene in rice leaves did not directly affect photosynthesis significantly; rather, it suppressed photosynthesis indirectly by stimulating respiration in the light (Fukayama et al., 2003). In transgenic rice leaves, it underwent activity regulation through protein phosphorylation in a manner similar to endogenous rice PEPC. It remained dephosphorylated and less active during the day time and became phosphorylated and more active in the night (Fukayama et al., 2003). Suppression of PEPC phosphorylation by pharmacological inhibition of *de novo* protein synthesis resulted in reduced photosynthetic rates in maize and sorghum also (Bakrim et al., 1993). In addition to N-terminal phosphorylation, C_4 PEPC enzymes also contain specific central domains and a typical serine in the C-terminal part that are both important for allosteric regulation. Furthermore, there is evidence for regulation of C_4 PEPC by tethering the protein to membranes.

The glycine decarboxylase complex (GDC) plays a critical role in the photorespiratory C_2 cycle of C_3 species by recovering carbon following the oxygenation reaction of Rubisco. In C_3 plants, GDC can be found in all photosynthetic cells, whereas in leaves of C$_3$–C$_4$ intermediate and C$_4$ species, its occurrence is restricted to BS cells (Engelmann et al., 2008). Loss of GDC from mesophyll cells (MCs) is considered as a key early step in the evolution of C$_4$ photosynthesis. Preferential reduction of GDCH in rice M cells by artificial microRNA (amiRNA) exhibited a photo-respiratory-deficient phenotype with stunted growth, accelerated leaf senescence; reduced chlorophyll, soluble protein, and sugars; and increased glycine accumulation in leaves (Lin et al., 2016). Gas exchange measurements also indicated an impaired ability to regenerate RuBP in photo-respiratory conditions and a significant reduction in the chloroplast area and coverage of the cell wall when grown in air. In another study, severely suppressed OsGDCH-RNAi plants (SSPs) were nonviable under ambient CO$_2$ concentration, although the plants could grow under high CO$_2$ condition (Zhou et al., 2013). The lack of functional GDC in the MCs has earlier shown reduced photorespiration in *M. arvensis* (Morgan et al., 1993). Promoter deletion studies in *Arabidopsis* identified two regions in the GLDPA promoter, one conferring repression of gene expression in mesophyll cells and one functioning as a general transcriptional enhancer. Subsequent analyses in transgenic *F. bidentis* confirmed that these two segments also fulfill the same function in the C$_4$ context (Engelmann et al., 2008). Hence, the BS cell-specific expression of four subunit proteins (P, L, T, and H) of glycine decarboxylase (GDC) is one of the most important targets (Bauwe and Kolukisaoglu, 2003).

Other enzymes that are target of modification include pyruvate pi dikinase (PPDK), NADP malate dehydrogenase, NADP-ME (Chi et al., 2004), NAD-ME, PEPCK, etc. The *cis*-elements that direct cell specificity in C$_4$ leaves are present in C$_3$ orthologues of genes recruited into C$_4$, probably facilitating parallel evolution of C$_4$ trait.

11.12.2 KRANZ ANATOMY

The wreath-like or concentric arrangement of M cells and the BS cells around the vascular bundles is called the Kranz (*German: wreath*) anatomy. BS cells are arranged internally close to the vascular tissue harboring Rubisco and the other Calvin-Benson cycle enzymes, while the M cells are arranged in the outer layer so that they can come in contact with the intercellular air space. The M and BS cells are effectively connected with each other by high density of plasmodesmata (Dengler and Nelson, 1999). Kranz anatomy is also marked with closer vein spacing mainly due to change in patterning of minor veins (Ueno et al., 2006). Though single cell C_4 photosynthesis through compartmentalization of organelles has also been observed in the Chenopodiaceae family (Voznesenskaya et al., 2002), plants with Kranz anatomy have the highest photosynthetic ability (Wang et al., 2013b). In fact, it has been predicted that rice needs to develop Kranz anatomy in order to effectively use the C_4 photosynthesis (Smillie et al., 2012; Bullis, 2015). Rice vein length per leaf, mesophyll thickness, and intercellular space volume are intermediate between those of most C_3 and C_4 grasses, which indicates that the introduction of Kranz anatomy into rice may not require radical changes in leaf anatomy, although deep lobing of chlorenchyma cells may constrain efforts to engineer C_4 photosynthesis into rice (Sage and Sage, 2009).

Initial analysis of six distinct rice mutant lines, termed altered leaf morphology (alm) mutants, for the architecture of their interveinal mesophyll cell arrangement did not provide any clue for Kranz anatomy (Smillie et al., 2012). A more systematic analysis of several TRIM mutants is in progress at the C_4 consortium. A genome-wide comparative transcriptome analysis of developmental trajectories in Kranz (foliar leaf blade) and non-Kranz (husk leaf sheath) leaves of maize was performed by Langdales's group (Wang et al., 2013b) to provide an insight into the genetic regulation of Kranz anatomy. The spatial and temporal co-expression analysis performed by them supports the partial involvement of SCARECROW/SHORTROOT regulatory network in the Kranz patterning that was proposed earlier. In fact, they have identified at least a 100 related genes, which are being functionally validated in rice through the transgenic method.

Several TFs belonging to the MYB family play an important role in both stomatal and non-stomatal responses by regulation of stomatal numbers and sizes, and metabolic components, respectively. Recently, the C_4 consortium has identified 11 transcription factors that express in cell-specific manner in C_4 plants such as sorghum, maize, and Setaria. These transcription factors that regulate the function of cascades of other target genes might play an important role in cellular function and are being functionally validated.

11.12.3   TRANSPORTERS

Efficient transportation of C_4 metabolites between M and BS cells is imperative to maintain high efficiency of C_4 photosynthesis. In NADP-ME type, the most efficient subtype of C$_4$ plants such as maize, the CO$_2$ is converted by carbonic anhydrase (CA) to bicarbonate, which is fixed by PEPC to generate OAA in the M cell cytoplasm. The OAA is transported to the chloroplast where it is reduced to malate. Then, malate is transported from M cell chloroplast to BS cell chloroplast where it gets decarboxylated by NADP-ME to pyruvate that is transported back to M cell chloroplast. Four maize proteins ZmpOMT1 and ZmpDCT1-3 have been predicted to be involved in transport of OAA and malate (Taniguchi et al., 2004). However, transporters involved in NADP-ME type C$_4$ photosynthesis are less characterized, though these are very important for incorporating C$_4$ photosynthesis through the transgenic approach. In the case of NAD-ME type of plants, four solute transporters have been documented in the chloroplast envelope membrane. The pyruvate: sodium symporter (BASS2) imports pyruvate and sodium into the chloroplast while the sodium: proton antiporter (NHD1) maintains the plastidial sodium homeostasis by exchanging the sodium with the cytosol for protons. The phosphoenolpyruvate/phosphate antiporter (PPT) counterexchanges PEP with inorganic ortho-phosphate (Pi). The OAA/malate antiporter DiT1 (dicarboxylate transporter 1) exchanges the OAA with malate at the MC chloroplast (Weber and Bräutigam, 2013). Comparative proteomics of chloroplast envelope has shown that C$_3$ and C$_4$ type chloroplasts have qualitatively similar but quantitatively very different chloroplast envelope membrane proteomes. In particular, translocators involved in the transport of triosephosphate and phosphoenolpyruvate as well as two outer envelope porins are much more abundant in C$_4$ plants (Brautigam et al., 2008). Several putative transport proteins have been identified that are highly abundant in C$_4$ plants but relatively minor in C$_3$ envelopes.

11.12.4 PLASTID BIOGENESIS

Land plants acquired their chloroplasts from endosymbiotic cyanobacterium. The consequent gene transfer between the chloroplast and nuclear genome and adaptive evolutions have resulted in interdependence between the organelle and nucleus genome for most of the complex genome signaling. Plastids have diversified in plants from their original function as chloroplasts to fulfill a variety of other roles such as metabolite biosynthesis and storage or purely to facilitate their own transmission according to the cell type that harbors them (López-Juez, 2007). Very little is known about how different plastid types differentiate, or about what mechanisms coordinate cell growth with plastid growth and division. Biogenesis of organelles and ultimately the development, metabolism, and evolution of higher plants may be the result of an intimate molecular cooperation among the nucleus/cytosol, plastids, and mitochondria (Herrmann et al., 1992). There might be a major, separate plastid and chloroplast "master switches," as indicated by the coordinated gene expression of plastid or chloroplast-specific proteins.

The higher amount of CO_2 to be re-fixed by Rubisco will need higher number of chloroplasts in BS cells. In C_4 rice plants, the chloroplast should mainly be present in BS cells for re-fixing the CO_2 and feeding it to the Calvin cycle. The size of the BS cells in C_4 plants is usually larger than C_3 plants so that it can accommodate more number of chloroplasts. In Kranz type C_4 species, spatial (or cell-specific) control of transcription of nuclear genes contributes to the development of dimorphic chloroplasts (Offermann et al., 2011). In maize, three different photosynthetic cells are observed. The C_3 M cells represent the ground state, while the specialized C_4 M and BS cells differentiate in response to light-induced positional signals (Rossini et al., 2001). The transcription factor Golden2 (G2) regulates plastid biogenesis in all photosynthetic cells during the C_3 stages of development and the differentiation of BS cell chloroplasts in C_4 leaf blades of maize while *ZmGLK1* transcripts accumulate mainly in C_4 M cells. Rossini et al. (2001) isolated G2-like (Glk) genes from both maize and rice, though the expression profiles of the rice *Glk* genes suggest that these genes may act redundantly to promote photosynthetic development in C_3 species. Compartmentalized function of a pair of *GLK* genes also regulates dimorphic chloroplast differentiation in sorghum (Wang et al., 2013a).

11.13 CONCLUDING REMARK

To increase the food production for ever-growing global population, crop yields must significantly increase. One of the avenues being recently explored is the improvement of photosynthetic capacity by engineering the C_4 photosynthetic pathway into C_3 crops like rice, which may drastically increase their yield. Crops with an enhanced photosynthetic efficiency would be able to utilize the solar radiation more efficiently, which can be translated into yield. Subsequently, this will help in producing more grain yield, reduce water loss, and increase nitrogen use efficiency especially in hot and dry environments. However, identifying the master regulators of the C$_4$ phenotype and their careful introduction can be much more difficult than predicted. With the ever-increasing atmospheric (CO$_2$), the advantage of C$_4$ plants over C$_3$ may also disappear as atmospheric (CO$_2$) nears 700 ppm. More elaborate engineering approaches are still in the experimental state and are being tested in model species. It is obvious that the establishment of C$_4$ rice will require more sophisticated gene transfer technologies. Gene pyramiding may be combined with modular cloning technologies such as Golden gate/Golden braid system and targeted genome editing technologies such as TALENs and CRISPR–Cas system. Once scientists solve the C$_4$ puzzle in a plant such as rice, the method can be extended to dramatically increase production of many other crops, including wheat, potatoes, tomatoes, apples, and soybeans (Bullis, 2015).

KEYWORDS

- C$_3$
- C$_4$
- chloroplast
- kranz anatomy
- photorespiration
- rice
- rubisco

REFERENCES

Bakrim, N., Prioul, J. L., Deleens, E., Rocher, J. P., Arrio-Dupont, M., Vidal, J., Gadal, P., & Chollet, R., (1993). Regulatory phosphorylation of C4 phosphoenolpyruvate carboxylase (a cardinal event influencing the photosynthesis rate in sorghum and maize). *Plant Physiol., 101,* 891–897.

Bauwe, H., & Kolukisaoglu, U., (2003). Genetic manipulation of glycine decarboxylation. J. *Exp. Bot., 54,* 1523–1535.

Bjorkman, O., Nobs, M., Pearcy, R., Boynton, J., & Berry, J., (1971). Characteristics of hybrids between C3 and C4 species of Atriplex. In: *Photosynthesis and Photorespiration,* Hatch, M., Osmond, C. & Slatyer, R. O., (eds.). (Wiley-Interscience Publishing, New York), pp. 105–119.

Brar, D. S., & Ramos, J. M., (2007). Wild species of Oryza: a rich reservoir of genetic variability for rice improvement. In: *Charting New Pathways to C4 Rice* (International Rice Research Institute (IRRI)),pp. 351–359.

Bräutigam, A., Kajala, K., Wullenweber, J., Sommer, M., Gagneul, D., Weber, K. L., Carr, K. M., Gowik, U., Mass, J., Lercher, M. J., Westhoff, P., Hibberd, J. M., & Weber, A. P., (2011). An mRNA blueprint for C4 photosynthesis derived from comparative transcriptomics of closely related C3 and C4 species. *Plant Physiol., 155*(1), 142–156.

Brautigam, A., Hoffmann-Benning, S., & Weber, A. P. M., (2008). Comparative proteomics of chloroplast envelopes from C3 and C4 plants reveals specific adaptations of the plastid envelope to C4 photosynthesis and candidate proteins required for maintaining C4 metabolite fluxes. *Plant Physiol., 148,* 568–579.

Brown, R. H., Bassett, C. L., Cameron, R. G., Evans, P. T., Bouton, J. H., Black, C. C., Sternberg, L. O., & Deniro, M. J., (1986). Photosynthesis of F1 hybrids between C4 and C3–C4 species of flaveria. *Plant Physiol., 82,* 211–217.

Brown, R. H., Byrd, G. T., Bouton, J. H., & Bassett, C. L., (1993). Photosynthetic characteristics of segregates from hybrids between flaveria brownii (C4 like) and flaveria linearis (C3-C4). *Plant Physiol., 101,* 825–831.

Bullis, K., (2015). Supercharged photosynthesis. *MIT Technol. Rev.iew,* February 2015. www.technologyreview.com/featuredstory/535011/supercharged-photosynthesis/

Burnell, J. N., (2011). Hurdles to engineering greater photosynthetic rates in crop plants: C4 Rice. In: *C4 Photosynthesis and Related CO_2 Concentrating Mechanisms*, Raghavendra, A. S., & Sage, R. F., eds., *Advances in Photosynthesis and Respiration.* (Springer Netherlands: Dordrecht), pp. 361–378.

Byrt, C. S., Grof, C. P. L., & Furbank, R. T., (2011). C4 plants as biofuel feedstocks: Optimising biomass production and feedstock quality from a lignocellulosic perspective. *J. Integr. Plant Biol., 53,* 120–135.

Chi, W., Yang, J., Wu, N., & Zhang, F., (2004). Four rice genes encoding NADP malic enzyme exhibit distinct expression profiles. *Biosci. Biotechnol. Biochem., 68,* 1865–1874.

Christin, P. A., Besnard, G., Samaritani, E., Duvall, M. R., Hodkinson, T. R., Savolainen, V., & Salamin, N., (2008). Oligocene CO_2 decline promoted C4 photosynthesis in grasses. *Curr. Biol., 18,* 37–43.

Dengler, N. G., & Nelson, T., (1999). Leaf structure and development in C4 plants. In: *C4 Plant Biology*, Sage, R., & Monson, R., (eds.). (Academic Press, San Diego), pp. 133–172.

Edwards, G., & Ku, M., (1987). Biochemistry of C3-C4 intermediates. In: *The Biochemistry of Plants – A Comprehensive Treatise*, Hatch, M. D., & Boardman, N. K., (eds.). (Academic Press, San Diego), pp. 275–325.

Edwards, G. E., Furbank, R. T., Hatch, M. D., & Osmond, C. B., (2001). What does it take to be C4 ? Lessons from the evolution of C4 photosynthesis. *Plant Physiol., 125,* 46–49.

Engelmann, S., Wiludda, C., Burscheidt, J., Gowik, U., Schlue, U., Koczor, M., Streubel, M., Cossu, R., Bauwe, H., & Westhoff, P., (2008). The gene for the P-subunit of glycine decarboxylase from the C4 species Flaveria trinervia: analysis of transcriptional control in transgenic Flaveria bidentis (C4) and Arabidopsis (C3). *Plant Physiol., 146,* 1773–1785.

Engler, C., Gruetzner, R., Kandzia, R., & Marillonnet, S., (2009). Golden gate shuffling: a One-Pot DNA shuffling method based on type IIs restriction enzymes., *PLoS One, 4,* e5553.

Fukayama, H., Hatch, M. D., Tamai, T., Tsuchida, H., Sudoh, S., Furbank, R. T., & Miyao, M., (2003). Activity regulation and physiological impacts of maize C(4)-specific phosphoenolpyruvate carboxylase overproduced in transgenic rice plants. *Photosynth. Res.,* *77,* 227–239.

Furbank, R., & Taylor, W., (1995). Regulation of photosynthesis in C3 and C4 plants: A molecular approach. *Plant Cell., 7,* 797–807.

Furbank, R. T., Hatch, M. D., & Jenkins, C. L. D., (2004). C4 Photosynthesis: Mechanism and regulation. In: *Advances in Photosynthesis and Respiration,* Leegood, R. C., Sharkey, T. D., & Von Caemmerer, S., (eds.). (Kluwer Academic Publishers, Dordrecht), pp. 435–457.

Gaj, T., Gersbach, C. A., & Iii, C. F. B., (2013). ZFN, TALEN, and CRISPR/Cas-based methods for genome engineering. *Trends Biotechnol., 31,* 397–405.

Gowik, U., & Westhoff, P., (2011). The path from C3 to C4 photosynthesis. *Plant Physiol., 155,* 56–63.

Hatch, M. D., (1999). C4 photosynthesis: a historical overview. In: *C4 Plant Biology,* Sage, R., & Monson, R., (eds.). (Academic Press, San Diego), pp. 17–46.

Heckmann, D., Schulze, S., Denton, A., Gowik, U., Westhoff, P., Weber, A. P. M., & Lercher, M. J., (2013). Predicting C4 photosynthesis evolution: Modular, individually adaptive steps on a mount fuji fitness landscape. *Cell., 153,* 1579–1588.

Herrmann, R. G., Westhoff, P., & Link, G., (1992). Biogenesis of plastids in higher plants. In: *Cell Organelles,* Herrmann, R. G., (ed.). (Springer Vienna), pp. 275–349.

Hibberd, J. M., & Quick, W. P., (2002). Characteristics of C4 photosynthesis in stems and petioles of C3 flowering plants. *Nature, 415,* 451–454.

Hibberd, J. M., Sheehy, J. E., & Langdale, J. A., (2008). Using C4 photosynthesis to increase the yield of rice-rationale and feasibility. *Curr. Opin. Plant Biol., 11,* 228–31.

Holaday, A., & Chollet, R., (1984). Photosynthetic/photorespiratory characteristics of C3−C4 intermediate species. *Photosynth. Res.,* 5, 307–323

Imaizumi, N., Ku, M. S. B., Ishihara, K., Samejima, M., Kaneko, S., & Matsuoka, M., (1997). *Plant Mol. Biol., 34,* 701–716.

John, C. R., Smith-Unna, R. D., Woodfield, H., Covshoff, S., & Hibberd, J. M., (2014). Evolutionary convergence of cell-specific gene expression in independent lineages of C4 grasses. *Plant Physiol., 165,* 62–75.

Kanai, R., & Edwards, G. E., (1999). The biochemistry of C4 photosynthesis. In: *C4 Plant Biology,* Sage, R., & Monson, R., (eds.). (Academic Press, San Diego), pp. 49–87.

Kellogg, E. A., (1999). Phylogenetic aspects of the evolution of C4 photosynthesis. In: *C4 Plant Biology,* Sage, R., & Monson, R., (eds.). (Academic Press, San Diego), pp. 411–444.

Ku, M. S., Wu, J., Dai, Z., Scott, R. A., Chu, C., & Edwards, G. E., (1991). Photosynthetic and photorespiratory characteristics of flaveria species. *Plant Physiol., 96,* 518–528.

Li, P., & Brutnell, T. P., (2011). Setaria viridis and Setaria italica, model genetic systems for the Panicoid grasses. *J. Exp. Bot., 62,* 3031–7.

Lin, H., Karki, S., Coe, R. A., Bagha, S., Khoshravesh, R., Balahadia, C. P., Ver Sagun, J., Tapia, R., Israel, W. K., Montecillo, F., De Luna, A., Danila, F. R., Lazaro, A., Realubit, C. M., Acoba, M. G., Sage, T. L., Von Caemmerer, S., Furbank, R. T., Cousins, A. B., Hibberd, J. M., Quick, W. P., & Covshoff, S., (2016). Targeted knockdown of GDCH in rice leads to a photorespiratory-deficient phenotype useful as a building block for C4 rice. *Plant Cell Physiol., 57*(5), 1–14 doi: 10.1093/pcp/pcw033.

López-Juez, E., (2007). Plastid biogenesis, between light and shadows. *J. Exp. Bot., 58,* 11–26.

Martins, P. K., Ribeiro, A. P., Cunha, B. A. D. B., Da, Kobayashi, A. K., & Molinari, H. B. C., (2015). A simple and highly efficient Agrobacterium-mediated transformation protocol for Setaria viridis. *Biotechnol. Reports, 6,* 41–44.

McCarty, D. R., & Meeley, R. B., (2009). Transposon resources for forward and reverse genetics in maize. In: *Handbook of Maize,* Bennetzen, J. L., & Hake, S., (eds.). (Springer New York, New York), pp. 561–584.

Molina, J., Sikora, M., Garud, N., Flowers, J. M., Rubinstein, S., Reynolds, A., Huang, P., Jackson, S., Schaal, B. A., Bustamante, C. D., Boyko, A. R., & Purugganan, M. D., (2011). Molecular evidence for a single evolutionary origin of domesticated rice. *Proc. Natl. Acad. Sci. USA, 108,* 8351–8356.

Monson, R., Edwards, G., & Ku, M., (1984). C3-C4 intermediate photosynthesis in plants. *Bioscience, 34 (9),* 563–574, https://doi.org/10.2307/1309599.

Moore, B. D., Franceschi, V. R., Cheng, S. H., Wu, J., & Ku, M. S., (1987). Photosynthetic characteristics of the C3-C4 intermediate Parthenium hysterophorus. *Plant Physiol., 85,* 978–983.

Morgan, C. L., Turner, S. R., & Rawsthorne, S., (1993). Coordination of the cell-specific distribution of the four subunits of glycine decarboxylase and of serine hydroxymethyltransferase in leaves of C3-C4 intermediate species from different genera. *Planta., 190,* 468–473.

Nelson, D., Lehninger, A., & Cox, M., (2013). *Lehninger Principles of Biochemistry* (6th ed.) New York, Freeman W. H.

Oakley, J. C., Sultmanis, S., Stinson, C. R., Sage, T. L., & Sage, R. F., (2014). Comparative studies of C3 and C4 Atriplex hybrids in the genomics era: physiological assessments. *J. Exp. Bot., 65,* 3637–3647.

Offermann, S., Okita, T. W., & Edwards, G. E., (2011). How do single cell C4 species form dimorphic chloroplasts? *Plant Signal. Behav., 6,* 762–765.

Rizal, G. Thakur, V., Dionora, J., Karki, S., Wanchana, S., Acebron, K., Larazo, N., Garcia, R., Mabilangan, A., Montecillo, F., Danila, F., Mogul, R., Pablico, P., Leung, H., Langdale, J. A., Sheehy, J., Kelly, S., & Quick, W. P., (2015). Two forward genetic screens for vein density mutants in sorghum converge on a Cytochrome P450 gene in the brassinosteroid pathway. *Plant J., 84*(2), 257–266.

Rossini, L., Cribb, L., Martin, D. J., & Langdale, J. A., (2001). The maize golden2 gene defines a novel class of transcriptional regulators in plants. *Plant Cell., 13,* 1231–1244.

Sage, R. F., Sage, T. L., & Kocacinar, F., (2012). Photorespiration and the Evolution of C_4 Photosynthesis. *Annu. Rev. Plant Biol., 63,* 19–47.

Sage, R. F., & Zhu, X. G., (2011). Exploiting the engine of C4 photosynthesis. *J. Exp. Bot., 62,* 2989–3000.

Sage, T. L., & Sage, R. F., (2009). The functional anatomy of rice leaves: Implications for refixation of photorespiratory CO2 and efforts to engineer C4 photosynthesis into rice. *Plant Cell Physiol., 50,* 756–772.

Saha, P., & Blumwald, E., (2016). Spike dip transformation of Setaria viridis. *Plant J., 86,* 89–101.

Sheehy, J., & Mitchell, P., (2011). Rice and global food security: the race between scientific discovery and catastrophe. In: *Access Not Excess – The Search for Better Nutrition,* Pasternak, C., (ed.). (Smith-Gordon), pp. 81–90.

Sheen, J., (1999). C4 gene expression. *Annu. Rev. Plant Physiol. Plant Mol. Biol., 50,* 187–217.

Smillie, I. R. A., Pyke, K. A., & Murchie, E. H., (2012). Variation in vein density and meso-phyll cell architecture in a rice deletion mutant population. *J. Exp. Bot., 63*, 4563–4570.

Taniguchi, Y., Nagasaki, J., Kawasaki, M., Miyake, H., Sugiyama, T., & Taniguchi, M., (2004). Differentiation of dicarboxylate transporters in mesophyll and bundle sheath chloroplasts of maize. *Plant Cell Physiol., 45*, 187–200.

Ueno, O., (2001). Environmental regulation of C3 and C4 differentiation in the amphibious sedge eleocharis vivipara. *Plant Physiolo., 127*, 1524–1532.

Ueno, O., Kawano, Y., Wakayama, M., & Takeda, T., (2006). Leaf vascular systems in C3 and C4 grasses: A two-dimensional analysis. *Ann. Bot., 97*, 611–621.

Ueno, O., Samejima, M., Muto, S., & Miyachi, S., (1988). Photosynthetic characteristics of an amphibious plant, Eleocharis vivipara: expression of C4 and C3 modes in contrast-ing environments. *Proc. Natl. Acad. Sci., 85*, 6733–6737.

Vicentini, A., Barber, J. C., Aliscioni, S. S., Giussani, L. M., & Kellogg, E. A., (2008). The age of the grasses and clusters of origins of C 4 photosynthesis. *Glob. Chang. Biol., 14*, 2963–2977.

Voet, D., Voet, J. G., & Pratt, C. W., (2013). *Principles of Biochemistry International Student Version*. In (Wiley Plus International), pp. 623–656.

Von Caemmerer, S., & Furbank, R. T., (2003). The C4 pathway: an efficient CO2 pump. *Photosynth. Res., 77*, 191–207.

Von Caemmerer, S., Quick, W. P., & Furbank, R. T., (2012). The development of C4 rice: Current progress and future challenges. *Science, 336*, 1671–1672.

Voznesenskaya, E., V., Franceschi, V. R., Kiirats, O., Artyusheva, E. G., Freitag, H., & Ed-wards, G. E., (2002). Proof of C4 photosynthesis without Kranz anatomy in Bienertia cycloptera (Chenopodiaceae). *Plant J., 31*, 649–662.

Wang, C., Guo, L., Li, Y., & Wang, Z., (2012). Systematic comparison of C3 and C4 plants based on metabolic network analysis. *BMC Syst. Biol., 6*, S9.

Wang, L. Czedik-Eysenberg, A., Mertz, R. A., Si, Y., Tohge, T., Nunes-Nesi, A., Arrivault, S., Dedow, L. K., Bryant, D. W., Zhou, W., Xu, J., Weissmann, S., Studer, A., Li, P., Zhang, C., LaRue, T., Shao, Y., Ding, Z., Sun, Q., Patel, R. V., Turgeon, R., Zhu, X., Provart, N. J., Mockler, T. C., Fernie, A. R., Stitt, M., Liu, P., & Brutnell, T. P., (2014). Compara-tive analyses of C4 and C3 photosynthesis in developing leaves of maize and rice. *Nat. Biotechnol., 32*, 1158–1165.

Wang, P., Fouracre, J., Kelly, S., Karki, S., Gowik, U., Aubry, S., Shaw, M. K., Westhoff, P., Slamet-Loedin, I. H., Quick, W. P., Hibberd, J. M., & Langdale, J. A., (2013a). Evolu-tion of GOLDEN2-LIKE gene function in C3 and C4 plants. *Planta., 237*, 481–95.

Wang, P., Kelly, S., Fouracre, J. P., & Langdale, J. A., (2013b). Genome-wide transcript anal-ysis of early maize leaf development reveals gene cohorts associated with the differen-tiation of C4 Kranz anatomy. *Plant J., 75*, 656–670.

Weber, A. P. M., & Bräutigam, A., (2013). The role of membrane transport in metabolic engi-neering of plant primary metabolism. *Curr. Opin. Biotechnol., 24*, 256–262.

Westhoff, P., & Gowik, U., (2010). Evolution of C4 photosynthesis–looking for the master switch. *Plant Physiol., 154*, 598–601.

Westhoff, P., Svensson, P., Ernst, K., Blasing, O., Burscheidt, J., & Stockhaus, J., (1997). Mo-lecular evolution of C4 phosphoenolpyruvate carboxylase in the genus flaveria. *Aust. J. Plant Physiol. 24*, 429.

Williams, B. P., Johnston, I. G., Covshoff, S., & Hibberd, J. M., (2013). Phenotypic landscape inference reveals multiple evolutionary paths to C4 photosynthesis. *Elife., 2*, 1–19.

Yeo, M. E., Yeo, A. R., & Flowers, T. J., (1994). Photosynthesis and photorespiration in the genus Oryza. *J. Exp. Bot., 45,* 553–560.

Young, J. N., Rickaby, R. E. M., Kapralov, M. V., & Filatov, D. A., (2012). Adaptive signals in algal Rubisco reveal a history of ancient atmospheric carbon dioxide. *Philos. Trans. R. Soc. B. Biol. Sci., 367,* 483–492.

Zhou, Q., Yu, Q., Wang, Z., Pan, Y., Lv, W., Zhu, L., Chen, R., & He, G., (2013). Knockdown of GDCH gene reveals reactive oxygen species-induced leaf senescence in rice. *Plant. Cell Environ., 36,* 1476–89.

Zhu, X. G., Long, S. P., & Ort, D. R., (2008). What is the maximum efficiency with which photosynthesis can convert solar energy into biomass? *Curr. Opin. Biotechnol., 19,* 153–159.

CHAPTER 12

MOLECULAR PHARMING (PHARMACEUTICALS): PRIMARY AND SECONDARY METABOLITES IN PLANTS

MIR ZAHOOR GUL,[1] MOHD YASIN BHAT,[2] ANIRUDH KUMAR,[3] and BEEDU SASHIDHAR RAO[1]

[1]Department of Biochemistry, University College of Sciences, Osmania University, Hyderabad, 500 007, Telangana, India

[2]Department of Plant Sciences, School of Life Sciences, University of Hyderabad, Hyderabad, 500 046, Telangana, India

[3]Department of Botany, Indira Gandhi National Tribal University, Amarkantak, 484 886, Madhya Pradesh, India

CONTENTS

12.1   INTRODUCTION

The concept of therapeutic use of plants and other botanicals is as old as early civilizations, and evolution of these medicines has occurred in parallel with the development of these civilizations. Although tremendous advances have been made in synthetic organic chemistry for manufacturing new and novel drugs, it is estimated that a quarter of the present day prescribed drugs are derived from plants (Lahlou, 2007; Ramawat and Goyal, 2008). During the last century, extensive research was conducted on the mechanism of drug action, and the development of new technologies has revolutionized the screening of natural products in discovering new drugs. The present state of the study suggests that the biological activity exhibited by any drug in the body is mediated by specific interactions of the drug molecules with biological macromolecules (proteins or nucleic acids in most of the cases). These interactions may trigger a particular response in the body cells, which ultimately leads to cellular homeostasis, and the body is finally restored to the normal state of health. The advent of modern molecular biology, particularly the DNA recombinant technology, has provided new impetus to the pharmaceutical industry for the development of bio-pharmaceutical products from nature through investigation of leads from the traditional systems of medicine. "Molecular farming" is an application of this technology involving the use of plants and, in some cases, animals, as a means to procure compounds of high therapeutic value for the benefit of humanity. Transformation of the organisms, for example, the development of genetically engineered plants by different techniques like *Agrobacterium*-mediated gene transformation approach or gene gun-induced transformation has revolutionized the production and development of useful products. Researchers, while taking advantage of these technological advancements, produce biopharmaceuticals products for varied purposes such as diagnostic and therapeutic applications as well as nutritional supplements. Through integrative approaches such as genetic modifications, by combining the various discovery tools and the new discipline of integrative biology, plants have emerged as a potentially new source of pharmaceutical products including antibodies, vaccines, blood substitutes, cytokines, growth factors, etc., etc. Plant-derived products have an advantage over those of mammalian origin as they are free from dangerous entities like mammalian viral vectors and other pathogens. Plants are also easy and cost-effective to cultivate, have high biomass production capability, and possess excellent gene to protein conversion speed, which makes them a suitable choice for the manufacture of the biotechnological products (Giddings et al., 2000; Breyer et al., 2012; Stryjewska et al., 2013). A considerable shift in the focus of research from basic sciences toward the commercial utilization of molecular farming has arisen in the recent past to produce pharmaceutical products that will benefit the mankind.

12.2   MOLECULAR FARMING IN PLANTS

Plant molecular farming refers to the development of genetic transformation of plants, i.e., introduction and expression of foreign genes, so that they accumulate non-native proteins (including pharmaceuticals and industrial proteins) and other secondary metabolites in their cells and tissues. The products derived therefrom can have an enormous therapeutic potential, for example, the large-scale production of recombinant proteins (Franken et al., 1997). The first successful attempt was the production of a recombinant plant-derived pharmaceutical protein (PDP), human serum albumin, initially produced in 1990 in transgenic tobacco and potato plants (Sijmons et al., 1990). Its ultimate purpose was to provide a safe and inexpensive means for the mass production of recombinant pharmaceutical proteins from plants. Plant molecular farming involves the growing of plants just like in agriculture to produce pharmaceutical or industrial compounds instead food, feed, or fiber. The valuable products that can be procured from such endeavors include drugs, vaccines, biodegradable plastics, and other industrially useful products. Molecular farming provides a broad range of products ranging from recombinant insulin expressed in bacteria to chimeric anti-tumor antibodies and multi-subunit protein complexes, which can prove to be a great boon for diabetic and cancer patients, respectively (Figure 12.1). Some of the products of molecular farming are commercially viable to be afforded by a common man. As the research in molecular farming has advanced, there have been technological breakthroughs on several levels, including transformation strategy, regulation of gene expression, protein accumulation, and the possibility of using a wide variety of crops as production platforms (Yusibov et al., 2002). The other strategies that are being adapted include the development of novel promoters, the improvisation of stability of target proteins and accumulation through manipulation of signaling pathways so that the target protein is compartmentalized into the intracellular milieu, and

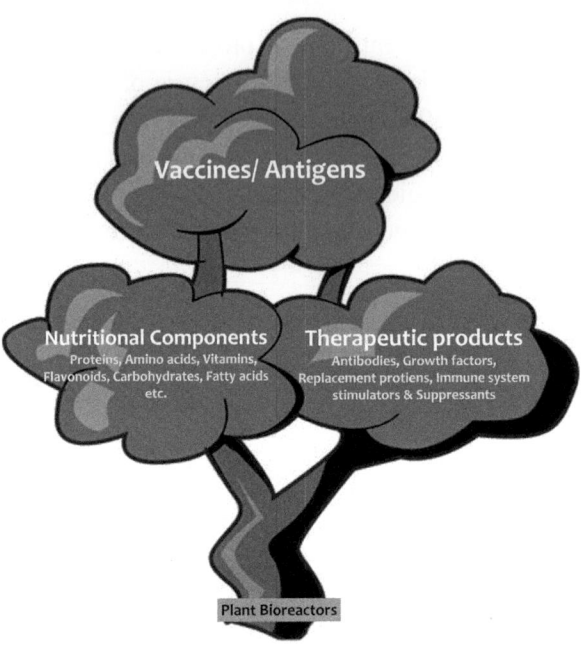

FIGURE 12.1   Molecular farming overview.

the improvement of downstream processing technologies (Menkhaus et al., 2004). The technological advancements will not only make different valuable products as a possibility but will also lead to the economic feasibility of these products. Further research into the development of this field may substantially reduce the production cost and thus reduce the market value of the products that are currently expensive, and consequently, a wider usage of these products by people across the globe may not be a remote possibility from now.

Plants usually produce large amounts of biomass, but the protein concentration level can still be elevated by using plant cell suspension culture in huge fermenters or by stably transforming and cultivating the plants in the field conditions (Kamenarova et al., 2005). Plants are a suitable choice for the production of complex mammalian proteins on a large scale by transformation. These expressed proteins resemble exactly regarding functionality and structure to the mammalian counterparts. The range of therapeutic proteins produced by plants varies from interleukins to recombinant antibodies. There have been successful attempts of production of transgenic plants that produce a recombinant protein in large quantities in any particular organ like seeds (Ma and Wang, 2012) for its long-term storage. Transgenic plants may be considered as bioreactors for the molecular farming of recombinant therapeutics comprising vaccines and diagnostics such as recombinant antibodies, plasma proteins, cytokines, and growth factors. That plants can be used for developing of recombinant pharmaceuticals was evident from the empirical results found between 1986 and 1990 when a rewarding expression of human growth hormone fusion protein (an interferon and human serum albumin) was obtained (Barta et al., 1986; De Zoeten et al., 1989; Sijmons et al., 1990). A breakthrough was achieved when successful expression of functional antibodies in plants was done in 1989 (Hiatt et al., 1989; Hiatt, 1990; Düring et al., 1990).

There is a shortage of biopharmaceuticals in the market owing to the high demand, which is further aggravated by the high cost and low efficiency of the current production systems like yeast, microbes. and animal cells (Jones, 2003; Knäblein, 2005). The expression of recombinant proteins in plants is a better alternative system than currently applied expression systems based on animal cells, yeast, or bacteria because of minimal production cost and lesser risk of pathogenic contamination (Jamal et al., 2009). It has led to an impetus in research for transgenic plants that can function as new generation bioreactors currently used in pharmaceutical companies, universities, and research institutes (Basaran and Rodríguez-Cerezo, 2008). The transgenic plants are known to be extremely versatile and capable of producing a broad range of proteins (Schillberg et al., 2003). The plants offer certain advantages for the purpose of molecular farming like high quantity of expression of recombinant proteins and secure storage of raw material. There are also least apprehensions regarding the contamination of these proteins with animal and human pathogens during downstream processing that has attracted biotechnologists toward plants for utilizing in molecular farming (Jamal et al., 2009; Breyer et al., 2012; Ma and Wang, 2012; Sparrow et al., 2013). Although transgenic animals, bacteria, and fungi are also used for the production of proteins, plants can be used to gain more economic profit (Horn et al., 2004). Preferably, higher plants are utilized for the production of proteins instead of animals due to the following reasons: (*i*) production cost is lesser than transgenic animals; (*ii*) management is easier than that for animals, as proficiency already exist for planting, harvesting, and processing of the plant material; (*iii*) plants are free from known human pathogen (such as virions); thus, there is no chance of contamination of the final product; (*iv*) higher plants generally synthesize proteins from eukaryotes with correct folding, glycosylation, and activity; (*v*) stability is greater than that of the animals because plant cells can store proteins to subcellular endo compartments that reduce degradation and therefore, increase stability (Horn et al., 2004; Obembe et al., 2011). For long-term storage, transgenic plants can also produce organs rich in a recombinant protein (Ma and Wang, 2012). By the dissemination of stably transformed plant lines in the field, a huge amount of biomass and protein can be produced (Kamenarova et al., 2005). Due to these reasons, plant molecular farming has become more attractive for modern biotechnologist, especially for plastid and chloroplast engineering (Horn et al., 2004; Breyer et al., 2012; Sparrow et al., 2013). Many common crops like rice, wheat, maize, banana, tomato, tobacco, potato, safflower, soy, canola, alfalfa, *Arabidopsis*, and oil rape seeds have been used scientifically for molecular farming. These plants were not selected on the random basis but are based on already existing knowledge of their pollination strategy, genetic, and other relevant aspects, which can be related to feasibility for conducting molecular farming experiments. *Nicotiana tabacum* is most commonly used as a model for the expression of recombinant proteins due to its capability to produce large quantities of green leaf materials per acre. But when the allocation of biopharmaceuticals is targeted toward any storage organ like a seed, *N. tabacum* is not a decent choice as it has tiny seeds and thus lesser storage capability. In such cases, it is always better to use other crops such as soybean and corn.

12.3   PLANT TRANSFORMATION SYSTEMS/EXPRESSION STRATEGIES

Plants provide an alternative to the use of animal expression system for recombinant protein production. Many such products have already been procured for plant transformation strategy, while many others are still in the pipeline (Fischer et al., 1999; Schillberg et al., 2003). The basic technological concept behind this is that genes for high-value proteins are expressed in crops that can be easily grown on a large scale, and the protein is then extracted and purified from the plant tissue. There are four ways of protein production from plants. Table 12.1 shows advantages and disadvantages of the systems mentioned above.

TABLE 12.1 Different Plant-Based Production Systems

System	Advantages	Disadvantages
Transgenic plants (Stable nuclear transformation of a crop species that will grow in the field or a green-house)	Yield, economy, scalability, establishment of permanent lines (when accumulated within plants) Containment, purification (when secreted from roots or leaves) Multiple gene expression, low toxicity, containment (transplas-tomic)	Production timescale, regula-tory compliance Scale yield cost production facilities Ab-sence of glycosylation, some evidence of horizontal gene transfer (transplastomic plants)
Virus-infected plants	Yield, timescale, mixed infections	High Cost
Agro-infiltrated leaves	Timescale	High Cost
Cell and tissue cultures	Timescale, containment, secretion into medium, purification, regula-tory compliance	High Cost

    a.   Plants for open field and/or greenhouse;
    b.   Stable transformation;
    c.   Transient transformation; and
    d.   Bioreactor-based systems.

12.3.1 PLANTS FOR OPEN FIELD AND/OR GREENHOUSE

Molecular farming approach has involved the utilization of a wide range of plants including both monocots and dicots like rice, banana, sugar beet, wheat, maize, *Arabidopsis*, tomato, pigeon pea, peanut, tobacco, potato, safflower, soy, white mustard, oilseed rape, etc. (Twyman et al., 2003, 2005). For the sake molecular farming, these plants are distinguished into dry seed crops (including cereals and grain legumes), leafy crops, fruits, vegetable crops, and oil crops.

Leafy crops are advantageous as they produce a large amount of bio-mass. However, the disadvantage is that the proteins are synthesized in an aqueous environment and are prone to rapid proteolytic degradation after harvesting. To overcome this, the material is processed as soon as possible after harvesting or transported in a dried frozen state, which adds to the production cost. The expression of proteins in vegetative tissues can also potentially interfere with plant metabolism. In the case of cereals and grain legumes (e.g., maize, rice), the protein allocation is directed to the seeds that allow long-term storage of the target protein, often at room temperature. By such methods, antibodies have been stored up for 3 years without any detectable loss of activity (Stoger et al., 2005a; Twyman et al., 2005). Downstream processing can, therefore, take place in batches of harvested material. This sort of processing is improved as no oxidizing substances are present compared to that in leafy crops like tobacco and alfalfa leaves. Currently, fruits and vegetable crops are being investigated mainly for the production of oral vaccines, because the raw plant material is edible. Likewise, in oil crops (e.g., safflower), proteins can be directed to oil bodies in the seeds that facilitate the downstream processing.

A concise description of the frequently used crops for molecular farming is given below:

12.3.1.1   Tobacco

One of the most abundantly used leafy crops is tobacco because of significant biomass yield and well-established procedures for the gene transfer and expression. The other benefit of this crop is that it is not used as a food material. However, the production of some toxic alkaloids and heterogeneous glycan structures produced in the tissues need to be considered before its usage on a wider scale. Among the different species of this plant, *Nicotiana tabacum* and *Nicotiana benthamiana* are abundantly used for molecular farming.

12.3.1.2   Maize

Maize also offers similar advantages like easy transformation and easy to scale up in the field. Among the cereal crops and grain legumes, it has the highest biomass content. The other benefits are that it has low maintenance and production cost. Apart from certain regulatory or environmental issues, maize is still considered by many developers to be the best available option for mass production of proteins.

12.3.1.3   Rice

Similar to maize, rice is easy to transform and manipulate in the laboratory. Field operations and scale-up can be easily handled. Producer prices for rice are significantly higher than that for maize. Because rice undergoes crop pollination, it is considered as one of the biologically safe options for the production of proteins.

12.3.1.4   Safflower

Safflower is a highly productive oilseed crop that can be easily transformed and scaled up in the production, and that has an advantage regarding confinement. It is a self-pollinating, and the seeds show minimal dormancy.

12.3.2   *STABLE TRANSFORMATION*

The dominant method used in molecular farming is the stable integration of the desired gene into the plant genome. Presently, many routine transformation protocols are available for a wide range of plants. The goal of stable integration of target gene incorporation is achieved using *Agrobacterium tumefaciens* or gene gun-mediated bombardment, whereby these specific traits are transferred to plants in which they were absent previously. Such transformed plants are stable over many generations and allow easy scale-up and low-cost production of the desired proteins for quite a long time. *Agrobacterium*-mediated gene transfer suffers from one drawback, i.e., it has a narrow host range and does not ordinarily infect the monocotyledons. But, rice can be transformed by *Agrobacterium* (Chan et al., 1993; Hiei et al., 1994, 1997), and efforts are being made to develop methods for transforming other monocots. For transforming plants, the desired gene is cloned into a binary vector that can be transferred between *Escherichia coli* and *Agrobacterium*. The transformed *Agrobacterium* itself delivers the target gene into the host cell genome. After the transformation is over, transformed cells are selected using a selectable marker in the expression vector. Depending upon the plant, it may take few months to a year for developing a full-fledged plant capable of being tested for the expression of the desired target protein.

The stable nuclear transformation has produced most of the recombinant proteins to date, and it is the most widely used process. This method can be manipulated to accumulate the protein of interest in the dry seeds of cereals, which allows long-term storage of the seed at room temperature without degradation of the protein (Horn et al., 2004). Moreover, this system has the high scale-up capacity, as crops like cereals are grown throughout the world around different biogeographical conditions. The main shortcomings of this process include a long production cycle of certain crops and intercrossing capability with other crops, which limits the public acceptance (Commandeur et al., 2003).

Another alternate strategy of transformation is by using DNA-containing organelle like plastids for transformation. In most of the plant species, the chloroplast is inherited from the maternal tissue and not through pollen. Thus, this strategy may provide the benefit of natural containment. The issue of transgene containment and prevention of its escape into the environment is becoming increasingly relevant due to extensive debates over the use of genetically modified (GM) crops. Stable plastid transformation has been archived in many plants like lettuce (Lelivelt et al., 2005) and soybean (Dubald et al., 2014), and the protocols have been established that could have potential both as a production and a delivery system for edible human therapeutic proteins. There is a keen interest in plastid-based recombinant protein production in nontoxic, edible plant species not only to minimize downstream protein processing costs but also to develop a combined production and delivery system for "edible" protein therapies. The plastid-based recombinant protein production is preferred in nontoxic or edible plant species not only to mitigate the downstream protein processing expenditures but also because it can provide a combined production and delivery system for edible protein therapies (Lelivelt et al., 2005) (Figure 12.2).

12.3.3 TRANSIENT EXPRESSION

This system depends on the capability of recombinant plant viruses such as tobacco mosaic virus (TMV) to infect tobacco plants and then transiently express a target protein in the plant tissue. The targeted protein will thus accumulate in the interstitial space between the cells, which is subsequently collected by centrifugation under vacuum condition. The transient production strategy is the fastest and secure method for plant molecular farming (Rybicki, 2010). It is primarily being used for early stages of the development process and in cases where there is a need for quickly obtaining gram amounts of the target protein, although it has been developed for scale-up and production. Transient gene expression strategy has certain advantages

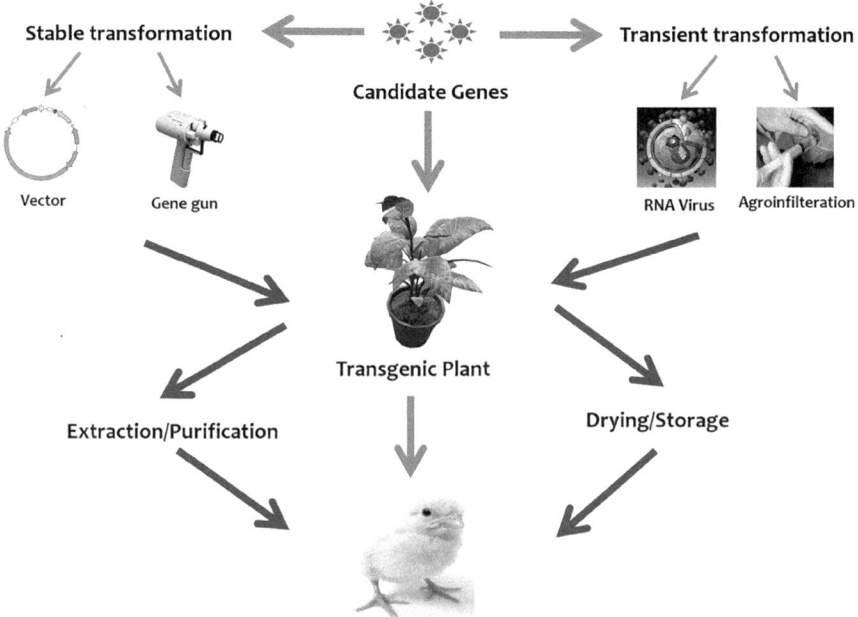

FIGURE 12.2 Transit and stable expression.

over the generation of stably transformed transgenic plants. The expressed protein in the case of transgenic plant requires a considerable amount of time before the target can be evaluated. In contrast to this, transient expression is rapid, flexible, and straightforward and often use either agrobacterial or viral vectors and subsequently results in protein expression that can be obtained in days (Vézina et al., 2009). This makes transient expression suitable for verifying that the gene product is functional before moving on to large-scale production in transgenic plants (Kapila et al., 1997). This platform is useful for a wide variety of custom proteins that are required in small quantities.

12.3.3.1   Agro-Infiltration Method

This method is based on a procedure that involves infiltration of a suspension of recombinant *Agrobacterium tumefaciens* into leaf tissues of tobacco. This facilitates the transfer of T-DNA to a large number of cells that express the transgene at very high levels without undergoing a stable transformation, contrary to transgenic crops. There are several instances where this method has been utilized for yielding clinical-grade biopharmaceuticals on a rapid and massive scale (Vézina et al., 2009; Pogue et al., 2010; Regnard et al., 2010).

12.3.3.2   Virus Infection Method

The virus infection method primarily relies on the capability of the plant viruses like TMV and potato virus (PV) to be used as vectors to insert foreign genes into the plant genome (Porta and Lomonossoff, 2002). The required target protein can thus be expressed in the plant tissue and thus can be obtained on a large scale. The virus infection method offers certain advantages over transgenic plants, e.g., virus infection takes few days or weeks to establish as compared to months spent for forming well-grown transgenic plants. The infection by the virus is also systemic in nature, which ultimately increases the yields of the required protein from a plant. The only disadvantage with this system is that the recombinant protein requires immediate processing because of tissue degradation and protein instability. Nevertheless, this method has thus become a source technology adapted by individual companies throughout the world. Large Scale Biology Corporation, a pharmaceutical company, has adopted the system for the production of idiotypic vaccines for the treatment of B-cell non-Hodgkin's lymphoma, which has successfully passed the phase I clinical testing (McCormick et al., 2008).

12.3.3.3   The Magnifection Technology

Both *Agrobacterium*-mediated stable transformation and viral vector-based transient infection method suffer from an inability to achieve high-level co-expression of two or more polypeptides necessary for the assembly of heterooligomeric proteins (Giritch et al., 2006). However, Icon Genetics, a German-based plant biotechnology company has developed a new expression system that fully utilizes advantages of a "deconstructed virus." The process called as "magnifection" is a scalable strategy for heterologous protein expression in plants and depends upon on transient amplification of viral vectors delivered to many areas of a plant by *Agrobacterium* by coat protein responsible for systemic movement of the noncompeting virus strains is removed and the systemic delivery of the resulting viral vectors to the entire plant using *Agrobacterium* as the vehicle of delivery and primary infection (Gleba et al., 2005). This method has been quite successful in enhancing amplification and ultimately leading to high-level co-expression of several polypeptides as an assemblage of functional hetero-oligomeric proteins at highly elevated levels up to 100-fold (Giritch et al., 2006). Further, advanced viral vectors are presently being planned by several platforms like Icon Genetics, Kentucky BioProcessing, Fraunhofer, and many others to facilitate the ease of manipulation, wider host range, speed, safety, low cost, and high quality and yield of pharmaceutical proteins required for clinical development (Rybicki, 2009; Komarova et al., 2010; Pogue et al., 2010).

12.3.4   BIOREACTOR-BASED SYSTEMS

Plant production platforms are also developed using plant-cell culture, moss, lemna, and algae, usually in fully contained bioreactors.

12.3.4.1   Plant Cell Culture System

Plant cell culture-based bioreactors from whole plants (e.g., tobacco, rice, carrots) is an alternative plant-based production platform to mammalian cells for the manufacture of biopharmaceuticals. It displays more potential than the traditional plant molecular farming using whole plants to produce pharmaceuticals (Schillberg et al., 2013; Magy et al., 2014; Raven et al., 2015). The system guarantees sterile *in vitro* conditions coupled with high-level containment, which is ideal for the production of high-purity pharmaceuticals, and cheaper and simpler downstream processing and purification (Kim et al., 2008; Franconi et al., 2010). They can be cultured in a very similar fashion as of mammalian cell lines in a sealed, sterilized container systems without human pathogens or soil contamination, but are less expensive to maintain than mammalian or microbial bioreactors as they require relatively simple synthetic nutrient growth medium to grow, and preferably, they allow the product to be secreted into the medium for purification, which further significantly lessens the cost of production. Moreover, the use of suspension cultured cells as an expression system might decrease heterogeneity of the protein and N-glycan due to the consistency in the size and type of cells (Liénard et al., 2007; De Muynck et al., 2009). Besides, the system is swift, as there is no prerequisite to regenerate and characterize transgenic plants as productive cell lines can be generated in a span of few months (Shaaltiel et al., 2007; Aviezer et al., 2009). The first Food and Drug Administration (FDA)-approved plant molecular farming-based pharmaceutical, taliglucerase alfa, used for the treatment of Gaucher's disease was produced in carrot cell suspension cultures (Fox, 2012). As Gaucher's disease is a very uncommon, the treatment with an orphan drug is very exorbitant ($200,000 US annually per patient for life). However, it reduces to $150,000/patient/year using the carrot cell production system. Around 20 recombinant proteins have been produced from plant cell culture systems in the recent past (Spök and Karner, 2008). Although economical, safer, easier to manipulate, and more rapid than most of the well-known conventional systems, suspension culture is still not the best production platform the plant system can offer, as the general yield and usability are slightly inadequate by the waning level of recombinant protein during the late stationary phase because of increased proteolytic activity (Corrado and Karali, 2009). Besides, the system is still limited to a small number of well-characterized plant cell lines [such as tobacco, rice carrot, or *Arabidopsis*] (Breyer et al., 2009).

12.3.4.2   Moss

At present, the use of moss, which is a non-seed plant, is being explored as a contender for the manufacture of pharmaceutical proteins in bioreactors (Huang and McDonald, 2012; Kircheis et al., 2012; Reski et al., 2015). The ability of moss cells to photosynthesize in culture expressively reduces the cost of culture nutrients. A moss variety, *Physcomitrella patens*, which is very vulnerable to transformation with recombinant DNA has been used as a model organism over more than three decades. As in yeast and plant cell suspension cultures, recombinant proteins can be planned to be secreted into moss culture medium, which assists downstream processing and purification of the recombinant protein. *Physcomitrella* has been used for proof-of-concept and feasibility studies of several biopharmaceuticals proteins, especially antibodies. With genetic manipulation, *P. patens* can also yield humanized form of a glycosylated protein, lewis Y-specific mAb MB314, thereby allowing for the elimination of unwanted side effects of the transformation. Some recombinant therapeutic proteins such as epidermal growth factor, α-galactosidase (Niederkrüger et al., 2014), α-amylase (Anterola et al., 2009), a glyco-optimized version of the antibody IGN311 (Orellana-Escobedo et al., 2015), a multiepitope fusion protein from the human immunodeficiency virus (Rubio-Infante et al., 2012), etc. have been produced in moss cultures.

The German company Greenovation Biotech has developed a *P. patens*-based commercial production system (Bryo-Technology) for large-scale, high-quality recombinant proteins. Up-scaling is being carried out by starting serial batteries of bioreactors of the same size and similar conditions. The yield, however, is 30 mg^{-1}/day range, and this relates to the yield of a typical fed-batch culture over 20 days of 600 mg^{-1}. Notwithstanding, the little yield the company is claiming corresponds to total savings in the range of 50% compared to mammalian cell culture because of the saving for purifications, quality control, and glycosylation.

12.3.4.3   Algal Bioreactors

Microalgae function as a primary natural source of valuable macromolecules comprising carotenoids, long-chain polyunsaturated fatty acids, and phycocolloids. Because algae are photoautotrophic, their meek necessities for growth make these primeval plants conceivably attractive bioreactor systems for the production of valued heterologous proteins. Some species are consistently transformed, and biopharmaceutical units have commenced discovering the possibilities of synthesizing recombinant therapeutic proteins in microalgae and the engineering of metabolic pathways to produce increased levels of desired entities. Algae can yield large-scale biomass within a very short period due to their short life cycle in conditions that makes it a desirable target for both increased production of natural compounds by metabolic engineering and exploitation as biological factories for the synthesis of novel high-value compounds. The downstream purification of recombinant proteins in algal species is parallel to yeast and bacterial systems and as a result is usually less expensive than the whole plant production systems (Walker et al., 2005). Among the single cell species, *Chlamomydomonas reinhardtii* is frequently used. Single cell algae can be grown under high density and large volumes. In the recent past, there has been a surge of interest in microalgal metabolites as a source of novel types of structures, including potential drugs that are not found in the higher plants (Shimizu, 1996; Cohen, 1999). Species of *Dunaliella, Hae-matococcus, Chlorella,* and *Euglena* produce carotenoids such as β-carotene, astaxanthin, and canthaxanthin, which are used as pigments in food products and cosmetics, as vitamin A supplements and health food products, and as feed additives for poultry, livestock, fish, and crustaceans (Borowitzka, 1988; Lorenz and Cysewski, 2000). Moreover, these systems have been explored to express a variety of biopharmaceutical and diagnostic recombinant proteins such as vaccines, enzymes, antibodies, interleukins, neurotrophic factors, and cholera toxin B unit (Lu et al., 2009; Gong et al., 2011; Rasala et al., 2011).

12.3.4.4   Duckweed (Lemna)

Duckweeds (Lemna) are small, free floating aquatic plants that grow on the surface of different water bodies such as ponds, lakes, and rivers. Lemna minor is used for food and feed in Asian countries and employed for waste-water management in North America. For plant molecular farming, Lemna is grown in water-based reactors. This plant has been genetically modified to produce 12 monoclonal antibodies together with small peptides and large multimeric enzymes (Fitzgerald, 2003; Gasdaska et al., 2003). Lemna has a biomass doubling time of 48–72 hours in control systems. Although Lemna might be attractive for reasons of containment, the rapid reproduction rate and the fact Lemna is flowering and produces pollen raises concerns. It might be tough to eradicate Lemna that has escaped, e.g., via wastewater. Thus, a strict containment system is considered pivotal (Dunwell and Ford, 2005) for this model system.

12.3.4.5   Organ Culture

Plant organs represent an attractive alternative to cell cultures for the production of secondary plant products. Two types of organs are considered for this objective: hairy root cultures and shoot cultures.

12.3.4.6   Hairy Root Cultures

Root cultures received substantial attention from biotechnologists for the production of secondary metabolites. These cultures can be sub-cultured and indefinitely further propagated on a synthetic medium without phyto-hormones and demonstrate impressive growth abilities due to the profu-sion of lateral roots (Chilton et al., 1982). This growth can be adapted to an exponential model when the number of generations of lateral roots becomes enormous (Hjortso, 1997). The specific growth rates are comparable, if not superior, to the parameters observed with undifferentiated cells. A major characteristic of hairy roots is that they can produce secondary metabolites concomitantly with growth. Hence, it is possible to obtain a continuous source of secondary compounds from actively growing hairy roots, unlike the usual results obtained with cell suspension cultures (Holmes et al., 1997). Cultured cells derived from the roots of higher plants produce secondary metabolites that have been used for in human healthcare since hoary past. *Agrobacterium*-transformed root cultures yield varied types of phytochemi-cals and other bioactive compounds. The production of Chinese folk medi-cine, Artemisinin, has been studied extensively in *A. rhizogenes*-transformed root cultures of *A. annua*, as an alternative of chloroquinone which is an anti-malarial drug. Polyacetylenes such as thiarubrine and terthienyl synthesized by *Calendula officinalis* are used as nematicides, phototoxins, antifungals, and antibacterials. Naphthaquinones like shikonis produced from Boragina-ceae have antimicrobial properties. Ginkgolides (diterpene) and forskolin (labdane diterpenoid) synthesized by *Ginkgo biloba* and *Coleus forskolin*, respectively, activate adenylate cyclase and are used in the treatment of asthma. An alkaloid, camptothecin, derived from *Camptothecin accuminata* finds its use in the treatment of cancers. *H. muticus* yields hyoscyamine in hairy root cultures after transformation with *A. rhizogenes*. Similarly, cocul-tures of hairy roots and other microorganisms have been found to produce unique secondary metabolites. For example, *Hordeum vulgare* roots infected with the vesicular-arbuscular mycorrhizae (VAM) fungi *Glomus intraradi-ces* induces accumulation of terpenoid glycoside. Likewise, *Catharanthus roseus* roots co-cultured with *Acaulospora scrobiculata* changed regulations of indole alkaloid biosynthesis (Hussain et al., 2012).

12.3.4.7 Shoot Cultures

As with roots, it is possible to cultivate aerial plant parts (shoots) for the production of secondary metabolites. Shoot cultures can be transgenic, the so-called shooty teratomas, if they are obtained after infection with *Agro-*

bacterium tumefaciens, or non-transgenic through the simple use of appropriate hormonal balance (Saito et al., 1985; Massot et al., 2000). Like organ cultures, shoots exhibit some comparable properties to hairy roots, namely genetic stability and good capacities for secondary metabolite production and also the possibility of gaining a link between growth and the production of secondary compounds (Bhadra et al., 1998; Massot et al., 2000). However, there are some differences in the metabolic pattern, as some syntheses are specifically located in either roots or shoots (Subroto et al., 1996). Other differences concern a somewhat slower growth rate as the fastest doubling time reported is approximately three days and also the necessity to expose shoot cultures to light, which can be a problem with large tank reactors made from steel (Heble, 1985).

12.3.4.8   Hydroponic Cultures

As a substitute for extracting the target proteins from plant tissues, the protein can be directed to the secretory pathway and recovered from root exudates (rhizosecretion) or leaf guttation fluid (phyllosecretion). This technology has, for instance, been used for producing antibodies (Raskin, 2000). Plants grown hydroponically are contained in a greenhouse setting and so have reduced chances of unintentional environmental release. Purification of the desired product is considerably easier because no tissue disruption is needed and the quantity of contaminating proteins is low.

12.4   PLANT-DERIVED RECOMBINANT PROTEINS

As the knowledge of several diseases increases through the sequencing of the human genome and advancement of medical research, many novel proteins that could be used for treatment have been identified. Several types of pharmaceuticals have been produced in plants. These comprise vaccines, antibodies, and biopharmaceuticals.

12.4.1   *PLANT-DERIVED VACCINE/ANTIGENS*

Vaccines are a significant tool to combat some of the world's fatal diseases and improve immunity against a particular disease. They are particularly important to use against diseases that are incurable once contracted. However, vaccines are often costly to produce and challenging to transport to far-off areas of the world owing to the need for refrigeration.

Plants offer mammoth potential as production platforms for vaccines and therapeutic proteins. Plant-derived vaccines were reckoned as substitutes, or at least supplements for, conventional vaccines and could resolve both these problems (Langridge, 2000; Streatfield, 2006). Vaccines and therapeutic proteins can be either expressed stably as transgenic plants or in a transient approach using agroinfiltration or by infection with recombinant virus expression vectors (Rybicki, 2009; Desai et al., 2010; Paul and Ma, 2011). Plant-derived vaccines and therapeutic proteins hold parallel biological activities as their mammalian-derived counterparts, unlike bacterial expression systems. Plant-derived vaccines have the twofold advantage of acting as the vaccine delivery vehicle as well as protect the vaccine protein from degradation by the harsh environment of the gastrointestinal tract so that it can reach the mucosal immune system more efficiently (Yusibo et al., 1999). Plant vaccines are also considered beneficial in terms of safety, versatility, and efficacy, as they are naturally free of microbial toxins and pyrogens or human and animal pathogens, including proviruses. On the other hand, oral immunization was believed to provide a significant advantage, and in the most passionate plans, plant-based vaccines were employed as edible vaccines. Besides, the progress of multicomponent vaccines is possible by insertion of multiple genetic elements or through cross-breeding of transgenic lines expressing antigens from various pathogenic organisms.

Usually, the production of plant-based vaccines reduces the overall cost, both in distribution and application particularly for developing countries. In the last 20 years, extensive research has revealed that plant-derived vaccines are attainable commodities. During the last two decades, some vaccine antigens have been expressed in plants, since the first vaccine from plants was reported 20 years ago (Rybicki, 2009; Tiwari et al., 2009). These include hepatitis B surface antigen, which has been expressed in transgenic potatoes (Richter et al., 2000), tomato (He et al., 2008), banana (Kumar et al., 2005), and tobacco cell suspension culture (Sojikul et al., 2003) and heat-labile enterotoxin B subunit (LTB) of *E. coli* expressed in potato tubers (Lauterslager et al., 2001), maize seed (Chikwamba et al., 2002), tobacco (Rosales- Mendoza et al., 2009), and soybean (Moravec et al., 2007). The cholera toxin B subunit (CTB) of *Vibrio cholera* has been expressed in several crops, including tobacco, tomato, and rice (Daniell et al., 2001a; Mishra et al., 2006; Nochi et al., 2007). Several plant-made vaccines for veterinary purposes including avian influenza, Newcastle disease, foot-and-mouth disease, and enterotoxigenic *E. coli* have been expressed in the plant (Lentz et al., 2010; Ling et al., 2010), and the hemagglutinin protein of rinderpest has been expressed in pigeon pea and peanut (Satyavathi et al., 2003). Others include the L1 protein of human papillomavirus types 11 and 16 (Biemelt et al., 2003; Maclean et al., 2007; Giorgi et al., 2010), the Norwalk virus capsid protein (Mason et al., 1996), the hemagglutinin protein of measles virus (Marquet-Blouin et al., 2003), and the H5N1 pandemic vaccine candidate (D'Aoust et al., 2010), all of which have been expressed in one or two of the following plants: tobacco, potato, and carrots.

Plant-produced antigens have been reported to induce immune responses, providing a shield against challenge in mouse model systems (Satyavathi et al., 2003; Streatfield and Howard, 2003). Several edible plant-produced antigens have been reported to induce protective immune responses in humans (Thanavala et al., 2005; McCormick et al., 2008) and mice (Daniell et al., 2001b; Tregoning et al., 2005). However, about 100-fold of the vaccine delivered by injection is required for oral administration (Streatfield and Howard, 2003). Notably, Franconi et al. (2006) and Massa et al. (2007) reported that immunization with plant-produced antigen protected the injected mice from tumor challenge with an E7-expressing tumor cell line, while Koya et al. (2005) reported that mice injected and immunized with chloroplast-derived anthrax protective antigen survived anthrax lethal toxin challenge. The Gal/GalNAc lectin of *Entamoeba histolytica* was found to be highly immunogenic in the injected mice (Chebolu and Daniell, 2007). There are various plant-produced vaccine candidates, which are at different stages of clinical trials, as summarized in Table 12.2. "In the present scenario, innovative plant based pharmaceutical grade protein production processes have attracted attention over the conventional approaches. This has contributed immensely towards the global public health (Buonaguro and Butler-Ransohoff, 2010)." It is evident that there are numerous opportunities to identify and develop low-cost plant-derived vaccine materials, including edible plant-based vaccines.

TABLE 12.2 Different Stages of Drug Development or Market Information of Many Plant-Derived Pharmaceutical Vaccines and Therapeutic Proteins for the Treatment of Various Diseases

Clinical Stage	Product/Plant Source	Target disorder/ disease	Organization
Phase 1	Hepatitis B antigen[V]/ Lettuce	Hepatitis	Thomas Jefferson University, USA
	Rhino RX[A]/Tobacco	Respiratory syncytial disease	Planet Biotech (USA)
	Vibrio cholerae[V]/Potato	Cholera	Arizona State University
	Capsid protein Norwalk virus[V]/Potato, Tomato	Diarrhea	Arizona State University
	Gastroenteritis virus (TGEV) capsid protein[V]/ Maize	Piglet gastroenteritis	ProdiGene, USA
	H5N1 vaccine candidate[V]/Tobacco	H5N1 pandemic influenza	Medicago, USA
	Lt-B vaccine[V]/Potatoes	Traveler's diarrhea	ProdiGene (USA)
	Fv antibodies[A]/Tobacco	Non-Hodgkin's lymphoma	Large Scale Biology, USA
	IgG (ICAM1)[A]/Tobacco	Common cold	Planet Biotechnology, USA
	Fibrinolytic drug[ThP]/ Duckweed	Blood clot	Biolex, USA
	Apolipoprotein[ThP]/ Safflower	Cardiovascular	SemBioSys, Canada
	Esherichia coli heat-labile toxin[V]/Potato, Corn	Diarrhea	ProdiGene (USA)/ Arizona State
Phase II	Hepatitis B antigen[V]/ Potato	Hepatitis	Arizona State University
	Cancer vaccine[V]/Tobacco	Non-Hodgkin's lymphoma	Large Scale Biology, USA
	Poultry vaccine[V]/Canola	Coccidiosis infection	Guardian Biosciences, Canada
	Lactoferon™(α-interferon)[ThP]/Duckweed	Hepatitis	Biolex, USA
	Glucocerebrosidase[ThP]	Gaucher's disease	Protalix (Israel)

(Source: Obembe, 2011)

12.4.2   PLANT-DERIVED ANTIBODIES

Antibodies or immunoglobulins (IgGs) are Y-shaped serum proteins produced by the immune system that binds to target molecules with a high level of specificity and are widely used for detection, prevention, and treatment of diseases. Monoclonal antibodies were only being used to treat diseases such as arthritis, cancer, and immune and inflammatory diseases. Passive immunization against pathogens has been reported to be obtained from recombinant antibodies as an alternative to combat to infectious diseases. This turn out to be an even more potential option with increasing microbial resistance to antibiotics and due to emergence and evolution of new potent pathogens (Casadevall, 1998). Although the demand for these antibodies is very high, the high cost of large-scale production is a drawback, which obstructs their way into healthcare market for treatment of infectious diseases. So, there is an utmost need of for cost-effective alternatives for large-scale production of antibodies (Stoger et al., 2005b). Although mammalian cells are used as a production system, it suffers from certain limitations like the complex nature of the product as well as the problem of post-translational modifications and proteolysis. Recent advances in molecular immunology and plant biotechnology have led to the possibility for large-scale production of recombinant monoclonal antibody (mAb) in plants as a safe and inexpensive alternative to mammalian-derived RIG (Ko et al., 2003). Plants now have potential as a virtually infinite source of mAbs, referred to by some as "plantibodies" (Figure 12.3). Studies have proven that plants are capable of expression of many different antigens, all of which have demonstrated immunogenicity when tested and some of them have been shown to provide protection in model and target animals.

Plants have been genetically modified to produce valuable proteins, thus showing enormous bioreactor potential for expressing and assembling mAbs for human and animal disease therapy (Daniell et al., 2001b; Ma et al., 2003). Plant expression systems offer a unique combination of advantages for the production of pharmaceutical antibodies for several reasons. They do not only provide economical production platforms, as plant-derived antibodies would cost just 0.1–1% of the production cost of mammalian culture and 2–10% of microbial systems (Chen et al., 2005), but they can also accumulate complex multimeric antibodies (Conrad and Fiedler, 1994) and therefore can be engineered for the production of varied tailor-made pharmaceutical proteins.

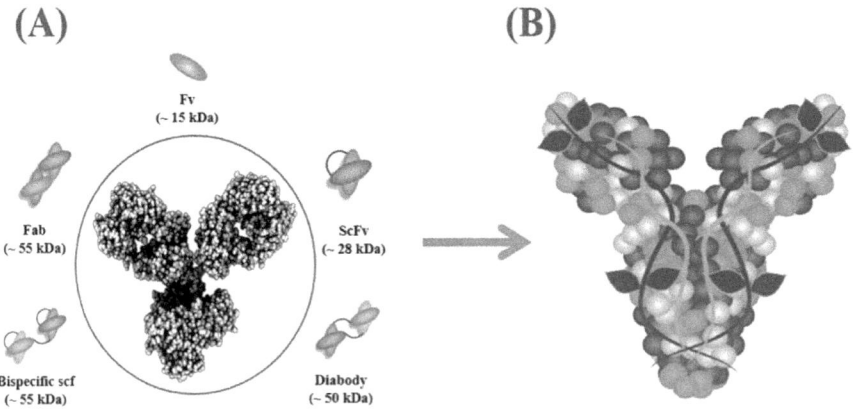

FIGURE 12.3 (A) IgG and its bioactive fragments; (B) Cloning of antibodies in plant system (plantibodies).

Two main approaches are being employed to produce biologically active whole antibodies in plants. Some transformation techniques have been employed to introduce antibody genes into plant cells. *Agrobacterium* and shotgun microprojectile bombardment-mediated transformation are the most suitable method at present. The vectors based on the tumor-inducing plasmid of *Agrobacterium tumefaciens,* which contain selectable plant markers, promoters upstream from polylinkers, and *E. coli* and *Agrobacterium* origins of replication, are usually quite huge, and the early approach to produce antibody in plants was to express each immunoglobulin chain separately in different plants and to introduce the two genes together in the progeny plant by cross-pollination of the individual heavy and light chain-expressing plants. Most groups have used the constitutive 35s promoter from cauliflower mosaic virus. With this promoter, a variety of cell types is likely to express the recombinant protein, even if there is an indication for differential expression related to developmental stage and tissue type (Ma et al., 1995).

The use of cross-pollination of individually transformed plants expressing antibody light or heavy chains involves two generations of plants to generate an antibody-producing plant (Hiatt et al., 1989; Ma et al., 1994). The yield of recombinant antibody is steadily high by this technique, between 1 and 5% of total plant protein. Others have used double-transformation techniques to introduce both the heavy- and light-chain genes into the same plant cell simultaneously (De Neve et al., 1993). Although reducing the time required to produce an antibody-secreting plant, the transformation may be less than optimal, and more screening may be necessary, because there can be individual variability of the expression of either gene. The yield of antibody was lower with this technique, amounting to only 0.055% of total soluble protein, and the efficiency of assembly also appeared to be reduced. An alternative strategy is to introduce the light- and heavy-chain genes on a single T-DNA (Düring et al., 1990; van Engelen et al., 1994). This is a more rapid technique and avoids the problem of segregation in the progeny plants, but the promoter and terminator need to be chosen carefully to coordinate expression of the two transgenes. Expression levels of up to 1.1% of total protein have been reported (van Engelen et al., 1994). Antibodies derived from plants have a multitude of applications, including binding to pathogenic organisms, binding to serum or body fluid effector proteins (e.g., interleukins), binding to tumor antigens to deliver imaging or anti-tumor agents, or binding to a cellular receptor site to up- or downregulate receptor function. The plant glycosylation arrangements vary from those in mammalian systems, and glycosylation is vital for antibody-facilitated activation of complement or the commencement of cellular immune responses (Russell, 1999; Breedveld, 2000). Plantibodies may transmit plant glycoproteins or maybe non-glycosylated due to genetically deleting glycosylation sites, but are incompetent of inducing the latter phenomena in both the cases (Breedveld, 2000). A chimeric secretory IgG/IgA antibody effective against a surface antigen of *Streptococcus mutans* has been expressed in tobacco and has been established to be effective against dental caries (Giddings, 2001). Soybeans can express a humanized anti-herpes simplex virus (HSV), which has been effective in preventing the transmission of vaginal HSV-2 in animals (Zeitlin et al., 1998). Rice and wheat expression systems can produce antibodies against the carcinoembryonic antigen, which may be useful for *in vivo* tumor imaging (Stoger et al., 2000). Finally, a plant viral vector has been used to produce a transiently expressed tumor-specific vaccine in tobacco for the treatment of lymphoma (McCormick et al., 1999).

12.4.3   OTHER THERAPEUTIC/NUTRACEUTICAL AGENTS

The third class of pharmaceuticals produced in plants is biopharmaceuticals. These are naturally occurring proteins that have medicinal use in humans.

A wide variety of other therapeutic agents has been derived from plants. The first therapeutic human protein to be expressed in plants was a human growth hormone (Barta et al., 1986). In 1990, human serum albumin, which is commonly isolated from blood, was produced in transgenic tobacco and potato for the first time (Sijmons et al., 1990). Since then, approximately 95 biopharmaceutical products have been approved by one or more regulatory agencies for the treatment of various human diseases including diabetes mellitus, growth disorders, neurological and genetic maladies, inflammatory conditions, and blood dyscrasias (Thomas, 2002). These include epidermal growth factor (Wirth et al., 2004; Bai et al., 2007); α-, β- and γ-interferons, which are used in treating hepatitis B and C (Sadhu and Reddy, 2003; Zhu et al., 2004; Arlen et al., 2007); erythropoietin, which promote red blood cell production (Weise et al., 2007; Musa et al., 2009); interleukin, which is used for treating Crohn's disease (Elias-Lopez et al., 2008; Fujiwara et al., 2010); insulin, which is used for treating diabetes (Nykiforuk et al., 2006); human glucocerebrosidase, which is used for the treatment of Gaucher's disease in genetically engineered carrot cells (Shaaltiel et al., 2007); and several others. Antimicrobial nutraceuticals such as human lactoferrin and lysozymes have now been successfully produced in several crops (Huang et al., 2008; Stefanova et al., 2008), and now commercially available only as fine chemicals. Other nutraceutical products at various stages of development are listed in Table 12.2. Plants could be used to produce large quantities of proteins that occur at low levels naturally. This would make some of the world's most expensive drugs much less costly to obtain.

12.5   RISKS AND CONCERNS

The plants utilized for molecular farming gives rise to issues like the utilization of GM crops. However, the difference lies in the fact that GM foods directly find their way into the food chain, while as the plants involved in molecular farming do not form the usual foodstuffs of the people. Molecular farming requires the plants that contain physiologically active compounds that accumulate in the plant's tissues. The concerns regarding GM foods have been discussed in many forms, and several arguments both in favor as well as against the GM foods have been put forward. Worldwide efforts are being made to assess the safety and efficacy of GM crops before recommendation for wider usage. Particular concerns should be addressed in a similar fashion and with the same seriousness about molecular farming in plants.

12.5.1 CROSS CONTAMINATION

There have been cases of cross-contamination in case of GM crops. One major issue regarding molecular farming is the possibility that the pharmaceuticals could potentially in future contaminate the general food supply. The risks can, however, be mitigating the use of vegetatively reproducing or self-pollinating plants for the production of pharmaceuticals. The idea of using cross-pollinating crops such as maize for such endeavors will not be a prudent one. The commercial varieties of banana that are sterile will be an ideal choice. The cross-pollinating ants should be assessed for risk–benefit ratio and if at all use must be limited to controlled production facilities, like the greenhouse to prevent any contamination during processing and distribution. This should not be a major hurdle because companies producing the pharmaceuticals will contract with a small number of growers and collect the products directly after harvest.

12.5.2 EFFECTS ON BIODIVERSITY

It is cause for grave research how the transformed plants are going to affect the physiology and behavior of wildlife that feeds on these plants. The intended proteins may be toxic to the animals, birds, etc., which may feed on such plants. Also, the root exudates of such plants may change in quality and composition, which may alter the soil microflora patterns that can be detrimental to some plant species.

Molecular farming could potentially occupy a significant portion of currently farmed land. It can lead to more requirement of land for growing such plants and thus may result in the conversion of natural habitats to cropland. In addition to this, extensive usage of pesticide for these crops can potentially be a threat to many insect and bird species. For example, tobacco is seen as an ideal candidate for molecular farming, but it is vulnerable to many diseases and therefore requires several applications of pesticides that can be harmful to biodiversity. Such questions need to be debated and discussed in detail, and steps be taken to avoid such causes of biodiversity loss (Fischer et al., 1999).

12.5.3 *EFFECT ON THE ACTIVITY OF PRODUCTS*

Another concern with pharmaceutical production in plants is whether plant-produced proteins will be active in humans and animals. Plant and mammalian post-translational protein modifications vary to a considerable extent, which may hamper the effectivity of these proteins produced in the plant-based system and used in human/animal system. Researchers have also demonstrated that certain mammalian enzymes can be engineered into plants, allowing plants to process mammalian proteins properly (Chrispeels and Faye, 1996). Such techniques can be useful if any pharmaceutical product is not shown to be effective due to the occurrence of plant-specific modifications.

12.6 CONCLUDING REMARKS

A sharp rise in pharmaceutical demands and complete information of the human genome have led to an upsurge of interest in plants as expression systems for the production of therapeutic entities. Plant molecular farming is being used to fulfill this increasing demand of recombinant proteins at an affordable cost and sufficient quantity, which is not possible in microbial and animal cell culture systems. Despite numerous benefits of this approach, however, the concerns that have been discussed must be adequately addressed. Although molecular farming offers an exciting alternative to pharmaceutical production, industry and government must proceed cautiously in this area to gain public acceptance.

KEYWORDS

- metabolites
- molecular farming
- nutraceutical
- pharmaceuticals
- therapeutics

REFERENCES

Anterola, A., Shanle, E., Perroud, P. F., & Quatrano, R., (2009). Production of taxa-4(5), 11(12)-diene by transgenic *Physcomitrella patens*. *Transgenic Res.*, *18*, 655–660.

Arlen, P. A., Falconer, R., Cherukumilli, S., Cole, A., Cole, A. M., Oishi, K. K., & Daniell, H., (2007). Field production and functional evaluation of chloroplast derived interferon-α2b. *Plant Biotechnol. J.*, *5*(4), 511–525.

Aviezer, D., Brill-Almon, E., Shaaltiel, Y., Hashmueli, S., Bartfeld, D., Mizrachi, S., Liberman, Y., Freeman, A., Zimran, A., & Galun, E., (2009). A plant-derived recombinant human glucocerebrosidase enzyme – a preclinical and phase I investigation. *PLoS One*, *4*(3), e4792.

Bai, J. Y., Zeng, L., Hu, Y. L., Li, Y. F., Lin, Z. P., Shang, S. C., & Shi, Y. S., (2007). Expression and characteristic of synthetic human epidermal growth factor (hEGF) in transgenic tobacco plants. *Biotechnol. Lett.*, *29*(12), 2007–2012.

Barta, A., Sommengruber, K., Thompson, D., Hartmuth, K., Matzke, M. A., & Matzke, A. J. M., (1986). The expression of a napoline synthase human growth hormone chimeric gene in transformed tobacco and sunflower callus tissue. *Plant Mol. Biol.*, *6*(5), 347–357.

Basaran, P., & Rodriguez-Cerezo, E., (2008). Plant molecular farming: opportunities and challenges. *Crit. Rev. Biotechnol.*, *28*(3), 153–172.

Bhadra, R., Morgan, J. A., & Shanks, J. V., (1998). Transient studies of light-adapted cultures of hairy roots of *Catharanthus roseus*: growth and indole alkaloid accumulation. *Biotechnol. Bioeng.*, *60*(6), 670–678.

Biemelt, S., Sonnewald, U., Galmbacher, P., Willmitzer, L., & Müller, M., (2003). Production of human papillomavirus type 16 virus-like particles in transgenic plants. *J. Virol.*, *77*(17), 9211–9220.

Borowitzka, M. A., (1988). *Vitamins and fine chemicals from microalgae.* In: *Micro-Algal Biotechnology,* Borowitzka, L. J., & Borowitzka, M. A., ed., Cambridge University Press, Cambridge, pp. 153–196.

Breedveld, F. C., (2000). Therapeutic monoclonal antibodies. *Lancet.*, *355,* 735–740.

Breyer, D., De Schrijver, A., Goossens, M., Pauwels, K., & Herman, P., (2012). Biosafety of molecular farming in genetically modified plants, In: *Molecular Farming in Plants: Recent Advances and Future Prospects,* Wang, A., Ma, S., ed., Springer, New York, pp. 259–274.

Breyer, D., Goossens, M., Herman, P., & Sneyers, M., (2009). Biosafety considerations associated with molecular farming in genetically modified plants. *J. Med. Plants Res.*, *3*(11), 825–838.

Buonaguro, F. M., & Butler-Ransohoff, J. E., (2010). Pharma Plant: the new frontier in vaccines. *Expert Rev. Vaccin.*, *9*(8), 805–807.

Casadevall, A., (1998). Antibody-based therapies as anti-infective agents. *Expert Opin. Emerg. Drugs*, *7*(3), 307–321.

Chan, M. T., Chang, H. H., Ho, S. L., Tong, W. F., & Yu, S. M., (1993). Agrobacterium-mediated production of transgenic rice plants expressing a chimeric alpha-amylase promoter/beta-glucuronidase gene. *Plant Mol. Biol.*, *22*(3), 491–506.

Chebolu, S., & Daniell, H., (2007). Stable expression of Gal/GalNAc lectin of *Entamoeba histolytica* in transgenic chloroplasts and immunogenicity in mice towards vaccine development for amoebiasis. *Plant Biotechnol. J.*, *5*(2), 230–239.

Chen, M., Liu, M., Wang, Z., Song, J., Qi, Q., & Wang, P. G., (2005). Modification of plant N-glycans processing: the future of producing therapeutic proteins in transgenic plants. *Med. Res. Rev.*, *25*(3), 343–360.

Chikwamba, R., McMurray, J., Shou, H., Frame, B., Pegg, S. E., Scott, P., Mason, H., & Wang, K., (2002). Expression of a synthetic E. coli heat-labile enterotoxin B sub-unit (LT-B) in maize. *Mol. Breed.*, *10*, 253–265.

Chilton, M. D., Tepfer, D. A., Petit, A., & Casse-Delbart, F., (1982). Tempe, J. *Agrobacterium* inserts T-DNA into the genome of the host plant root cells. *Nature*, *295*, 432–434.

Chrispeels, M. J., & Faye, L., (1996). The production of recombinant glycoproteins with defined non immunogenic glycans. In: *Transgenic Plants: a Production System for Industrial and Pharmaceutical Proteins.* Owen, M. R. L., & Pen, J., ed., John Wiley & Sons, Chichester, UK, pp. 99–113.

Cohen, Z., (1999). *Chemicals From Microalgae.* Taylor & Francis Ltd., London, pp. 409.

Commandeur, U., Twyman, R. M., & Fischer, R., (2003). The biosafety of molecular farming in plants. *AgBiotech Net.*, *5*, ABN110.

Conrad, U., & Fiedler, U., (1994). Expression of engineered antibodies in plant cells. *Plant Mol. Biol.*, *4*, 1023–1030.

Corrado, G., & Karali, M., (2009). Inducible gene expression systems, and plant biotechnology. *Biotechnol. Adv.*, *27*(6), 733–743.

Daniell, H., Lee, S. B., Panchal, T., & Wiebe, P. O., (2001a). Expression of the native cholera toxin B subunit gene and assembly as functional oligomers in transgenic tobacco chloroplasts. *J. Mol. Biol.*, *311*(5), 1001–1009.

Daniell, H., Streatfield, S. J., & Wycoff, K., (2001b). Medical molecular farming: production of antibodies, biopharmaceuticals, and edible vaccines in plants. *Trends Plant Sci.*, *6*(5), 219–226.

D'Aoust, M. A., Couture, M. M., Charland, N., Trepanier, S., Landry, N., Ors, F., & Vezina, L. P., (2010). The production of hemagglutinin-based virus-like particles in plants: a rapid, efficient and safe response to pandemic influenza. *Plant Biotechnol. J.*, *8*(5), 607–609.

De Muynck, B., Navarre, C., Nizet, Y., Stadlmann, J., & Boutry, M., (2009). Different subcellular localization and glycosylation for a functional antibody expressed in *Nicotiana tabacum* plants and suspension cells. *Transgenic Res.*, *18*(3), 467–482.

De Neve, M., De Loose, M., Jacobs, A., Van Houdt, H., Kaluza, B., Weidle, U., Van Montagu, M., & Depicker, A., (1993). Assembly of an antibody and its derived antibody fragment in *Nicotiana* and *Arabidopsis. Transgenic Res.*, *2*, 227–237.

De Zoeten, G. A., Penswick, J. R., Horisberger, M. A., Ahl, P., Schultze, M., & Hohn, T., (1989). The expression, localization, and effect of a human interferon in plants. *Virology*, *172*(1), 213–222.

Desai, P. N., Shrivastava, N., & Padh, H., (2010). Production of heterologous proteins in plants: strategies for optimal expression. *Biotechnol. Adv.*, *28*, 427–435.

Dubald, M., Tissot, G., & Pelissier, B., (2014). Plastid transformation in soybean. *Methods in Mol. Biol.*, *1132*, 345–354.

Dunwell, J., & Ford, C. S., (2005). *Technologies for biological containment of GM and Non-GM crops. Project Report Defra.*, pp. 275.

Düring, K., Hippe, S., Kreuzaler, F., & Schell, J., (1990). Synthesis and self-assembly of a functional monoclonal antibody in transgenic *Nicotiana tabacum. Plant Mol. Biol.*, *15*(2), 281–293.

Elias-Lopez, A. L., Marquina, B., Gutierrez-Ortega, A., Aguilar, D., Gomez-Lim, M., & Hernandez-Pando, R., (2008). Clinical and experimental immunology transgenic tomato expressing interleukin-12 has a therapeutic effect in a murine model of progressive pulmonary tuberculosis. *Clin. Exp. Immunol.*, *154*(1), 123–133.

Fischer, R., Emans, N., Schuster, F., Hellwig, S., & Drossard, J., (1999). Towards molecular farming in the future: using plant-cell-suspension cultures as bioreactors. *Biotechnol. Appl. Biochem.*, *30*, 109–112.

Fitzgerald, D. A., (2003). Revving up the Green Express. Companies explore the use of transgenic plants for economical, large scale protein expression. *The Scientist Magazine*, *17*, 14.

Fox, J. L., (2012). First plant-made biologic approved. *Nat. Biotechnol.*, *30*, 472.

Franconi, R., Demurtas, O. C., & Massa, S., (2010). Plant-derived vaccines and other therapeutics produced in contained systems. *Expert Rev Vaccine*, *9*(8), 877–892.

Franconi, R., Massa, S., Illiano, E., Mullar, A., Cirilli, A., Accardi, L., Di Bonito, P., Giorgi, C., & Venuti, A., (2006). Exploiting the plant secretory pathway to improve the anticancer activity of a plant-derived HPV16 E7 vaccine. *Int. J. Immunopathol. Pharmacol.*, *19*(1), 187–197.

Franken, E., Teuschel, U., & Hain, R., (1997). Recombinant proteins from transgenic plants. *Curr. Opin. Biotech.*, *8*, 411–416.

Fujiwara, Y., Aiki, Y., Yang, L., Takaiwa, F., Kosaka Tsuji, N. M., Shiraki, K., & Sekikawa, K., (2010). Extraction and purification of human interleukin-10 from transgenic rice seeds. *Protein Expr. Purif.*, *72*(1), 125–130.

Gasdaska, J. R., Spencer, D., & Dickey, L., (2003). Advantages of therapeutic protein production in the aquatic plant *Lemna*. *BioProcess J.*, *2*(2), 49–56.

Giddings, G., (2001). Transgenic plants as protein factories. *Curr. Opin. Biotechnol.*, *12*, 450–454.

Giddings, G., Allison, G., Brooks, D., & Carter, A., (2000). Transgenic plants as factories for biopharmaceuticals. *Nature Biotechnol.*, *18*, 1151–1155.

Giorgi, C., Franconi, R., & Rybicki, E. P., (2010). Human papillomavirus vaccines in plant. *Expert. Rev. Vaccines*, *9*(8), 913–924.

Giritch, A., Marillonnet, S., Engler, C., Van Eldik, G., Botterman, J., Klimyuk, V., & Gleba, Y., (2006). Rapid high yield expression of full-size IgG antibodies in plants coinfected with noncompeting viral vectors. *Proc. Natl. Acad. Sci. USA*, *103*(40), 14701–1406.

Gleba, Y., Klimyuk, V., & Marillonnet, S., (2005). Magnifection – a new platform for expressing recombinant vaccines in plants. *Vaccine*, *23*(17–18), 2042–2048.

Gong, Y., Hu, H., Gao, Y., Xu, X., & Gao, H., (2011). Microalgae as platforms for production of recombinant proteins and valuable compounds: Progress and prospects. *J. Ind. Microbiol. Biotechnol.*, *38*, 1879–1890.

He, Z. M., Jiang, X. L., Qi, Y., & Luo, D. Q., (2008). Assessment of the utility of the tomato fruit-specific E8 promoter for driving vaccine antigen expression. *Genetica*, *133*(2), 207–214.

Heble, M. R., (1985). Multiple shoot cultures: a viable alternative *in vitro* system for the production of known and new biologically active plant constituents. In: *Primary and Secondary Metabolism of Plant Cell Cultures*, Neumann, K. H., Barz, W., Reinhart, E. J., ed., Springer, pp. 281–289.

Hiatt, A., (1990). Antibodies produced in plants. *Nature*, *344*, 469–470.

Hiatt, A., Cafferkey, R., & Bowdish, K., (1989). Production of antibodies in transgenic plants. *Nature*, *342*, 76–78.

Hiei, Y., Komari, T., & Kubo, T., (1997). Transformation of rice mediated by *Agrobacterium tumefaciens*. *Plant Mol. Biol.*, *35*(1–2), 205–218.

Hiei, Y., Ohta, S., Komari, T., & Kumashiro, T., (1994). Efficient transformation of rice (*Oryza sativa* L.) mediated by Agrobacterium and sequence analysis of the boundaries of the T-DNA. *Plant J.*, *6*(2), 271–282.

Hjortso, M. A., (1997). Mathematical modelling of hairy root growth. In: *Hairy Roots, Culture and Applications*, Doran, P. M., ed., Harwood Academic Publishers: Amsterdam, pp. 169–178.

Holmes, P., Li, S. L., Green, K. D., Ford-Lloyd, B. V., & Thomas, N. H., (1997). Drip-tube technology for continuous culture of hairy roots with integrated alkaloid extraction. In: *Hairy Roots, Culture and Applications*. Doran, P. M., ed., Harwood Academic, pp. 201–208.

Horn, M. E., Woodard, S. L., & Howard, J. A., (2004). Plant molecular farming: Systems and products. *Plant Cell Rep.*, *22*(10), 711–720.

Huang, N., Bethell, D., Card, C., Cornish, J., Marchbank, T., Wyatt, D., Mabery, K., & Playford, R., (2008). Bioactive recombinant human lactoferrin, derived from rice. *In Vitro Cell Dev. Biol.*, *44*(10), 464–471.

Huang, T. K., & McDonald, K. A., (2012). Bioreactor systems for *in vitro* production of foreign proteins using plant cell cultures. *Biotechnol. Adv.*, *30*, 398–409.

Hussain, M. S., Fareed, S., Ansari, S., Rahman, M. A., Ahmad, I. Z., & Saeed, M., (2012). Current approaches toward production of secondary plant metabolites. *J. Pharm. Bioallied. Sci.*, *4*(1), 10–20.

Jamal, A., Ko, K., Kim, H. S., Choo, Y. K., Joung, H., & Ko, K., (2009). Role of genetic factors and environmental conditions in recombinant protein production for molecular farming. *Biotechnol. Adv.*, *27*(6), 914–923.

Jones, D. N., Kroos, N., Anema, R., Van Montfort, B., Vooys, A., Van der Kraats, S., Van der Helm, E., Smits, S., Schouten, J., Brouwer, K., Lagerwerf, F., Van Berkel, P., Opstelten, D. J., Logtenberg, T., & Bout, A., (2003). High-level expression of recombinant IgG in the human cell line per.c6. *Biotechnol. Prog.*, *19*(1), 163–168.

Kamenarova, K., Abumhadi, N., Gecheff, K., & Atanassov, A., (2005). Molecular farming in plants: an approach of agricultural biotechnology. *J. Cell Mol. Bio.*, *4*, 77–86.

Kapila, J., De Rycke, R., Van Montagu, M., & Angenon, G., (1997). An *Agrobacterium* mediated transient gene expression system for intact leaves. *Plant Sci.*, *122*(1), 101–108.

Kim, T. G., Baek, M., Lee, E. K., Kwon, T. H., & Yang, M. S., (2008). Expression of human growth hormone in transgenic rice cell suspension culture. *Plant Cell Rep.*, *27*(5), 885–891.

Kircheis, R., Halanek, N., Koller, I., Jost, W., Schuster, M., Gorr, G., Hajszan, K., & Nechansky, A., (2012). Correlation of ADCC activity with cytokine release induced by the stably expressed, glyco-engineered humanized Lewis Y-specific monoclonal antibody MB314. *MAbs*, *4*, 532–541.

Knablein, J., (2005). Plant-based expression of biopharmaceuticals. In: *Encyclopedia of Molecular Cell Biology and Molecular Medicine*. Meyers, R. A., ed., Weinheim, Wiley-VCH Verlag GmbH & Co., KGaA, pp. 386–407.

Ko, K., Tekoah, Y., Rudd, P. M., Harvey, D., Dwek, R. A., Spitsin, S., Hanlon, C. A., Rupprecht, C., Dietzschold, B., Golovkin, M., & Koprowski, H., (2003). Function, and glycosylation of plant-derived antiviral monoclonal antibody. *Proc. Natl. Acad. Sci. USA*, *100*, 8013–8018.

Komarova, T. V., Baschieri, S., Donini, M., Marusic, C., Benvenuto, E., & Dorokhov, Y. L., (2010). Transient expression systems for plant-derived biopharmaceuticals. *Expert. Rev. Vaccines*, *9*(8), 859–876.

Koya, V., Moayeri, M., Leppla, S. H., & Daniell, H., (2005). Plant-based vaccine: mice immunized with chloroplast-derived anthrax protective antigen survive anthrax lethal toxin challenge. *Infect. Immun.*, *73*(12), 8266–8274.

Kumar, G. B. S., Ganapathi, T. R., Revathi, C. J., Srinivas, L., & Bapat, V. A., (2005). Expression of hepatitis B surface antigen in transgenic banana plants. *Planta*, *222*(3), 484–493.

Lahlou, M., (2007). Screening of natural products for drug discovery. *Expert Opin. Drug Discov.*, *2*(5), 697–705.

Langridge, W. H., (2000). Edible vaccines. *Sci. Am.*, *283*, 66–71.

Lauterslager, T. G. M., Florack, D. E. A., Van der Wal, T. J., Molthoff, J. W., Langeveld, J. P., Bosch, D., Boersma, W. J., & Hilgers, L. A., (2001). Oral immunization of naïve and primed animals with transgenic potato tubes expressing LT-B. *Vaccine*, *19*(17–19), 2749–2755.

Lelivelt, C. L., McCabe, M. S., Newell, C. A., Desnoo, C. B., Van Dun, K. M., Birch-Machin, I., Gray, J. C., Mills, K. H., & Nugent, J. M., (2005). Stable plastid transformation in lettuce (Lactuca sativa L.). *Plant Mol Biol.*, *58*(6), 763–774.

Lentz, E. M., Segretin, M. E., Morgenfeld, M. M., Wirth, S. A., Santos, M. J. D., Mozgovoj, M. V., Wigdorovitz, A., & Bravo-Almonacid, F. F., (2010). High expression level of a foot and mouth disease virus epitope in tobacco transplastomic plants. *Planta*, *231*(2), 387–395.

Lienard, D., Sourrouille, C., Gomord, V., & Faye, L., (2007). Pharming and transgenic plants. *Biotechnol. Ann. Rev.*, *13*, 115–147.

Ling, H. Y., Pelosi, A., & Walmsley, A. M., (2010). Current status of plant-made vaccines for veterinary purposes. *Expert. Rev. Vaccin.*, *9*, 971–982.

Lorenz, R. T., & Cysewski, G. R., (2000). Commercial potential for *Haematococcus* microalgae as a natural source of astaxanthin. *Trends Biotechnol.*, *18*, 160–167.

Lu, Y., & Oyler, G. A., (2009). Green algae as a platform to express therapeutic proteins. *Discov. Med.*, *8*, 28–30.

Ma, J. K. C., Drake, P. M., & Christou, P., (2003). The production of recombinant pharmaceutical proteins in plants. *Nat. Rev. Genet.*, *4*, 794–805.

Ma, J. K. C., Hiatt, A., Hein, M., Vine, N. D., Wang, F., Stabila, P., Van Dolleweerd, C., Mostov, K., & Lehner, T., (1995). Generation and assembly of secretory antibodies in plants. *Science*, *268*, 716–719.

Ma, J. K. C., Lehner, T., Stabila, P., Fux, C. I., & Hiatt, A., (1994). Assembly of monoclonal antibodies with IgG1 and IgA heavy chain domains in transgenic tobacco plants. *Eur. J. Immunol.*, *24*(1), 131–138.

Ma, S., & Wang, A., (2012). Molecular farming in plants: An overview. In: *Molecular Farming in Plants: Recent Advances and Future Prospects*. Springer, pp. 1–20.

Maclean, J., Koekemoer, M., Olivier, A. J., Stewart, D., Hitzeroth, I. I., Rademacher, T., Fischer, R., Williamson, A. L., & Rybicki, E. P., (2007). Optimization of human papillomavirus type 16 (HPV-16) L1 expression in plants: comparison of the suitability of different HPV-16L1 gene variants and different cell- compartment localization. *J. Gen. Virol.*, *88*(5), 1460–1469.

Magy, B., Tollet, J., Laterre, R., Boutry, M., & Navarre, C., (2014). Accumulation of secreted antibodies in plant cell cultures varies according to the isotype, host species and culture conditions. *Plant Biotechnol. J.*, *12*, 457–467.

Marquet-Blouin, E., Bouche, F. B., Steinmetz, A., & Muller, C. P., (2003). Neutralizing immunogenicity of transgenic carrot (*Daucus carota* L.) – derived measles virus hemagglutinin. *Plant Mol. Biol., 51,* 459–469.

Mason, H. S., Ball, J. M., Shi, J. J., Jiang, X., Estes, M. K., & Arntzen, C. J., (1996). Expression of Norwalk virus capsid protein in transgenic tobacco and potato and its oral immunogenicity in mice. *Proc. Natl. Acad. Sci. USA, 93*(11), 5335–5340.

Massa, S., Franconi, R., Brandi, R., Muller, A., Mett, V., Yusibov, V., & Venuti, A., (2007). Anti-cancer activity of plant-produced HPV16 E7 vaccine. *Vaccine, 25*(16), 3018–3021.

Massot, B., Milesi, S., Gontier, E., Bourgaud, F., & Guckert, A., (2000). Optimized culture conditions for the production of furanocoumarins by micropropagated shoots of *Ruta graveolens. Plant Cell Tiss. Org., 62,* 11–19.

McCormick, A. A., Kumagai, M. H., Hanley, K., Turpen, T. H., Hakim, I., Grill, L. K., Tuse, D., Levy, S., & Levy, R., (1999). Rapid production of specific vaccines for lymphoma by expression of the tumor-derived single-chain Fv epitopes in tobacco plants. *Proc. Natl. Acad. Sci. USA, 96,* 703–708.

McCormick, A. A., Reddy, S., Reinl, S. J., Cameron, T. I., Czerwinkski, D. K., Vojdani, F., Hanley, K. M., Garger, S. J., White, E. L., Novak, J., Barrett, J., Holtz, R. B., Tusé, D., & Levy, R., (2008). Plant produced idiotype vaccines for the treatment of non-Hodgkin's lymphoma: Safety and immunogenicity in a phase I clinical study. *Proc. Natl. Acad. Sci. USA, 105*(29), 10131–10136.

Menkhaus, T. J., Bai, Y., Zhang, C., Nikolov, Z. L., & Glatz, C. E., (2004). Considerations for the recovery of recombinant proteins from plants. *Biotechnol. Prog., 20*(4), 1001–10014.

Mishra, S., Yadav, D. K., & Tuli, R., (2006). Ubiquitin fusion enhances cholera toxin B subunit expression in transgenic plants and the plant-expressed protein binds GM1 receptors more efficiently. *J. Biotechnol., 127*(1), 95–108.

Moravec, T., Schmidt, M. A., Herman, E. M., & Woodford-Thomas, T., (2007). Production of *Escherichia coli* heat labile toxin (LT) B subunit in soybean seed and analysis of its immunogenicity as an oral vaccine. *Vaccine, 25,* 1647–1657.

Musa, T. A., Hung, C. Y., Darlington, D. E., Sane, D. C., & Xie, J., (2009). Overexpression of human erythropoietin in tobacco does not affect plant fertility or morphology. *Plant Biotechnol. Rep., 3,* 157–165.

Niederkruger, H., Dabrowska-Schlepp, P., & Schaaf, A., (2014). Suspension culture of plant cells under phototrophic conditions. In: *Industrial Scale Suspension Culture of Living Cells,* Meyer, H. P., Schmidhalter, D. R. ed., Wiley-VCH Verlag GmbH & Co. KGaA: Weinheim: Germany, pp. 259–292.

Nochi, T., Takagi, H., Yuki, Y., Yang, L., Masumura, T., Mejima, M., Nakanishi, U., Matsumura, A., Uozumi, A., Hiroi, T., Morita, S., Tanaka, K., Takaiwa, F., & Kiyono, H., (2007). Rice-based mucosal vaccine as a global strategy for cold chain and needle-free vaccination. *Proc. Natl. Acad. Sci. USA, 104*(26), 10986–10991.

Nykiforuk, C. L., Boothe, J. G., Murray, E. W., Keon, R. G., Goren, H. J., Markley, N. A., & Moloney, M. M., (2006). Transgenic expression and recovery of biologically active recombinant human insulin from *Arabidopsis thaliana* seeds. *Plant Biotechnol. J., 4*(1), 77–85.

Obembe, O. O., Popoola, J. O., Leelavathi, S., & Reddy, S. V., (2011). Advances in plant molecular farming. *Biotechnol. Adv., 29*(2), 210–222.

Orellana-Escobedo, L., Rosales-Mendoza, S., Romero-Maldonado, A., Parsons, J., Decker, E. L., Monreal-Escalante, E., Moreno-Fierros, L., & Reski, R., (2015). An Env-derived multi-epitope HIV chimeric protein produced in the moss *Physcomitrella patens* is immunogenic in mice. *Plant Cell Rep.*, *34*, 425–433.

Paul, M., & Ma, J. K., (2011). Plant-made pharmaceuticals: leading products and production platforms. *Biotechnol. Appl. Biochem.*, *58*, 58–67.

Pogue, G. P., Vojdani, F., Palmer, K. E., Hiatt, E., Hume, S., Phelps, J., Long, L., Bohorova, N., Kim, D., Pauly, M., Velasco, J., Whaley, K., Zeitlin, L., Garger, S. J., White, E., Bai, Y., Haydon, H., & Bratcher, B., (2010). Production of pharmaceutical-grade recombinant aprotinin and a monoclonal antibody product using plant-based transient expression systems. *Plant Biotechnol. J.*, 8(5), 638–654.

Porta, C., & Lomonossoff, G. P., (2002). Viruses as vectors for the expression of foreign sequences in plants. *Biotechnol. Genet. Eng. Rev.*, *19*, 245–291.

Ramawat, K. G., & Goyal, S., (2008). The Indian herbal drugs scenario in global perspectives. In: *Bioactive Molecules and Medicinal Plants*, Ramawat, K. G., Merillon, J. M., ed., Springer, Berlin Heidelberg, New York, pp. 323.

Rasala, B. A., & Mayfield, S. P., (2011). The microalga *Chlamydomonas reinhardtii* as a platform for the production of human protein therapeutics. *Bioeng Bugs*, *2*, 50–54.

Raskin, I., (2000). Methods for recovering polypeptides from plants and portions thereof. US Patent 6096546.

Raven, N., Rasche, S., Kuehn, C., Anderlei, T., Klockner, W., Schuster, F., Henquet, M., Bosch, D., Buchs, J., Fischer, R., & Schillberg, S., (2015). Scaled-up manufacturing of recombinant antibodies produced by plant cells in a 200-L orbitally-shaken disposable bioreactor. *Biotechnol. Bioeng.*, *112*, 308–321.

Regnard, G. L., Halley-Stott, R. P., Tanzer, F. L., Hitzeroth, I. I., & Rybicki, E. P., (2010). High level protein expression in plants through the use of a novel autonomously replicating geminivirus shuttle vector. *Plant Biotechnol. J.*, *8*, 38–46.

Reski, R., Parsons, J., & Decker, E. L., (2015). Moss-made pharmaceuticals: From bench to bedside. *Plant Biotechnol. J.*, *13*, 1191–1198.

Richter, L. J., Thanavala, Y., Arntzen, C. J., & Mason, H. S., (2000). Production of hepatitis B surface antigen in transgenic plants for oral immunization. *Nat. Biotechnol.*, *18*, 1167–1171.

Rosales-Mendoza, S., Alpuche-Solis, A. G., Soria-Guerra, R. E., Moreno-Fierros, L., Martinez-Gonzalez, L., Herrera-Diaz, A., & Korban, S. S., (2009). Expression of an *Escherichia coli* antigenic fusion protein comprising the heat labile toxin B subunit and the heat stable toxin, and its assembly as a functional oligomer in transplastomic tobacco plants. *Plant J.*, *57*(1), 45–54.

Rubio-Infante, N., Govea-Alonso, D. O., Alpuche-Solís, A. G., García-Hernandez, A. L., Soria-Guerra, R. E., Paz-Maldonado, L. M., Ilhuicatzi-Alvarado, D., Varona-Santos, J. T., Verdin-Teran, L., Korban, S. S., Moreno-Fierros, L., & Rosales-Mendoza, S., (2012). Chloroplast-derived C4V3 polypeptide from the human immunodeficiency virus (HIV) is orally immunogenic in mice. *Plant Mol. Biol.*, *78*, 337–349.

Russell, D. A., (1999). Feasibility of antibody production in plants for human therapeutic use. *Curr. Topics Microbiol.*, *240*, 119–138.

Rybicki, E. P., (2010). Plant-made vaccines for humans and animals. *Plant Biotechnol. J.*, *8*, 620–637.

Rybicki, E. P., (2009). Plant-produced vaccines: promise and reality. *Drug Discov. Today*, *14*, 16–24.

Sadhu, L., & Reddy, V. S., (2003). Chloroplast expression of His-tagged GUS-fusions: a general strategy to overproduce and purify foreign proteins using transplastomic plants as bioreactors. *Mol. Breed.*, *11*, 49–58.

Saito, K., Murakoshi, I., Inze, D., & Van Montagu, M., (1985). Biotransformation of nicotine alkaloids by tobacco shooty teratomas induced by a Ti plasmid mutant. *Plant Cell Rep.*, *7*(8), 607–610.

Satyavathi, V. V., Prasad, V., Khandelwal, A., Shaila, M. S., & Sita, G. L., (2003). Expression of hemagglutinin protein of rinderpest virus in transgenic pigeon pea [*Cajanus cajan* (L.) Millsp.] plants. *Plant Cell Rep.*, *21*, 651–658.

Schillberg, S., Fischer, R., & Emans, N., (2003). Molecular farming of antibodies in plants. *Nuturwissenschaften*, *90*(4), 145–55.

Schillberg, S., Raven, N., Fischer, R., Twyman, R. M., & Schiermeyer, A., (2013). Molecular farming of pharmaceutical proteins using plant suspension cell and tissue cultures. *Curr. Pharm.*, *19*, 5531–5542.

Shaaltiel, Y., Bartfeld, D., Hashmueli, S., Baum, G., Brill-Almon, E., Galili, G., Dym, O., Boldin-Adamsky, S. A., Silman, I., Sussman, J. L., Futerman, A. H., & Aviezer, D., (2007). Production of glucocerebrosidase with terminal mannose glycans for enzyme replacement therapy of Gaucher's disease using a plant cell system. *Plant Biotechnol. J.*, *5*(5), 579–590.

Shimizu, Y., (1996). Microalgal metabolites: a new perspective. *Ann. Rev. Microbiol.*, *50*, 431–465.

Sijmons, P. C., Dekker, B. M., Schrammeijer, B., Verwoerd, T. C., Van den Elzen, P. J., & Hoekema, A., (1990). Production of correctly processed human serum albumin in transgenic plants. *BioTechnol. (NY)*, *8*(3), 217–221.

Sojikul, P., Buehner, N., & Mason, H. S., (2003). A plant signal peptide hepatitis B surface antigen fusion protein with enhanced stability and immunogenicity expressed in plant cells. *Proc. Natl. Acad. Sci. USA*, *100*, 2209–2214.

Sparrow, P., Broer, I., Hood, E. E., Eversole, K., Hartung, F., & Schiemann, J., (2013). Risk assessment and regulation of molecular farming-a comparison between Europe and US. *Curr. Pharm. Des.*, *19*(31), 5513–5530.

Spok, A., Karner, S. Basaran, P., & Rodríguez-Cerezo, E., (2008). Plant molecular farming: opportunities and challenges. *Crit. Rev. Biotechnol.*, *28*(3), 153–172.

Stefanova, G., Vlahova, M., & Atanassov, A., (2008). Production of recombinant human lactoferrin from transgenic plants. *Biol. Plant*, *52*, 423–428.

Stoger, E., Ma, J. K., Fischer, R., & Christou, P., (2005a). Sowing the seeds of success: pharmaceutical proteins from plants. *Curr. Opin. Biotechnol.*, *16*(2), 167–173.

Stoger, E., Sack, M., Nicholson, L., Fischer, R., & Christou, P., (2005b). Recent progress in plant antibody technology. *Curr. Pharm. Des.*, *11*, 2439–2457.

Stoger, E., Vaquero, C., Torres, E., Sack, M., Nicholson, L., Drossard, J., Williams, S., Keen, D., Perrin, Y., Christou, P., & Fischer, R., (2000). Cereal crops as viable production and storage systems for pharmaceutical scFv antibodies. *Plant Mol. Biol.*, *42*, 583–590.

Streatfield, S. J., (2006). Mucosal immunization using recombinant plant-based oral vaccines. *Methods*, *38*, 150–157.

Streatfield, S. J., & Howard, J. A., (2003). Plant-based vaccines. *Int. J. Parasitol.*, *33*, 479–493.

Stryjewska, A., Kiepura, K., Librowski, T., & Lochyński, S., (2013). Biotechnology and genetic engineering in the new drug development. Part I. DNA technology and recombinant proteins. *Pharmacol. Rep.*, *65*(5), 1075–1085.

Subroto, M. A., Kwok, K. H., Hamill, J. D., & Doran, P. M., (1996). Coculture of genetically transformed roots and shoots for synthesis, translocation, and biotransformation of secondary metabolites. *Biotechnol. Bioeng.*, *49*(5), 481–494.

Thanavala, Y., Mahoney, M., Pal, S., Scott, A., Richter, L., Natarajan, N., Goodwin, P., Arntzen, C. J., & Mason, H. S., (2005). Immunogenicity in humans of an edible vaccine for hepatitis B. *Proc. Natl. Acad. Sci. USA, 102*(9), 3378–3382.

Thomas, J. A., (2002). Biotechnology: Safety evaluation of biotherapeutics and agribiotechnology products. In: *Biotechnology and Safety Assessment,* Thomas, J. A., & Fuchs, R. L., Ed., Academic Press, New York, pp. 347–384.

Tiwari, S., Verma, P. C., Singh, P. K., & Tuli, R., (2009). Plants as bioreactors for the production of vaccine antigens. *Biotechnol. Adv., 27,* 449–467.

Tregoning, J. S., Clare, S., Bowe, F., Edwards, L., Fairweather, N., Qazi, O., Nixon, P. J., Maliga, P., Dougan, G., & Hussell, T., (2005). Protection against tetanus toxin using a plant-based vaccine. *Eur. J. Immunol., 35*(4), 1320–1326.

Twyman, R. M., Schillberg, S., & Fischer, R., (2005). Transgenic plants in the biopharmaceutical market. *Expert Opin. Emerg. Drugs, 10*(1), 185–218.

Twyman, R. M., Stoger, E., Schillberg, S., Christou, P., & Fischer, R., (2003). Molecular farming in plants: host systems and expression technology. *Trends Biotechnol., 21*(12), 570–578.

Van Engelen, F. A., Schouten, A., Molthoff, J. W., Roosien, J., Salinas, J., Dirkse, W. G., Schots, A., Bakker, J., Gommers, F. J., Jongsma, M. A., et al., (1994). Coordinate expression of antibody subunit genes yields high levels of functional antibodies in roots of transgenic tobacco. *Plant Mol. Biol., 26,* 1701–1710.

Vézina, L. P., Faye, L., Lerouge, P., D'Aoust, M. A., Marquet-Blouin, E., Burel, C., Lavoie, P. O., Bardor, M., & Gomord, V., (2009). Transient co-expression for fast and high-yield production of antibodies with human-like N-glycans in plants. *Plant Biotechnol. J., 7*(5), 442–455.

Walker, T. L., Purton, S., Becker, D. K., & Collet, C., (2005). Microalgae as bioreactors. *Plant Cell Rep., 24,* 629–641.

Weise, A., Altmann, F., Rodriguez-Franco, M., Sjoberg, E. R., Bäumer, W., Launhardt, H., Kietzmann, M., & Gorr, G., (2007). High-level expression of secreted complex glycosylated recombinant human erythropoietin in the *Physcomitrella* Delta-fuc-t Delta-xyl-t mutant. *Plant Biotechnol. J., 5*(3), 389–401.

Wirth, S., Calamante, G., Mentaberry, A., Bussmann, L., Lattanzi, M., Baranao, L., & Bravo-Almonacid, F., (2004). Expression of active human epidermal growth factor (hEGF) in tobacco plants by integrative and non-integrative systems. *Mol. Breed., 13,* 23–35.

Yusibov, V., Hooper, D. C., Spitsin, S. V., Fleysh, N., Kean, R. B., Mikheeva, T., Deka, D., Karasev, A., Cox, S., Randall, J., & Koprowski, H., (2002). Expression in plants and immunogenicity of plant virus-based experimental rabies vaccine. *Vaccine, 20*(25–26), 3155–3164.

Yusibov, V., Shivprasad, S., Turpen, T. H., Dawson, W., & Koprowski, H., (1999). Plant viral vectors based on tobamoviruses. *Curr. Top. Microbiol. Immunol., 240,* 81–94.

Zeitlin, L., Olmsted, S. S., Moench, T. R., Co, M. S., Martinell, B. J., Paradkar, V. M., Russell, D. R., Queen, C., Cone, R. A., & Whaley, K. J., (1998). A humanized monoclonal antibody produced in transgenic plants for immunoprotection of the vagina against genital herpes. *Nat. Biotechnol., 16,* 1361–1364.

Zhu, Z., Hughes, K. W., Huang, L., Sun, B., Liu, C., & Li, Y., (2004). Expression of human alpha-interferon cDNA in transgenic rice plants. *Plant Cell Tiss. Org. Cult., 36,* 197–204.

CHAPTER 13

PLANT-BASED EDIBLE VACCINES: AN INNOVATIVE CONCEPT TO IMMUNIZATION

RAJEEV SHARMA,[1] NISHI MODY,[1] SONAL VYAS,[2] and SURESH P. VYAS[1]

[1] Drug Delivery Research Laboratory, Department of Pharmaceutical Sciences, Dr. H. S. Gour Vishwavidyalaya, Sagar, M.P., 470003, India

[2] Department OF Pathology, Index Medical ColLEGE, Hospital & Research Centre, Indore (M.P.), India, E-mail: spvyas54@gmail.com; rajeevs472@gmail.com; nishimody@gmail.com; sonalvyas07@gmail.com

CONTENTS

13.1   INTRODUCTION

Vaccines have emerged as a promising medical intervention for immunopro-phylaxis, as a cost-effective approach particularly in developing countries. Injectable vaccines pose some difficulties, viz., inconvenient administration, need for regulated temperature from the producer site to the user end, and affordability in terms of cost. Hence, plant-derived edible vaccines were discovered as an attractive alternative as they do not face any of these lim-itations. The idea of edible vaccines came from a group of philanthropic organizations led by the World Health Organization (WHO) who launched the children's vaccine initiative to set goals for developing vaccines that are safe, heat stable, orally administered, and widely accessible in 1990. They suggested that these goals can be fulfilled by producing subunit vaccines in edible tissues of transgenic crop plants (Mason et al., 1992). Plant-derived vaccines offer increased safety, envision low-cost programs for mass vac-cination, and propose a wider use of vaccination for human and veterinary use (Landridge, 2000). Introduction of modern molecular and biotechno-logical methods have resulted in the development of vaccines that comprise recombinant proteins derived from pathogenic viruses, bacteria, or parasites (generally, these proteins are produced not by the pathogens themselves, but by expression of the gene encoding the protein in a "surrogate organism") (Lal et al., 2007). In the late 90's, it was found that green plants can also be employed as "surrogate production organism" to generate antigens of human pathogens (including HBsAg). These proteins can induce priming and can boost the immune response in humans upon oral administration. In addi-tion, unlike almost all other cell lines used for the production of vaccines, components of plant cells have always been an important part of the normal human diet. Plants, therefore, offer significant new opportunities for mak-ing safe and effective oral vaccines (Giddings et al., 2000; Rybicki et al., 2012, 2014). Various parts of plants (seeds, leaves, chloroplasts, and fruits) can also be used as vehicles for the production of edible vaccine. This will not only increase the cost-effectiveness but also the safety of the vaccines as compared to biologicals produced in animal tissues (Mor et al., 1998; Streatfield, 2001; Carter and Langridge, 2002; Ma et al., 2003; Arntzen et al., 2005; Twyman, 2005; Castle et al., 2006). In general, the term "plant-derived edible vaccines" refers to antigens produced in edible parts of a plant, such as fruits, leaves, roots, or tubers. Vaccines expressed in nonedible plant tissues may also be orally administered after processing. Also, not all plant-derived vaccine antigens are edible; they may be effective only when administered by the appropriate method (topically; intramuscularly, i.m.; intraperitoneally, i.p.; or subcutaneously, s.c.) (Mason and Arntzen, 1995; Walmsley and Arntzen, 2003). Transgenic plants expressing antigens from pathogenic microorganisms offer many advantages as low-cost production systems and effective for vaccine. This new technology might contribute to global vaccine programs and might have a dramatic impact on healthcare in developing countries. The economic and technical benefits offered by plant-derived vaccines propose these vaccines as ideal substitutes for traditional vaccines (Fischer et al., 2004; Koprowski, 2005; Streatfield, 2006).

13.2 ADVANTAGES OF EDIBLE VACCINES

Traditional or subunit vaccines are often associated with some difficulties like affordability, safety, and storage, particularly in developing countries. The concept of edible vaccine can help to meet these challenges as they do not require cold chains for storage. They are also safer to use as compared to their animal counterpart because the production of recombinant proteins in plants minimizes potential human or animal pathogen and toxic contamination in form of toxins. Plants are potentially more economical than industrial fermentation or bioreactor facilities, and the amounts of protein produced by plants are comparable to those produced by industrial approaches; thus, they can be easily used for oral vaccines as the dosage required is generally high. Purification and cost-associated limitations with it could also be eliminated using plant-derived vaccines. The potential for administration by oral route and the convenient storage when the vaccines are produced in seeds and tubers could dramatically reduce the costs associated with syringes and needles and the requirement for cold storage. Elimination of needles in most vaccination programs will also reduce the risk associated with reuse of needles in some developing countries. Seed and tubers are simple, cheap and convenient storage organs for relatively simple, cheap antigen storage. Plants offer enough flexibility that the recombinant proteins can be targeted to intracellular compartments where they are more stable. The procedures involved in the production of edible vaccines largely rely on established molecular and genetic manipulation protocols, which are generally available in developing countries;

ultimately, the cost of production will be reduced 100–1000 times compared with that of traditional vaccines. The different advantages are listed in Figure 13.1 (Streatfield et al., 2001; Streatfield and Howard, 2003; Warzecha and Mason, 2003; Kirk et al., 2005; Raney, 2006).

13.3 MECHANISM OF ACTION

Mucosal surface lining the respiratory, digestive, and urino-reproductive tracts are the common portals for the entry of pathogens, and thus, these collectively form the largest immunologically active tissue in the body. Nasal and oral are the most effective route for mucosal immunization as the muco-

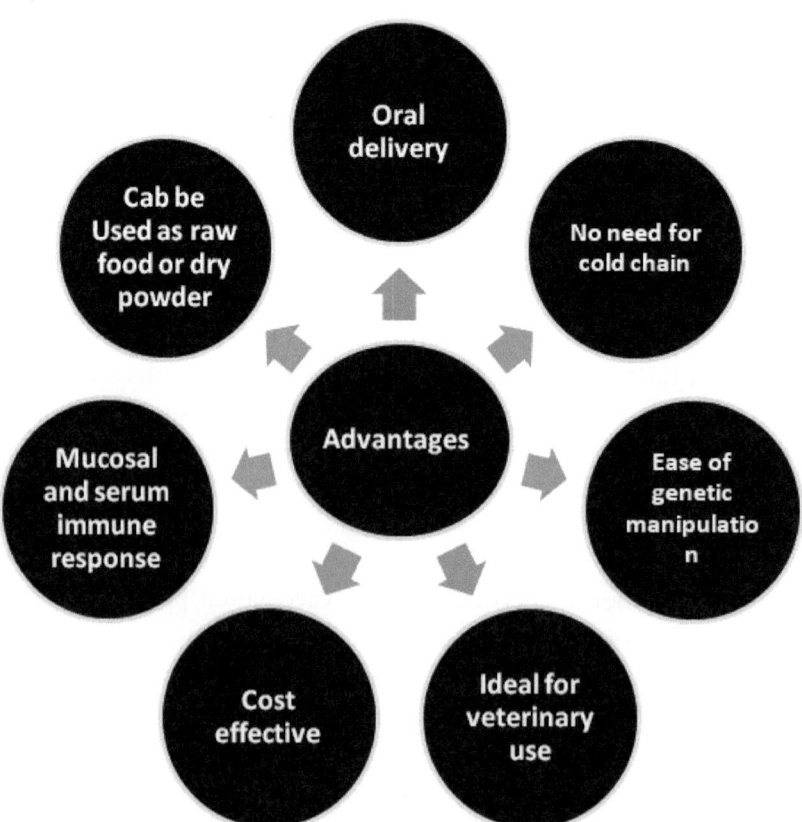

FIGURE 13.1   Advantages of edible vaccines.

sal immune system is the first line of defense and the most effective site for vaccination against pathogens. The aim of oral vaccine is to stimulate both mucosal and humoral immunity against pathogens. Edible vaccines upon ingestion with oral route undergo mastication followed by degradation of plant cell in the intestine as a result of action of digestive enzyme on them. Peyer's patches (PP) are an enriched source of IgA-producing plasma cells and have the potential to populate mucosal tissue and serve as mucosal immune effector site. The breakdown of edible vaccine occurs near PP, which consist of 30–40 lymphoid nodules on the outer surface of the intestine and also contain follicles from which the germinal center develops after antigenic stimulation. These follicles act as a site for the penetration of antigens in intestinal epithelium. The antigen then comes in contact with M-cells. M-cells express class-2 major histocompatibility complex molecules and transport the antigens across the mucous membrane, and these endocytosed antigens can activate B-cells within these lymphoid follicles. The activated B-cells leave the lymphoid follicles and migrate to diffuse mucosal associated lymphoid tissue where they differentiate into plasma cells that secrete the IgA class of antibodies. These sIgA (secretory antibodies) antibodies are transported across the epithelial cells into secretions of the lumen where they can interact with antigens present in the lumen as shown in Figure 13.2 (Crippes et al., 2001).

13.4   CHOICE OF PLANTS AND PLANT TISSUES

Selection of appropriate plant material is the crucial task for producing transgenic plant vaccines. Plant materials that can both be suitable for extensive storage and oral delivery and express high levels of a chosen vaccines are most suitable candidates for the production of transgenic vaccines The choice of plant usually depends on its capacity for transformation or infectivity with genetically engineered viruses, and to date, scientists have often made this decision based on the system with which they are most competent. The choice of the plant species (tissue in which the protein accumulates) is important and is usually determined through how the vaccine is to be applied in the future. This limitation is overcome in nonedible plants by vaccine antigen extraction and purification. Some plants such as potato, tomato, and tobacco are more agreeable to tissue culture and transformation, and others such as corn, soybean, and wheat are relatively more

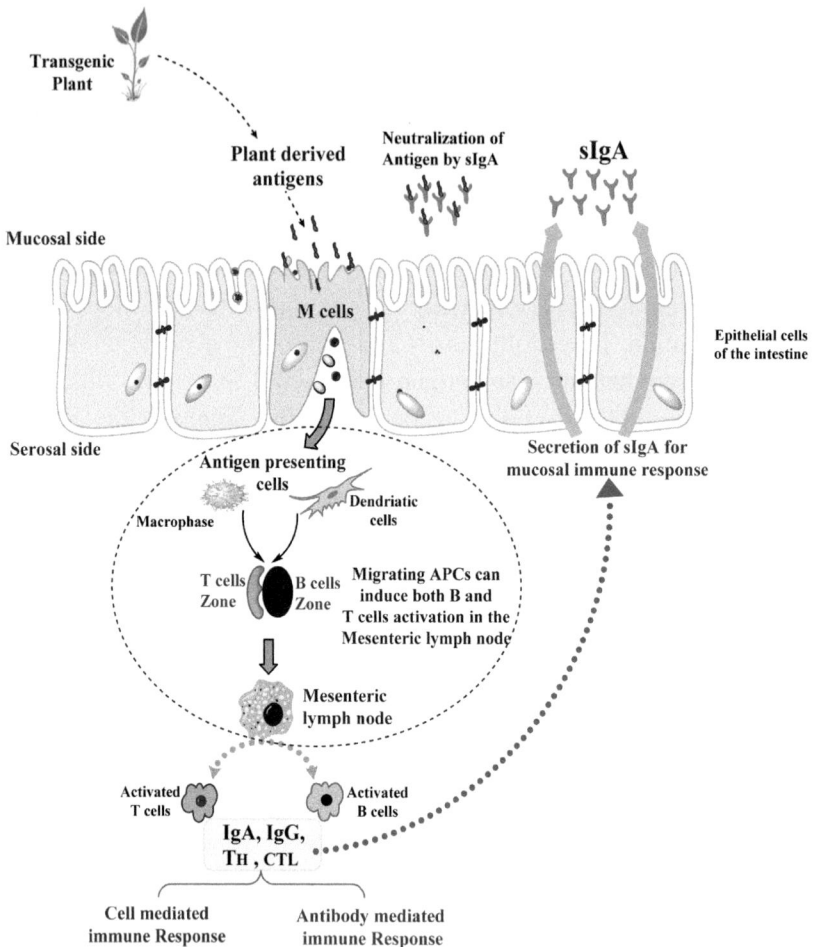

FIGURE 13.2   Mechanism of immune response of edible vaccines.

recalcitrant to *in vitro* manipulation. The desired method of administering the recombinant protein also plays an important role. Various plant platforms have been demonstrated for the production of recombinant proteins in plants, including leafy crops, cereals, legume seeds, oilseeds, fruits, vegetables, higher plant tissue and cell cultures, hydroponic systems, algae, and halobios (Table 13.1) (Tacket, 1999; Mason et al., 2002; Walmsley and Arntzen, 2003).

13.4.1 VEGETABLES

Vegetables are good candidate for the production of edible vaccines as many of them are free of toxicant, full of nutrients, appetizing, and fresh edible. Potato *(S. tuberosum),* tomato *(Lycopersicon esculentum),* and carrot *(Daucuscarota* subsp. *sativus*) have already been reported to successfully express vaccine candidates. LT-B, CT-B, HBsAg, and Norwalk virus capsid proteins are example of some of the antigens that are produced from transgenic potato. Most comprehensive antigen production and immunological studies, including human clinical trials, were carried out with transgenic potato-derived antigens (Mason et al., 1998; Youm et al., 2007). However, potato is not highly palatable and requires cooking, which may result in denaturation of the antigens. In recent years, tomato has emerged as a new expression system because the transformation system, industrial greenhouse culture, and processing of tomatoes are well established. Because of these advantages, tomatoes have appeared as an attractive choice. Antigen genes encoding HBsAg, malarial antigen, HIV gag, and rabies capsid proteins have been successfully transformed to tomato (Ruf et al., 2001; Chowdhury et al., 2007). Carrot is another good alternative for tissue culture and transgenic technology. It was reported that high level of recombinant protein expressions was observed in proplastids of cultured carrot cells. Oral delivery of therapeutic proteins via edible carrots preserved the structural integrity of their target proteins, as no cooking is needed. In addition to these plants as noted above, more and more vegetable transgenic systems such as lettuce *(Lactuca sativa),* celery cabbage *(Brassica rapa* var. *pekinensis),* and cauliflower *(B. oleracea*var *botrytis)* are now under the process of establishment, but low expression level is still a problem in these plant expression systems (Tacket and Mason, 1999; Walmsley and Arntzen, 2003; Koprowski, 2005).

13.4.2 FRUITS

When the idea of edible vaccines was first articulated, people envisioned that a fruit like bananas *(Musa acuminate)*, which are highly palatable raw, would be the ideal system for edible vaccine production. However, inefficient transformation, inadequate information on gene expression, fruit-specific promoters, and the difficulty and expense of cultivation in the greenhouse

TABLE 13.1   Different Plants Used for the Generation of Edible Vaccines

Plants categories	Name of Plant	Advantage	Disadvantage	Examples
Vegetable	Potato	Efficient transformation system, Clonally propagated, low potential for outcrossing in field, Tuber-specific promoters available,Industrial tuber processing well established	Relatively low tuber protein content,Unpalatable in raw form; cooking may cause denaturation of antigen	Cholera toxin, rota viruses, Colonization factor antigen, Epstein virus
	Tobacco	Facile, efficient transformation system Abundant material for protein characterization	Toxic alkaloids incompatible with oral delivery, Potential for outcrossing in field	Hepatitis B virus, Rabies virus, E. coli, Tetanus
	Tomato	Efficient transformation system, Fruit is edible raw,Fruit-specific promoters available, Industrial green-house culture and fruit processing well established	Low fruit protein content, Due to acidic nature it is incompatible with some antigens, No *in vitro*model system to test fruit expression	Norwalk virus
Fruits	Papaya	Cultivated widely in developing countries inexpensive fruit, Eaten raw by infants and adults, low potential for out crossing in field	High cultivation space requirement, Very expensive in green house, low gene expression	*Taeniasolium*
	Banana	Does not need cooking, Proteins not destroyed even if cooked, Inexpensive, Grown widely in developing countries	Trees take 2–3 years to mature and transformed trees take about 12 months to bear fruit.Fruits spoil rapidly after ripening and contain very low protein content so unlikely to produce large amounts of recombinant proteins	Cholera bacterium

TABLE 13.1 (Continued)

Plants categories	Name of Plant	Advantage	Disadvantage	Examples
Crops	Alfa-alfa	Relatively efficient transformation system, High protein content in leaves, Ideal system for animal vaccines, uncooked leaves can be eat	Potential for out crossing in field, Deep root system problematic for cleaning field	FMDV, Rota virus
	Maize	Production technology widely established, High protein content in seeds, Stable protein in stored seeds, Well-suited for animal vaccines	Low transformation, Potential for outcrossing in field for some species	*E. coli*
	Rice	Commonly used in baby food because of low allergenic potential, High expression of proteins/antigens, Easy storage/transportation, Expressed protein is heat stable	Grows slowly, Requires specialized glasshouse conditions	Cholera toxin B subunit (CTB)

have hampered pharmaceutical production of this crop (Mason et al., 2002; Trivedi et al., 2004). Papaya *(Carica papaya)* is a widespread tropical and semi-tropical fresh edible fruit considered as another ideal plant species for vaccine production. Embryogenic callus of papaya was genetically transformed with different constructions including the reporter *GUS* gene by biolistic techniques. Sciutto et al. (2002) reported that inserted transgenes were identified in papaya; this strategy will allow vaccine production on a large scale to evaluate possibilities of plant systemic and oral immunization in the near future (Carter et al., 2002; Sciutto et al., 2002).

13.4.3   CROPS

Alfalfa *(Medicago sativa)* is a perennial crop with strong regenerative capacity and propagation, which continuously produces large clonal populations by stem cuttings in a limited period of time for up to 5 years after establishment. Alfalfa is a palatable crop with high leaf protein content and high annual yields of green leaf tissue. In this crop, pharmaceuticals could be administered in fresh or dried leaf or as leaf extract. Several proteins have been produced in alfalfa, including foot-and-mouth disease virus (FMDV) antigens. Maize *(Zea mays)* was investigated as a commercial platform by several companies (e.g., ProdiGene Inc.) for producing a range of pharmaceutical proteins such as recombinant antibodies, vaccine candidates, and enzymes. Similar to maize, some proteins have been expressed in rice *(Oryza sativa* L.) at very high levels by using constitutive and endosperm-specific promoters. Cereal crops are also ideal plant species because the endosperms of such crops are full of soluble proteins and can be easily separated from the whole plant, which enhances the antigen concentration and reduces the oral doses. Seed legumes offer an attractive option for edible vaccines, because production of antigens in seeds makes for convenient administration of edible vaccines and simplified preparation of antibodies. These crops are particularly suitable for the production of edible vaccines for humans and livestock, because many are natural components of food and livestock feeds (Hood et al., 2002; Nicholson et al., 2003; Streafield et al., 2003; Dus et al., 2005).

13.4.4   ALGAE AND OTHER

Although the biotechnological processes based on transgenic microalgae are still in their infancy, researchers are considering the potential of microalgae as green cell factories to produce value-added metabolites and heterologous proteins for pharmaceutical application. There are five species which have been transformed: *Chlamydomonas reinhartii, Phaeodactylum tricornutum, Amphidinium carterae, Symbiodinium microadriaticum,* and *Cylindrotheca fusiformis.* Tobacco is also easily cultured and transformed plants, and thus has been used as a test system for many antigens. Tobacco produces large volumes of green tissue, and several crops can be harvested per year by cutting the foliage. As a consequence, the research of plant-derived recombinant proteins have been developed from the earth to oceans. Using algae as a bioreactor to produce oral vaccines can solve many problems (e.g., low expression content, toxicity, and long-growth period), which can hardly be overcome by higher organisms (Featherstone, 1996; Hood and Jilkat, 1999; Sun et al., 2003; Fuhrmann et al., 2004; Leon-Baiiares, 2004).

13.5   TECHNIQUES FOR PRODUCTION OF PLANT-DERIVED VACCINES

Gene expression in plants can be classified into two categories viz., transient expression and stable expression. In transient expression, the foreign gene is expressed only for a few days after being introduced into the cell and does not get integrated in the genome. In stable expression, the foreign gene is integrated in the genome and will be passed to the progeny. In transient expression, DNA encoding the antigen of interest is introduced into the plant cell, followed by its recognition by the transcription machinery, finally leading to the expression. On the other hand, stable transformation involves introduction of a foreign gene into the plant cell and its stable integration into the plant genome. The choice of expression system for recombinant protein production in plants depends largely on the aim and the scale of the project (Hansen and Wright, 1999; Sharma et al., 2005). Transient expression is useful for research in order to validate a new technology or to produce recombinant proteins on a large scale. For large-scale production, however, a transgenic plant system may be more convenient, but will require longer periods for development and optimization and scale-up of tissue culture, selection and transformation conditions (Horn et al., 2004; Twyman et al., 2005). Stable expression of a foreign gene involves the production of transgenic plants or transgenic cell cultures. Stable transformation offers multigenerational stable expression that largely depends upon the species. Although the transformation of chloroplasts is also possible for some species, in general, the transformation method involves nuclear transformation. Moderate to low expression of the recombinant protein is observed by this method, but the level of expression can be increased by controlling a number of parameters like use of an optimized codon and strong promoters for plant expression, translation-enhancing leader sequences at the 5' end, polyadenylation signals at the 3' end, and microsomal retention sequences. However, stable transformation requires efficient transformation techniques and meticulous selection and breeding of the high-expressing transgenic lines (Hansen and Wright, 1999; Lorence and Verpoorte, 2004; Horn et al., 2004; Sharma et al., 2005). The DNA sequence coding for a protective antigen and the construction of an expression "cassette" suitable for plant transformation are the two important factors that affect the production of edible vaccines. Generally, transformation methods can be classified in "indirect transformation," where gene transfer is mediated by a bacterium or virus, or "direct transformation," where different techniques such as particle bombardment or electroporation of protoplasts is used to physically introduce "naked" DNA into the plant cell (Sharma et al., 2005). In direct transformation, there are two processes, namely viral vector-based and *Agrobacterium*-based transient amplification. The level of gene expression in both the cases relies on the vector-based delivery of transgenes and the vector could be a bacterium or virus. *Agrobacterium*-mediated transformation may be the choice for a large number of species for which well-established protocols are available (Hansen and Wright, 1999). For some species, however, particularly important crops like maize and soybean, transformation through particle bombardment is usually employed (Bidney et al., 1992). Another good alternate for high-level gene expression is the transformation of chloroplasts, because the chloroplast genome is present in high copy numbers per chloroplast, which in turn are present in large number per cell. Transplastomic plants potentially can express very high levels of foreign protein. Viral vectors are generally more appropriate for expressing products in leaf tissue of high biomass-producing plants like tobacco, potato, etc. (Daniell et al., 2005). The various steps involved in the production of edible vaccine in transgenic plants are shown in Figure 13.3.

13.5.1 BACTERIAL VECTOR

Plants do not contain any naturally occurring plasmid DNA molecules. However, a bacterial plasmid, the tumor inducing (Ti) plasmid of the soil microorganism *Agrobacterium tumefaciens,* has been comprehensively used to introduce genes into plant cells. A large number of plants species can be transfected by using *Agrobacterium*-based plasmid vectors by capitalizing on a natural bacterial system to introduce DNA into the nuclear genome of plants (Figure 13.4). This tumor-inducing (Ti) plasmid is large (~200 kbp) and carries a number of genes that are required for the infection process

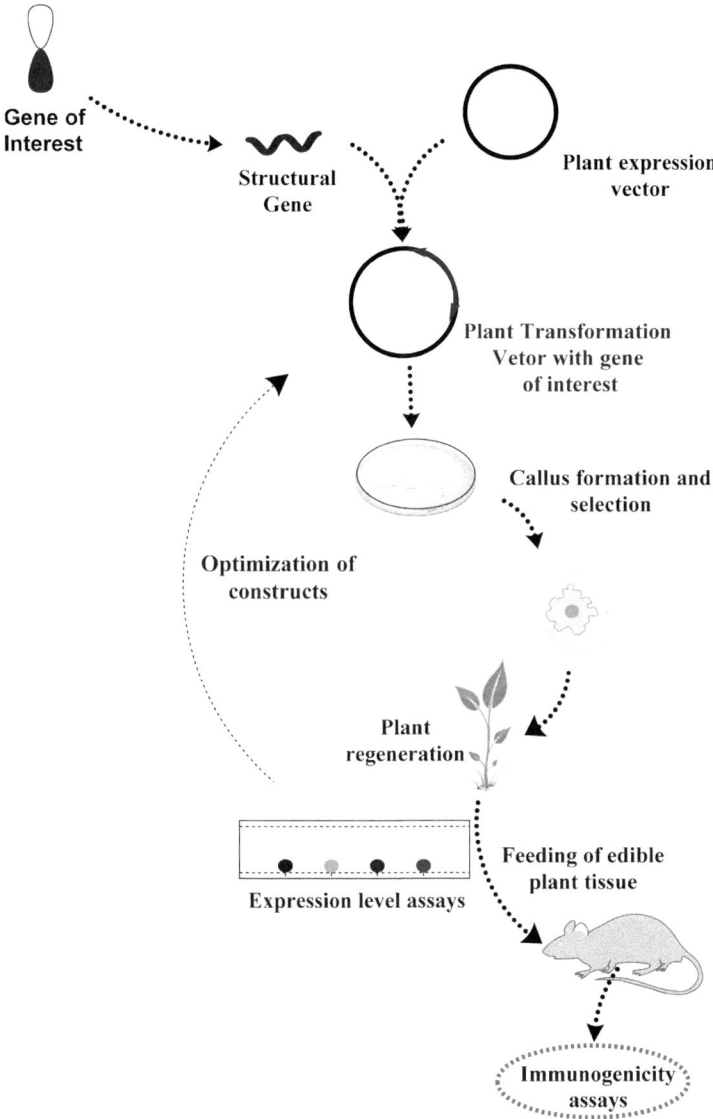

FIGURE 13.3 Production of edible vaccine in transgenic plants involves various steps like cloning of interest genes into a plant expression vector, callus formation, recognition and stimulation of immune response, etc.

(Binns and Thomashaw, 1988; Hooykaas and Shilperoort, 1992; Cheng and Fry, 2000; Escobar and Dandekar, 2003). After infection, part of the Ti plasmid, called the T-DNA, becomes integrated into the plant genome

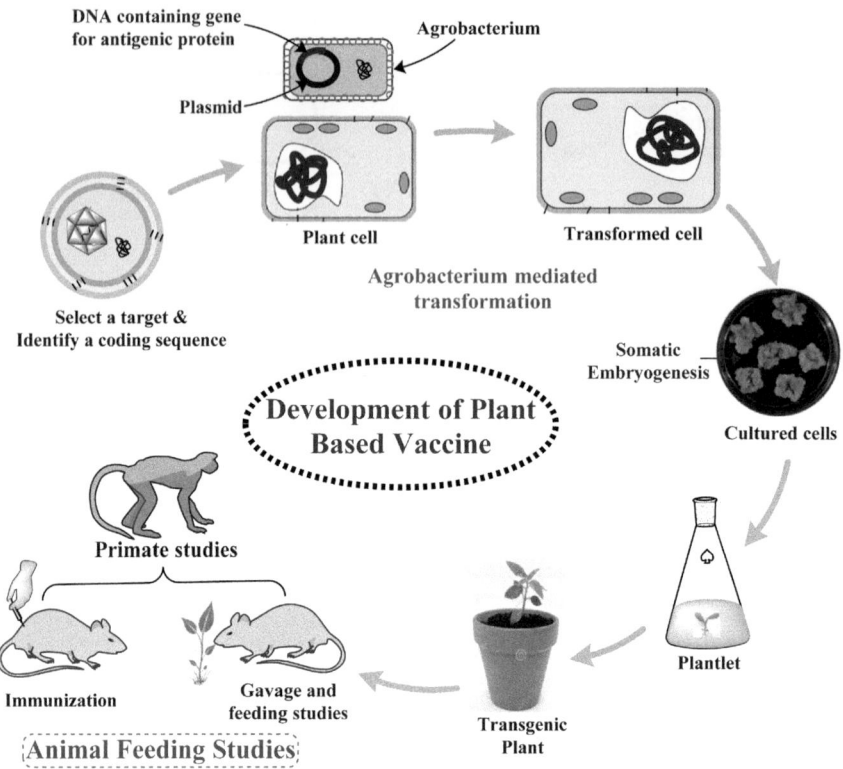

FIGURE 13.4   Schematic representation of *Agrobacterium-based* generation of edible vaccine.

at an apparently random position through non-homologous recombination. T-DNA, approximately 23 kbp in size, contains not only the genes responsible for the cancerous properties of the transformed cells (e.g., those controlling the production of the plant hormones auxin and cytokinin to stimulate cell division and growth) but also those responsible for the synthesis of opines, which are amino acid derivatives. Within the Ti plasmid, the T-DNA is flanked by two 25 bp imperfect direct repeats, known as the left and right border sequences. These sequences are necessary for the integration of T-DNA into the plant genome, and any DNA sequence located between them will be integrated into the plant genome. Once integrated into the plant, the T-DNA is stably maintained and is passed on to daughter cells. In addition to the T-DNA, the Ti plasmid also carries the genes needed for opine utilization, a series of approximately 35 genes required for virulence (vir) and an *Agrobacterium* origin of replication (ori). Because the T-DNA fragment becomes integrated into the plant genome, the Ti plasmid offers an excellent mechanism for the introduction of novel or manipulated genes into the plants (Sheng and Citovsky, 1996; Hellens and Mullineaux, 2000; Opabode, 2006). Apart from *A. tumefacien, another agrobacterium species, Agrobacterium rhizogenes* has also been used as a vector for plant transformation. *A. rhizogenes* stimulates root hair formation in many dicotyledonous plant species, carries a large extrachromosomal element, called Ri (root-inducing) plasmid that functions in a manner analogous to the Ti plasmid of *A. tumefaciens.* Transformation using *A. rhizogenes* has developed analogously to that of *A. tumefaciens* and has been successfully utilized to transform, for example, alfalfa, *Solanum nigrum* L.

Two basic strategies have been devised for inserting new sequences into the T-DNA region (Zupan, 1995; Hamilton, 1997; Klee, 2000).

- ***Co-integration vectors***. The insertion of new DNA into a Ti plasmid results from the recombination of a small vector plasmid, for example, an *E. coli* vector, and a Ti plasmid harbored in *Agrobacterium*. The recombination takes place through a homologous region present in both the plasmids.
- ***Binary vectors***. The T-DNA does not need to be physically associated with the virulence genes in order to become integrated into the plant genome. Therefore, the disarmed T-DNA can be cloned into a small plasmid and manipulated appropriately. The plasmid can then be transformed into *Agrobacterium* that possesses a Ti plasmid that still contains all the virulence genes, but from which the T-DNA has been removed. The virulence proteins provide all the function required to integrate the T-DNA into the plant genome (Bevan, 1988; Hellens et al., 2000).

Leaf tissue is most commonly used for transfecting plants using *Agrobacterium*. Small discs are taken from the leaf by using a paper punch and incubated with the recombinant *Agrobacterium*. Bacterial invasion occurs at the "wounded" edges of the disk. Other alternative approaches used for the introduction of DNA via *Agrobacterium* into plant cells are protoplast transformation and biolistic gene gun approaches. In protoplast transformation, genetic material can be transferred to the protoplastic state of cells through chemical, electrophoretic, or liposomal means. Although DNA inserted into cell in this manner in incapable of independent replication, it may become randomly integrated into one of the plant chromosomes through a process of non-homologous recombination. The biolistic gene gun approach is applicable to the introduction of DNA into a wide range of cells (Bidney et al., 1992). In this approach, the DNA is transferred to the target cell by firing of DNA-coated metal particles, generally tungsten or gold. Pressurized gas, gun powder, or electrical discharges can be used in order to accelerate the process.

13.5.2   VIRAL VECTORS

During their systemic infection of host plants, viruses often produce a large number of genome copies, as well as high titers of some of their encoded proteins. They do not integrate into the host genome and therefore can be considered as natural episomal vectors. The possibility of manipulating the viral genome *in vitro* allowed the exploitation of viruses as vectors for expressing foreign genes, taking advantage of the fact that a huge amplification of the inserted gene can be achieved (Scholthof et al., 1996). Potential features that make virus vectors attractive for protein production include: easy genetic manipulation based on small viral genome sizes and the availability of cDNA clones, potential knowledge of regulatory elements involved in transcription and translation allowing optimization of expression levels of recombinant product, short cycle times of approximately 2-3 weeks in plants, wide host range that may allow the infection of various species using the same vector construct, and high yields of recombinant product based on the highly efficient transcription from some viral promoters. Plants have many natural viral pathogens (Baulcombe, 1999; Yusibov, 1999). Plant virus-based vectors have been developed by fusion of proteins or protein domains to the plant viral coat protein (CP) and presentation on the surface of the assembled virus particle or independent expression of foreign genes inserted into the viral genome. Fusion of peptides to the plant viral CP has been exploited to develop subunit vaccines of human or animal pathogens. Plant virus CP that was fused to immunogenic protein domains of the pathogens was assembled in infected plant cells into virus particles presenting the foreign epitope on the surface (Porta et al., 1996; 2003). The plant virus expression system has single-stranded DNA (geminiviruses), double-stranded nonintegrating DNA (pararetroviruses), and/or plus-sense

RNA viruses. Within the past three decades, several genetic manipulation techniques have been developed to overcome the limitations in the use of viral vectors. Reverse transcription of viral RNA into cDNA allowed their insertion into plasmid vectors for molecular cloning. *In vitro* transcription systems facilitated the synthesis of infectious RNA transcripts from full-length cDNA clones. In addition, the ability to clone viral genes into *A. tumefaciens* enabled this bacterium to be used as a vector to deliver viral genomes into the plant cell's nucleus, thus allowing the virus to be transcribed, replicate, and move throughout the plant. A plant virus expression vector can develop by inserting the gene of interest into the fully functional wild-type virus vector. These vectors are essentially fully functional viruses that have retained infectivity, are relatively stable, and have ability to move systemically within their host and produce infectious virus particles. Recently, some modifications have been made on the fully functional viral approaches by substituting existing viral genes (non-functional genomic sequence) with foreign genes. The extreme evolution of this concept lead to deconstructed viral vectors missing several components of the original virus and usually delivered to the plant by independent constructs. Two full virus vectors, the potato virus X (PVX) and the cowpea mosaic virus (CMV), and two deconstructed vectors, based upon bean yellow dwarf virus (YDV) and tobacco mosaic virus (TMV) have been used to produce protein antigen in plants. Compared to other transient gene expression methods, the expression level is maintained for longer periods due to the elevated number of transcripts as the virus replicates and spreads throughout the plant (Chapman et al., 1992; Twyman et al., 2005). Viral vectors also present several limitations and disadvantages. As a transient expression system, viral vectors are not integrated into the genome and must be repeatedly inoculated so that infected plants can express the desired protein. This implies extra work for inoculation and a need for a supply of plants suitable for being efficiently infected. Some viral vectors like the PVX vector have a limited host range. The choice for the host is an important aspect, as many *Solanaceae* species, commonly used as host for PVX and TMV, are rich in alkaloids that may hamper the use of the infected plants as a raw material for feeding as an oral vaccine (Greene and Allison, 1994; Awram et al., 2002; Scholthof et al., 2002).

13.5.3   CHLOROPLAST TRANSFORMATION

Genetic engineering involving chloroplast transformation offers several unique advantages over nuclear transformation, including very high expression levels of the recombinant proteins/antigen. The chloroplast, like the mitochondrion, possesses its own genome. The chloroplast genome is a single circular double-stranded DNA molecule, which, in the model dicotyledonous plant *Arabidopsis thaliana*, is composed of approximately 155 kbp of DNA containing 87 protein-coding genes. Similar to bacterial genes, chloroplast genes are also arranged in operons. This permits for the insertion of multiple foreign genes into the chloroplast that can be expressed from the same polycistronic transcript. Chloroplasts can also process eukaryotic proteins, including, vaccine antigens, interferons, and human somatotropin with correct folding of subunits and formation of disulfide bridges. Chaperones present in chloroplasts enable the correct folding and assembly of complex multi-subunit proteins. As a result, the expression of genes inserted into the chloroplast can greatly exceed that from a gene in the nucleus. The foreign gene that is being introduced into the chloroplast must be flanked by sequences homologous to the chloroplast genome itself. Chloroplast genes are themselves transcribed by chloroplast-specific promoters and use chloroplast-specific termination signals. Therefore, foreign genes to be expressed in the chloroplast must be placed between suitable promoter and terminator sequences. Additionally, a selectable marker is used to identify transformed cells (Daniell et al., 2005; Streatfield, 2006). Furthermore, several challenges in nuclear genetic engineering could be eliminated, including position effect due to site-specific integration of transgenes by homologous recombination. The problem of gene silencing both at the transcriptional and translational levels has not been observed in transgenic chloroplasts in spite of high levels of translation.

13.5.4   DIRECT GENE TRANSFER

Several so-called direct gene transfer procedures have been developed to transform plants and plant tissues without the use of an *Agrobacterium* intermediate. In the direct transformation of protoplasts, the uptake of exogenous genetic material into a protoplast may be enhanced by use of a chemical agent or electric field. The exogenous material may then be integrated into the nuclear genome (Paszkowski et al., 1984; Lorz et al., 1985; Potrykus et al., 1985). Introduction of DNA into protoplasts of *N. tabacum* is affected by treatment of the protoplasts with an electric pulse in the presence of the appropriate DNA in a process called electroporation (Fromm, 1987; Shillito, 1987). DNA viruses have been used as gene vectors. A cauliflower mosaic virus carrying a modified bacterial methotrexate-resistance gene is used to infect a plant. The foreign gene is systematically spread in the plant. The advantages of this system are the ease of infection, systematic spread within the plant, and multiple copies of the gene per cell. Liposome fusion has also been shown to be a method for transformation of plant cells. Protoplasts are brought together with liposomes carrying the desired gene. As the membranes merge, the foreign gene is transferred to the protoplast. Polyethylene glycol (PEG)-mediated transformation has been carried out in *N. tabacum*, a dicot, and *Lolium multiflorum*, a monocot (Negrutiu et al., 1987). Alternatively, exogenous DNA can be introduced into cells or protoplasts by microinjection. A solution of plasmid DNA is injected directly into the cell with a finely pulled glass needle. A more recently developed procedure for direct gene transfer involves bombardment of cells by microprojectiles carrying DNA. In this procedure called particle acceleration, tungsten or gold particles coated with the exogenous DNA are accelerated toward the target cells. At least transient expression has been achieved in onion. This procedure has been utilized to introduce DNA into Black Mexican sweet corn cells in suspension culture and maize immature embryos and into soybean protoplasts (Reich, et al., 1986; Klein et al., 1987; McCabe et al., 1988; Stomp et al., 1991).

13.6   IMMUNOLOGICAL CONSEQUENCES AGAINST PLANT-BASED IMMUNIZATION

In recent years, noteworthy advances have been made in the development of innovative plant-derived vaccines using recombinant/genetic transformation technology. Plant-derived new generation vaccines have shown the ability to harness the body's immune system to kill infections and seem to be an excellent tool for mass prevention and spread of diseases (Canizares et al., 2005; Streatfield et al., 2007). Today, a number of new plant species are generated by using modern transformation technologies for better expression of recombinant proteins, including rice, potato, soybean, corn, banana, tobacco, tomato, etc. (Figure 13.5).

Scale-up production of transgenic plants at cheaper cost has been presented for tobacco, soybean, and corn using established agricultural techniques (Hood and Jilka, 1999). Another exciting possibility is using plants for mucosal delivery of the expressed antigens, which cause stimulation of both systematic and mucosal immune response (Mason et al., 2002; Walmsley and Arntzen, 2003). This concept of oral vaccination has significant implications for the induction of mucosal immune response, which is the main barrier against hazardous pathogenic microbes present in food or water (Haq et al., 1995; Streatfield, 2005). Mucosal surfaces are the most common sites that serve as entry ports for pathogens to initiate their infectious processes. Immunologists are continuously working to ascertain protection through these surfaces as they constitute the first line of defense against infection. To guard against infections, these surfaces are protected by a complex system of defenses that comprise the mucosal immune system. It has been well established that mucosal-associated lymphoid tissues (MALT) is consisted of immunocytes, including plasma cells, B and T lymphocytes than any other tissue in the body. Gut-associated lymphoid tissues (GALT), nasal-associated lymphoid tissues (NALT), bronchus-associated lymphoid tissues (BALT), genital tract, ocular tissues, salivary glands, and mammary glands are the most common constituent of the mucosal immune system. MALT mainly governs the secretion of sIgA (secretory

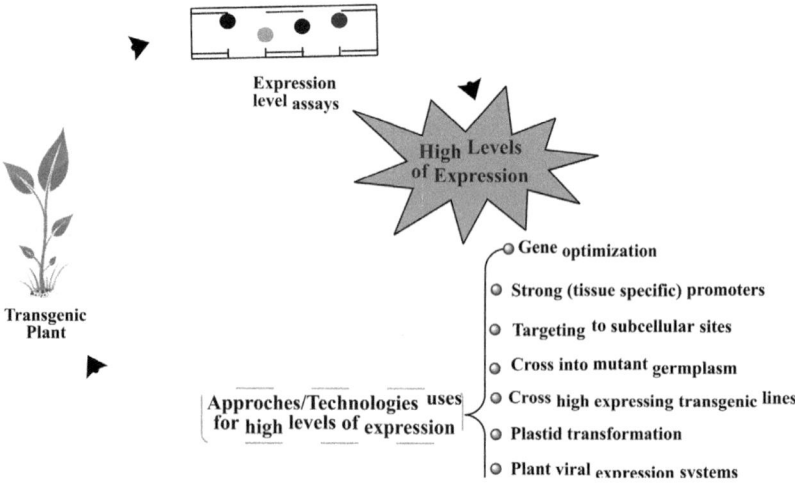

FIGURE 13.5 Various technologies to boost levels of expression in transgenic plants.

antibody) across the mucosal surface in Th$_2$-dependent antigen-specific reactions (Kiyono and Fukuyama, 2004). These sites are in intimate and constant contact with the external environment, contain mucocilliary epithelium, possess secretory components and/or sIgA in the epithelium and lamina propria, and contain organized lymphoid follicles in the subepithelial regions. After *in vivo* stimulation in the GALT, these tissues participate in circulation of antigen reactive sIgA, B lymphocytes, and specifically sensitized T cells to other distant sites (McGhee et al., 1992; Shalaby, 1995; Ogra, 2003). This network of mucosal sites responding against a single antigen after exposure to any mucosal site is known as common mucosal immune system. Transgenic plants that express foreign/therapeutic proteins of pharmaceutical or industrial value represent economical alternative to fermentation-based expression systems or ideal vehicles to produce recombinant proteins and orally deliver protective antigens. The stimulation of the immune response from edible plant-derived vaccines depends upon the antigen interaction with mucosal tissues and its ability to cross the epithelial barrier and to be captured and presented by GALT. GALT consists of the organized lymphoid follicles, PP, appendix, tonsils, and mesenteric lymph nodes. From the mouth to the anus, lymphoid follicles are covered by specialized antigen sampling cells, termed microfold (M) cells. The lymphoid follicles mainly consist of B-lymphocytes but also have helper T cells. The most prominent aggregates of the lymphoid follicles are present in the small intestine. These follicles are termed as PP and contain specialized B and T cell areas. The tonsils and adenoids are also part of GALT. PP is considered as unique inductive sites of the immune response initiated in the intestine. Indeed, the dome epithelium of PP comprise a phagocytic epithelial cell type, i.e., the M cell, specialized in the attachment of particulate antigens, including microorganisms and latex beads as well as macromolecules and translocation of these antigens for delivery to antigen presenting cells (APCs) present in the underlying lymphoid compartment. In addition, M-cells display a poorly organized brush border with short irregular microville, which constitute the physical barriers preventing antigen attachment to villus epithelium. Furthermore, there is ample experimental evidence in various edible plant-based vaccines systems that initiation of the immune response requires antigen presentation to T cells in an organized lymphoid compartment, such as the PP rather than in a diffuse lymphoid compartment such as the lamina propria of the intestinal villi. To induce gastrointestinal immune responses, antigen must be transported from the intestinal mucosa into PP. Antigen sampling cells or

M cells are capable of transporting macromolecules, particles, or microorganisms directly into intestinal PP. M cells are different from intestinal absorptive cells as they lack brush border. In addition, M cells contain a large number of endocytic vesicles to uptake and transport lumen contents across the epithelial layer. A unique feature of M cells is the presence of an intraepithelial pocket containing lymphocytes and a few macrophages, which can interact with the transported antigen or microorganisms. Subsequent to the transport of soluble antigens by M cells, dendritic cells are probably one of the major cells involved in initiating cellular immune responses in PP (Neutra et al., 1992; Ogra, 2003). Humoral immune response is mainly characterized by the production of sIgA by B-lymphocytes. However, IgE responses are also associated with transmucosal exposure to antigens, which are allergenic. B cells are localized in a central region of PP beneath the layer of M cells. Follicular dendritic cells present antigen after trapping to B cells localized around follicular dendritic cells. Primed T helper cells, which migrate through this region, can interact with B cells associated with follicular dendritic cells. Differentiation and maturation of sIgA responses occurs after stimulated B cells leave the PP, and their activation is pivotal in both regulatory and effector activities of the immune system. Most of the T cells are localized around B cell follicles in PP (Parrott, 1976). T cells are also just beneath the M cell layer, where antigen presentation by macrophages and dendritic cells occur. Activation of CD4+ T cells and cytotoxic T lymphocytes takes place within PP followed by migration of these cells to other sites like lamina propria or intestinal epithelium. The activation of T helper cells and secretion of Th_2-like cytokines contributes to the development of mucosal IgA response. Various sites associated with the common mucosal system are shown in Figure 13.6 (Bhatia and Dahia, 2015).

13.7 FACTORS AFFECTING THE EFFICACY OF EDIBLE PLANT-BASED VACCINES

A critical issue of edible plant-based vaccine development is the compatibility of the antigen with cooking and loss of antigenicity of antigens at acidic pH. in oral vaccine research, the major concern not only the harsh gut milieu of the GIT; oral immune tolerance and loss of structural integrity of antigens are also important issues. The major challenge of the edible vegetable-based vaccine approach will be to deliver antigens that look like vaccines, in the context of vegetable that looks like food. The edible vaccine

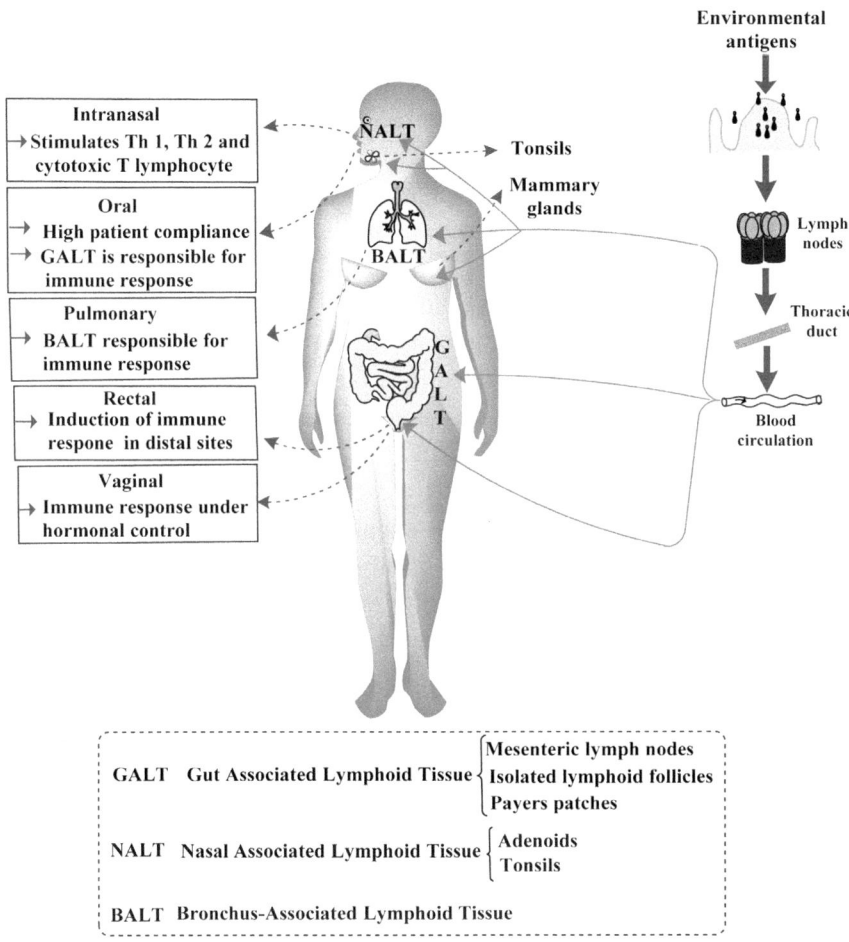

FIGURE 13.6   Various sites of the common mucosal system.

must stimulate an immunogenic response to the vaccine protein and main-tain the tolerogenic response to the food vehicle. Furthermore, in the case of oral route, low concentration of the vaccine antigens in plant requires the consumption of large amounts for immunization or strategies to increase recombinant protein levels. Hence, there is always need of adjuvant to increase the immunogenicity of antigens in edible vaccines (Chargelegue et al., 2001; Lauterslager et al., 2003; Mei et al., 2006). Currently, various approaches have been developed to protect antigens from digestive enzymes like encapsulation of the antigens in a protective coat (Van et al., 1990; Ryan et al., 1997) or use of attenuated bacterial strains of *Vibrio cholerae* and *Salmonella* as delivery vehicles; the attenuated strains have the benefits that they tend to direct the antigen to the surface of the GI tract where it will be available for sampling by the mucosal immune system. On the other hand, the encapsulation approach is an inherently safer means of antigen presentation and can be achieved using biocompatible and biodegradable polymers (Eldridge et al., 1991), liposomes (Jackson et al., 1990), proteosomes (Mallett et al., 1995), or a product derived from transgenic plants expressing the antigens. Immunogenicity, targeting, and immunomodulatory properties of plant-derived vaccines can be altered by modification of expression vector. For example, the antigens can be expressed as a fusion with a molecule that can deliver it to the intestinal cell surface receptors. Fusions with the receptor binding subunit of the heat-labile toxin of enterotoxigenic strains for *E. coli* (LTB) or to the receptor binding subunit of cholera toxin (CTB) follow this approach. These toxin subunits have been shown to act as immune modulators for co-delivered proteins. For example, LTB tends to bias a Th_2 response. Such fusion approaches are starting to be tested in plants. The most straightforward approach to orally co-deliver antigens expressed in plant material would be possible to express the protein adjuvant in a separate plant line and then blend the antigens and adjuvant material prior to administration. This would allow a set ratio of antigen to adjuvant to be ensured, whereas the expression of antigen and adjuvant in the same line may result in some batch-to-batch variation in the ratio. Mutant forms of bacteria can be used to reduce the toxicity and to retain the adjuvanticity. These altered toxins include active site mutants LT-K 63 and LT-R72 and hinge cleavage mutants LT-G192. Another factor, which is important in the efficacy of the vaccine, is the palatability of the delivery vehicles. Fruits such as banana and tomato, which are eaten raw, retain the immunogenicity of the delivered antigen, but in the case of potato, which is boiled to make it palatable, the immunogenicity of the delivered antigen is reduced (Arakawa et al., 1998; Mason et al., 1998; Sala et al., 2003).

13.8 APPLICATION OF EDIBLE PLANT-DERIVED VACCINES

Plants offer a suitable protein production and oral delivery option for subunit vaccines to combat a wide range of diseases. Various research groups have investigated a wide range of protective antigens from microbial and viral pathogens in plants. There is no limit to the number and range of antigens that can be produced in plants if the DNA sequences coding for the appropriate genes are available (Mor et al., 1998; Streatfield and Howard, 2003; Teli and Timco, 2004). Many of these have entered preclinical trials, wherein the immunogenicity is usually evaluated using an animal model (Streatfield, 2006, 2007). Edible plant-based vaccines represent a novel, more effective, safer platform against various infectious/hazardous diseases. Recently, numerous chimeric plant viruses have emerged as promising, novel, and attractive vector systems for the expression of foreign epitopes to be used as immunogens for the development of innovative plant-derived vaccines (Lico et al., 2012). The early focus was on human pathogens, but in the last few years, attention has also given to animal pathogens (e.g., Newcastle and foot and mouth disease). Plant-derived pharmaceuticals for active and passive immunization and auto-tolerance have proven to be immunologically active and effective in animals and humans. Edible plant-based vaccines provide a promising example of a novel approach that combines the innovations in plant biology, biotechnology, and medical science to create affordable biopharmaceuticals.

13.8.1 HEPATITIS B

Hepatitis B virus (HPV) is one of the most common human pathogen and causes of chronic viremia, liver cirrhosis, chronic hepatitis, persistent infection, liver necrosis, etc. (Michel and Tiollais, 2010). Because immunization is the only known method to prevent the diseases of the HPV, any attempt to reduce its infection requires the availability of large quantities of vaccine. Current vaccines consisting of the HBV surface antigen (HBsAg) produced in recombinant yeast is commercially available for parenteral vaccination. However, in developing nations, the cost of production, distribution of the resulting vaccine, and administration cost is too high. The second major hurdle is the need of trained personnel to administer injections. Thus, there are clear needs for the development of inexpensive new hepatitis B vaccines that would allow vaccination of large segments of the population. Therefore, various research programs have been initiated to express a hepatitis vaccine in transgenic plants. Edible plant-derived vaccines allow oral delivery of an antigen while overcoming the cost concerns described above. The production of recombinant proteins in plants is inexpensive compared to other expression systems (Daniell et al., 2001; Hood et al., 2002; Pniewski, 2013). HBsAg was the first reported viral antigen to be produced in transgenic plants. Mason et al. (1992) expressed HBsAg in tobacco and demonstrated the plant-derived antigen formed virus-like particles (VLPs). The particles were physically similar to those derived from human serum and retained antigenicity. Mice immunized with partially purified, tobacco-derived hepatitis B VLPs responded in a manner similar to those immunized with yeast-derived rHBsAg (commercial vaccine). Potato tubers have received considerable attention. Transgenic potatoes expressing HBsAg at higher levels should allow assessment of the immunogenic responses of the antigen by oral delivery. Potato-derived HBsAg is currently being tested in human clinical trials using the orally delivered antigen as a booster for the commercial vaccine (Richter et al., 2000). A second criterion for a plant-based hepatitis B vaccine is that the antigen concentration should be uniform, allowing for even dosing of subjects. Huang and coworker (2006) assembled hepatitis B core antigen (HBc or HBcAg) into capsid particles by using a novel plant viral expression system (*Nicotiana benthamiana*) leaves. HBc has been extensively studied for its production in various expression systems and for the use of HBc particles for high-density, immunogenic presentation of foreign epitopes. Studies showed rapid, high-level (up to 7% of total soluble protein) production of HBc in plant leaves by using a novel plant viral expression system. The plant-derived HBc (p-HBc) self-assembles into VLPs that are morphologically and physically similar to *E. coli*-derived HBc (e-HBc)-VLPs and induced strong humoral and mucosal antibody responses in mice.

13.8.2 MALARIA

Malaria is one of the most prevalent disease caused by protozoan parasites (Genus *Plasmodium*) throughout the world, with increasing morbidity and mortality rate each year of over 300 to 500 million new cases of infection, resulting in 1.5 to 2.7 million deaths worldwide. The mortality and morbidity rate of malaria has increased because of the development of resistance of anopheles mosquito vectors to various insecticides, increasing drug resistance strain of malarial parasites (Abdel-Hameed and Gregson, 2005). Although this disease can be controlled by chemotherapeutic antimalarial agents, because of the above-mentioned reasons, there is an urgent need for the development of a successful malaria vaccine. Although various antigenic candidate expressed on the surface of the parasite have been identified and tested, a successful vaccine is still an immense challenge for immunologists (Targett, 1989). Currently, three antigens are being investigated for the development of a plant-derived malaria vaccine, namely merozoite surface protein (MSP) 4 and MSP 5 from *Plasmodium falciparum*, and MSP 4/5 from *P. yoelli* (Mishra et al., 2005). Chowdhury and Bagasra (2012) hypothesized the development of an edible malaria vaccine using transgenic tomatoes to carry different subunit antigens. Oral immunization of individuals against 2–3 antigens and against each stage of the lifecycle of the multistage parasites would certainly provide an affordable and safe alternative. In the same manner, Wang et al. (2004) investigated oral immunization of mice with recombinant MSP 4, MSP 4/5, and MSP1, co-administered with cholera toxin B subunit as a mucosal adjuvant and evaluated the induced antibody responses effective against blood-stage parasite. However, one of the major limitations is expression of antigens in transgenic plants are too low and high level of expression is essential to confer meaningful immunity. With the advancement of techniques, a dramatic increase in the level of antigenic expression in plants can be achieved using transgenic technology. The use of this approach can improve antigenic expression in plants and can induce strong immune response in susceptible populations like those with moderate food intake. In addition to high levels of antigen anticipated to be necessary, it is likely that strong adjuvants will also be required. Therefore, suitable adjuvants have to be identified and tested. Since the inception of these concepts more than a decade ago, several developments have been made to date, including pioneering human clinical trials.

13.8.3 CHOLERA TOXIN

In developing countries, cholera is devastating infectious diarrheal disease that is the leading cause of death among children. *Vibrio cholerae* and enterotoxigenic *E. coli* (ETEC) are two causal agents of bacterial diarrhea. Two examples of enterotoxins are the heat labile toxin (LT) and the *V. cholera* cholera toxin (CT). LT and CT consist of two subunits, the A subunit (LTA or CTA) and the B subunit (LTB or CTB). Therefore, the critical challenge for cholera toxin prevention lies in the establishment of affordable alternatives, including effective oral vaccine development that provide long-lasting immune protection after single immunization or convenient and readily available vaccines for frequent administration to people living in cholera-prevalent areas. Based on previous studies and facts, it is well established that cholera is controlled more effectively through oral immunization rather than parental (Waldor, 1996b). Plant-derived edible vaccines may provide exactly such a tool because they are safe, nutritious, and easy to administer. The first human clinical trials for a transgenic, plant-derived antigen were planned and performed after receiving approval from US Food and Drug Administration in 1997. Oral immunization with or CTB results in the production of antigen-specific sIgA that prevents the binding of LTB and or CTB to the epithelial cells and hence interfere with the toxic effect of LT and CT. Mason et al. (1998) demonstrated expression and correct assembly of heat-labile enterotoxin LTB by potato tissues and tested the plant-derived vaccine in animal feed trials. The potato-immunized mice produced both serum IgG and gut mucosal IgA at a high level comparable to those in mice given bacteria-expressed LTB. Tobacco and potato tissues can express enterotoxins, especially LTB. Transgenic potatoes constitutively expressing a synthetic bacterial diarrhea vaccinogen (LTB) were orally delivered to human volunteers in Phase I/ II clinical trials. Administration of plant-derived recombinant LTB (rLTB) to mice produced LTB-specific serum IgG and mucosal IgA (Rigano et al., 2003). In another study, Hamorsky and coworkers (2013) developed a rapid, robust, and scalable bioproduction system in plants for a CTB variant, aglycosylated CTB (pCTB). This system allowed for the accumulation of pCTB at >1 g per kg of fresh leaf of tobacco-related plants within 5 days, which accounts for over 1000 doses of original CTB included in the WHO prequalified vaccine "Dukoral." The integrity of pCTB was subsequently analyzed using biochemical, biophysical, and immunological experiments, and it was finally concluded that the plant-derived protein is feasible as a cholera vaccine antigen. The plant-expressed protein behaved like native CTB in terms of effects on cytokine levels and T-cell proliferation, indicating the suitability of plant expression systems for the production of bacterial antigens, which could be used as edible vaccine (Jani et al., 2004).

13.8.4   ANTHRAX

The current available human licensed vaccine for anthrax (protective antigen; PA) is produced by a cell-free filtrate from cultures of non-encapsulated attenuated strain of *Bacillus anthracis* that contains trace amounts of lethal factor (LF) and edema factor (EF), which are toxic and cause adverse side effects. Although current available PA-based vaccines proved to be protective in various animal models and humans, some characterization issues such as safety, stability, reactogenicity, expanded dosage schedule, and route of administration remain a matter of concern. In order to overcome the above-mentioned problems related with the current available vaccines, plant-derived vaccines offer safe and affordable alternative in which vaccine antigens can be successfully expressed and devoid of bacterial contaminants (Ivins et al., 1998; Fellows et al., 2001; Pittman et al., 2001; Wang et al., 2005). Plant-derived recombinant PA is free of LF and EF and is easy to produce, without the need for costly fermenters. In an effort to produce anthrax vaccine in large quantities and free of extraneous bacterial contaminants, PA was expressed in transgenic tobacco chloroplasts by inserting the *pagA* gene into the chloroplast genome. Further, s.c. immunization of mice with partially purified chloroplast-derived or *B. anthracis*-derived PA with adjuvant yielded high IgG titers, and both groups of mice showed 100% survival when challenged with lethal doses of toxin. The results showed high levels of expression without using fermenters and significant immunoprotection by the plant-derived anthrax vaccine antigen (Koya et al., 2005). In the same manner, Gorantala et al. (2011) successfully expressed *B. anthracis* PA gene [PA (dIV)] in plants and *E. coli* [rPA(dIV)] and further evaluated the protective response in mice by intraperitoneal (i.p.) and oral immunizations with or without adjuvant. The obtained results showed that plant-derived *Bacillus anthracis* PA-based vaccine could stimulate better protective immune response as compared to current available vaccine upon i.p. or oral immunization. Recently, the same research group also studied the expression of the PA gene in Indian mustard by *Agrobacterium*-mediated transformation and in tobacco by plastid transformation. Oral and i.p. immunization experiments with plant PA in the murine model showed high serum PA-specific IgG and IgA antibody titers. PA-specific mucosal immune response was significantly noted in orally immunized groups. (Gorantala et al., 2014). Today, numerous investigations have proved the immunogenic and immunoprotective properties of plant-derived *B. anthracis* PA; thus, prospective investigations need to focus on the development of more efficient edible vaccine against anthrax (Aziz et al., 2002; Azhar et al., 2005; Kamarajugadda, 2006; Masri et al., 2006; Streatfield, 2006).

13.8.5 PLAGUE

Plague is caused by *Yersinia pestis*, a gram-negative bacterium, and human infections are often acquired through an infected flea vector. There are three different forms of plague, of which bubonic plague is the most common. The current available vaccine is a killed whole vaccine containing formaldehyde-killed *Y. pestis* strain 195/P cells, which is moderately effective against bubonic plague and ineffective against septicemic and pneumonic plague. In addition, it requires a course of vaccination over 6 months and carries a significant risk of transient and severe side effects (Russell et al., 1995). Alvarez et al. (2006) reported the production of an economical alternative plague vaccine candidate and found that it elicited immune response in mice. In this study, the F1–V-antigen fusion protein was expressed in transgenic tomato plants that were then molecularly characterized. The immunogenicity of the oral plant-made vaccine was tested in BALB/c mice primed s.c. using bacterial-produced F1–V and boosted orally with freeze-dried, powdered, F1–V transgenic tomato fruit. The results showed generation of high IgG1 titre level in serum and mucosal IgA in fecal pellets. Several plant-made subunit vaccines have been evaluated for immunogenicity against *Y. pestis*. CaF1 and LcrV are the most effective antigens identified so far against *Y. pestis*. F1 is a capsular protein located on the surface of the bacterium with antiphagocytic properties. The fusion protein of capsular protein with V antigen has been shown to be more immunogenic when the mice were challenged with the bacteria. When the F1–V fusion gene was expressed in transgenic chloroplasts, the expression levels obtained were as high as 14.8% of the total soluble protein (Daniell, 2006; Alvarez et al., 2006; 2010; Streatfield, 2006; Mett et al., 2007). However, there is still a clear need to provide alternative, economical vaccines more suited to the large-scale immunization of populations. Such a vaccine would be ideally administered noninvasively and promote a much better mucosal immunity against the infection.

13.8.6   TUBERCULOSIS

Tuberculosis (TB) caused by the bacterium *Mycobacterium tuberculosis* is the leading fatal infectious disease in the developing world. Because the portal of entry and primary site of infection of *M. tuberculosis* is the respiratory track, mucosal vaccination should have inherent advantage to induce a protective immune response. Currently, the only available TB vaccine is the Bacille Calmette-Guerin (BCG), which is an attenuated vaccine derived from *M. bovis*. Although BCG has been widely used for decades, its efficacy has been shown to be highly variable, and it fails to protect against lung infections, including adult pulmonary TB. Over the last decade, intensive research activities have led to significant progress in the development of TB vaccine, reflected by a myriad of subunit vaccine candidates and DNA vaccines, from which a dozen have been assessed in clinical trials (Mendoza et al., 2015). Among the efforts performed, the use of plant-derived vaccines has been explored by various research groups. Rigano et al. has developed the transgenic plant vaccines against *tuberculosis* in which TB antigen ESAT-6 was fused to the B subunit of *E. coli* LTB and expressed in the *A. thaliana* plant. The plant-made LTB-ESAT-6 fusion protein induced antigen-specific responses from $CD4^+$ cells and increased IFN-γ production, indicating a Th_1 response. In addition, a Th_2 response was also induced in the PP. This is the first report of an orally delivered, subunit tuberculosis vaccine priming an antigen-specific, Th_1 response (Rigano et al., 2006). Numerous research groups have explored this concept demonstrating that the use of minimally processed plant biomass may confer immunity when used as oral vaccines in test animal models. Therefore, it has been proposed that oral vaccines consisting of gelatin capsules containing freeze-dried biomass derived from transformed plants expressing the target antigen are convenient means to provide mucosal as well as systemic immune response (Davoodi-Semiromi et al., 2010; Kwon et al., 2013; Govea-Alonso et al., 2014). Four oral plant-derived vaccines have been evaluated in clinical trials thus far, reflecting the potential to induce specific humoral responses in humans with at least modest immunogenicity or capacity to serve as a boost in previously immunized individuals (Tacket et al., 1998, 2000; Yusibov et al., 2002; Thanavala et al., 2005).

13.8.7 HUMAN IMMUNODEFICIENCY VIRUS TYPE 1 EPITOPES

Human immunodeficiency virus (HIV) is one of the most devastating infectious diseases globally, and thousands of people continue to be infected with HIV every day. Vaccination is the most efficient method for protection against viral diseases; an effective HIV vaccine requires both B cell and T cell epitopes to stimulate an efficient CD4+ T cell-mediated immune response and to elicit cytotoxic lymphocytes and protecting antibodies. In light of this, plants have gained much interest as affordable and attractive production system for the development of potential HIV vaccines (Hefferon, 2013). To date, varied HIV antigens have been produced in various types of plant using a variety of expression systems, and some plant derived anti-HIV MAbs have progressed to Phase 1 trial [Pharma plant news June 2011]. These antigens consist of envelope proteins gp120 and 160, a transmembrane protein gp41, and protease genes. In 1995, the expression of an epitope of the HIV-1 virus was demonstrated for the first time in the plant cell. A translational fusion of a gp120 epitope to the CP of TMV resulted in the expression of virus particles consisting of the CP and CP fusion protein. Initial attempts were made to generate HIV Gag virus like particles in plants using either the native gene or a plant codon usage optimized version of the Gag gene proved to be unsuccessful. In due course, Gag was produced using a plant chloroplast-based expression system proved to be satisfactory in terms of expression levels and immunogenicity, suggesting that this may be the much better approach for the development of plant-derived HIV vaccine (Hefferon, 2013). Recently, HIV-1 Tat protein has been explored as a prospective vaccine candidate. Various studies suggested that HIV-1 Tat should be considered as an important component of potential HIV vaccines. Tat is not only a key regulator of HIV-1 replication in infected cells, but it is also an extracellular immunomodulator, which increases the efficiency of virus dissemination and promotes AIDS progression. Some studies suggested that TMV-based transient expression system for the production of HIV-1 Tat protein in an experimental plant *Nicotiana benthamiana*, and in an edible plant, spinach, at levels of up to 300 µg of Tat per 1 g of leaf tissue. Feeding these Tat-producing spinach leaves to mice primed them for antibody production following a subsequent immunization with a plasmid expressing Tat and resulted in significantly higher anti-Tat antibody levels compared to that in control mice (Horn et al., 2003; Canizares et al., 2005;

Karaseva et al., 2005; Webster, 2005). In another approach, Rubio et al. (2012) generated a synthetic gene encoding C4V3, a recombinant protein, and it was highly expressed in tobacco chloroplasts. The C4V3 recombinant protein was demonstrated to cross-react with both an anti-C4V3 rabbit serum and with sera from HIV-positive patients. When orally administered, the plant-derived C4V3 in BALB/c showed generation of significant CD4+ T-cell proliferation responses and both systemic as well as mucosal antibody responses. To date, various techniques have been developed to improve the expression levels and production platform in plants, but there is still a need to provide alternative, economical vaccines more suited to the large-scale immunization of populations.

13.8.8   FOOT AND MOUTH VIRUS VP1 COAT PROTEIN

Foot and mouth disease (FMD), caused by a highly variable RNA virus, is one of the most economically and socially devastating diseases affecting all cloven-hoofed animals; it produces vesicles in the mouth, on the skin between and above the hoofs, and on the teats of lactating animals worldwide. Currently, inactivated vaccines are used as a major tool in FMD eradication programs in Europe as well as in other parts of the world. However, these vaccines have a number of limitations such as limited shelf-life, threat of virus escape from the manufacturing sites, propagation of virulent virus, and requirement of a booster injection after 4–12 months. In addition, some other challenges such as inadequate safety, sterility, cost-effectiveness, easy delivery, and long-lasting immunity against multiple serotypes are associated with conventional (inactivated or attenuated) vaccines. In this scenario, transgenic vaccines in plants are attractive alternatives to conventional FMD vaccines. Plants can be grown efficiently at large scale and easily delivered. However, the production of FMD chimeric plant-based vaccines from local isolates is a real challenge worldwide (Rodriguez et al., 2011; Ali Saeed et al., 2015). Various studies have demonstrated the structural capsid protein VP1 to carry critical epitopes responsible for the induction of neutralizing antibodies has been also successfully expressed as an immunogenic antigen in *Arabidopsis thaliana*, *alfalfa*, and *potato* plant. In addition, fast engineering and short cycle times for recombinant protein production have great advantages in vaccine development. The VP1 capsid protein of FMDV can be expressed in transgenic *A. thaliana* plant and after successful immunization, it elicits strong and protective antibody responses against virulent FMDV in experimental hosts (Dus Santos, 2005).

13.9 CONCLUDING REMARKS

A vast amount of evidence has accumulated in the last decade in favoring of plant-derived antigens as effective, safe, and inexpensive means of acquiring or delivering a vaccine. The success of human trials with edible transgenic plant-derived vaccines and successful results in animal studies have demonstrated the potential of plant-derived vaccines to expand the armament of preventative medicine. Edible plant-derived vaccines hold great potential, especially in developing countries where transportation cost, poor refrigeration, and needle use complicates the vaccine administration. Edible vaccines, in particular, might overcome some of the difficulties of production, distribution, and delivery associated with traditional vaccines. Significant challenges are still to be overcome before vaccine crops can become a reality. However, while access to essential healthcare remains limited in much of the world and the scientific community is struggling with complex diseases such as HIV and malaria, plant-derived vaccines represent an appetizing prospect.

KEYWORDS

- **edible vaccines**
- **immunoprophylaxis**
- **innovative concept**
- **microalgae**
- **plasmid**
- **transformation**
- **vector**

REFERENCES

Abdel-Hameed, A. A., (2005). Antimalarial drug resistance in the Eastern Mediterranean Region. *East Medit Health J.*, *9*(4), 492–508.

Alvarez, M. L., & Cardineau, G. A., (2010). Prevention of bubonic and pneumonic plague using plant-derived vaccines. *Biotech Adv.*, *28,* 184–196.

Alvarez, M. L., Pinyerd, H. L., Crisantes, J. D., Rigano, M. M., Pinkhasov, J., Walmsley, A. M., Mason, H. S., & Cardineau, G. A., (2006). Plant-made subunit vaccine against pneumonic and bubonic plague is orally immunogenic in mice. *Vaccine*, *2414,* 2477–2490.

Arakawa, T., Chong, D. K., & Langridge, W. H., (1998). Efficacy of a food plant-based oral cholera toxin B subunit vaccine. *Nat Biotechnol.*, *16,* 292–297.

Arntzen, C. J., Mason, H. S., Tariq, H. A., & Clements, J. D., (2002). *Oral immunization with transgenic plants*. US Pat No. 6395964.

Arntzen, C., Plotkin, S., & Dodet, B., (2005). Plant-derived vaccines and antibodies: Potential and limitations. *Vaccine*, *23,* 1753–1756.

Awram, P., Gardner, R. C., Forster, R. L., & Bellamy, A. R., (2002). The potential of plant viral vectors and transgenic plants for subunit vaccine production. *Adv. Virus Res.*, *58,* 81–124.

Azhar, A. M., Singh, S., Anand, K. P., & Bhatnagar, R., (2005). Expression of protective antigen in transgenic plants: a step towards edible vaccine against anthrax. *Biochem. Biophy. Res. Cummun.*, *2993,* 345–51.

Aziz, M. A., Singh, S., Kumar, P. A., & Bhatnagar, R., (2002). Expression of protective antigen in transgenic plants: a step towards edible vaccine against anthrax. *Biochem. Biophy. Res. Cummun.*, *299,* 345–351.

Baulcombe, D. C., (1999). Fast forward genetics based on virus-induced gene silencing. *Curr. Opin. Plant Biol.*, *2,* 109–113.

Bevan, D., (1998). Binary Agrobacterium vectors for plant transformation. *Nucleic Acids Res.*, *12,* 8711–8721.

Bhatia, S., & Dahiya, R., (2015). Edible vaccines, modern applications of plant biotechnology in pharmaceutical sciences. Elsevier, ISBN: 9780128024980.

Bidney, D., Scelonge, C., Martich, J., Burus, M., Sims, L., & Huffman, G., (1992). Microprojectile bombardment of plant tissues increased transformation frequency of Agrobacterium tumefaciens. *Plant Mol Biol.*, *18,* 301–313.

Binns, A. N., & Thomashaw, M. F., (1998). Cell biology of Agrobacterium infection and transformation of plants. *Ann. Rev. Microbiol.*, *42,* 575–606.

Canizares, M. C., Lomonossoff, G. P., & Nicholson, L., (2005). Development of cowpea mosaic virus-based vectors for the production of vaccines in plants. *Exp. Rev. Vaccines*, *45,* 687–697.

Carter, J. E., & Langridge, W. H. R., (2002). Plant-based vaccines for protection against infectious and autoimmune diseases. *Crit Rev Plant Sci.*, *21,* 93–109.

Castle, L. A., Wu, G., & McElroy, D., (2006). Agricultural input traits: past, present and future. *Curr. Opin. Biotechno.*, *17,* 105–112.

Chapman, S., Kavanagh, T., & Baulcombe, D., (1992). Potato virus X as a vector for gene expression in plants. *Plant J.*, *2,* 549–557.

Chargelegue, D., Obregon, P., & Drake, P. M., (2001). Transgenic plants for vaccine production: expectations and limitations. *Trends in Plant Sci.*, *6,* 495–496.

Cheng, M., & Fry, J. E., (2000). An improved efficient *Agrobacterium*-mediated plant transformation method. *Int. Patent publ. WO.*, 0034/491.

Chowdhury, K., & Bagasra, O., (2007). An edible vaccine for malaria using transgenic tomatoes of varying sizes, shapes and colors to carry different antigens, *Medical Hypotheses*, *68*, 22–30.

Cripps, A. W., Kyd, J. M., & Foxwell, A. R., (2001). Vaccines and mucosal immunisation. *Vaccine*, *19*, 2513–2515.

Daniell, H., (2006). Production of biopharmaceuticals and vaccines in plants via the chloroplast genome. *Biotechnol J.*, *110*, 1071–9.

Daniell, H., Chebolu, S., Kumar, S., Singleton, M., & Falconer, R., (2005). Chloroplast-derived vaccine antigens and other therapeutic proteins. *Vaccine*, *23*, 1779–1783.

Daniell, H., Streatfield, S. J., & Wycoff, K., (2001). Medical molecular farming: production of antibodies, biopharmaceuticals and edible vaccines in plants. *Trends Plant Sci.*, *6*, 219–26.

Davoodi-Semiromi, A., Schreiber, M., Nalapalli, S., et al., (2010). Chloroplast-derived vaccine antigens confer dual immunity against cholera and malaria by oral or injectable delivery. *Plant Biotechnol J.*, *82*, 223–42.

Dus, S. M. J., & Wigdorovitz, A., (2005). Transgenic plants for the production of veterinary vaccines. *Immunol Cell Biol.*, *83*, 229–238.

Eldridge, J. H., Staas, K., Meulbroek, J. A., McGhee, R., Tice, T. R., & Gilley, R. M., (1991). Biodegradable microspheres as a vaccine delivery system. *Mol. Immunol.*, *28*, 287–294.

Escobar, M. A., & Dandekar, A. M., (2003). Agrobacterium tumefaciens as an agent of disease. *Trends Plant Sci.*, *8*, 380–386.

Featherstone, C., (1996). Vaccines by agriculture. *Mole Med.*, 278–281.

Fellows, P. F., Linscott, M. K., Ivins, B. E., Pitt, M. L., Rossi, C. A., Gibbs, P. H., et al., (2001). Efficacy of a human anthrax vaccine in guinea pigs, rabbits, and rhesus macaques against challenge by Bacillus anthracis isolates of diverse geographical origin. *Vaccine*, *19*, 3241–7.

Fischer, R., Stoger, E., Schillberg, S., Christou, P., & Twyman, R. M., (2004). Plant-based production of biopharmaceuticals. *Curr. Opin. Plant Biol.*, *7*, 152–158.

Fromm, M. E., Wu, R., & Grossman, L., (1987). (eds.). In: *Methods in Enzymology*. Academic Press, Orlando, Fla, 153, 307.

Fuhrmann, M., (2004). Production of antigens in chlamyclomonasreinhardtii: green microalgae as a novel source of recombinant proteins. *Methods Mol Med.*, *94*, 191–195.

Giddings, G., Allison, G., Brooks, D., & Carter, A., (2000). Transgenic plants as factories for biopharmaceuticals. *Nat. Biotechnol.*, *18*, 1151–5.

Gorantala, J., Grover, S., Goel, D., Rahi, A., JayadevMagani, S. K., Chandra, S., & Bhatnagar, Ro., (2011). A plant based protective antigen [PAdIV,] vaccine expressed in chloroplasts demonstrates protective immunity in mice against anthrax. *Vaccine*, *292*, 4521–4533.

Gorantala, J., Grover, S., Rahi, A., Chaudhary, P., Rajwanshi, R., Bhalla, N. S., & Bhatnagar, R., (2014). Generation of protective immune response against anthrax by oral immunization with protective antigen plant-based vaccine. *J. Biotech.*, *176*, 10.

Govea-Alonso, D. O., Rybicki, E., & Rosales-Mendoza, S., (2014). Plant-based vaccines as a global vaccination approach: current perspectives. In: *Genetically Engineered Plants as a Source of Vaccines Against Wide Spread Diseases -An Integrated View*. Springer, Science Business Media, LLC, *233* Spring Street, New York, NY, 10013, USA.

Greene, A. E., & Allison, R. F., (1994). Recombination between viral RNA and transgenic plant transcripts. *Science*, *263*, 1423–1425.

Gregson, A., & Plowe, C. V., (2005). Mechanisms of resistance of malaria parasites to antifolates. *Pharmacol Rev.*, *571*, 117–145.

Hamilton, C. M., (1997). A binary-BAC system for plant transformation with high-molecular-weight DNA. *Gene.*, *200*, 107–116.

Hamorsky, K. T., J. Calvin, K., Lauren J., Bennett, B. K. J., Kajiura, H., Fujiyama, K., & Matoba, N., (2013). Rapid and scalable plant-based production of a cholera toxin b subunit variant to aid in mass vaccination against cholera. *Outbreaks PLoSNegl Trop Dis.*, *73*, e2046.

Hansen, G., & Wright, M. S., (1999). Recent advances in the transformation of plants. *Trends Plant Sci.*, *4*, 226–231.

Haq, T. A., Mason, H. S., Clements, J. D., & Arntzen, C. J., (1995). Oral immunization with a recombinant bacterial antigen produced in transgenic plants. *Science*, *268*, 714–716.

Hefferon, K. L., (2013) Applications of plant-derived vaccines for developing countries. *Trop Med. Surg. 1*, 106. doi: 10.4172/2329–9088. 1000106.

Hefferon, K. L., (2013). Applications of plant-derived vaccines for developing countries. *Trop Med. Surg.*, *1*, 1.

Hellens, R., Mullineaux, P., & Klee, H., (2000). A guide to *Agrobacterium* binary Ti vectors, *Trends in Plant Science*, *5*, 1–4.

Hood, E. E., & Jilkat, J. M., (1999). Plant-based production of xenogenic proteins. *Curr. Opin. Biotech.*, *10*, 382–386.

Hood, E. E., Woodard, S. L., & Horn, M. E., (2002). Monoclonal antibody manufacturing in transgenic plants–myths and realities. *Curr. Opin. Biotechnol.*, *13*, 630–635.

Hooykaas, P. J J., & Shilperoort, R. A., (1992). Agrobacterium and plant genetic engineering. *Plant Mol. Biol.*, *19*, 15–38.

Horn, M. E., Pappu, K. M., Bailey, M. R., Clough, R. C., Barker, M., Jilka, J. M., Howard, J. A., & Streatfield, S. J., (2003). Advantageous features of plant-based systems for the development of HIV vaccines. *J. Drug Target*, *118*(10), 539–545.

Horn, M. E., Woodard, S. L., & Howard, J. A., (2004). Plant molecular farming: Systems and products. *Plant Cell Report*, *22*, 711–720.

Huang, Z., Santi, L., LePore, K., Kilbourne, J., Arntzen, C. J., & Mason, H. S., (2006). Rapid, high-level production of hepatitis B core antigen in plant leaf and its immunogenicity in mice. *Vaccine*, *24*, 2506–2513.

Ivins, B. E., Pitt, M. L., Fellows, P. F., Farchaus, J. W., Benner, G. E., & Waag, D. M., (1998). Comparative efficacy of experimental anthrax vaccine candidates against inhalation anthrax in rhesus macaques. *Vaccine*, *16*, 1141–1148.

Jackson, S., Mestecky, J., Childers, N. K., & Michalek, S. M., (1990). Liposomes containing anti-idiotypic antibodies: an oral vaccine to induce protective secretory immune responses specific for pathogens of mucosal surfaces. *Infect. Immun.*, *58*, 1932–1936.

Jani, D., Singh, N. K., Bhattacharya, S., Meena, L. S., Singh, Y., Upadhyay, S. N., Sharma, A. K., & Tyagi, A. K., (2005). Studies on the immunogenic potential of plant-expressed cholera toxin B subunit. *Plant Cell Rep.*, *22*, 471–477.

Kamarajugadda, S., & Daniell, H., (2006). Chloroplast-derived anthrax and other vaccine antigens: their immunogenic and immunoprotective properties. *Expert Rev. Vaccines*, *56*, 839–849.

Karaseva, A. V., Candice, W. S. F., Rich, A., Shon, K. J., Zwierzynski, I., Hone, D., Koprows-ki, H., & Reitz, M., (2005). Plant based HIV-1 vaccine candidate: Tat protein produced in spinach, *Vaccine, 23,* 1875–1880.

Kirk, D. D., & McIntosh, K., (2005). Social acceptance of plant-made vaccines: Indications from a public survey. *AgBio Forum., 84,* 228–234.

Kiyono, H., & Fukuyama, S., (2004). NALT-versus Peyer's-patch-mediated mucosal immu-nity. *Nat. Rev. Immunol., 4*(9), 699–710.

Klee, H., (2000). A guide to Agrobacterium binary Ti vectors. *Trends in Plant Science, 5,* 446–451.

Klein, T. M., Wolf, E. D., & Wu, R., & Sanford, J. C., (1987). High velocity microprojectiles for delivery of nucleic acids into living cells. *Nature, 327,* 70–73.

Koprowski, H., (2005). Vaccines and sera through plant biotechnology. *Vaccine, 23,* 1757–1763.

Koya, V., Mahtab, M., & Stephen, H., (2005). Leppla, and henry daniell, plant-based vaccine: Mice immunized with chloroplast-derived anthrax protective antigen survive anthrax lethal toxin challenge. *Infect Immun., 73*(12), 8266–8274.

Kwon, K. C., Verma, D., & Singh, N. D., (2013). Oral delivery of human biopharmaceuticals, autoantigens and vaccine antigens bioencapsulated in plant cells. *Adv. Drug Deliv. Rev., 65*(6), 782–99.

Lal, P., Ramachandran, V. G., Goyal, R., & Sharma, R., (2007). Edible vaccines: Current status and future. *IJMM, 25,* 93–102.

Landridge, W., (2000). Edible vaccines. *Scientifi. Am., 283,* 66–71.

Lauterslager, T. G., Stok, W., & Hilgers, L. A., (2003). Improvement of the systemic prime/oral boosts strategy for systemic and local response. *Vaccine, 21*(13–14), 1391–1399.

Leon-Baiiares, R., Gonzilez-Ballester, D., Galvin, A., & Fernindez, E., (2004). Transgenic microalgae as green cell-factories. *TIBTECH., 22,* 45–52.

Lico, C., Santi, L., Twyman, R. M., Pezzotti, M., &Avesani, L., (2012). The use of plants for the production of therapeutic human peptides. *Plant Cell Reports, 31,* 439–451.

Lorence, A., & Verpoorte, R., (2004). Gene transfer and expression in plants. *Methods Mol. Biol., 267,* 329–350.

Lorz, H. B., & Baker, J. S., (1985). Gene transfers to cereal cells mediated by protoplast transformation. *Mol. Gen. Genet., 199,* 178–182.

Ma, Y., Lin, S., Gao, Y., Li, M., Zhang, J., & Xia, N., (2003). Expression of ORF2 partial gene of hepatitis E virus in tomatoes and immunoactivity of expressed product. *World J. Gastroenterorol., 9,* 2211–2215.

Mallett, C. P., Hale, T. L., Kaminski, R. W., Larsen, T., Orr, N., Cohen, D., & Lowell, G. H., (1995). Intranasal or intragastric immunization with proteosome-Shigella lipopolysac-charide vaccines protects against lethal pneumonia in a murine model of Shigella infec-tion. *Infect. Immun., 63,* 2382–2386.

Mason, H. S., & Arntzen, C. J., (1995). Transgenic plants as vaccine production systems, *TIBTECH, 13,* 388–392.

Mason, H. S., Haq, T. A., Clements, J. D., & Arntzen, C. J., (1998). Edible vaccine protects mice against Escherichia coli heat-labile enterotoxin LT: potatoes expressing a syn-thetic LT-B gene. *Vaccine, 16,* 1336–1343.

Mason, H. S., Lam, D. M. K., & Arntzen, C. J., (1992). *Proc. Natl. Acad. Sci., USA,* 11745–11749.

Masri, S. A., Rast, H., Hu, W. G., Nagata, L. P., Chau, D., Jager, S., & Mah, D., (2007). Cloning and expression in E. coli of a functional Fab fragment obtained from single human lymphocyte against anthrax toxin. *MolImmunol.*, *44*(8), 2101–1206.

McCabe, D. E., Swain, W. F., Martinell, B. J., & Christou, P., (1988). Stable transformation of soybean Glycine max, by particle acceleration. *Bio-Technology.*, *6*, 923–926.

McGhee, J. R., Mestecky, J., Dertzbaugh, M. T., Eldridge, J. H., Hirasawa, M., & Kiyono, H., (1992). The mucosal immune system: from fundamental concepts to vaccine development. *Vaccine*, *10*(2), 75–88.

Mei, H., Tao, S. U., Yuan-Gang, Z., & Zhi-Gang, A., (2006). Research advances on transgenic plant vaccines. *Acta. Genetica. Sinica.*, *334*, 285–293.

Mendoza, S. R., Huerta, R. R., & Angulo, C., (2015). An overview of tuberculosis plant-derived vaccines. *Expert Rev. Vaccines Early Online*, 1–13.

Mett, V., Lyons, J., Musiychuk, K., Chichester, J. A., Brasil, T., Couch, R., Sherwood, R., Palmer, G. A., Streatfield, S. J., & Yusibov, V., (2007). A plant-produced plague vaccine candidate confers protection to monkeys. *Vaccine*, *25*(16), 3014–3017.

Michel, M. L., & Tiollais, P., (2010). Hepatitis B vaccines: Protective efficacy and therapeutic potential. *Pathol. Biol.*, *58*, 288–295.

Mishra, N., Gupta, P. N., Khatri, K., Goyal, A. K., & Vyas, S. P., (2007). Edible vaccines: a new approach to oral immunization. *Indian J. Biotechnol.*, 283–294.

Mor, T. S., Gomez-Lim, M. A., & Palmer, K. E., (1988). Perspective: edible vaccines: a concept coming of age. *Trends in Microbiology*, *6*, 449–453.

Negrutiu, I., Shilito, R., Potrykus, I., Biasini, G., & Sala, F., (1987). Hybrid gene in the analysis of transformation conditions. *Plant Mol. Biol.*, *8*, 363–373.

Neutra, M. R., & Kraehenbuhl, J. P., (1992). M cell-mediated antigen transport and monoclonal IgA antibodies for mucosal immune protection. *Adv. Exp. Med. Biol.*, *327*, 143–150.

Nicholson, L., Gonzalez-Melendi, P., Van Dolleweerd, C., Tuck, H., Pemn, Y., Ma, J. K. C., Fischer, R. Christou, P., & Stoger, E., (2003). A recombinant multimeric immunoglobulin expressed in rice shows assembly-dependent subcellular localization in endosperm cells. *Plant Biotech. J.*, *3*, 115–127.

Ogra, P. L., (2003). Mucosal immunity: some historical perspective on host-pathogen interactions and implications for mucosal vaccines. *Immunol Cell Biol.*, *811*, 23–33.

Opabode, J. T., (2006). Agrobacterium-mediated transformation of plants: emerging factors that influence efficiency. *Biotechnology and Molecular Biology Review*, *1*, 12–20.

Parrott, D. M., (1976). The gut as a lymphoid organ. *Clin. Gastroenterol.*, *52*, 211–128.

Paszkowski, J., Shillito, R. D., Saul, M., Mandak, V., Hohn, T., Hohn, B., & Potrykus, I., (1984). Direct gene transfer to plants. *EMBO J.*, *3*, 2717.

Pharma-Planta launches a pivotal phase I clinical trial of a plant-derived microbicidal protein, (2011). *Plant pharma news*, [http://www.pharma planta.net/index.php?pg=50].

Pittman, P. R., Gibbs, P. H., Cannon, T. L., & Friedlander, A. M., (2001). Anthrax vaccine: short-term safety experience in humans. *Vaccine*, *20*, 972–978.

Pniewski, T., (2013). The twenty-year story of a plant-based vaccine against hepatitis B: Stagnation or promising prospects *Int. J. Mol. Sci.*, *14*, 1978–1998.

Porta, C., Spall, V. E., Findlay, K. C., Gergerich, R. C., Farrance, C. E., & Lomonossoff, G. P., (2003). Cowpea mosaic virus-based chimaeras: Effects of inserted peptides on the phenotype, host range, and transmissibility of the modified viruses. *Virology.*, *310*, 50–63.

Porta, C., Spall, V. E., Lin, T., Johnson, J. E., & Lomonossoff, G. P., (1996). The development of cowpea mosaic virus as a potential source of novel vaccines. *Intervirology.*, *39*, 79–84.

Potrykus, I., Paszkowski, J., Saul, M. W., Petruska J., & Shillito, R. D., (1985). DNA transfer to protoplasts. *Mol. Gen. Genet.*, *1*, 169.

Raney, T., (2006). Economic impact of transgenic crops in developing countries. *Curr. Opin. Biotechnol.*, *17*, 174–178.

Reich, T. J., Iyer V. N., Scobie, B., & Miki, B. L., (1986). Efficient transformation of alfalfa protoplasts by the intranuclear microinijection of Ti plasmids. *Bio-Technology*, *4*, 1001–1004.

Richter, L. J., Thanavala, Y., Arntzen, C. J., & Mason, H. S., (2000). Production of hepatitis B surface antigen in transgenic plants for oral immunization. *Nat. Biotechnol.*, *18*, 1167–1171.

Rigano, M. M., Dreitz, S., Kipnis, A. P., Izzo A. A., & Walmsley, A. M., (2006). Oral immunogenicity of a plant-made, subunit, tuberculosis vaccine. *Vaccine*, *245*, 691–695.

Rigano, M. M., Sala, F., Arntzen, C. J., & Walmsley, A. M., (2003). Targeting of plant-derived vaccine antigens to immunoresponsive mucosal sites, *Vaccine*, *21*, 809–811.

Rodriguez, L. L., & Gay, C. G., (2011). Development of vaccines toward the global control and eradication of foot-and-mouth disease. *Exp. Rev. Vaccines.*, *10*, 377–387.

Rubio-Infante, N., Govea-Alonso, D. O., Alpuche-Solís, A. G., Garcia-Hernandez, A. L., Soria-Guerra, R. E., et al., (2012). A chloroplast-derived C4V3 polypeptide from the human immunodeficiency virus HIV, is orally immunogenic in mice. *Plant Mol. Biol.*, *78*, 337–349.

Ruf, S., Hermann, M., Berger, I. J., Carrer, H., & Bock, R., (2001). Stable genetic transformation of tomato plastids and expression of a foreign protein in fruit. *Nat. Biotech.*, *19*, 870–875.

Russell, P., Eley, S. M., Hibbs S. E., Manchee, R. J., Stagg, A. J., & Titball, R. W., (1995). A comparison of plague vaccine, UPS and EV76 vaccine induced protection against Yersinia pestis in a murine model. *Vaccine*, *13*, 1551–1556.

Ryan, E. T., Butterton, J. R., Smith, R. N., Carroll, P. A., Crean, T. I., & Calderwood, S. B., (1997). Protective immunity against *Clostridium difficile* toxin A induced by oral immunization with a live, attenuated Vibrio cholerae vector strain. *Infect Immun.*, *65*, 2941–2949.

Rybicki, E. P., (2014). Plant-based vaccines against viruses. *Virol J.*, *11*, 205.

Rybicki, E. P., Chikwamba, R., Koch, M., Rhodes, J. I., & Groenewald, J. H., (2012). Plant made therapeutics: an emerging platform in South Africa. *Biotechnol. Adv.*, *30*, 449–459.

Saeed, A., Kanwal, S., Arsad, M., Ali, M., Shaikh, R. S., & Abubakar, M., (2015). Foot-and-mouth disease: overview of motives of disease spread and efficacy of available vaccines. *J. Anim. Sci. Technol.*, *57*, 10.

Sala, F., Rigano, M. M., Barbante, A., Basso, B., Walmsley, A. M., & Castiglione, S., (2003). Antigen production in transgenic plants: strategies, gene constructs and perspectives. *Vaccine*, *21*, 803–808.

Scholthof, H. B., Scholthof, K. B., & Jackson, A. O., (1996). Plant virus gene vectors for transient expression of foreign proteins in plants. *Annu. Rev. Phytopathol.*, *34*, 299–3231.

Scholthof, K. B. G., Mirkov, T. E., & Scholthof, H. B., (2002). Plant viral gene vectors: biotechnology applications in agriculture and medicine. In: Setlow, J. K. ed., *Genetic Engineering: Principles and Methods*, vol. *24*. Kluwer Academic/Plenum Publishers, New York, pp. 67–85.

Sciutto, E., Fragoso, G., Manoutcharian, K., Gevorkian, G., Rosas-Salgad, G., Hem d ndez-Gonzalez, M. Herrera-Estrella, L., Cabrera-Ponce, J. L., Lapez-Casillas, F., Gonzalez-Bonilla, C., Santigago-Machuca, A., Sanchez, R. J., Goldbaum, F., Alujac, A., & Larraldea, C., (2002). New approaches to improve peptide vaccine against porcine *Taeniasolium Cysticercosis*. *Arch. Med. Res.*, *33*, 371–378.

Shalaby, W. S., (1995). Development of oral vaccines to stimulate mucosal and systemic immunity: barriers and novel strategies. *Clin. Immunol. Immuno. Pathol.*, *742*, 127–34.

Sharma, K. K., Bhatnagar P., & Thorpe, T. A., (2005). Genetic transformation technology: Status and problems. *In Vitro Cell DevBiol Plant*, *41*, 102–112.

Sheng, O. J., & Citovsky, V., (1996). Agrobacterium-plant cells DNA transport: have virulence proteins will travel. *Plant Cell*, *8*, 1699–1710.

Shillito, R. D., & Potrykus, I., (1987). In: *Methods in Enzymology*, Wu, R., Grossman, L. eds., Academic Press, Orlando, Fla., *153*, 283.

Stomp, A. M., Weissinger, A., & Sederoff, R. R., (1991). Transient expression from microprojectile-mediated DNA transfer in pinustaeda. *Plant Cell Reports*, *104*, 187–191.

Streatfield, S. J., & Howard, J. A., (2003). Plant-based vaccines. *Int. J. Parasitol.*, *33*, 479–493.

Streatfield, S. J., (2006). Mucosal immunization using recombinant plant-based oral vaccines. *Methods*, *38*, 150–157.

Streatfield, S. J., (2007). Approaches to achieve high-level heterologous protein Heterologous protein production in plants. *Plant Biotechnology Journal*, *5*, 2–15.

Streatfield, S. J., Jilka, J. M., Hood, E. E., Turner, D., D., Bailey, M. R., & Mayor, J. M., (2001). Plant based vaccines unique advantages. *Vaccine*, *19*, 2742–2748.

Streatfield, S. J., (2005). Oral hepatitis B vaccine candidates produced and delivered in plant material. *Immunology and Cell Biology*, *83*, 257–262.

Sun, M., Qian, K. X., Su, N., Chan, H. Y., Liu, J. X., & Shen, G. F., (2003). Foot-and-mouth disease virus VP1 protein fused with cholera toxin B subunit expressed in *Chlamydomonas reinhardtii* chloroplast. *Biotechnol. Lett.*, *25*, 1087–1092.

Tacket, C. O., & Mason, H. S., (1999). A review of oral vaccination with transgenic vegetables. *Microbes. Infect.*, *l*, 777–783.

Tacket, C. O., Mason, H. S., & Losonsky, G., (2000). Human immune responses to a novel Norwalk virus vaccine delivered in transgenic potatoes. *J. Infect. Dis.*, *1821*, 302–305.

Tacket, C. O., Mason, H. S., & Losonsky, G., (1998). Immunogenicity in humans of a recombinant bacterial antigen delivered in a transgenic potato. *Nat. Med.*, *45*, 607–609.

Targett, G. A. T., (1989). Status of malaria vaccine research, *J. Royal Society Med.*, *17*, 52–56.

Teli, N. P., & Timko, M. P., (2004). Recent developments in the use of transgenic plants for the production of human therapeutics and biopharmaceuticals. *Plant Cell, Tissue and Organ Culture*, *79*, 125–145.

Thanavala, Y., Mahoney, M., & Pal, S., (2005). Immunogenicity in humans of an edible vaccine for hepatitis B. *Proc. Natl. Acad. Sci. USA*, *1029*, 3378–82.

Trivedi, P. K., & Nath, P., (2004). MaExpl, an ethylene-induced expansin from ripening banana fruit. *Plant*, *167*, 1351–1358.

Twyman, R. M., Schillberg, S., & Fischer, R., (2005). Transgenic plants in the biopharmaceutical market. *Expert. Opin. Emerg. Drugs*, *10*, 185–218.

Van, D. V. L., Herrington, D. A., Murphy, J. R., Wasserman, S. S., Formal, S. B., & Levine, M. M., (1990). Specific immunoglobulin a secreting cells in peripheral blood of humans following oral immunization with a bivalent Salmonella typhi–Shigellasonnei, vaccine or infection by pathogenic *S. sonnei*. *Infect. Immun.*, *58*, 2002–2004.

Waldor, M. K., & Mekalanos, J. J., (1996). "Progress toward live-attenuated cholera vaccines," In: *Mucosal Vaccines*, Kiyono, H., Ogra, P. L., & McGhee, J. R., eds., Acedemic press, *San Diego*.

Walmsley, A. M., & Arntzen, C. J., (2003). Plant cell factories and mucosal vaccines, *Curr. Opin. Biotechnol., 14,* 145–150.

Wang, J. Y., & Roehrl, M. H., (2005). Anthrax vaccine design: strategies to achieve comprehensive protection against spore, bacillus, and toxin. *Med. Immunol., 4,* 4.

Wang, L., Goschnick, M. W., & Coppel, R. L., (2004). Oral immunization with a combination of *Plasmodium yoeliimerozoite* surface proteins 1 and 4/5 enhances protection against lethal malaria challenge, *Infect. Immunol., 72,* 6172–6175.

Warzecha, H., & Mason, H. S., (2003). Benefits and risks of antibody and vaccine production in transgenic plants. *J. Plant Physiol., 1607,* 755–764.

Webster, D. E., Thomas, M. C., Pickering, R., Whyte, A., Dry, I. B., Gorry, P. R., & Wesselingh, S. L., (2005). Is there a role for plant-made vaccines in the prevention of HIV/ AIDS? *Immunol. Cell Biol., 833,* 239–247.

Youm, J. W., Won, Y. S., Jeon, J. H., Ryuc, C. J., Choi, Y. K., Kimb, H. C., Kime, B. D., Joung, H., & Kim, H. S., (2007). Oral immunogenicity of potato-derived HBsAg middle protein in BALB/c mice. *Vaccine, 25,* 577–584.

Yusibov, V., Hooper, D. C., & Spitsin, S. V., (2002). Expression in plants and immunogenicity of plant virus-based experimental rabies vaccine. *Vaccine, 20*(25–26), 3155–3164.

Yusibov, V., Shivprasad, S., Turpen, T. H., Dawson, W., & Koprowski, H., (1999). Plant viral vectors based on tobamoviruses. *Curr. Top. MicrobiolImmunol., 240,* 81–94.

Zupan, J. R., & Zambryski, P. C., (1995). Transfer of T-DNA from Agrobacterium to the plant cell. *Plant Physiology., 107,* 1041–1047.

CHAPTER 14

BREEDING FOR NUTRITIONAL QUALITY ENHANCEMENT IN CROPS: BIOFORTIFICATION AND MOLECULAR FARMING

RAHUL PRIYADARSHI,[1] HITENDRA KUMAR PATEL,[2] and ANIRUDH KUMAR[3]

[1]ICAR–Indian Institute of Rice Research, Hyderabad – 500030, India

[2]CSIR–Centre for Cellular and Molecular Biology, Hyderabad – 500007, India

[3]Department of Botany, Indira Gandhi National Tribal University (IGNTU), Amarkantak–484886, Madhya Pradesh, India, E-mail: singhanir@gmail.com

CONTENTS

ABSTRACT

Micronutrient deficiency presents a considerable threat to achieve nutritional security globally. The intensity of the challenge becomes evident from the fact that nearly half of the global population is affected severely by micronutrient malnutrition. Among major micronutrient deficiencies, the most prevalent concern is related to iron (Fe), zinc (Zn), and vitamin A. Various scientific strategies have been suggested to address the challenge of micronutrient deficiency, of which developing nutrient dense food crops remains the most attractive in terms of cost-effectiveness, sustainability, and environmental safety. This strategy known as crop biofortification can be implemented using modern biotechnology and plant breeding approaches. The development of genetically engineered "golden rice" enriched with beta carotene illustrates a well-established example of bio-fortification via biotechnological techniques. In a similar way, breeding for nutritionally rich crop genotypes also offers a potential method to facilitate nutritional enhancement. In this article, we provide an overview on the current status of crop biofortification along with future projections toward the expansion of biofortified produce.

14.1 INTRODUCTION

Global population is increasing rapidly and has been anticipated to increase from current 7.2 billion to 9.6 billion by 2050 (Gerland et al., 2014). Green revolution enabled feeding the humungous population growth. Presently, however, nearly half of the global population is not able to access healthy and balanced foods. In developing nations, a large proportion of the population can afford only staple food such as rice, maize, and wheat. However, nutritious foods like fresh fruits, vegetables, fish, meat, and milk are also required for improving the livelihood of poor people. The lack of essential nutrients is affecting almost one-third of the world population, in particular children below the age of 5 years. Based on a recent report (Maternal and Child Nutrition, 2015), the growth and development of one out of four children worldwide are severely affected by malnutrition. This has led to the creation of a scenario where people are suffering from macro and micronutrient deficiencies despite having surplus of food production, a phenomenon popularly referred to as "Hidden Hunger." There are 49 nutrients required for the proper functioning of the human immune system and good health. At present, greater than 3 billion peoples are suffering from hidden hunger, and it is worth noting that 300,000 people succumb to acute malnutrition across the world globe each year (FAO, 2009). The multiple effects of malnutrition become apparent in the form of physical and mental development, reduced immunity, increased fatigueness/weakness, infertility, and muscle dystrophy. This might also lead to death in case of severity (Stein, 2010). The adverse impact of malnutrition can be seen directly by observing individual health condition, and indirectly, these cause a great loss to the nation. The most common and conspicuous reason that underlies malnutrition is the lack of essential micronutrients in the regular diet (Von Grebmer et al., 2014). Unfortunately, most of the food crops either contain very minute amount of the micronutrient or even completely lack these nutrients. In addition, processes such as milling further reduces the fiber content and nutritional substances such as vitamin A, folic acid, lysine, iron (Fe), zinc (Zn), selenium (Se), and antioxidants. Nevertheless, these substances remain crucial to allow normal growth and development. In recent years, malnutrition has emerged as one of the biggest health problems in African and Asian countries. One of the promising methods to knob the problem of malnutrition is to enhance essential nutrient content in edible part of the staple crops by using modern breeding and biotechnology approaches, the methods collectively defined as biofortification. Concerning breeding nutritionally rich crops, genetic variations for micronutrient content has been observed in many crops. This implies a tremendous scope to improve staple food crops with high values of micronutrients through breeding programs (Hirschi, 2009). In order to enhance the nutritional superiority in the staple food, the genetic and metabolic engineering has to be undertaken as a tool to enhance the quantity of the desired compound to a certain level and to reduce the quantity of competitive or inhibitory compounds. Efforts are needed to uncover novel metabolic pathways for extended amount of micronutrient production, which indeed may serve the purpose to greater extent.

In conjunction, a paradigm change is needed in the legislative regulatory mechanism in order to accept new technologies that hold the potential to maximize the promising opportunities for better nutrition. Nutrition is a solution for sustainable economic development. The three most severe forms of diseases growingly surfacing due to nutrient deficiency are blindness (vitamin A deficiency), anemia (Fe deficiency), and goiter (I deficiency). Some other minerals and vitamins such as Zn, Ca, Se, proteins, folate, carotenoids, fluoride, chlorine, cobalt, etc. are also responsible to cause severe diseases. This article highlights the importance of biofortification and its role in eradication of malnutrition. The present status of micronutrient-rich staple crops, available techniques, strategies, and future directions for developing biofortified crops have also been discussed.

14.2 MODERN TECHNIQUES TO MEASURE MICRONUTRIENT CONCENTRATION

A broad range of techniques exists allowing the precise assessment of micronutrient concentration. Among these, X-ray fluorescence (XRF) spectrometry stands to be the most important that is employed to assess the concentration of micronutrients in particular genotypes. By using this technique, plants having low or high content of micronutrient can be efficiently discriminated. Similarly, X-ray fluorescence spectrometer is another user-friendly and non-destructive technique in which given samples are irradiated with X-ray, and the produced fluorescence is then used to define the quantity and nature of compounds contained in the samples. Compared to atomic absorption spectrometry (AAS) and inductively coupled plasma mass spectrometry (ICP-MS), the quantification of Fe and Zn of plant samples can be carried out with less effort and more accuracy. All the three techniques, AAS, ICP-MS, and XRF, essentially demand an absolutely dust-free and temperature-controlled environment. All the techniques described above can be used to determine the Fe content; however, differences were often reported while performing quantification. The preference for XRF also stems from its user-friendly protocol.

14.3 MAJOR STRATEGIES TO ENABLE BIOFORTIFICATION

Micronutrient fortified crops can be developed using plant breeding and modern biotechnology approaches. This technology has many advantages over conventional time-consuming and labor-intensive technology. Staple food is consumed in greater quantities by a large fraction of the society in the developing and low-income countries. Therefore, developing staple food with micronutrient richness may serve the purpose to a greater extent of meeting demands of quality food. It is expected that biofortified food will reach to majority of the low-income families in near future. Another benefit of biofortification is that the entire process demands one-time investment. Once micronutrient-rich crop is developed, the seeds can be distributed across the globe. Interestingly, biofortification is completely different from other methods of making food nutrient rich. In this technology, the plant is developed to make the nutrient-rich seed, which can be used for generations. Interestingly, there is no intermediate step like adding micronutrient during preparation, which makes it cost effective and affordable by even economically weaker sections of the society. The different approaches that are opted to facilitate biofortification of crops are indicated in Figure 14.1.

14.3.1 CONVENTIONAL PLANT BREEDING

Conventional plant breeding remains the most rewarding method to permit crop improvement, which can be deployed to develop food crops remarkably enriched with micronutrients. In breeding technique, existing

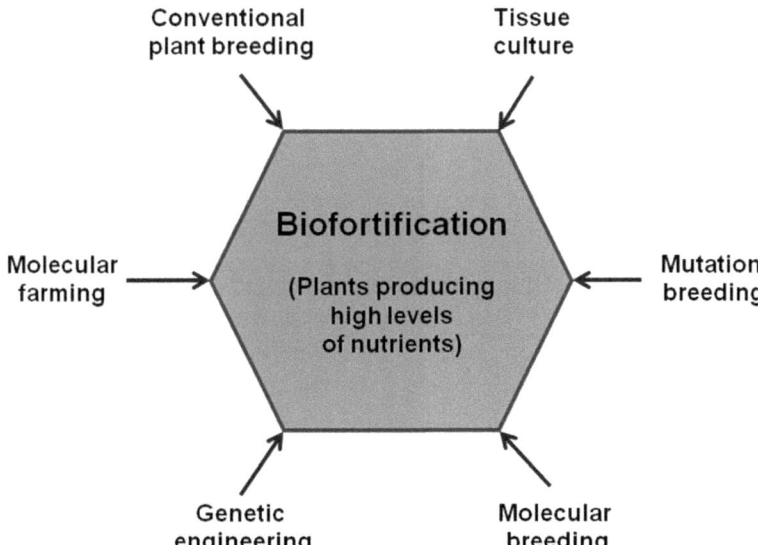

FIGURE 14.1   Different approaches of biofortification.

genotypes are improved through exploiting the genetic resources carrying specific genes available in crossable gene pool, and the technique is capable of accumulating a greater amount of accessible micronutrients. Most of the biofortification work such as the development of beta-carotene, Fe, and Zn-enriched crop is undertaken through plant breeding techniques (Qaim et al., 2007).

The different component traits responsible for enhancing micronutrient density in edible plant parts are reported to be stable across the environment (Hefferon, 2015). Hence, improved traits become more prominent in subsequent generation. Furthermore, it allows for the improvement of crop with enriched nutritional quality as well as eating and cooking quality and physio-chemical traits. The major drawback that largely hampers the efficacy of breeding method is its reliance upon the genetic variability existing among the cross-compatible genotypes. Notwithstanding this, traditional plant breeding approach is considered as the most suitable approach in developing and underdeveloped countries in view of the social, economic, and ethical problems.

14.3.2 TISSUE CULTURE

Tissue culture technique was pioneered by the German botanist Gottlieb Haberlandt in 1902. The tissue culture work in India was initiated by Dr. Panchanan Maheshwari in 1950. The commercial use of this technique was started in 1960 with the orchid multiplication by Dr G. M. Morel. Subsequently, a whole plant was developed from a single cell by using tissue culture technique. Now, this technique is popularly being used in many area of plant biology such as production of disease-free potato, banana, cassava, etc. plant material to enhance the commercial value. Tissue culture in combination with embryo rescue has become more useful in transporting genes from wild relatives to cultivated one, which is not possible by conventional plant breeding. Therefore, genetic variability can be improved using tissue culture technique. Tissue culture technique is mostly used to propagate root and tuber plants such as potato and cassava. New rice for Africa (NERICA), a cross product of Asian and African rice with greater produce and tolerance to water stress, has been developed successfully using this technology.

14.3.3 MUTATION BREEDING

Mutation breeding is used to improve the grain quality and agronomic traits of many crops. The genetic base is expanded through introducing mutations by irradiation or chemical mutagenesis. Presently, more than 3,200 mutants have been created in 210 plant species across 70 countries of the world (Mutant Varieties Database; https://mvd.iaea.org/). Among these, 1,950 are in Asia, 955 are in Europe, 68 are in Africa, and 250 are in Latin and North America. The majority of the European and American mutants contain mutations in flowers; however, the main products of mutants in Asia are topmost food crops such as wheat, rice, maize, and soybeans (Unnevehr et al., 2007). This technique has many benefits over genetic engineering and conventional approaches, such as cost-effective, quick, ubiquitously applicable, transferrable, non-hazardous, and environmentally friendly.

14.3.4 MOLECULAR BREEDING

Molecular breeding is a modern technique that refers to the combined application of plant biotechnology and breeding for crop improvement. It is based on the genetic breeding process, but insertion of genes is also carried out in order to meet the desired characteristic in targets plants. This led to change in the DNA content of plants to some extent and therefore faces regulatory resistance during release. In the age of high-throughput genotyping and next generation sequencing (NGS), genomics and molecular are expanding at a pace that was never expected previously. It reduces the costs of screening plant tissue material because it uses tightly linked marker for screening of gene of interest, which is supposed to be attached to unique DNA fragment. Molecular breeding can be further used as an approach to speed up and monitor the efforts to move a particular trait of interest in a plant. This technique reduces the duration of a typical crop-development process of varieties by years. The marker can be detected in the early stage of plant development, such as tissue of new seedlings. In this way, the presence or absence of the trait of interest can be determined at an early age. Both public and private institutions are using this technology. Plant breeders can pile several genes of different traits or pyramiding of different genes of the single trait of interest into one genotype. Interestingly, some recessive trait identification has become possible, which was not possible through conventional breeding or other existing techniques.

14.3.5 GENETIC ENGINEERING TO DEVELOP NUTRIENT-RICH CROP GENOTYPES

All plant species differ in their genetic ability to accumulate and deposit the available minerals and nutrients. In the context, genetic engineering (GE) can be undertaken to improve the efficiency of accumulation and deposition in food crops (Borg et al., 2009). It is faster, and the expression of genes in genetically modified organism is often stable. The introduction of new traits into genotypes takes fewer generations than conventional plant breeding approaches (Bouis, 2010). Plants produced using GE are known as genetically modified organisms (GMOs) or transgenic and face the hurdles of GMO regulations. Furthermore, social, ethical, and health issues have always been associated with this technology (Seralini et al., 2007). The precursor of vitamin A from a daffodil plant was transferred using this technique for the "Golden Rice" development. This trait was absent in rice plants, and it could not be transferred by conventional breeding approach because daffodils cannot be crossed with rice plants (Paine et al., 2005). Molecular breeding and GE also have benefits over conventional breeding because they make it tranquil to grow crops with many nutritional traits of interest. High-lysine corn already exists, but molecular breeding and marker-assisted selection can further make it possible to improve corn with more than three other essential amino acids.

14.3.6 MOLECULAR FARMING

The term molecular farming or metabolic engineering collectively represent the molecular and genetics techniques that involve synthesis of commercial products in plants. Metabolic engineering of transgenic plants is gaining importance in recent years as an attractive alternative to animals and microorganisms for the manufacture of biologically important products. This technology is very useful for the commercial-scale production of phytochemicals and proteins through genetic modification of crops. The majority of the third world country cannot pay for the expensive therapeutics produced by the current methods. Hence, we must produce new cheaper therapeutics than existing drugs. This technique could put forward a worthwhile alternative for the increasing need of biopharmaceuticals. Therapeutics developed using plant is nontoxic, inexpensive, easily stored, and can be profusely produced. Moreover, regulations related to the bio-safety of this emerging field is also easier (Ahmad, 2014). It offers an approach to deal with some of the existing difficulties through engineering plants in a way such that they synthesize additional products that can alleviate malnutrition to some magnitude.

14.4   STATUS OF BIOFORTIFICATION OF FOOD CROPS

14.4.1   BIOFORTIFIED RICE

Nearly three million preschool-aged children face eye-related problems caused by vitamin A deficiency per year. About half million of them become permanently blind, while the remaining are reported to die. The precursor molecule of vitamin A is β-carotene, which is not found in the major staple food such as rice (Christou et al., 2004). Therefore, people depending heavily on rice-based diets always remain at the risk of vitamin A deficiency. The mechanism of β-carotene synthesis is active only in leaves. In view of the above, golden rice project (GRP) was initiated to introduce the β-carotene biosynthesis pathway into rice endosperm. The addition of two genes, a bacterial phytoene desaturase (*crtI*) from bacterium *Erwinia uredovora* and a plant phytoene synthase (*psy*) from daffodil, into the rice genome turned on the synthesis of β-carotene in rice grains (Tang et al., 2009). According to Paine et al. (2005), golden rice engineered with β-carotene is capable to produce up to 37 µg/g β-carotene of dry rice. The name golden rice was given due to its grain color attained owing to the greater abundance of β-carotene. Golden rice producing β-carotene in polished rice ensures that consumers are free from vitamin A deficiency (Potrykus, 2010). The most recent variant of the golden rice technology, which is known as GR2, exploits two provitamins A pathways, where the phytoene synthesis gene of daffodil (used in GR1) was substituted with an analogous maize gene in GR2 (Shumskaya et al., 2013).

Reported to be naturally low in rice, the Fe content is restricted to the aleurone layer. Furthermore, milling and polishing cause a substantial decrease in its concentration, extending up to 80% (Brinch et al., 2007). The levels of Fe and Zn are reported to range from 6.3–24.4 µg/g and 13.5–58.4 µg/g, respectively, in diverse rice genotypes (Babu, 2013). Rice biofortification for Fe is based on overexpression of the Fe storage protein such as ferritin. The ferritin gene has been sequenced and isolated from various plant species including French bean, pea, soybean, and maize. Apart from playing a role in the synthesis of Fe protein complexes such as ferredoxin and cytochromes, ferritin also protects the cells from free radicals damage generated by interactions of iron and dioxygen. Interestingly, the overexpression of the ferritin gene alone in rice ensure up to 3-fold increase in Fe concentration, whereas ferritin and nicotianamine synthase when expressed together in rice have been reported to cause 6-fold enhancement (Goto et al., 1999; Lucca et al., 2001; Wirth et al., 2009).

GE of rice crop has also been carried out in order to tackle undernourishment including folate deficiency and essential amino acids. For instance, transgenic rice expressing *Arabidopsis* GTP-cyclohydrolase I (GTPCHI) and amino deoxychorismate synthase (ADCS) was reported to increase folate biosynthesis (Storozhenko et al., 2007). In a similar manner, transgenic rice expressing essential amino acid was also developed using RNAi-based silencing technology.

More recently, biofortified rice with 30% higher Zn content has also been developed and released for cultivation in Bangladesh (Chowdhury, 2014; Harvest Plus, 2014a). Zn-biofortified rice store Zn content in the endosperm in place of the aleurone layer; thus, it is not lost in polishing processes. Phytic acid (PA) is known as a notorious inhibitor of zinc and other minerals, which impedes its absorption in the gastrointestinal tract. PA is found to be present in rice, wheat, and corn which causes serious nutritional consequences, and research is in progress to reduce its content (Brnić et al., 2014). Interestingly, mutation breeding has led to the creation of low PA mutants in cereal crops including wheat, rice, and maize (Howe et al., 2006).

14.4.2 BIOFORTIFIED MAIZE

Maize is one of the primary staple foods in sub-Saharan Africa (SSA). It is consumed by 1.2 billion plus people of SSA and Latin America. Still, more dependence on maize as a diet increases undernourishment and vitamin deficiency, which has led to increase in diseases such as kwashiorkor and night blindness. Therefore, biofortification of maize is being carried out with β-carotene and several other essential micronutrients in order to promote healthy wellbeing. Biofortified maize, which has a high amount of three vitamins, namely ascorbate, folate, and β-carotene in the endosperm, has been developed through metabolic engineering (Mugode et al., 2014). The quantity of folate, ascorbate, and β-carotene was reported to enhance by 2-, 6-, and 169-fold, respectively, in triple vitamin-fortified maize as compared to traditionally grown crops. According to Gannon et al. (2014), high β-carotene maize is as good as artificial supplements. High yielding β-carotene orange maize variety was released in Zambia in 2012. A high impact on nutrition was shown by biofortified maize (De Moura et al., 2014). In multivitamin corn, three different metabolic pathways have been amended concurrently (Zhu et al., 2008; Naqvi et al., 2009, 2011a, 2011b). The speedy development of nutritionally improved GM crops may save millions of lives around the globe and may trim down the terrible impact of undernourishment in the poorest areas of the world (Zhu et al., 2007).

The opaque 2 and kernel modifiers are collectively known as quality protein maize (QPM). It contains 3.5 g of lysine per 100 g of protein. Normally, maize kernel contains 90% starch and 10% protein. Out of this, 70% are composed of various kinds of alcohol-soluble prolamins known as zeins. Polymerase chain reaction-based tightly linked microsatellite markers for the opaque2 gene and phenotypic selection markers for kernel virtuousness have been developed at the International Maize and Wheat Improvement Center (CIMMYT), Mexico. The conversion of normal maize lines into QPM line using these markers is simple, rapid, accurate, efficient, and cost-effective (Gupta et al., 2009). Some of the markers utilized extensively are umc1066, phi057, and phi112. Surprisingly, 20-fold upsurge in the tocotrienol level was observed when the Barley HGGT (homogentisic acid geranylgeranyl transferase) gene was overexpressed in maize seeds. HGGT catalyzes an analogous reaction to HPT, which is very specific for GGDP (geranylgeranyl diphosphate), whereas HPT uses PDP (phytyl diphosphate) as its prenyl substitute. Transgenic plants study expressing the HGGT gene suggests that HGGT has a strong affinity for GGDP substrate rather than PDP. HGGT also leads to the accumulation of vitamin E antioxidants such as tocotrienols, principally as γ-tocotrienols in transgenic tobacco calli and *Arabidopsis* leaves (Cahoon et al., 2003).

14.4.3 BIOFORTIFIED WHEAT

The worldwide second largest production of cereal crop is wheat (*Triticum* sp.). It contributes 28% of the world's food requirement. Wheat biofortification has been achieved by several ways including the employment of fertilizers and growth promoting agents of plant in cultivation practices, conventional plant breeding methods, and advanced biotechnology approaches. The foliar and soil application of iron, nitrogen, zinc, and fertilizers have been found to be one of the important approach that has led to enrichment of wheat grains with iron and zinc content (Kutman et al., 2010; Aciksoz et al., 2011; Zhang et al., 2012). Biofortification using microbial inoculation have also been found promising for wheat. The inoculation of rhizobacteria and cyanobacteria improved the wheat grains containing higher amount of protein and micronutrients, such as zinc, iron, copper, and manganese (Rana et al., 2012). As part of conventional breeding programs, several wheat breeding lines have been screened and used for zinc and iron biofortification (Rawat et al., 2009). The QTLs associated with zinc, iron, and protein biofortification have also been identified (Xu et al., 2012). Furthermore, biotechnology-based RNAi technology has been used for reducing the levels of gliadins in wheat. More concentration of gliadins in wheat has led to celiac disease. Likewise, the use of GE helps in increasing the level of limited free lysine in wheat. It is an essential amino acid. For the segment of population that is gluten sensitive or intolerant, and for those who utilize wheat as a staple food, biofortified wheat provides more options and helps in nutritional food security (Hefferon, 2015).

14.4.4 BIOFORTIFIED PEARL MILLET

Pearl millet (*Pennisetum glaucum*) cultivation covers an approximate 31 million hectares worldwide. More than 30 countries of the globe including Africa, Latin America and arid and semi-arid tropical and subtropical regions of Asia cultivate pearl millet. It is a very nutritious and a major staple cereal food for poor farming communities. In terms of nutritional content/quality as well as digestibility, pearl millet is closest to maize and is generally superior to sorghum. Condensed polyphenols are not highly present in Pearl millet. The content of tannins present in sorghum decreases digestibility (Dykes et al., 2006). Grain of Pearl millet is affluent in essential micronutrients such as iron and zinc, thus giving it a complete amino acid profile over crops like sorghum. Considering all these qualities, it is a key contributor to dietary protein, iron, and zinc intake among a diversity of rural populations in India and SSA (Kodkany et al., 2013).

Recently, a biofortified pearl millet using conventional plant breeding approach has been produced by International Crops Research Institute for the Semi-Arid Tropics (ICRISAT), Andhra Pradesh, India. It contain up to 90 μg iron/g when compared to the standard pearl millet containing 36-50 μg iron/g (Rai et al., 2013). A high-iron content variety ICTP-8203Fe of pearl millet was developed from ICTP 8203, an open-pollinated variety. It was again developed by ICRISAT and released in 2012 as Dhanshakti in Maharashtra, India. ICRISAT has also developed the biofortified pearl millet hybrid ICMH 1201 (Shakti-1201) that was released in 2014. Currently, several hybrid lines are under national trials in India, and some of them are expected to be released for cultivation soon.

14.4.5   BIOFORTIFIED SORGHUM

Sorghum is a chief cereal crop and fifth most produced crop worldwide, after maize, rice, wheat, and barley. It is the main food for more than 500 million people and cultivated over 30 countries across the world. Its farming is more common in tropical and subtropical regions of the globe, which is a region of hot climate. Moreover, sorghum is a major source of carbohydrates for local diets of over 300 million people in SSA. However, it lacks a key nutrient: vitamin A. In order to enhance the micronutrient contents and their bioavailability, Africa Harvest Biotechnology Foundation International (Africa Harvest) along with DuPont Pioneer and Bill and Melinda Gates Foundation (BMGF) initiated the Africa Biofortified Sorghum (ABS) project. They developed transgenic sorghum that has reduced phytate (35–80%), improved protein profile, and elevated levels of provitamin A (up to 21 ppm). Transgenic plants are at present in greenhouse trials and predicted to be released in 2018. Sorghum has poorer comparative bioaccessibility in transgenic varieties. However, the higher content of carotenoids in the sorghum grain has led to improve the overall availability of carotenoids and provitamin A. According to bioavailability studies, when phytate levels are reduced in sorghum, zinc and iron absorption improves by 30–40% and 20–30%, respectively (Kumar et al., 2013).

14.4.6   BIOFORTIFIED TOMATOES

Anthocyanins having excellent antioxidant properties are present in many foods and play a significant role in reducing the threat of some vital diseases, including cancer. Very few people consume the suggested quantity of fruits and vegetables. The improvement of the level of micronutrient content of tomato holds the possible solution to recover human health. Tomato is a key crop worldwide and rich in nutrients like lycopene, anthocyanins, ascorbic acid, and folate. Still, the levels of these phytonutrients are measured sub-optimal in tomato. By improving the level of such health-promoting compounds in tomatoes, consumers will consume increased nutritional components, which is especially significant when the daily suggested standards are not being met. Keeping that in mind, researchers have used both metabolic engineering and conventional breeding approach to construct novel tomato lines that carry useful genes, which help in the improvement of micronutrient content in tomato fruit. Metabolic engineering in tomato has been achieved by one of these four techniques; (i) modification of a key-rate limiting step: this was used to alter the level of ascorbic acid and β-carotene, (ii) heterologous gene expression: this was used to alter the level of ascorbic acid and folate, (iii) silencing of a critical pathway: this was used for anthocyanins, and (iv) increase in the level of transcription factors: this was used to alter anthocyanins and polyphenols levels. Even though metabolic engineering is fast and easy to apply than conventional breeding approach, due to regulatory reasons, conventional breeding approach is more favorable than metabolic engineering. Conventional breeding in tomato is based on the availability of natural genome variations. In the tomato genetic studies, polygenic trait phytonutrients in tomato plant, for example, anthocyanins, lycopene, ascorbic acid, and phenols, which are governed by many genes, have been identified and introduced to cultivated tomato varieties through conventional breeding (Raiola et al., 2014). A summary of tomato biofortification by metabolic engineering and conventional breeding approach is presented in Table 14.1.

14.4.7   BIOFORTIFIED BEAN

The common bean is a legume and vital source of micronutrients across the world, especially in parts of Latin America and Eastern Africa, which is useful for more than 300 million people (Harvest Plus Iron-bean, 2009).

TABLE 14.1 A Summary of Tomato Biofortification by Metabolic Engineering and Conventional Breeding Approach (Adopted and Modified from Raiola et al., 2014)

Approach	Techniques	Gene	Pathway	Natural sources	Potential health benefits
Metabolic engineering	Modification of a key-rate limiting step	GDP-mannose-3,5'-epimerase	Ascorbate	Citrus fruits, leafy vegetables, etc.	Antioxidant, lowering hyper tension, and reduces free radicals, reduces risk of cancer.
		monodehydroascorbate reductase (Mdhar)			
		dehydroascorbate reductase (Dhar)			
	Heterologous gene expression	*Erwinia uredovora*-phytoene desaturase (*crtl*)	β-Carotene	Orange, yellow vegetables, spinach, etc.	Antioxidant, reduces free radicals, reduces risk of cancer.
		Arabidopsis-arabinono-1,4lactone oxidase (ALO)	Ascorbate	Citrus fruits, leafy vegetables, etc.	Antioxidant, lowering hyper tension, and reduces free radicals, reduces risk of cancer.
		Yeast-derived-GDP mannose pyrophosphorylase (GPase)			
		Bacterial GTP-cyclohydrlase I (GCHI)	Ascorbate		
		Arabidopsis-Aminodeoxchorismate synthase (ADCS)	Folate	Vegetables, fruits and fruit juices, nuts, etc.	Encourages cell and tissue growth, reduces miscarriage risk and neural tube defects in the fetus.
		Erwinia herbicola lycopene b-cyclase	Carotenoid	Carrot, various fruits, green vegetables. etc.	Antioxidant, reduces free radicals damage, protects against cancers.
		Grape stylbene synthase	Resveratrol	Grapes, berries, peanuts, etc.	Antioxidant, neutralize free radicals,control aging, etc.

TABLE 14.1 (Continued)

Approach	Techniques	Gene	Pathway	Natural sources	Potential health benefits
	Silencing of a critical pathway	De-etiolated1 (DET-)1	Anthocyans	Berries, wine, grapes, etc.	Moderates hypertension, pyrexia, liver illnesses, dysentery and diarrhoea.
	Over expression of transcription factors	Delia (Del) and Rosea1 (Ros1)	Anthocyans		Moderates hypertension, pyrexia, liver illnesses, dysentery and diarrhoea.
		AtMYB12	Anthocyans	Grape, cacao, fruits, vegetables, etc.	Antioxidants, protect cardiovascular system and skin from uv radiation, etc.
	Association mapping QTL	29 SSR and 15 morpho physiological traits	Anthocyans	Berries, wine, grapes, red/ purple vegetables, etc.	Moderates hypertension, pyrexia, liver illnesses, dysentery and diarrhoea.
Conventional breeding	QTL linkage mapping	*Lyc 7.1* and *lyc 12.1*	Lycopene	Tomato, watermelon, papaya, etc.	Antioxidant, control cancer and cadiovascular diseases.
		Variation during the plant development	Ascorbate, polyphenols, and soluble solid	Berries, grapes, cherry, olive, sorghum, maize, etc	Prevent cancer, reduces high blood pressure, heart disease and diabetes.
			Total phenols	Grape, cacao, fruits and vegetables	Antioxidants, protect cardiovascular system and skin from uv radiation, etc.

The iron concentration in beans is high, and it can be further improved by biofortification approaches. The main limitation of the improvement of iron content in bean by the biofortification process is low absorption capacity of iron. Because of the presence of polyphenol(s) (PP) and phytic acid (PA), the absorption capacity of iron is low in bean. Biofortified bean varieties with high iron content have been distributed to combat malnutrition in Uganda, the Democratic Republic of Congo, and Rwanda. The initial data show that biofortified beans have enhanced the availability of iron content for women in Rwanda (Haas et al., 2005).

14.4.8   BIOFORTIFIED ORANGE FLESH SWEET POTATO

One of the most essential food security crop for small farming households, mainly in Africa is sweet potato. By tradition, in Africa, yellow or white varieties of sweet potato are cultivated as the main source of food, but they either lack or have very minute quantities of vitamin A. In Africa, a biofortified orange flesh sweet potato (OFSP) was introduced to provide sufficient content of vitamin A. It contains high quantity of β-carotene, a building block for vitamin A. It is a drought-tolerant and high-yielding variety. According to an analysis, following the boiling of potato in meal preparation, 75% of the β-carotene content is retained (Harvest Plus, 2014a). Improved nutritional impacts (higher vitamin A and β-carotene status) have been widely recognized among consumers for these biofortified OFSP (Van Jaarsveld et al., 2005; Hotz et al., 2012; Jamil et al., 2012). Thirty-one varieties of orange flesh sweet potatoes were released in eight African countries since 2009 (Harvest Plus 2014b).

14.4.9   BIOFORTIFIED CASSAVA

Cassava (*Manihot esculenta*), a starch-rich root crop, is consumed by more than 700 million people worldwide, and of these, 250 million of African population rely on cassava as their staple food. The daily fixed diet constraint based on cassava provides less than 30% protein and about 10–20% of zinc, iron, and vitamin A. The plan of Bio Cassava Plus (BC+) has been introduced in Africa that employed modern biotechnologies techniques to deliver genetically modified cassava with improved levels of iron, zinc, vitamin A, and protein. This approach is helpful for improving health of Africans (Sayre et al., 2011). Cassava rich with β-carotene has been produced through introduction of two transgenes: *crtB* phytoene-synthase gene isolated from *Erwinia* and 1-deoxyxylulose-5-phosphate synthase (*dxs*) gene isolated from *Arabidopsis*. Cassava varieties rich with β-carotene are called yellow or golden cassava. Consuming yellow cassava has been shown that significantly high level of vitamin A in children (Talsma, 2014). These biofortified cassava varieties were at first released in 2013 in Nigeria, and at present, it was planted by more than 500,000 farmers (Harvest Plus, 2014b).

14.4.10   BIOFORTIFIED POTATO

Potato (*Solanum tuberosum*) is a chief non-cereal food crop. It lacks two important sulfur-containing amino acids cysteine and methionine, and essential amino acids lysine and tyrosine. Sequentially, for the improvement of essential amino acids in potato, the *amA1* seed albumin gene was isolated and introgressed from *Amaranthus hypochondriacus* (Chakraborty et al., 2010). The amA1 protein is non-allergenic with high content of essential amino acids.

14.4.11   SOYBEANS WITH IMPROVED OILS

Soybeans (*Glycine max*) are an essential crop rich in oil and protein content. The high-oleic soybean variety was developed by Pioneer Hi-Bred. It is stable at even high cooking temperatures and is better for consumer's health. The modified oil lacks trans fats and having 20% less saturated fat than the wild parent. This variety was developed by introgression of a segment of coding region of omega-6 desaturase gene-1 (FAD2-1) into the soybean genome. Its insertion downregulates the function of native omega-6 desaturase gene and produce high level of oleic oil. In USA and Canada, field and oil testing is still ongoing and commercialization is anticipated soon. The genetically engineered stearidonic acid (SDA) omega-3 soybeans that are rich in omega-3 fatty acids (heart-healthy fatty acids) have developed by Monsanto Company, in collaboration with Solae. Oil from SDA soybeans has the potential to benefit consumers because the availability of good dietary sources of omega-3 fatty acids. Use of SDA soybean oils may reduce the pressure on another source of the omega-3 fatty acids, i.e., fish populations that are currently being overharvested. This variety is still under field trials. and its commercialization plan has not been announced yet.

14.4.12 BIOFORTIFIED CHICKPEA

Chickpea (*Cicer arietinum* L.) is a key source of carbohydrates, proteins, vitamins, and various micronutrients. It is the second-most essential pulse crop subsequent to common bean and staple diet of millions of people of developing countries. High concentrations of selenium (Se) are stored in the grain of chickpea. The concentration of selenium-methionine in chickpea seeds is increased either through soil or a foliar application of selenium-containing fertilizer which is used as a foliar application (Poblaciones et al., 2014). The genetics of iron and zinc content in chickpea seeds was performed by genome-wide association analysis with the help of single nucleotide polymorphisms (SNPs). It is advocated that the level of zinc and iron significantly varies in chickpea germplasm. The high zinc and iron germplasm lines would be used as donors for the development of chickpea varieties with better micronutrient content (zinc and iron) through breeding techniques.

14.4.13 BANANA BIOFORTIFICATION

Bananas are the staple food in Uganda and are as a key source of starch. Like many other starchy staple foods such as rice, maize, etc., banana also contains low amount of vitamin A, iron, and protein and thus have the low level of nutrients. Micronutrient deficiencies related to bananas as the staple food is not common in the highlands of Kenya and Tanzania as well as Burundi, Uganda, Rwanda and eastern areas of the Democratic Republic of Congo. Cultivated bananas are nearly sterile; hence, genetic improvement of bananas for nutrient enhancement through conventional breeding has been a major challenge, and genetic engineering was used as an alternative strategy for the biofortification of banana. The banana project "Banana21" was jointly started by the researchers from the Australia and the National Agricultural Research Organization (NARO), Queensland University of Technology (QUT), Uganda, to improve the micronutrient content of East African Highland bananas (EAHB) in 2005. The project expenditure was totally covered by Bill and Melinda Gates Foundation. Phytoene synthase and ferritin genes were introduced from Asupina banana and soybean, respectively, in order to produce transgenic banana producing high amounts of vitamin A and iron, respectively. These iron and vitamin A transgenic banana lines were introduced to field trials in 2008 in Australia at Queensland and later, in 2010 in Uganda and Rwanda (http://www.banana.go.ug/index.php/banana-biofortification). A super banana with enhanced vitamin A developed at Queensland University in Australia has undergone clinical human feeding trials on the Iowa State University campus in Ames, Iowa, in 2015. This biofortified banana is targeted to release in West African countries, where banana is a staple food.

14.4.14   BIOFORTIFIED ORANGES

Blood oranges are a rich source of vitamin C, carotenoids, and anthocyanin pigments. These constituents are very important for good health. The anthocyanin pigment of blood oranges is only expressed when the oranges have been exposed to a period of cold conditions during fruit development. This was reported by scientists of the John Innes Centre; they reported a method that can bypass the "cold" trigger. A gene responsible for anthocyanin synthesis has been identified in blood orange called *Ruby*. It functions as MYB transcription activator. The expression of *Ruby* is found to be controlled by retrotransposon. The production of anthocyanin depends on cold condition in blood orange. Therefore, the activation of retrotransposon is considered to be associated with cold stress (Butelli et al., 2012). This result supported that blood oranges with complete anthocyanins could be grown in more commercial orange growing areas like their normal orange cousins.

14.4.15   BIOFORTIFIED FEED CROPS

Adequate and balanced feed for domestic animals and poultry birds has been a long-time problem. The development of biofortified food crop, considering the requirement of livestock animals and birds may improve the existing situation. These crops include alfalfa, maize, soybean, oats, barley etc. (Hefferon, 2015). Alfalfa (*Medicago sativa*) is a perennial flowering plant and cultivated as the main forage crop in many countries. Molecular biology and GE have played a pivotal role in the development of genetically modified animal feed with improved content of amino acids like methionine and cysteine. Two bacterial genes encoding for aspartate kinase and adenylyl sulfate reductase were cloned and introduced into alfalfa plants. These transgenic alfalfa plants produced 30% and 60% more amounts of cysteine and methionine, respectively. In genetically modified alfalfa plants, the contents of some other amino acids like lysine and aspartate were also found to be improved (Tong et al., 2014).

The current trend of animal feed is mostly grain based. Therefore, research in the area of grain-based feed has showed considerable improvement. Transgenic maize crop expressing the cell wall invertase gene has been developed, which has shown great improvement in grain quality and yield. This experiment can be incorporated in other important animal grain-based feed (Li et al., 2013). One of the important tasks is to remove glucan (anti-nutrient compound) from the major forage crops and feeds. It is a polysaccharide that prevents the absorption of important micronutrients in the intestine during digestion. In order to preclude the action of glucan, transgenic maize expressing endo-β-1,3-1,4-glucanase (Bgl7AM) from *Bispora* sp. has been developed. MEY-1 was found to produce a high level of β-glucanase Bgl7AM, which is stable in the normal environment of the digestive tract at pH 1.0–8.0. The simplification of the feed processing procedure through these transgenic plants makes the feed more amenable for livestock consumption (Zhang et al., 2013).

14.4.16 MOLECULAR PHARMING IN PLANTS

Production of proteins or small molecules for medical application by pharmaceutical crops is also at the forefront (Ma et al., 2003, 2005; Ramessar et al., 2008; Yao et al., 2015). Molecular biology technique can be used to produce required protein in food crop, popularly known as molecular farming/pharming (MF). This is not new concept; in 1986, the very first human growth promoting hormone was produced in tobacco and sunflower by using molecular biology technique (Barta et al., 1986). Subsequently, hepatitis B antigen (HBsAg) was expressed in transgenic tobacco and was found to be very similar to recombinant technology-derived HBsAg of yeast and human serum (Mason et al., 1992). To date, more than 20 proteins produced in plants using MF techniques are in clinical trials, while more than 100 proteins are in the pipeline of expression in different plant species. Some representative examples include ZMApp for Ebola virus, PRX-102 for Fabry disease, vaccinePfs25 VLP for malaria, HAI-05 as H5N1 vaccine, P2G12 antibody for HIV, DPP4-Fc respiratory disease and MERS-CoV corona virus have been expressed in tobacco will be helpful in plant molecular pharming. Similarly, for the treatment of vitamin B12 deficiency, recombinant intrinsic factor was expressed in *A. thaliana* (Yao et al., 2015). Many of these proteins are at different stages of clinical trials, but at least one ELELYSO™ (taliglucerase alfa) (Protalix BioTherapeutics, Karmiel, Israel) has received approval for human treatment by the US Food and Drug Administration (FDA) in 2012. The treatment of Gaucher Disease (GD) has been done by enzyme replacement therapy using ELELYSO™; a recombinant taliglucerase alfa protein synthesized in carrot tissues.

Even though, there are some production challenges of plant molecular farming (PMF) in plants, such as purification and hurdles with downstream processing, low yield, and plant glycosylation, yet plants systems are promising in order to achieve the production of pharmaceutical proteins on a large scale when compared to other systems like mammalian cells or bacteria. The naturally present pigment anthocyanins in many plants act as detoxifying agents in the human body. Plant biologists are now looking at ways to upsurge the amounts of anthocyanins in vegetables and fruits.

14.5 CONCLUDING REMARKS

Malnutrition is one of the major problems that obstruct the pace of attaining food/nutritional security worldwide. Concerning biofortification, conventional breeding and modern biotechnology hold immense relevance to crop modification in order to address the nutritional security problems. Biofortification allows improvement in the nutritional quality of food crop through modern scientific interventions. While creating biofortified genotypes, the agronomic traits are kept unaffected as an essential prerequisite. Owing to its cost-effectiveness and sustainability, this approach has attracted the attention of scientific community for crop improvement of many species. In the coming 10 to 20 years, the world population will reach 8 to 9 billion. Along these lines, the growing problem of malnutrition in the third world is likely to exacerbate. The challenge of producing food crops with higher micronutrient value is getting intense based on to the requirement of the undernourished population. In the modern genomics era, biofortification is technically feasible; it offers sustainable long-term, nutritional security for the world population. This approach also contributes to equitable development in society to address socioeconomic and socio-political need. This technology is an important approach to tackle malnutrition. Besides creating awareness and public acceptance of biofortified food, appropriate public distribution system and increasing access to market would play a fundamental role in reaching the projected goal of food and nutritional security.

ACKNOWLEDGMENTS

Authors gratefully acknowledge the persons who are directly or indirectly involved in the completion of this chapter.

KEYWORDS

- **biofortification**
- **breeding**
- **genetic engineering**
- **micronutrients**
- **molecular farming**
- **nutrition**

REFERENCES

Aciksoz, S. B., Yazici, A., Ozturk, L., & Cakmak, I., (2011). Biofortification of wheat with iron through soil and foliar application of nitrogen and iron fertilizers. *Plant and Soil, 349*(1), 215–225.

Ahmad, K., (2014). Molecular farming: strategies, expression systems and bio-safety considerations. *Czech. J. Genet Plant Breed, 50*(1), 1–10.

Babu, V. R., (2013). Importance and advantages of rice biofortification with iron and zinc. *SAT eJournal/ E. Journal. Icrisat.Org, 11,* 1–6.

Barta, A., Sommergruber, K., Thompson, D., Hartmuth. K., Matzke, M. A., & Matzke, A. J. M., (1986). The expression of a nopaline synthase – human growth hormone chimaeric gene in transformed tobacco and sunflower callus tissue. *Plant Mol. Biol., 6,* 347–357.

Borg, S., Brinch-Pedersen, H., Tauris. B., & Holm, P. B., (2009). Iron transport, deposition and bioavailability in the wheat and barley grain. *Plant and Soil*, *325*(1), 15–24.

Bouis, H. E., & Welch, R. M., (2010). Biofortification-a sustainable agricultural strategy for reducing micronutrient malnutrition in the global south. *Crop Sci.*, *50*, 20–32.

Brinch, P. H., Borg, S., Tauris, B., & Holm, P. B., (2007). Molecular genetics approaches to increasing mineral availability and vitamin content of cereals. *J. Cereal Sci.*, *46*, 308–326.

Brnic, M., Wegmuller, R., Zeder, C., Senti, G., & Hurrell, R. F., (2014). Influence of phytase, EDTA, and polyphenols on zinc absorption in adults from porridges fortified with zinc sulfate or zinc oxide. *J. Nutr.*, *144*, 1467–1473.

Cahoon, E. B., Hall, S. E., Ripp, K. G., Ganzke, T. S., Hitz, W. D., & Coughlan, S. J., (2003). Metabolic redesign of vitamin E biosynthesis in plants for tocotrienol production and increased antioxidant content." *Nat. Biotechnol.*, *21*, 1082–1087.

Chakraborty, S., Chakraborty, N., Agrawal. L., Ghosh, S., Narula, K., Shekhar, S., Kash, S., Naik, P., Pandec, P. C., Chakrborti, S. K., & Datta, A., (2010). Next-generation protein-rich potato expressing the seed protein gene AmA1 is a result of proteome rebalancing in transgenic tuber. *PNAS*, *107*(41), 17533–17538.

Chowdhury, M., (2014). Address of ICN2 by Ms. Matia chowdhury, honorable minister for agriculture, government of people's republic of bangladesh rome, Italy. http://www.Harvestplus.org/sites/default/files/Bangladesh%20 Statement%20at%20ICN2.pdf.

Christou, P., & Twyman, R. M., (2004). The potential of genetically enhanced plants to address food insecurity. *Nut. Research Rev.*, *17*, 23–42.

De Moura, F., Palmer, A., Finkelstein, J., Haas, J., Murray-Kolb, L., Wenger, M., Birol, E., Boy, E., & Peña-Rosas, J. P., (2014). Are Biofortified Staple Food Crops Improving Vitamin A and Iron Status in Women and Children? New Evidence from Efficacy Trials, *Adv Nutr.*, *5*, 56–570.

Dykes, L., & Rooney, L. W., (2006). Sorghum and millet phenols and antioxidants. *J. Cereal Sci.*, *44*, 236–251.

Food and Agriculture Organization of the United Nations (FAO), (2009). http://www.fao.org/home/en/.

Gannon, B., Kaliwile, C., Arscott, S., Schmaelzle, S., Chileshe, J., Kalungwana, N., Mosonda, M., Pixley, K., Masi, C., & Tanumihardjo, S. A., (2014). Biofortified orange maize is as efficacious as a vitamin A supplement in Zambian children even in the presence of high liver resources of vitamin A: a community-based, randomized placebocontrolled trial. *Am. J. Clin. Nutr.*, *100*, 1541–50.

Garland, P., Raftery, A. E., Sevcikova, H., Li, N., Gu, D., Spoorenberg, T., Alkema, L., Fosdick, B. K., Chunn, J., Lalic, N., Bay, G., Buettner, T., Heilig, G. K., & Wilmoth, J., (2014). World population stabilization unlikely this century. *Science*, *346*, 234–237.

Goto, F., Yoshihara, T., Shigemoto, N., Toki, S., & Takaiwa, F., (1999). Iron fortification of rice seed by the soybean ferritin gene. *Nat. Biotechnol.*, *17*, 282–286.

Gupta, H. S., Agrawal, K., Mahajan, V., Bisht, G. S., Kumar, A., Verma, P., Srivastava, A., Saha, S., Babu, R., Pan, M. C., & Mani, V. P., (2009). Quality protein maize for nutritional security rapid development of short duration hybrids through molecular marker assisted breeding. *Curr. Sci.*, *96*(2), 1–12.

Haas, J., Beard, J., Murray-Kolb, L., Del Mundo, A., Felix, A., & Gregorio, G., (2005). Iron-biofortified rice improves the iron stores of nonanemic Filipino women. *Jou. Nutr.*, *135*, 2823–30.

Harvest Plus, Kigali declaration on biofortified nutritious foods. Second global conference on biofortification, kigali, rwanda. http://biofortconf. ifpri.info/files/2014/04/Kigali-Declaration-on Biofortified-Nutritious-Foods-April-9–20142.pdf.

Harvest Plus, Biofortification progress brief: August 2014. Washington, DC. http://www.harvestplus.org/sites/default/files/Biofortification_Progress_BriefsAugust2014_WEB_2.pdf.

Harvest Plus Iron-bean, 2009. [(accessed on 01 March 2016)]. Available online: http://www.harvestplus.org/sites/default/files/HarvstPlus_Bean_Strategy.pdf.

Haberlandt, G., (1902). Culturversuche mit isolierten Pflanzenzellen. S*itz-Ber. Mat. Nat. Kl. Kais. Akad. Wiss. Wien., 111*, 69–92.

Hefferon, K. L., (2015). Nutritionally enhanced food crops, progress and perspectives. *Int J Mol Sci., 16*(2), 3895–3914.

Hirschi, K. D., (2009). Nutrient biofortification of food crops. *Annu. Rev. Nutr., 29*, 401–421.

Hotz, C., Loechl, C., Lubowa, A., Tumwine, J. K., Ndeezi, G., Nandutu Masawi, A., Baingana, R., Carriquiry, A., de Brauw, A., Meenakshi, J. V., & Gilligan, D. O., (2012). Introduction of β-carotene rich orange sweet potato in rural Uganda resulted in increased vitamin A intakes among children and women and improved vitamin A status among children. *Jou. Nutr., 142,* 1871–1880.

Howe, J. A., & Tanumihardjo, S. A., (2006). Carotenoid-biofortified maize maintains adequate vitamin a status in Mongolian gerbils. *Jou. Nutr., 136,* 2562–2567.

Jamil, K., Brown, K., Jamil, M., Peerson, J., Keenan, A., Newman, J., & Haskell, M., (2012). Daily Consumption of Orange-Fleshed Sweet Potato for 60 Days Increased Plasma β-Carotene Concentration but Did Not Increase Total Body Vitamin A Pool Size in Bangladeshi Women. *Jou. Nutr., 142*(10), 1896–1902.

Kodkany, B., Bellad, R., Mahantshetti, N., Westcott, J., Krebs, N., Kemp, J., & Hambidge, K. M., (2013). Biofortification of pearl millet with iron and zinc in a randomized controlled trial increases absorption of these minerals above physiologic requirements in young children. *J. Nutr., 143,* 1489–1493.

Kumar, A. A., Anuradha, K., & Ramaiah, B., (2013). Increasing grain Fe and Zn concentration in sorghum: progress and way forward. *SAT Ejournal, Ejournal. Icrisat. Org, 11,* 1–5.

Kutman, U. B., Yildiz, B., Ozturk, L., & Cakmak, I., (2010). Biofortification of durum wheat with zinc through soil and foliar applications of nitrogen. *Durum Wheat Pasta Symposium, 87*(1), 1–9.

Li, B., Liu, H., Zhang, Y., Kang, T., Zhang, L., Tong, J., Xiao, L., & Zhang, H., (2013). Constitutive expression of cell wall invertase genes increases grain yield and starch content in maize. *Plant Biotechnol J., 11*(9), 1080–1091.

Lucca, P., Hurrell, R., & Potrykus, I., (2001). Genetic engineering approaches to improve the bioavailability and the level of iron in rice grains. *Theor Appl Genet., 102,* 392–397.

Ma, J. K. C., Barros, E., Bock, R., Christou, P., Dale, P. J., Dix, P. J., Fischer, R., Irwin, J., Mahoney, R., Pezzotti, M., Schillberg, S., Sparrow, P., Stoger, E., & Twyman, R. M., (2005). Molecular farming for new drugs and vaccines. Current perspectives on the production of pharmaceuticals in transgenic plants. *EMBO Rep., 6,* 593–599.

Mason, H. S., Lam, D. M. K., & Arntzen, C. J., (1992). Expression of hepatitis B surface antigen in transgenic plants. *Proc. Natl. Acad. Sci., 89,* 11745–11749.

Morel, G., (1963). Le culture *in vitro* du m´erist`eme apical de certaines orchid´ees. *Comptes Rendus de l'Academie des Sciences Paris., 256,* 4955–4957.

Mugode, L., Há, B., Kaunda, A., Sikombe, T., Phiri, S., Mutale, R., Davis, C., Tanumihardjo, S. A., & De Moura, F. F., (2014). Carotenoid retention of biofortified provitamin a maize (*Zea mays* L.) after Zambian traditional methods of milling, cooking and storage. *J. Agric. Food Chem.*, *62*, 6317–6325.

Naqvi, S., Ramessar, K., Farre, G., Sabalza, M., Miralpeix, B., Twyman, R. M., Capell, T., Christou, P., & Zhu, C., (2011a). High value products from transgenic corn. *Biotechnol. Adv.*, *29*, 40–53.

Naqvi, S., Zhu, C., Farre, G., Ramessar, K., Bassie, L., Breitenbach, J., Perez Conesa, D., Ros, G., Sandmann, G., Capell, T., & Christou, P., (2009). Transgenic multivitamin corn through biofortification of endosperm with three vitamins representing three distinct metabolic pathways. *Proc. Natl. Acad. Sci.*, *106*, 7762–7767.

Naqvi, S., Zhu, C., Farre, G., Sandmann, G., Capell, T., & Christou, P., (2011b). Synergistic metabolism in hybrid corn indicates bottlenecks in the carotenoid pathway and leads to the accumulation of extraordinary levels of the nutritionally important carotenoid zeaxanthin. *Plant Biotechnol. J.*, *9*, 384–393.

Paine, J. A., Shipton, C. A., Howells, R. M., Kennedy, M. J., & Vernon, G., (2005). Improving the nutritional value of Golden Rice through increased pro-vitamin A content. *Nat. Biotechnol.*, *23*, 482–487.

Poblaciones, M. J., Rodrigo, S., Santamaría, O., Chen, Y., & McGrath, S. P., (2014). Agronomic selenium biofortification in Triticum durum under Mediterranean conditions: From grain to cooked pasta. *Food Chem.*, *146*, 378–384.

Potrykus, I., (2010). Lessons from the 'Humanitarian Golden Rice' project: regulation prevents development of public good genetically engineered crop products. *Nat. Biotechnol.*, *27*, 466–472.

Qaim, M., Stein, A. J., & Meenakshi, J. V., (2007). Economics of bio-fortification. *Agr. Econ-Blackwell.*, *37*, 119–133.

Rai, K. N., Yadav, O. P., Rajpurohit, B. S., Patil, H., T., Govindaraj, M., Khairwal, I. S., Rao, A. S., Shivade, H., Pawar, V. Y., & Kulkarni, M. P., (2013). Breeding pearl millet cultivars for high iron density with zinc density as an associated trait. *J. of SAT Agric. Res.*, *11*, 1–7.

Raiola, A., Rigano, M. M., Calafiore, R., Frusciante, L., & Barone, A., (2014). Enhancing the health-promoting effects of tomato fruit for biofortified food. *Mediators of Inflamm.*, 1–16.

Ramessar, K., Sabalza, M., Capell, T., & Christou, P., (2008). Corn plants: an ideal production platform for effective and safe molecular pharming. *Plant Sci.*, *74*, 409–419.

Rana, A., Joshi, M., Prasanna, R., Shivay, Y. S., & Nain, L., (2012). Biofortification of wheat through inoculation of plant growth promoting rhizobacteria and cyanobacteria. *Eur. J. Soil Biol.*, *50*, 118–126.

Rawat, N., Tiwari, V. K., Singh, N., Randhawa, G. S., Singh, K., Chhuneja, P., & Dhaliwal, H. S., (2009). Evaluation and utilization of Aegilops and wild Triticum species for enhancing iron and zinc content in wheat. *Genet. Resour. Crop Evol.*, *56*, 53.

Sayre, R., Beeching, J. R., Cahoon, E. B., Egesi, C., Fauquet, C., Fellman, J., Fregene, M., Gruissem, W., Mallowa, S., Manary, M., Dixon, B. M., Mbanaso, A., Schachtman, D. P., Siritunga, D., Taylor, N., Vanderschuren, H., & Zhang, P., (2011). The BioCassava plus program: biofortification of cassava for sub-Saharan Africa. *Annu. Rev. Plant Biol.*, *62*, 251–272.

Séralini, G. E., Cellier, D., & Vendomois, J. S. D., (2007). New analysis of a rat feeding study with a genetically modified maize reveals signs of hepatorenal toxicity. *Arch. Environ. Contam. Toxicol.*, *52*(4), 596–602.

Shumskaya, M., & Wurtzel, E. T., (2013). The carotenoid biosynthetic pathway: Thinking in all dimensions. *Plant Sci., 208,* 58–63.

Stein, A. J., (2010). Global impacts of human malnutrition. *Plant Soil, 335,* 133–154.

Storozhenko, S., De Brouwer, V., Volckaert, M., Navarrete, O., Blancquaert, D., Zhang, G. F., Lambert, W., & Van Der Straeten, D., (2007). Folate fortification of rice by metabolic engineering. *Nat. Biotechnol., 25*(11), 1277–9.

Talsma, E., (2014). Yellow cassava: efficacy of provitamin a rich cassava on improvement of vitamin a status in Kenyan schoolchildren. Dissertation for Wageningen University, Netherlands. http://library.wur.nl/WebQuery/wurpubs/454759.

Tang, G., Qin, J., Dolnikowski, G. G., Russell, R. M., & Grusak, M. A., (2009). Golden Rice is an effective source of vitamin A. *Am. J. Clin. Nutr., 89,* 1776–1783.

Tong, Z., Xie, C., Ma, L., Liu, L., Jin, Y., Dong, J., & Wangm, T., (2014). Co-expression of bacterial aspartate kinase and adenylylsulfate reductase genes substantially increases sulfur amino acid levels in transgenic alfalfa (Medicago sativa L.). *PLoS One., 9,* 883–910.

Unnevehr, L., Pray, C., & Paarlberg, R., (2007). Addressing micronutrient deficiencies: Alternative Interventions and Technologies. *AgBio. Forum., 10*(3), 124–134.

Van Jaarsveld, P., Faber, M., Tanumihardjo, S., Nestel, P., Lombard, C., & Spinnler Benadé, A., (2005). β-Carotene–rich orange-fleshed sweet potato improves the vitamin A status of primary school children assessed with the modified-relative-dose-response test. *Am. J. Clin. Nutr., 81*(5), 1080–1087.

Von Grebmer, K., Saltzman, A., Birol, E., Wiesmann, D., Prasai, N., Yin, S., Yohannes, Y., Menon, P., Thompson, J., & Sonntag, A., (2014). Global hunger index, international food policy research institute, Washington, DC, USA.

Wirth, J., Poletti, S., Aeschlimann, B., Yakandawala, N., Drosse, B., Osorio, S., Tohge, T., Fernie, A. R., Günther, D., Gruissem, W., & Sautter, C., (2009). Rice endosperm iron biofortification by targeted and synergistic action of nicotianamine synthase and ferritin. *Plant Biotechnol. J., 7*(7), 631–44.

Xu, Y., An, D., Liu, D., Zhang, A., Xu, H., & Li, B., (2012). Molecular mapping of QTLs for grain zinc, iron and protein concentration of wheat across two environments. *Field Crops Res., 138,* 57–62.

Yao, J., Weng, Y., Dickey, A., & Wang, K. Y., (2015). Plants as factories for human pharmaceuticals: applications and challenges. *Int. J. Mol. Sci., 16*(12), 28549–28565.

Zhang, Y., Xu, X., Zhou, X., Chen, R., Yang, P., Meng, Q., Meng, K., Luo, H., Yuan, J., Yao, B., & Zhang, W., (2013). Overexpression of an acidic endo-β-*1,* 3–1, 4-glucanase in transgenic maize seed for direct utilization in animal feed. *PLoS One, 8*(12), 1–8.

Zhang, Y. Q., Sun, Y. X., Ye, Y. L., Karim, M. R., Xue, Y. F., Yan, P., Meng, Q. F., Cui, Z. L., Cakmak, I., Zhang, F. S., & Zou, C. Q., (2012). Zinc biofortification of wheat through fertilizer applications in different locations of China. *Field Crops Res., 125,* 1–7.

Zhu, C., Naqvi, S., Breitenbach, J., Sandmann, G., Christou, P., & Capell, T., (2008). Combinatorial genetic transformation generates a library of metabolic phenotypes for the carotenoid pathway in corn. *Proc. Natl. Acad. Sci., 105,* 18232–18237.

Zhu, C., Naqvi, S., Gomez-Galera, S., Pelacho, A. M., Capell, T., & Christou, P., (2007). Transgenic strategies for the nutritional enhancement of plants. *Trends Plant Sci., 12,* 548–555.

PART V

RECENT TECHNIQUES IN ADVANCED MOLECULAR PLANT BREEDING

CHAPTER 15

TERMINATOR TECHNOLOGY (GURT TECHNOLOGY)

D. R. MEHTA[1] and A. K. NANDHA[2]

[1]Associate Professor, Department of Genetics and Plant Breeding, Junagadh Agricultural University, Junagadh–362001, Gujarat, India

[2]PhD Scholar, Department of Biotechnology, Junagadh Agricultural University, Junagadh–362001, Gujarat, India

CONTENTS

15.1   INTRODUCTION

Biotechnology techniques are being applied to plants to produce plant materials with improved composition, functional characteristics, or organoleptic properties. Genetic modifications have produced fruits that can ripen on the vine for better taste and yet have a longer shelf life through delayed pectin degradation (Gross, 1988; Bennett et al., 1989) or altered responses to the plant hormone ethylene (Bleecker, 1989). A recently patented method for preventing plants from producing viable seeds could have a serious implication for farming systems and even biodiversity and food security especially in developing countries.

The ability to modify plant genomes and introduce genes for a specific desired trait into a desired plant gave rise to an array of experiments on several commercially important crops. Several biotech companies developed seeds for plants that are disease-resistant, pest-resistant, herbicide-resistant, or that give high yields. In order to protect their intellectual property rights (IPR), a new technology known as the Genetic Use Restriction Technology (GURT) or Terminator Technology was born. The seeds generated through this technology are called terminator seeds or suicide seeds. The technology was developed under a cooperative research and development agreement between the Agricultural Research Service of the United States Department of Agriculture and Delta and Pine Land Company (the largest cotton seed producing company in the world) in the 1990s; it is not yet commercially available, but tests are currently being conducted in greenhouse in the United States (Anon., 1998a).

The characteristic feature of terminator seeds is their ability to generate plants that give rise to sterile seeds. In simple terms, a farmer buys the seeds and sows them to reap a good harvest of a crop that is genetically modified (GM) to possess a desirable trait. But, the new seeds that are formed in these genetically modified (GM) crops are sterile and cannot be used for the next season. They have to buy the seeds again to grow the next season crop. Advancement in this technology is the development of a genetically engineered crop that yields sterile seeds, but the desired trait that has been engineered will be functional only when an inducer chemical is administered. This inducer chemical needs to be purchased from the respective company. Thus, the farmers may not save the seeds from his harvest but needs to purchase the inducer every year (Dutfield, 2007).

A huge controversy surrounds this technology since its proposition by Monsanto in the 1990s, with claims that it is a new way to rob farmers and make profits. But the proponents argue that they need to make up for the costs incurred in developing the advance technology. Apart from the protection of intellectual property, the biggest advantage conferred by such a technology is that it prevents the genetically altered trait from spreading to wild plants, which is one of the risks involved in the use of GM crops. Nevertheless, the agri-business giant agreed not to commercialize terminator seeds. More than 50 GURT patents are held by universities, governments, and private firms in the US (Shi, 2006).

The technology was discussed during the 8th Conference of the Parties to the United Nations Convention on Biological Diversity in Curitiba, Brazil, during March 20–31, 2006 (Table 15.1).

15.2   TYPES OF GURTS

There are two types of GURTs; variety GURTs (V-GURTs) and trait-specific GURTs (T-GURTs). V-GURTs (also called terminator genes) can render seeds sterile and prevent unauthorized use of genetic material and self-supply of commercial seeds by the farmers. V-GURTs are potential technological solutions to the IPR market failures in developing countries. GURTs could significantly enhance the innovator's ability to capture rents from their innovation, which may result in more investment in plant breeding R&D and higher yielding seeds that are suitable for local growing conditions. T-GURTs involve the external application of inducers to trigger the expression of some specific traits. For example, traits of tolerance to salt or heavy metals could be switched on or off. Traits can be activated at the time of purchase or activators can be purchased for later use (Jefferson et al., 1999). The key to T-GURTs is to switch on or off some target characteristics of a plant through inducible, promoters regulating the expression of the transgenes (FAO, 2001). T-GURTs could be used in all crops to enhance certain value-added traits or to protect the plants in extreme harsh weather.

15.2.1   V-GURTS

V-GURTS are more criticized and even more strongly opposed. Also known as the "terminator technology," this technology is a result of a change in the

TABLE 15.1   Brief History and Progress of Terminator Technology

No.	Year	Event
1	1994	Biotech companies obtain patents on Genetic Use Restriction Technology (GURT)
2	March 03, 1998	Patent on terminator seeds granted
3	March 30, 1998	USDA is strong supporter of terminator technology
4	May 11, 1998	Monsanto announce intent to purchase Delta & Pine Land company
5	1998	Former chairman of FAO Council, Dr. M.S. Swaminathan, speaks out against terminator seeds
6	June 12, 1998	Delta and Pine Land executive: Poor farmers 'locked' in cycle of saving obsolete seed varieties
7	September 14, 1998	Clinton administration clamps down on scientists' research into terminator technology
8	October 10, 1998	Indian government bans terminator seeds
9	October 30, 1998	International research network condemns terminator technology
10	December 2, 1998	Indian government summarizes threat posed by terminator seed
11	January 1999	Dozens of biotech companies developing terminator technologies
12	April 7, 1999	Maurice F. Strong, former secretary general of UNCED speaks out against terminator seeds
13	June 15 – 21, 1999	Biodiversity convention's scientific body caves in under US pressure in passing terminator resolution
14	June 21, 1999	Panama says it's opposed to terminator technology
15	June 24, 1999	Rockefeller foundation president, Dr. Gordon Conway, tells Monsanto to end its research into terminator technology
16	July 20, 1999	USDA and Delta & Pine Land secure new patent for improvements in terminator genetic seed sterilization technology
17	October 4, 1999	Monsanto pledges not to commercialize terminator seeds
18	December 21, 1999	USDA official dismayed at Monsanto decision not to commercialize terminator
19	January 14, 2000	Ghana opposed to terminator technology
20	February 8, 2000	Dr. Jacques Diouf, FAO Director-General against terminator technology
21	May 2000	Scientists call for 5-year ban on GM crops
22	May 15- 26, 2000	Biodiversity convention adopts its scientific advisory body's recommendation on terminator and related technology

TABLE 15.1   (Continued)

No.	Year	Event
23	September 26-28, 2000	UN food and agriculture organization's ethics panel voices strong concerns about the development of GM crops
24	July 6, 2001	USDA and Delta & Pine Land conclude negotiations on terminator patent
25	January 10, 2003	UPOV concludes that terminator technology could be bad for farmers
26	February 19-21, 2003	Biodiversity Convention Expert Panel Considers Impact of Terminator Technology on Small Farmers, Indigenous Peoples, and Local Communities
27	March 13-April 11, 2003	US and Seed Industry Strong Arms UPOV into Removing Criticisms of Terminator Technology from Document
28	November 10-14, 2003	SBSTTA Recommends that Biodiversity Convention Forgoe Action on Terminator Seed Report
29	Before February 7, 2005	Canadian Government will propose lifting of De Facto Ban on Terminator Seeds
30	February 7, 2005	Biodiversity Convention Scientific Body Upholds Recommendation to Ban Terminator Seeds
31	February 11, 2005	Percy Schmeiser speaks out against Terminator Seeds at UN meeting
32	September 26-27, 2005	Andean and Amazon Indigenous Leaders denounce Terminator Seeds
33	January, 2006	Granada: Ad-hoc working group on Art. 8J CBD accept a recommendation paper for Curitiba with "case by case assessment"
34	January 23- 27, 2006	Rich Countries and Seed Company Lobbyists Seek to Undermine Terminator Ban at Meeting of Biodiversity Convention Working Group
35	March 16, 2006	European Parliament Resolution Backs Terminator Ban
36	March 18, 2006	Andean and Amazon Indigenous Groups Call on Syngenta to Abandon Patent on GM Potatoes
37	March, 2006	Curitiba: a lot of protest (over all the international ban terminator campaign) Result: The moratorium remains without limitation
38	March 20, 2006	India: Prime minister, Manmohan Singh sign 500,000 petition denouncing Terminator Technology
39	March 23, 2006	Members of Convention on Biological Diversity Uphold ban on Terminator Seeds
40	March 23-27, 2004	Peasant farmers demonstrate at Convention on Biological Diversity

genetic makeup of a plant cell, whereby plants regenerated from this cell will develop non-viable seeds, which will not germinate in the next generation. The seed of the next generation is sterile, which means that farmers who use it would not be able to save ("brown bag") seeds from this crop to plant in the next vegetation season.

The USDA/D&PL technology involves inserting three transgenes (toxin gene, site-specific recombinase gene, and recombinase repressor gene) into the plant DNA. The genes are connected so that (a) the repressor gene prevents the recombinase gene from functioning, (b) the recombinase gene, if it functions, allows the toxin gene to activate, and (c) the toxin gene produces a toxin that kills the embryo in the seed, so that the seed cannot germinate. The seed producer can control the system by spraying the first-generation seed with a regulator. The regulator then inactivates the repressor gene. Because the repressor gene does not function, the recombinase gene is allowed to do its job, as in step (b) above. If the seed producer wishes to protect the intellectual property embedded in the seeds, he sprays the seeds with the regulator before delivery to the farmers (Shi, 2006).

V-GURTS mechanism relies on the transfer of a combination of three genes: (i) a gene coding for a toxic substance (terminator or lethal gene), which is linked to a blocking sequence preventing the activation of the terminator gene; (ii) a recombinase gene (CRE/LOX). This gene contains the information for a protein that cuts the blocking sequence linked to the toxic gene; and (iii) a repressor gene that contains nucleotide sequence for coding the protein that suppresses the recombinase gene and that could be controlled by an external stimulus (Gupta, 1998). Once the chemical trigger is released, the repressor gene is switched off allowing the recombinase gene to be switched on, which in turn removes the blocking sequence from the terminator gene, resulting in lethal gene expression (Figure 15.1). Toxins produced by the activated terminator gene destroy the embryo, thereby rendering the seeds sterile. This, however, leaves all the other aspects of plant growth unaffected, because the toxic effects stimulated by chemical treatment only occur during the later stages of embryo development and, therefore, cannot adversely affect final yields (Lehmann, 1998).

At least three V-GURT strategies can be distinguished (FAO, 2001). In strategy-I, the seed is fertile by default. The activator is used to induce a disrupter gene that results in sterility in the next generation. The breeder treats the seeds with a chemical inducer when sold; thus, the second-gen-

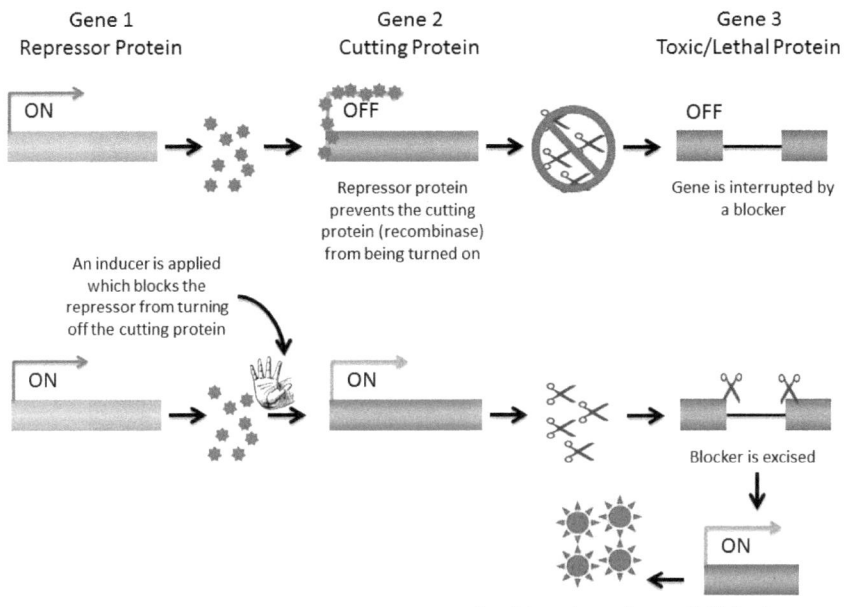

FIGURE 15.1 Switching on and off of gene responsible for GURT technology.

eration seed is fit for consumption but infertile. The USDA/D&PL technology mentioned above follows strategy-I. In strategy-II, the seed is infertile by default; the breeder applies a chemical in all generations to inactivate the disrupter gene that causes the sterility, but not before selling the seeds. Strategy-III focuses on crops produced vegetatively such as roots and tuber crops. Growth can be prevented during storage and it helps to extend shelf life. In this case, the blocking gene is expressed by default until being suppressed by the application of a chemical (Table 15.2).

There are three distinctive strategies on which VGURTs can be based. The first one, described in patent application by Delta & Pine Land/USDA, is based on plant cell containing a gene disruptor that makes embryo formation impossible (FAO, 2001). The second strategy, such as in Zeneca application, is based on reversed process, which means that, in order to suppress activity of nondormant disruptor gene chemicals need to be used. The third application is only applied in the production of vegetatively propagated species to prevent unwanted growth during storage (e.g., root and tuber crops,

TABLE 15.2 Different V-GURTs Used to Prevent Spread of Specific Transgenes

Sr. No.	Trait	Tested Species	Natural (N)or Transgene (T)	Development stage	References
1	**Reduced admixture**	**Seed**			
	Seed shattering	*Arabidopsis*, Rice, *Brassica*	Natural and Transgene	Mature pods	Konishi (2006)
	Seed size	*Brassica*, *Arabidopsis*, Tobacco	Natural and Transgene	Mature seed	Ma et al. (2002); Jofuku et al. (2005); Slade et al. (2006)
2	**Eliminate transgenics**	**Plant**			
	Seed color	Maize, *Brassica*	Natural and Transgene	Mature seed	Krebbers et al. (1999); Mahmood et al. (2006)
	Seed sterility	Tobacco, *Brassica*	Natural and Transgene	Germination	Olive et al. (1999); Schernthaner et al. (2003)
	Conditional lethality	Tobacco, *Arabidopsis*	Transgene	Vegetative growth	Schernthaner et al. (2003) Fabijanski et al. (2006); Nasholm (2006);Erikson et al. (2004)
	Reduced fitness	*Arabidopsis*, Tobacco, B*rassica*	Natural and Transgene	Vegetative growth	Ait-ali et al. (2003);Al-Ahmad and Gressel (2005); Al-Ahmad et al. (2006); Gressel (1999)
3	**Reduce gene movement**	**Pollen**			
	Flowering time	*Brassica*, *Arabidopsis*, Wheat, Barley	Natural and Transgene	Time to flowering	Lemmetyinen et al. (2004);Fu, et al. (2005);Tadege, et al. (2001);Lough, et al. (2006); Fischer, et al. (2006)
	Physical flower structure	Birch	Transgene	Flowering	Lemmetyinen et al. (2004)

TABLE 15.2   (Continued)

Sr. No.	Trait	Tested Species	Natural (N)or Transgene (T)	Development stage	References
	Male sterility	Tobacco, *Brassica*, Maize	Natural and Transgene	Pollen viability	Perez, et al. (2006); Luo, et al. (2005); Burgess, et al. (2002); Kuvshinov, et al. (2005); Kuvshinov, et al. (2001); Hodges, et al. (1999); Fabijanski, et al. (1998); Paul,(2006); Cigan,(2000)
	Maternal inheritance	Soybean, *Arabidopsis*, Tobacco	Natural and Transgene	Chloroplast inheritance	Grevich & Daniell,(2005) Daniell et al. (2005); Kumar, et al.; (2004)

(Reprinted with permission from Hills, Melissa J. et al., (2007) Genetic use restriction technologies (GURTs): strategies to impede transgene movement, Trends in Plant Science , 12 , 4 , 177 – 183.© 2007 Elsevier.)

ornamentals), where the gene that inhibits growth provides variety, and a second gene, which can be made active through the application of a chemical, can restore this growth (Visser et al., 2001).

In addition to this, there are hybrids that are always fertile but have significantly lower performance in second and every other generation. However, hybrids might be infeasible, or ineffective for many self-fertilizing crops such as rice, wheat, soybean, cotton, and horticultural crops. These crops are likely to be the target application of V-GURTS. T-GURTS may in principle be applied to virtually all the crops to protect the enhanced trait.

15.2.2   T-GURTS

T-GURTS are not intended to affect the viability of seeds, but to control the traits. This is achieved by regulating gene expression, which means switching off or on one or more genes conferring a single trait, through the application of chemical inducers (Pendleton, 2004) or by certain conditions such as environmental factor like heat, cold, and increased precipitation. Traits of interest that could be controlled by T-GURTS include male sterility, pest resistance, stress tolerance, nutrient production, seed germination, or flower development (Gupta, 1998).

There are two possible ways in which T-GURTS can be designed either constructing a chain of genes, in which a toxin gene can be programmed to delete/prevent the expression of a "trait gene," or by activating a trait gene by farmers' will when applying an activator compound. The system can be designed so that subsequent generations of the seeds will contain the trait gene, but in an inactive state. For example, insecticidal genes (e.g., Bt toxin gene) under the control of an inducible promoter could remain inactivated until an insect pest outbreak justified the application of a chemical to induce the formation of gene products toxic to insects. The inducer chemical would, of course, be under the control and licensing of the seed company; however, the ultimate "trigger" of this technology could be under the control of producers (Pendleton, 2004).

15.3   MECHANISM INVOLVED IN TERMINATOR TECHNOLOGY

Terminator technology is a transgenic approach that uses a ribosomal inactivating protein (RIP) gene – Gene X (lethal gene), expressed under the control of a tissue specific – late embryogenesis activity (LEA) promoter. The RIP gene is activated by the action of a recombinase (produced by Gene Y) that is activated by an inducer chemical (tetracycline), and its expression is controlled by a repressor (Gene Z). The RIP's expression kills the embryo, and the saved seed fails to germinate. This technology allows the plants to produce good normal seed that can be used for food purposes cannot propagate itself (Sarkar and Ghosh, 2001).

The patented method for terminator technology is based on a gene that produces a protein that is toxic to the plant and therefore, does not allow the seed to germinate. One such gene indicated in the patent is the ribosomal inactivating protein (RIP) gene, which if expressed, does not allow protein synthesis to take place. The gene is placed under the control of LEA promoter permitting RIP to express only during late embryogenesis, thus affecting only the embryo development. This gene (RIP gene) will not express in the first generation, because its expression is blocked through the use of a spacer or a blocking sequence between the promoter and the lethal RIP gene. On either side of the spacer, specific excision sequences are placed that are recognized by a recombinase enzyme (CRE/LOX system from a bacteriophage), whose function is to excise the spacer or the blocking sequence. The "second" gene encoding recombinase is placed behind another promoter/operator, specific for a repressor encoded by the third gene, which is a repressor gene. The genetic elements, as above, differ for pure lines and hybrid seed production systems (Crouch, 1998; Gupta, 1998).

Plant cells are genetically modified by introducing the strips of DNA, and plants are regenerated through tissue culture methods. Because the promoter is active only during a certain stage of seed formation, the lethal gene has no chance of being expressed. When the first-generation plants start producing seeds, the blocking sequence is firmly placed to prevent the lethal gene for being active. The first-generation seeds are, therefore, formed without any trouble. When the first-generation seeds mature, they are exposed to a certain chemical. This chemical can repress the protein by the third strip of DNA and prevent it from repressing the promoter attached to the recombinase enzyme. With this repression removed, the cells of the mature seed produce recombinase. The recombinase promptly removes the excision and blocking sequences in the first-strand of DNA (Mukherjee and Kumar, 2014).

Although the promoter and lethal gene are brought together, the lethal gene is not expressed because the promoter has been chosen to be active only at an earlier stage of seed development, which is not safely passed. As a result, these seeds can be sold to farmers, and they germinate properly to produce healthy plants. However, these second-generation plants carry the promoter and lethal gene, bidding their time to spring into action. A time comes when the second-generation plants start producing seeds. At the stage when the promoter becomes active, the lethal gene is activated and the chemicals it produces disrupts the process of seed formation. As a result, the second-generation seeds will not be fertile.

15.4 TECHNOLOGY FOR PURE LINES SEED PRODUCTION

In self-pollinated crops, where pure lines are used as cultivars, the promoter for the above recombinase gene is repressor specific and is used by a specific repressor protein encoded by the third gene, which is a repressor gene. The repressor gene can be switched off using a stimulus in the form of a chemical, a heat shock, etc. Due to the stimulus, the repressor gene is switched off and CRE/LOX (recombinase gene) is switched on. This will lead to the production of recombinase enzyme, which will excise the blocking sequence or the spacer which blocked the expression of the terminator gene (e.g., RIP). Due to excision of the blocking sequence, the promoter will come to lie adjacent to the terminator gene, which will now be expressed and will kill the developing embryos.

In actual commercial production and sale of seeds of pure lines as above, the crop for commercial seed production will be grown by the breeders or seed companies without chemical treatment. The harvested seeds from this crop will be treated with the chemical, before it is sold to the farmer, so that in crops grown by the farmers, the toxic substance is produced at the time of embryo development, and the embryos die or fail to develop. A seed thus produced will carry the endosperm, but not the embryo, so that it can be used or sold as grain but cannot be used for sowing. Tetracycline is one such chemical trigger that can be used for treatment of the seeds before selling them in the market (Gupta, 1998).

15.5   VERMINATOR: AN ALTERNATIVE TO "TERMINATOR"

Responding to the controversy over terminator technology, recently, Astra Zeneca of UK indicated that it will seek patents in more than 50 countries for its improved plant germplasm invention (Anon., 1998b). They came up with a more farmer-friendly technology. The verminator seed prevents suicide but germinates only when it is exposed to Zeneca's chemical trigger. The technology uses a rodent fat gene (brown adipose tissue of *Rattus rattus*) to block the normal growth and development of transgenic plants and forces the plants to remain unproductive or susceptible to biotic stress through genetic programming. The technique has been dubbed as "verminator" by RAFI and appears to be wider and more flexible than the "terminator," though intended to serve the same purpose. The plant remains unproductive unless they sprayed with Zeneca's chemical at a particular cropping stage which allows them to grow normal plant. This way, the company succeeds not only in selling seeds but also in making profits through its associated chemical business (Sarkar and Ghosh, 2001).

Although such technology may not yet be commercially available, it would only be a matter of time before such a technology might find itself into widespread use. Before they do, it is important that every aspect and methods surrounding the technology be extensively studied and tested in order to safeguard and protect the interests of the general population instead of just serving the interests of a privileged few. Table 15.3 describes few applicants on which GURT technology has been targeted now.

TABLE 15.3   Targeted for GURT Technology (Visser et al., 2001)

No.	Crops	Trait examples	Current status
1	Wheat	Nutrient quality; taste; yield; disease resistance; drought avoidance; cold tolerance	Staple crop; increased R&D expected
2	Rice	–	Staple crop; increased R&D expected
4	Maize	–	Staple crop/specialty products; gene flow containment desirable
5	Soybean	Nutrient quality; feed quality	Increased R&D expected
6	Cotton	Agronomic traits; color	Increased R&D expected
7	Oil crops	Fatty acid composition	Sunflower, olive, oil palm; Canola: gene flow containment (variety)
8	Horticultural crops	Quality traits	V-GURDs for non-hybrids
9	Plantation crops	Agronomical traits	Coffee, banana
10	Cattle	Meat quality; feed conversion efficiency	Specialty products (pharmaceuticals)
11	Trees	Environmental concerns; health benefits; lignin content	Eucalyptus, Populus, Pinus, Acacia
12	Fish and other aquatic species	Environmental concerns; yield; low-temperature tolerance; disease resistance	Salmonids, crap, tilapia, crustaceans, molluses

(Reprinted from Visser, B.; van der Meer, I.; Louwaars, N.; Beekwilder, J. & Eato, D. (2001), "The impact of 'terminator' technology." Biotechnology and Development Monitor, No. 48, p. 9-12.)

15.6   MOTIVES FOR GURTS

The application of GURTs can be inspired by different, not mutually exclusive motives. These motives fall in the following categories.

15.6.1   PROPERTY PROTECTION INTERESTS

Breeding companies wish to safeguard their investments in improved varieties, whether produced by classical breeding or genetic engineering. GURTs may offer a better biological insurance against the free use of genetic innovations than patents, plant breeder's rights or licenses. Hybrid seed, stemming from two genetically different paternal and maternal lines, offer a certain protection of germplasm against reproduction by the farmers, because the seeds normally result in different and heterogeneous plants in the next generation.

For cross-fertilizing crops, this prevents straightforward regeneration of elite varieties. For many self-fertilizing crops, hybrids are not feasible or effective. Therefore, these crops represent the primary targets for the use of V-GURTs in order to obtain biological property protection. Crops concerned include rice, wheat, soybean and cotton, and horticultural crops as well as ornamentals that are vegetatively multiplied. T-GURTs to protect high value-added traits in newly released commercial varieties may be applied to virtually all the crops.

15.6.2   ENVIRONMENTAL INTERESTS

GURTs can be used for the environmental containment of entire transgenic varieties (V-GURT) or of specific transgenes contained in transgenic varieties (T-GURT). In both cases, the focus will be most likely on species that may establish themselves in ecological niches, and/or for which wild relatives exist in involved crop production systems. T-GURTs can be used to contain traits with suspected health risks for the farmer or consumer or traits that may somehow form a threat to biodiversity or other aspects of the environment. An example is a pathogen resistance gene that may outcross to weedy relatives and thus increase the fitness of those weedy relatives.

15.6.3   PRODUCTION INTERESTS

It can be in the producer or farmer's interest to restrict the expression of a trait, for example, to a specific phase in the development of the plants or animals or during biotic or abiotic stress. Stress response engages intense resources and may diminish the quality of the processed product and is only desirable when really needed (Visser et al., 2001)

15.7   IMPACT GROUPS AND GURTS (SWANSON AND GOSCHL, 2000)

There are three basic assumptions with which the economic impacts by GMO and GURT, at the country level, will be assessed (Figure 15.2). Firstly, assume that GURTs will have the effect of increasing appropriability of the value from plant breeding activities, given that the country has biotechno-

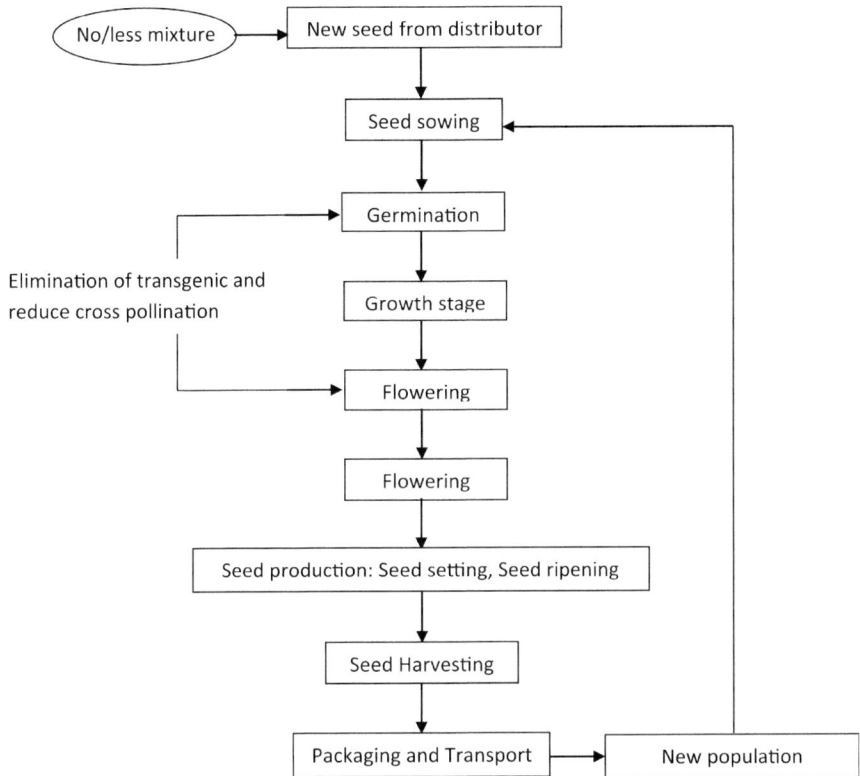

FIGURE 15.2   GURT impedes a specific mechanism of gene flow.

logical (BT) capability. Secondly, assume that GURTs will have positive impacts on the agriculture-based productivity in those countries with a) BT capability and b) significant land area dedicated to GURT crop varieties. Finally, assume that when a country is not BT capable, then the key to the impacts of GURT technologies will be the rate and extent of the diffusion of innovation to that country.

Based on these assumptions, developing countries are categorized into five distinct groups as given. **Group A** comprises those countries with exis-tent of BT capabilities. These are the developing countries that will produce GURT-based crops themselves. **Group B** includes those with an incipient biotech sector. This may be the result of either (a) an already existent but immature sector or (b) a good prospect for the development of GURT capac-ity through foreign investment and ease of regulation. In other words, these are the countries with good potential to catch up to group A.

Group C countries are likely to convert their agriculture to GURT-based systems and thus reap the productivity benefits from that technology, although they will be unable to develop domestic GURT capacity for the time being. They are countries with a built-in tendency toward GURT crops, and a moderately high rate of diffusion of innovations from other countries' plant breeding sectors. **Group D** features those countries that are currently highly dependent on public R&D spending and may lack the liquidity to adopt risky new technology. This means that these countries are in danger of suffering disadvantages from GURTs, particularly by virtue of a slow-down in the rate of agricultural productivity growth. They are "slow diffusion" countries, in that, innovations from other countries' plant breeding sectors do not necessarily confer productivity benefits in these countries.

Group E consists of those countries with a small amount of land in crops likely to be targeted by GURTs. Although these countries are not likely to benefit from this biotechnology in any way, they are also not likely to find themselves in a worse position than before. These are countries that will face a small loss in R&D rents, but whose level of productivity growth – already low – will be unaffected by GURTs.

In the United States, new plant varieties created through genetic engineering are subject to protection under three statutory systems: (i) the Plant Protection Act (covering asexually reproducing new plant varieties); (ii) the Plant Variety Protection Act (covering sexually reproducing new plant varieties); and (iii) the Patent Statute. The most generous of these three regimes is the last one. Until the 1980s, it was not clear that the Patent Statute covered living things and plants in particular. Since then, the developers of new plant varieties and specifically of new varieties created through genetic engineering have relied whenever possible on patent law (Overwalle, 1999).

The legal protections available to plant breeders in Europe resemble those available in the United States, but are somewhat less generous. Beginning in the 1940s, several European countries adopted special statutes providing protection for new plant varieties. The formation in 1961 of the *Union pour la Protection des Obtentions Vegetales* (UPOV) resulted in the repeal of these statutes and their replacement with a uniform system of "Breeders' Rights." The UPOV has been amended several times, and the contours of its successive iterations vary. The protections it provides to

plant breeders roughly resemble patent law – with three significant exceptions. (1) It provides no protection for novel methods of developing plants.
(2) It provides less protection than patent law for "equivalent" plant varieties. And (3) it permits member countries to provide an exemption for
"farmer's rights"; in other words, farmers may be allowed to reuse seeds
produced by protected crops (Foster, 1998). The 1973 European Patent
Convention is less hospitable to the protection of plants than the American patent statute (Foster, 1998). At present, it is more difficult for plant
breeders to avail themselves of European patent protection than to obtain
American patent protection.

Compliance by developing countries with their obligations under
TRIPS has been slow and uneven – for reasons aptly summarized at the
recent Global Conference on Biotechnology. In many developing countries, the politicians who negotiated and signed the agreement are no longer in power. Establishing systems of the sort required by TRIPS will be
costly. Finally, the benefits conferred upon developing countries of a vigorous intellectual property regime are far from obvious. These circumstances
make it likely that many developing countries will fail to meet their obliga-
tion under TRIPS to comply with article 27 by January 1, 2000. Even those
countries that do meet the formal deadline may fail to provide effective
enforcement of the new provisions. Whether threats of trade sanctions by
the United States will be sufficient to overcome their reluctance to comply
with the agreement remains to be seen (Anon., 1999). The bottom line is
that we can expect that the levels of effective intellectual property protection for GM plants will remain lower – and perhaps dramatically lower – in
developing countries than in the United States and Europe for some time
to come.

15.8 EFFECT OF TERMINATOR TECHNOLOGY ON VARIOUS LEVELS

Terminator technology leads to various beneficial paths to plant breeders,
private sectors, and at certain level to the government sector too, but the
coin always has two sides. On the bright side, this technology may provide
the world their next revolution, while on the dark side of using terminator
technology, it will adversely affect at various levels such as farmers, environment, and agro-biodiversity.

15.8.1 EFFECT OF TERMINATOR TECHNOLOGY ON THE FARMERS

The greatest concern about terminator technology, especially as it affects poor farmers, is that they would be unable to maintain commercial varieties from their own seed stock and would be forced to return to the seed provider. This will translate into nonavailability or lack of seed inputs to the farmers. This will greatly affect the level of agricultural production and the farmer's income, thus undermining food and social security. This will tantamount to a technology that imprisons target users, which is an issue of social concern (Yusuf, 2010).

For quite long, the farmers in developing countries have the culture of seed sharing and exchange among themselves with the traditional landraces and available crop varieties. However, with terminator seeds, this is not possible, and it will greatly hamper this beneficial and sustainable culture of the local farmers who mostly practice subsistence agricultural production. Also, courtesy of this culture traditional landraces possessing desirable characteristics and adaptability to the local environment such as disease and pest resistance, drought tolerance, *etc.* acquired over time compared to modern varieties are exchanged among the farmers, and this cycle of seed exchange will be impaired by terminator technology. More complicating is the fact that IPR prohibits the farmers from saving seed and undertaking their own breeding programs and prohibit plant breeders from using the material to create new generation of varieties adapted to specific regions of growing conditions (Guebert, 2001). The regulatory process, in fact, may not answer most questions about the environmental and human health risks of commercial production of transgenic crops (Anon, 2003).

15.8.2 EFFECT OF TERMINATOR TECHNOLOGY ON ENVIRONMENTS

Farmers may worry that their use of GM seeds will create "**super weeds**" or "**superbugs**" that, over time, become resistant to GM seeds and crops and to other herbicides and pesticides. There is some research that suggests that weeds and bugs could possibly evolve into resistant organisms. Gene movement from crop to weed through pollen transfer has been demonstrated for GM crops when the crop is grown near a closely related weed species (Kruft,

2001). There is great concern that GM plant species are able to mate with their wild-type counterparts and pass on their genetic material to naturally occurring species (Smyth, 2002). This transfer of genetic material from a GMO into the genome of another plant is called "**genetic drift**." (Repp, 2000). If left uncontrolled, genetic drift could result in GM traits entering the environments and possibly mutating or hybridizing to produce "subsequent generations of plants with unforeseen properties." Similarly, insects have, in the past, developed a resistance to pesticides. A recent study documented a decreased susceptibility in pests to the use of BT as a sprayed pesticide (Baron, 2008). Many GMOs are designed to be drought or pest resistant, which allows them to survive in environments where wild-type plants cannot. This can result in GMOs creating a decline in biodiversity by dominating ecological niches usually filled by the less vigorous and hardy wild-type species.

However, such constructs may still survive, and exceptions to this rule can be imagined. First, the trait could be induced by a range of related compounds or triggered by naturally occurring events, such as steroids, pests, and disease infestations. Second, the receiving organisms could already contain a GURT construct controlled by related inducers so that the inducer would activate both constructs. Such undesirable effects might be avoided by choosing highly specific inducers. A third and more problematic scenario is the outcrossing of GURT constructs that exhibit negative control of the involved trait. This means that a trait is expressed unless it is blocked by the application of an inducer. In the case of outcrossing, the chemical compound would most likely not be applied to the organism that newly integrated the construct and the trait would be continuously suppressed, when not desired. Such outcrossing could affect not only domesticates but also wild relatives. Although individual elements of a GURT construct may outcross independently and exhibit certain effects, the likelihood of such events is very low, because these assume both a recombination event and an outcrossing event in most cases. In other words, such elements will only outcross separately from the other elements of the same construct, once they have been physically separated from each other in the genome (Visser et al., 2001).

15.8.3 EFFECT ON AGRO-BIODIVERSITY

Agro-biodiversity refers to the variety of plants and animals that are used in the agricultural systems worldwide. The world's agro-biodiversity depends heavily on seed saving, selecting, and re-planting. This practice has resulted in crop varieties that are adapted to the local environments, soil, and local pests. This technique has also resulted in creating new crop varieties that fetch more money in the market, for example, Basmati rice of India and Pakistan. Introducing "terminator seeds" will replace the age-old practice of seed saving and can lead to the loss of traditional seed varieties.

15.9 SOCIO-ECONOMIC IMPACT OF TERMINATOR TECHNOLOGY

The agricultural system of the world is very diverse. In developed countries like Canada, farms tend to big business-like operations, known as intensive farming operations. In developing countries, the farming system tends to be much smaller and include a lot of subsistence farmers – these are people who farm to feed their families and survive rather than making money.

Terminator technology can be good for the intensive farming operations in the developed world. These farms produce high-value produce (crops that fetch more money in the market such as seedless fruits, unblemished vegetables, etc.) and rarely save seeds for replanting, making it less vulnerable to terminator technology. But, medium, low and subsistence farming practices dominate the agricultural systems of the developing world. There are nearly *1.4 billion farmers* around the world engaged in these farming systems. These farming practices rely heavily on saved seeds and use it for replanting. If "terminator seeds" are introduced in these systems, it will replace the existing seeds and force the farmers to buy seeds every season, which poor farmers from developing countries cannot afford (Somalinga, 2007). Eaton et al. (2002) summarized potential economic benefits, costs, and risks of GURT, which are given in Table 15.4.

15.10 MERITS AND DEMERITS OF TERMINATOR TECHNOLOGY

The terminator technology seems to have certain *advantages* and *disadvantages*, depending on the kind of stand that an individual has on genetic modification. Some say that this technology may be able to prevent the accidental spread of GM crops to its wild relatives and help prevent any adverse effects in biodiversity. And yet, many opponents think that the terminator technology

TABLE 15.4 Potential Economic Benefits, Costs, and Risks of GURT (Eaton et al., 2002)

Categories	Benefits	Cost	Risks
Farmers	Increased productivity from improved genetic inputs due to increased research and development (R&D) investment	Increased input costs from seed purchase	Misuse of monopoly powers by breeders Reduced seed security and access to genetic improvements (marginalized farmers)
Breeders (especially private sector)	Increased share of research benefits from new products	Increased cost for access to genetic resources of other breeders	—
Governments	Reduced investment requirements in breeding Fewer enforcement costs for plant variety protection (PVP)	Complementary R&D investment requirements Establishment and enforcement of new regulatory requirements	—
Society	Increased agricultural productivity		Reduced genetic diversity in fields

(Reprinted with permission form Eaton, D., et al., (2002). Economic and policy aspects of 'terminator' technology. Biotech. Develop. Monit., 49, 19–22.)

may lead to farmers getting more dependent on seed companies to provide seeds to plant. In a way, this can be seen as a means that can affect food security.

15.10.1 MERITS OF TERMINATOR TECHNOLOGY

1. The most obvious benefit of the new technology is that it would enable the developers of new plant varieties to make a profit selling seeds in countries that currently lack intellectual property protections for plants.
2. Opportunities for such profits should encourage biotechnology firms to develop plant varieties suitable for cultivation in developing countries and then to make those varieties available to farmers.
3. Terminator technology is more than a potential substitute for IPRs; from the standpoint of the private firms, it is even better than IPRs. It is plainly superior (from the firms' standpoint) to "sui generis" plant-protection statutes styled on the UPOV, for the obvious reason

that it overrides the traditional entitlement of farmers to reuse seeds produced by protected plant varieties.

4. In developing countries, the practical impediments to effective enforcement of restrictive licenses are much more severe. By contrast, terminator technology makes it simply impossible for farmers to reuse the seeds. Second, patent law in many countries limits in various ways (e.g., through the "patent misuse" doctrine) the kinds of license agreements that patentees may extract from their customers.

5. By employing terminator technology, the firms can escape those restrictions. These two differences suggest that the incentive effects of terminator technology will be even stronger than that of IPRs (whether they give rise to unacceptable adverse side-effects will be considered below).

6. By discouraging arbitrage, terminator technology increases the ability of seed suppliers to engage in price discrimination and other forms of precise marketing. To be sure, the technology does not eliminate arbitrage altogether; farmers who bought seeds at low prices could, instead of planting them, resell them to other farmers able and willing to spend more.

7. These technologies were developed primarily as a means of protecting companies holding patented crop varieties from the unauthorized use of seeds saved from earlier crops. Patent protection laws are the standard approach for this purpose, but are difficult to enforce; GURTs are intended to create built-in patent protection. Some promoters claim that, with better patent protection, more commercial effort can be expended to improve varieties of minor crops, with gains from added value and increased yields.

8. Transgenic plants have genetically contaminated other species, and many groups and individuals are alarmed about the unpredictable effects of genetically altered crops on wild plant populations. Promoters believe terminator technology would reduce (though it would not eliminate) this threat, because wild plants that are pollinated by terminator crops will produce (largely) sterile seeds. However, it should be noted that genetic engineering, like all technologies, will have a certain failure rate, which has not at this time been precisely determined; its success as a biosafety tool cannot be

guaranteed (Bosselmann, 1996; Bai, 1997; Sherwood, 1997; Anon, 2006; Danial, 2010).

15.10.2 DEMERITS OF TERMINATOR TECHNOLOGY

1. The potential disadvantage of terminator technology that looms largest in contemporary debate is the danger that farmers will become dependent on Western biotechnology companies for their supplies of seeds. The companies, it is feared, will then take advantage of that situation to charge farmers exorbitant prices.

2. Most poor farmers depend on saving seeds from previous crops and would not be able to afford commercial seeds each year or the chemicals required.

3. The net effects will be further impoverishment of farmers in the developing world and increase the regrettable flow of revenues from the developing countries to the developed world.

4. A more subtle variation on this theme is the risk that the biotechnology companies will use their economic leverage to trap farmers in a cycle of economic dependency analogous to the "crop lien" system that prevailed in the American in the late 19th century.

5. In some areas, poor farmers may be unable to pay for the seeds at the start of the growing season. The seed companies will thus be tempted to provide the farmers the seeds they need in exchange for a promise to pay the companies higher prices at the time of harvest – a promise the farmers will likely have trouble keeping. After a few repetitions of this arrangement, farmers would find themselves hopelessly in debt.

6. The pollen from terminator plants could contaminate and kill seeds of other nearby plants. Thus, neighboring crop seeds in the first generation could be rendered sterile, unknowingly to the farmers harvesting them.

7. It may have a negative impact on the ability of wild populations to maintain them. Wild plant populations could be reduced or endangered.

8. Treatments used to activate trait technology in seeds or plants could be ecologically damaging in various ways. The antibiotic tetracycline, for instance, has been suggested as one such gene switching

substance, but increasing its use in the environment could add to the growing problem of anti-microbial resistance in disease causing bacteria.

9. With commercialization of these technologies, the genetic diversity of the world's major food crops will be narrowed, thus increasing their vulnerability to disease and insects and reducing local crop adaptation to local conditions.

10. The TPS patent notes that the ribosome inhibitor is the preferred agent for killing seeds because it is not lethal to people who might eat the seeds. However, the patent authors suggest that "any lethal gene should be acceptable" in cases where there is no use for the seeds. They suggest that diphtheria toxin or any cytotoxic protein could be used for the purpose. Seed companies and government regulators may want to consider the possible impact on seed-eating birds, insects, and microorganisms (Ranson and Stch, 1997; Tansakul and Burt, 1999; Ewing, 2002; Anon, 2006).

15.11   WHAT IS TO BE DONE FOR TERMINATOR TECHNOLOGY?

The substantial potential benefits of terminator technology, as discussed above, suggest that it would be unfortunate if the seed companies were compelled either through legal prohibitions or popular opinion to eschew use of it altogether. On the other hand, developing countries would be wise to regulate its usage so as to avoid the dangers sketched of demerits (Rakoff, 1983).

1. Manipulation of the rules of patent law is an unpromising way of seeking to prevent the technology from being used in exploitative ways. If anything, denying a patent to this technology thus enabling many biotechnology companies (rather than just one) to use it in marketing their genetically altered seeds would increase rather than decrease the dangers associated with it.

2. More promising would be mandatory disclosure rules – analogous to those imposed on all residential mortgage transactions in the United States. Sellers of genetically altered seeds containing "terminator" genes would be required to explain to each customer the implications of the new technology – and, in particular, the fact that the customers would need to purchase new seeds each year. This strategy could go

some distance toward reducing the information asymmetries between sellers and buyers, but would likely not eliminate altogether the risks cataloged in demerits (Rakoff, 1983).

3. More promising would be price controls. A developing country could permit the sale and use within its jurisdiction of seeds containing "terminator" genes, but only on the condition that the seller not exceed price ceilings set by an administrative tribunal. American copyright law (although not patent law) contains many such arrangements. Generally speaking, they work well in balancing the interests of innovators and consumers. A similar strategy could enable developing countries simultaneously to create incentives for the development of new plant varieties while curbing exploitative use of the resultant economic levers.

4. Finally, more elaborate regulatory mechanisms might be effective. For example, a developing country could permit the sale and use within its jurisdiction of seeds containing the terminator gene, but only on the condition that each seller agrees in all future years to continue to provide the seeds to the original purchasers at the same price (adjusted for inflation). Under such an arrangement, the seller could charge a higher price to new customers, but not to existing customers. Such a system would shield farmers from many of the exploitative tactics predicted by the critics of the technology (Foster, 1998).

5. It is impossible, without knowing a good deal more about the economic and social conditions in a specific country, to determine which regulatory apparatus would work best. But the general idea should be clear enough: A sensible response to the advent of terminator technology would permit – but regulate – its use.

15.12 CONCLUDING REMARKS

Our agriculture is also fully dependent on this practice of the farmers. But due to introduction of terminator seed, farmers will not be able to save surplus seed for future cultivation; rather, they have to buy seed again from the seed companies because it stops enhanced yielding. Although such technology may not yet be commercially available, it would only be a matter of time before such a technology might find itself into widespread use. Therefore, this terminator technology is a direct threat to our agriculture and our food security. Before they do, it is important that every aspect and methods surrounding the technology be extensively studied and tested in order to safeguard and protect the interests of the general population instead of just serving the interests of a privileged few. Now we can either ban or impose restrictive legal regime to this terminator technology. We can enforce the draft Biodiversity Act and draft Plant Variety Protection Act containing provision for farmers' rights in line with fair and equitable sharing of the benefits from the utilization of genetic resources. Then, we will be able to protect our food security and agriculture within TRIPS mandate. Terminator technology may have both positive and negative impacts on the world's agricultural system. In developed countries like Canada, terminator technology will not have much impact on farmers and the way they farm. But in developing and low-income countries, terminator technology might be harmful to the farmers. Moreover, technical aspects of terminator technology design still need to be fine-tuned. These aspects need to be perfected before introducing terminator technology in the farms worldwide. Apart from this, as with any other GMO, the impact of introducing terminator technology on the world's biodiversity is not yet known.

KEYWORDS

- risks of terminator technology
- terminator technology
- T-GURTS
- verminator
- V-GURTS

REFERENCES

Ait-ali, T., et al., (2003). Flexible control of plant architecture and yield via switchable expression of *Arabidopsis gai. Plant Biotechnol. J.*, *1,* 337–343.

Al-Ahmad, H., & Gressel, J., (2005). Mitigation using a tandem construct containing a selectively unfit gene precludes establishment of *Brassica napus* transgenes in hybrids and backcrosses with weedy *Brassica rapa. Plant Biotechnol. J.*, *3,* 178–192.

Al-Ahmad, H., et al., (2006). Mitigation of establishment of *Brassica napus* transgenes in volunteers using a tandem construct containing a selectively unfit gene. *Plant Biotechnol. J.*, *4,* 7–21.

Anonymous, (2003), Advisory Committee on Biotechnology and 21st Century Agriculture (AC21). Summary, First Plenary Meeting of AC21, pp. 21.

Anonymous, (2006), Genetic use restriction technologies (gurts) or terminator technology. *Fact Sheet Series on Innovative Technologies*, pp. 1–4.

Anonymous, (1998a), United States Patent and Trademark Office, [Available at http://patft. uspto.gov/netacgi/nphParser?Sect1=PTO1&Sect2=HITOFF&d=PALL&p=1&u=/ netahtml/srchnum.htm&r=1&f=G&l=50&s1=5, 723, 765. WKU & OS=PN/5, 723, 765&RS=PN/5, 723, 765].

Anonymous, (1998b), Seed industry consolidation: Who own whom? *RAFI Communique*, pp. 1–4.

Anonymous, (1999), *World Intellectual Property Report 13*, pp. 310.

Bai, J. B., (1997). Comment, protecting plant varieties under TRIPS and NAFTA: Should utility patents be available for plants? *Texas International Law Journal, 32*, 139.

Barron, J. A., (2008). Genetic use restriction technologies: Do the potential environmental harms outweigh the economic benefits? *Georgetown Inter. Environ. Law Rev., 20*, 271–286.

Bennet, A. B., et al., (1989). Tomato fruits polygalacturonase, gene regulation and enzyme function In: *Biotech. Food Quality*, Kings. S. D., Bills. D. D., Quatrano, R., (eds.) Boston, Mass: Butterworths Publishing, pp. 167–180.

Bleecker, A. B., (1989). Prospects for the use of genetic engineering in the manipulation of ethylene biosynthesis and action in higher plants. In: *Biotech. Food Quality*. Kings S., D., Bills, D. D., Quatrano, R., (eds.). Boston, Mass: Butterworths Publishing, pp. 159–166.

Bosselmann, K., (1996). Plants and politics: The international legal regime concerning biotechnology and Biodiversity. *Colorado Journal of International Environmental Law and Policy, 7*, 111.

Burgess, D. G., et al., (2002). A novel, two-component system for cell lethality and its use in engineering nuclear male-sterility in plants. *Plant J., 31*, 113–125.

Cigan, A., (2000). Reversible nuclear genetic system for male sterility in transgenic plants, *US Patent 6, 072*, 102.

Crouch, M. L., (1998). How the terminator terminates, occasional paper, Edmonds Institute, Washington, University of Indiana, Department of Biology, pp. 1–8

Daniel, D., (2010). Seeds of hope: how new genetic technologies may increase value to farmers, *Rutgers Comp. Tech. Law J., 36*, 250–285.

Daniell, H., et al., (2005). Breakthrough in chloroplast genetic engineering of agronomically important crops. *Trends Biotechnol., 23*, 238–245.

Dutfield, G., (2007). Social and economic consequences of genetic use restriction technologies in developing countries AB international. *Agril. Biotech. Intellec. Prop.*, pp. 294.

Eaton, D., et al., (2002). Economic and policy aspects of 'terminator' technology. *Biotech. Develop. Monit., 49*, 19–22.

Erikson, O., et al., (2004). A conditional marker gene allowing both negative and positive selection in plants. *Nat. Biotechnol., 22*, 455–458.

Ewing, J., (2002). Agricultural biotechnology: is the international regulation of transgenic agricultural plants for the birds (and the Bees)? *Suffolk Transnational Law Rev., 25*, 617, 634.

Fabijanski, S., et al., (2006). Methods and constructs for plant transformation, *US Patent 7, 112, 721*.

Fabijanski, S., et al., (1998). Molecular methods of hybrid seed production, *US Patent 5, 728, 558*.

FAO, (2001), "Potential Impacts of Genetic Use Restriction Technologies (GURTs) on agricultural biodiversity and agricultural production systems." CGRFA/WG-PGR- 1/01/7.

Fischer, R., et al., (2006). Methods for modulating floral organ identity, modulating floral organ number, increasing of meristem size, and delaying flowering time, *US Patent 7, 109,* 394.

Foster, G. K., (1998). Opposing forces in a revolution in international patent protection: the U. S. and India in the uruguay round and its aftermath, *UCLA Journal of International and Foeign Affairs, 3,* 283.

Fu, D., et al., (2005). Large deletions within the first intron in VRN-1 are associated with spring growth habit in barley and wheat. *Mol. Genet. Genomics, 273,* 54–65.

Gressel, J., (1999). Tandem constructs: preventing the rise of super weeds. *Trends Biotechnol., 17,* 361–366.

Grevich, J., & Daniell, H., (2005). Chloroplast genetic engineering: recent advances and future perspectives. *Crit. Rev. Plant Sci., 24,* 83–107.

Gross, K. C., (1988). Cell wall dynamics. In: *Biotech Food Quality. Boston Mass,* Kungs, S. D., Bills, D. D., Quatrano, R., (eds.). Butterworths Publishing, pp. 143–158.

Guebert, A., (2001). Supreme Court blesses plant patents: Bye-bye binrun seed. *The Land (Minnesota),* pp. 3.

Gupta, P. K., (1998). The terminator technology for seed production and protection: Why and how? *Current Sci., 75,* 1319–1323.

Hills, Melissa J. et al., (2007) Genetic use restriction technologies (GURTs): strategies to impede transgene movement, Trends in Plant Science, 12, 4, 177–183.

Hodges, T., et al., (1999). Method for the production of hybrid plants, US Patent 5,929,307.

Jefferson, R. A., et al., (1999). Genetic use restriction technologies: technical assessment of the set of new technologies which sterilize or reduce the agronomic value of second generation seed, as exemplified by U.S. Patent No. 5,723,765 and WO 94I03619. *Convention on Biological Diversity.* (UNEP/CBD/SBTTA/4/9/Rev. 1).

Jofuku, K. D., et al., (2005). Control of seed mass and seed yield by the floral homeotic gene. *Proc. Natl. Acad. Sci. USA, 102,* 3117–3122.

Konishi, S., (2006). An SNP caused loss of seed shattering during rice domestication. *Sci., 312,* 1392–1396.

Krebbers, E., et al., (1999). Use of anthocyanin genes to maintain male sterile plants, *US Patent 5,* 880, 331.

Kruft, D., (2001). Impacts of genetically-modified crops and seeds on farmers. Colorado State University, *Transgenic Crops: An Introduction and Resource Guide, 25.*

Kumar, S., et al., (2004). Stable transformation of the cotton plastid genome and maternal inheritance of transgenes. *Plant Mol. Biol., 56,* 203–216.

Kuvshinov, V., et al., (2005). Double recoverable block of function – a molecular control of transgene flow with enhanced reliability. *Environ. Biosafety Res., 4,* 103–112.

Kuvshinov, V., et al., (2001). Molecular control of transgene escape from genetically modified plants. *Plant Sci., 160,* 517–522.

Lehmann, V., (1998). Patent on seed sterility threatens seed saving. *Biotech. Dev. Monit., 35,* 6–8.

Lemmetyinen, J., et al., (2004). Prevention of the flowering of a tree, silver birch. *Mol. Breed., 13,* 243–249.

Lough, T., et al., (2006). Control of floral induction, *US Patent 7,* 071, 380.

Luo, H., et al., (2005). Controlling transgene escape in GM creeping bentgrass. *Mol. Breed., 16,* 185–188.

Ma, Q. H., et al., (2002). Increased cytokinin levels in transgenic tobacco influence embryo and seedling development. *Funct. Plant Biol.*, *29*, 1107–1113.

Mahmood, T., et al., (2006). Molecular markers for seed color in *Brassica juncea*. *Genome*, *48*, 755–760.

Mukherjee, S., & Kumar, N. S., (2014). Terminator gene technology- their mechanism and consequences. *Sci. Vis.*, *14*(1), 51–58.

Nasholm, T., (2006). Selective plant growth using D-amino acids, US Patent 7,105,349.

Oliver, M., et al., (1999). Control of plant gene expression, US Patent 5,977,441.

Overwalle, G., (1999). "Patent protection for plants: a comparison of American and European approaches," *Journal of Law and Technology*, *39*, 143.

Overwalle, G., (1999). Patent protection for plants: acomparison of American and European approaches. *J. Low Tech.*, pp. 143.

Paul, W., (2006). Tapetum-specific promoters, US Patent 7,078,587.

Pendleton, C. N., (2004). The peculiar case of "terminator" technology: agricultural biotechnology and intellectual property protection at the crossroads of the third green revolution. *Biotech. Law Rep.*, *23*, 1–29.

Perez, P., et al., (2006). Use of male sterility to prevent transgene spread in plants, US Patent 7,112,719.

Rakoff, T. D., (1983). Contracts of adhesion: An essay in reconstruction. *Harvard Law Review*, *96*, 1173.

Ranson, R., & Sutch, R., (1977). One Kind of Freedom, *Explorations in Economic History*, *38*, 1–5.

Repp, R., (2000). Biotech pollution: assessing liability for genetically modified crop production and genetic drift, *Idaho Law Rev.*, *36*, 585, 587.

Sarkar, C. K. G., & Ghosh, S. K., (2001). Bio-wealth protection: Genetic and legislative approaches, *Financial Daily*, BLFeedback@thehindu.co.in.

Schernthaner, J. P., et al., (2003). Control of seed germination in transgenic plants based on the segregation of a two-component genetic system. *Proc. Natl. Acad. Sci. USA*, *100*, 6855–6859.

Sherwood, R. M., (1997). The TRIPS Agreement: Implications for developing countries. *Idea*, *37*, 491.

Shi, G., (2006). Intellectual property rights, genetic use restriction technologies (GURTs), and strategic behavior. *Selected Paper Prepared for Presentation at the American Agricultural Economics Association Annual Meeting*, Long Beach, California, pp. 31.

Slade, A., et al., (2006). Compositions and methods for modulation of plant cell division, US Patent 7,056,739.

Smyth, S., (2002). Liabilities and economics of transgenic crops, *Natural Biotech.*, *20*, 537–537.

Somalinga, V., (2007). Terminator technology for GM crops. http://www.genomebc.ca/education/teachers/articles/terminator-technology/.

Swanson, T., & Goschl, T., (2000). Genetic use restriction technologies (GURTs): Impacts on developing countries. *Int. J. Biotechnol.*, *2*(1/2/3), 56–84.

Tadege, M., et al., (2001). Control of flowering time by FLC orthologues in *Brassica napus*. *Plant J.*, *28*, 545–553.

Tansakul, R., & Burt, P., (1999). People power vs the gene giants, *Bangkok Post.*, *17*.

Visser, B., et al., (2001). The impact of 'terminator' technology. *Biotech. Dev. Monit.*, *48*, 9–12.

Yusuf, M., (2010). Ethical issues in the use of the terminator seed technology. *Afri. J. Biotech.*, *9*(52), 8901–8904.

CHAPTER 16

OMICS IN PLANT BREEDING

D. R. MEHTA[1] and RAMESH KUMAR[2]

[1]Associate Professor, Department of Genetics and Plant Breeding, Junagadh Agricultural University, Junagadh–362001, Gujarat, India

[2]PhD Scholar, Department of Genetics and Plant Breeding, Junagadh Agricultural University, Junagadh–362001, Gujarat, India

CONTENTS

16.1   INTRODUCTION

From the time when agriculture is believed to begin, in approximately 10,000 BC, people have consciously or instinctively selected plants with improved characteristics for cultivation of subsequent generations. Plant breeding became a science only after the rediscovery of Mendel's laws in 1900. It was only in the late 19th century that the Mendel uncovered the secrets of heredity, thus giving rise to genetics, the fundamental science of plant breeding.

Scientists added a few more pieces to the puzzle that was becoming this new science in the first half of the 20th century by concluding that something inside the cells was responsible for heredity. This hypothesis generated answers and thus consequent new hypotheses, leading to the continuing accumulation of knowledge and progress in the field. For example, the genetic recombination in 1910, double helix structure of DNA in 1953, recombinant DNA technology in 1972, and DNA sequencing in 1975 (Table 16.1).

The importance of determining the entire genome sequence of crop plant and other organisms were recognized more than three decades ago and was an important first step in ushering the field of genomics. It was also apparent at that time that the goal of deciphering the complete genome was achievable with existing sequencing techniques and by developing large-scale cloning and mapping strategies.

The "omics" approach integrates genome, transcriptome, proteome, metabolome, physiognome, and phenome data into a single data set and can lead to the identification of unknown genes and their regulatory networks involved in metabolic pathways of interest. This will also help in understanding the genotype–phenotype relationship. Given the importance of omics, subsequent paragraphs will discuss the several tools and methods currently in use or having great scope for future use in crop breeding. The role of fields of omics, the inter-relationships among them, and their corresponding biological processes can be visualized by the omics, as shown in Figure 16.1.

16.2   APPROACHES IN OMICS

Omics is not a single word, but it contains a group of tools and methods that allowed us to address the complex global biological systems that underlie various plant functions. These approaches are as follow:

1. Genomics
2. Transcriptomics
3. Proteomics
4. Metabolomics
5. Physiognomics
6. Phenomics

TABLE 16.1 Chronology of Major Advances in Genetics and Biotechnology Relevant to Plant Breeding

Year	Historical landmark
1809–1882	Develops the theory of natural selection.
1866	Concluded that traits are determined by particulate factors (now called genes) which carry hereditary information.
1868	Fredrich Miescher isolated phosphate rich chemicals from nuclei of white blood cells and called these as nuclein (now known as DNA).
1884-85	E. Strasburger and Walther Flemming showed that nuclei contain chromosomes.
1900	Rediscovery of Mendel's principles of heredity.
1910	The effects of genetic recombination in *D. melanogaster* demonstrates that genetic factors (genes) are located on chromosomes.
1913	Alfred Henry Sturtevant devised the principle for constructing a genetic linkage map.
1928	Frederick Griffith discovered genetic transformation of a bacterium and called the agent responsible the "transforming principle."
1941	George Beadle and Edward Tatum demonstrate that a gene produces a protein.
1944	Elucidates the process of genetic recombination by studying satellite chromosomes and genetic crossing-over related to linkage groups in chromosomes 8 and 9 of maize.
1953	Watson and Crick, proposed the double helix structure of the DNA molecule.
1957	Developed biochemical markers based on the expression of isoenzymes.
1967	U.E. Loening introduced Polyacrylamide – Gel Electrophoresis.
1969	Discovery of restriction enzymes.
1972	Recombinant DNA technology begins with the first cloning of a DNA fragment.
1973	Performed the first genetic engineering experiment on a microorganism, the bacterium *E. coli.*
1977	Sanger et al. (1975) developed DNA sequencing by the enzymatic method.
	Maxam and Gilbert (1977) developed DNA sequencing by chemical degradation.
	F. Bolivar and co workers constructed the plasmid vector pBR322.
1980	Develop the RFLP (Restriction Fragment Length Polymorphism) technique.
1983	The first transgenic plant is produced, a variety of tobacco into which a group of Belgian scientists introduced kanamycin antibiotic resistance genes.
1985	Genentech the first biotech company to launch its own biopharmaceutical product, human insulin produced in cultures of *E. coli* transformed with a functional human gene.

TABLE 16.1 (Continued)

Year	Historical landmark
1985	The first plant with a resistance gene against Lepidoptera was produced.
1986	The first field trial of transgenic plants was conducted in Ghent, Belgium.
1987	The first plant tolerant to a herbicide, glyphosate, was created.
1987	Mullis et al. (1987) identified thermostable Taq DNA polymerase enzyme, which enabled the automation of PCR.
1988	The first transgenic cereal crop, Bt maize, was developed.
1990	Rafalski et al. (1990) develop the first genotyping technique using PCR, RAPD (Random Amplified Polymorphism DNA).
1990	New tools for NCBI sequence alignment are created (BLAST – Basic Local Alignment Search Tool).
1994	The first permit was issued for the commercial cultivation and consumption of a GMO, the Flavr Savr tomato.
1995	Patrick O. Brown developed microarray technology.
1996	Ian Wilmut Cloned Dolly from an adult using the technique of nuclear transfer the first mammal to be cloned.
1997	The first plant containing a human gene, human protein c-producing tobacco, produced.
2000	Sequencing of a prokaryotic organism, the bacterium *E. coli*.
2003	The first eukaryotic genome sequence that of the human is released by two major independent research groups in the US.
2005	Next Generation Sequencing was used as a tool to unravel whole genomes quickly.
2011	Second and third-generation large-scale sequencing systems were developed.
2012	Technologies that control the temporal and spatial expression of genes were used in genetic transformation and the exclusion of auxiliary genes.

(Source: Borem and Fritsche-Neto, 2013)

16.3 GENOMICS

Genomics consists of the development of large-scale analyses of not only structural but functional features of genomes, allowing the discovery of evolutionary and functional dynamics in crop plants. Presently, genomics provides plant breeders with a new set of tools and techniques that allow analysis of the whole genome thereby facilitating the direct study of the genotype and its relationship with the phenotype (Tester and Langridge, 2010). Advances in genomics can also contribute to crop improvement in two general ways. First, a better understanding of the biological mechanisms can

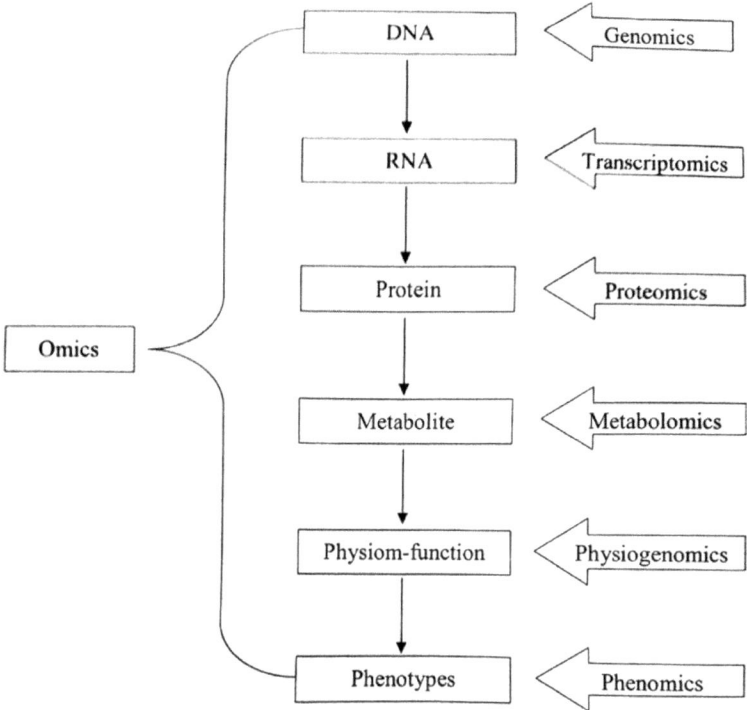

FIGURE 16.1   Inter-relationships among the fields of omics.

lead to new or improved screening methods for selecting superior genotypes more efficiently. Second, new knowledge can improve the decision-making process for more efficient breeding strategies (Varshney et al., 2005).

16.3.1 DNA SEQUENCING

The determination of base sequence of DNA is called DNA sequencing. DNA sequencing became feasible due to availability of restriction enzymes, development of electrophoresis techniques, gene cloning, and PCR techniques that make available large number of copies of DNA fragments required for sequencing.

Two methods, a chemical (Maxam and Gilbert, 1977) and an enzymatic (Sanger et al., 1977) method, of sequencing were developed; these two methods are popularly termed as first-generation DNA sequencing methods. Soon the second- or next-generation DNA sequencing (NGS) methods were developed, which used PCR for *in vitro* cloning in the place of *in vivo* cloning and are much faster and cheaper (Pandey et al., 2008; Shendure and Ji, 2008; Edwards, 2013). At present, the third-generation DNA sequencing (TGS) methods are becoming commercially available; these methods sequence single DNA molecules without any cloning (Schadt et al., 2010). As the fourth-generation sequencing technique, nanopores have the potential to become a label-free, rapid, and low-cost DNA sequencing technology (Feng et al., 2015).

16.3.2   CHAIN TERMINATION METHOD

This was most the widely used method of DNA sequencing at that time. In this method, the chain termination process is used to stop DNA synthesis selectively. Basic requirement for sequencing by this method is (1) DNA sequence, (2) primer, (3) four deoxynucleotide bases (dATP, dGTP, dCTP, dTTP), and (4) low concentration of a chain terminating nucleotide (dideoxynucleotide).If a dideoxynucleotide is added to the end of a chain, it will block subsequent extension of that chain as the dideoxynucleotides have no 3'-OH. By using dideoxythymidine triphosphate (ddTTP), dideoxycytidine triphosphate (ddCTP), dideoxyadenosine triphosphate (ddATP), and dideoxyguanosine triphosphate (ddGTP), each labeled with a dye that fluoresces a different color, as chain terminators in a DNA synthesis reaction, a population of nascent fragments will be generated that includes chains with 3' termini at every possible position (Figure 16.2). The entire nascent fragment that terminates with dideoxythymidine will fluoresce one color; those that terminate with dideoxycytidine will fluoresce a second color; chains that terminate with dideoxyadenosine will fluoresce a third color; and those that terminate with dideoxyguanosine will fluoresce a fourth color. At the time of reaction in the reaction tube, the ratio of deoxynucleotide:dideoxynucleotide is kept at approximately 100:1, so that the probability of termination at a given base in the nascent chain is about 1/100. This yields a population of fragments terminating at all potential termination sites within a distance of a few hundred nucleotides from the original primer terminus.

After the DNA chains generated in the reaction are denatured and separated by polyacrylamide capillary gel electrophoresis, their positions in the

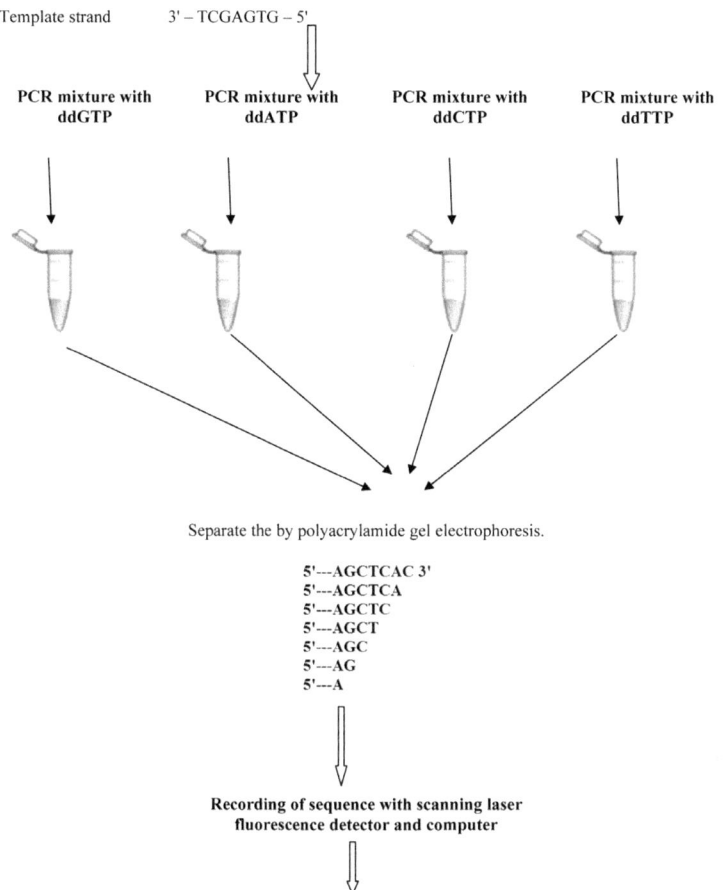

Template strand   3' – TCGAGTG – 5'

PCR mixture with ddGTP **PCR mixture with ddATP** **PCR mixture with ddCTP** **PCR mixture with ddTTP**

Separate the by polyacrylamide gel electrophoresis.

5'---AGCTCAC 3'
5'---AGCTCA
5'---AGCTC
5'---AGCT
5'---AGC
5'---AG
5'---A

Recording of sequence with scanning laser fluorescence detector and computer

DNA sequence computer printout

FIGURE 16.2 Sequencing DNA by the 2,3-dideoxynucleoside triphosphate chain-termination procedure.

gel are detected by a fluorescence detector and recorded on a computer. The shortest chain moves through the gel first, and each chain thereafter is one nucleotide longer than the preceding one. The dideoxynucleotide at the end of each chain will be determined by the fluorescence. Thus, the sequence of the longest newly synthesized chain can be determined by simply reading the sequence of fluorescence peaks from the shortest DNA chain to the longest DNA chain.

The genomes of *Arabidopsis* (Arabidopsis Genome Initiative, 2000) and rice (International Rice Genome Sequencing Project, 2005) were completed using automatic Sanger sequencing method. Since 2005, a new generation of technologies has emerged (Varshney et al., 2009), allowing the generation of an even larger amount of data per sample by reaction and revolutionizing the genomics landscape, viz., 454 pyrosequencing, ABI SOLID, Illumina genome analyzer, Helicos, Ion torrent, and SMRT. All these technologies perform DNA sequencing on platforms capable of generating several million base outputs in a single run. Among the NGS, the Roche 454 FLX and Solexa/Illumina are used worldwide.

16.3.3   454 PYROSEQUENCING

In the 454 pyrosequencing approach, genomic DNA is first fragmented into smaller pieces, and short adaptors are then ligated to blunt ends of the single-stranded fragments (Figure 16.3). These adaptor-flanked fragments are then mixed with small 28-μm streptavidin-coated beads whose surfaces have been covered with short sequences complementary to one of the ligated adaptors. Fragments are then hybridized to their corresponding beads in such a way that each bead carries a unique fragment. Fragment amplification must be performed for intensifying the light signal that is required for the precise detection of added bases by a CCD camera (Ansorge, 2009). Amplification of DNA fragments is performed through emulsion PCR and then loaded onto a Pico Titer Plate device so that each single bead is deposited in a well with the dimensions that can contain only one bead (Shendure, et al., 2005). The wells are then filled with smaller beads (1 μm) that carry immobilized enzymes required for pyrosequencing (sulfurylase and leuciferase) and also help beads to deposit in the wells gravitationally (Kircher and Kelso, 2010). Pico Titer Plates are centrifuged to ensure that the enzymes are in close contact with the DNA beads (Nowrousian, 2010). The loaded plates are placed in a 454 FLX instrument in which dNTPs and buffers are delivered to the wells of the plate. When a nucleotide is added to the growing chain, a signal is recorded by a CCD camera. The signal strength is proportional to the number of nucleotides. Signal strength can be precisely detected only for less than 10 consecutive nucleotides; after that, the signal declines rapidly (Mardis, 2008). As a result of this drawback, insertion and deletion error rates are the most frequently observed errors in this method.

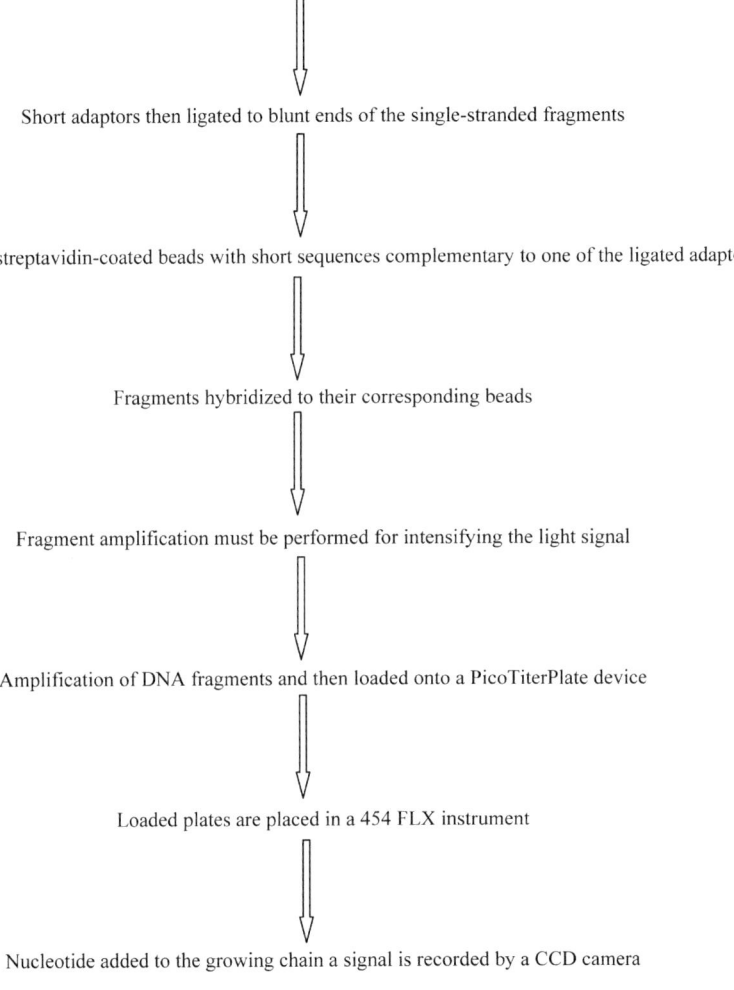

Genomic DNA fragmented into smaller pieces

Short adaptors then ligated to blunt ends of the single-stranded fragments

Mixed with streptavidin-coated beads with short sequences complementary to one of the ligated adaptor

Fragments hybridized to their corresponding beads

Fragment amplification must be performed for intensifying the light signal

Amplification of DNA fragments and then loaded onto a PicoTiterPlate device

Loaded plates are placed in a 454 FLX instrument

Nucleotide added to the growing chain a signal is recorded by a CCD camera

FIGURE 16.3   A schematic representation of 454 pyrosequencing technology.

16.3.4 BRIDGE SEQUENCING [ILLUMINA (SOLEXA) GENOME ANALYZER]

The Genome Analyzer was first introduced by Solexa (San Diego, CA, USA) and then further developed by Illumina (Nejad et al., 2013). In Illumina sequencing, a DNA library can be constructed by any method that generates adaptor-flanked fragments of several hundred base pairs in length

(Shendure, et al., 2005). During the annealing step, the extension product from one bound primer forms a bridge to the other bound primer, and its complementary strand is synthesized; it is therefore also called **bridge sequencing technology.** The amplified fragments form clusters, each of which contains approximately one million copies of the initial fragment. Moreover, each cluster contains both forward and reverse strands of the original sequences, but to have homogenous populations of strands in each cluster, which can be sequenced precisely without the interference of the complementary strand, one of the strands must be removed before initiating the sequencing process (Kircher and Kelso, 2010). The sequencer adopts the technology of sequencing by synthesis, which includes four reversible blocked nucleotides, which transiently block the extension of nucleotides, and the imaging step is carried out. After the acquisition of images in each cycle, the 3'-OH blocking group is chemically removed, and another cycle is subsequently started. The steps of adding nucleotides and other PCR reagents and capturing image is continued for a specific number of times and finally generates 100 bp reads.

16.3.5 LIMITATIONS OF THE NGS METHODS

1. Generate short-length reads that are not easy to assemble as genome sequences.
2. Sequencing is based on synthesis or hybridization reaction that uses as template millions of copies.
3. Each NGS platform provides its own software package for signal acquisition.
4. Sample preparation takes several days, which often involves additional equipment costs, chemicals and physical space.
5. Large amounts of data are generated, which created difficulties for their storage, analysis, and management.

The third-generation sequencing methods (Helicos Genetic Analysis System, Single-Molecule Real-Time Technology) are based on single DNA molecules, and they do not suffer from the above limitations. The single-molecule techniques used by nanopore-based fourth-generation DNA Sequencing technology allow us to further study the interaction not only between nucleic acid and protein but also between protein and protein. This technique opens a new door to molecular biology investigation at the single molecule scale.

16.3.6 COMPARATIVE GENOMICS

It is the study of similarities and differences of the structure and functions of the hereditary information among taxa. In plants, the evolution of small but essential segments in the genome occurs slowly, allowing the recognition of common intragenic regions between species that diverged a long time ago, as well as similar gene arrangements on the chromosomes, showing a synteny among genomes (Moore et al., 1995). A comparison of genomes at the sub-centimorgan level establishes that gene collinearity is maintained (Chen et al., 1997). The variations in the levels of collinearity and synteny are given by diverse factors, such as duplications and chromosomal segmentations, transposable element mobility (transposon and retrotransposon), gene deletion, and local rearrangements (Paterson et al., 2000). *Arabidopsis thaliana* is a model dicot species, but rice is largely used in comparative genome research. In monocot species, rice has economical and cultural value and is considered a model species because its genome is the smallest among the cereals that have economic values.

16.4 TRANSCRIPTOMICS

Transcriptomics is the field of molecular biology that studies the transcriptome: the complete set of transcripts in a cell, tissue, or organism, which includes the messenger RNA (mRNA) and noncoding RNA (ncRNA) molecules (Morozova et al., 2009). Transcriptome study helps the researchers in determining which sets of genes are turned on or off in a particular condition and to quantify the changes in gene expression among different biological conditions. With the advent of high-throughput sequencing technologies, it is also possible to map transcripts onto the genome, thus obtaining valuable information about gene structure, splicing patterns, and other transcriptional modifications (Wang and Snyder, 2009).

Numerous methods have been developed to identify and quantify the transcriptome. Such methods have evolved from the candidate gene-based detection of a few transcripts using northern blotting to high-throughput techniques, which detect thousands of transcripts simultaneously, such as microarrays and next-generation sequencing technologies (e.g., RNA-Seq) (Morozova et al., 2009). Methods of studying the transcriptome can also be divided into sequencing-based approaches (expressed sequence tags sequencing, serial analysis of gene expression, massive parallel signature sequencing, and RNA-seq) and hybridization-based approaches (microarray technology and tiling-arrays).

16.4.1   CONSTRUCTION OF CDNA LIBRARIES FOR EST SEQUENCING

Expressed sequence tags (EST) was first proposed by Adams et al. (1991). ESTs are short-sequence reads derived from complementary DNA (cDNA) sequences by partial sequencing. The information generated by ESTs was the main resource used to identify gene transcripts and find out gene expression levels in a given biological system.

Construction of a cDNA library and the generation of ESTs involve: the isolation of total RNA and purification of mRNAs; synthesis of cDNA using a reverse transcriptase enzyme; cloning of cDNA fragments; and sequencing of randomly selected clones using the Sanger method (Sanger et al., 1977) (Figure 16.4).

Both the 3' and 5' ends of a cDNA clone could be sequenced, resulting in ESTs ranging in size from 100 to 700 bp (Nagaraj et al., 2006). After generating ESTs, the next step is EST sequence analysis, which can be divided into pre-processing (generation of high-quality ESTs), clustering/assembly of ESTs on the basis of their sequence similarity, and EST annotation.

16.4.2   SUPPRESSION SUBTRACTIVE HYBRIDIZATION (SSH)

Suppression subtractive hybridization (SSH) is a method used for separating cDNA molecules that distinguish two related samples. In particular, the SSH technique can be applied to study transcriptomics, as it is a comparative method that examines the relative abundance of transcripts of a sample of interest in relation to a control sample (Lukianov et al., 1994; Diatchenko et al., 1996; Gurskaya et al., 1996). SSH is based on the principle of PCR suppression, which allows the amplification of desired sequences and at the

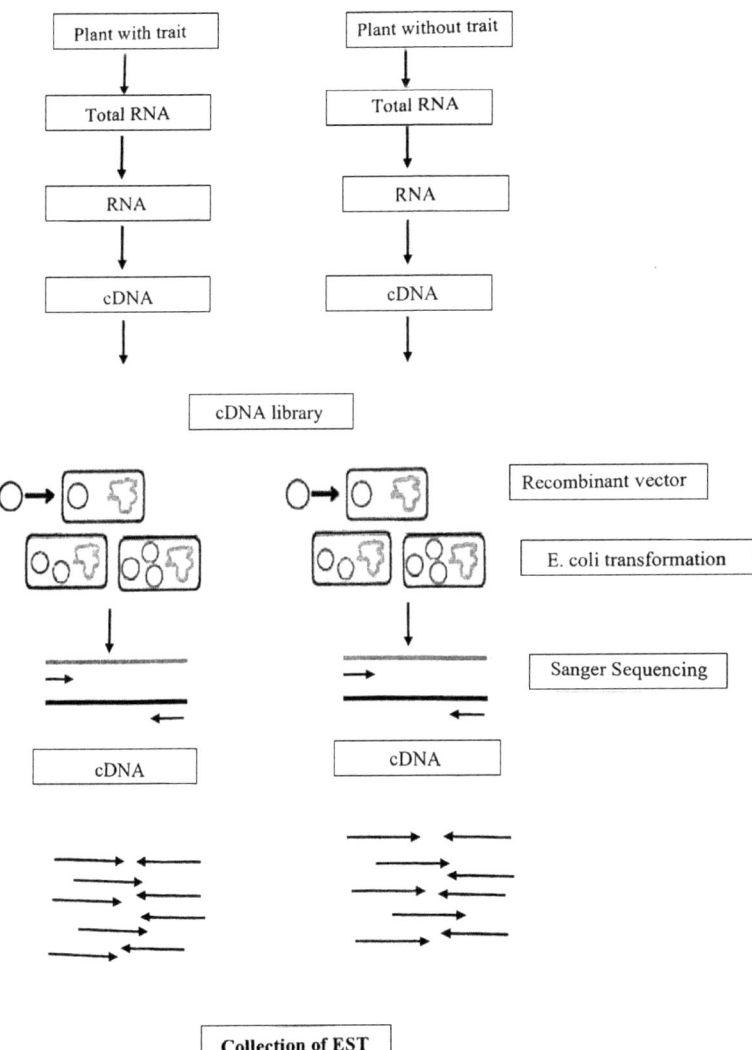

FIGURE 16.4   Overview of construction of cDNA libraries and EST sequencing.

same time suppresses the amplification of undesirable ones. In addition, SSH combines normalization and subtraction during the hybridization step, which removes cDNAs that are common between the sample of interest and control samples and normalizes cDNAs with different concentrations. Normalization occurs because during hybridization, more abundant cDNA molecules can anneal faster than less abundant cDNA fragments that remain single-stranded (Lukyanov et al., 2007). Finally, the method retains only differentially expressed or variable sequence transcripts that were present in the tester (Desai et al., 2001). These techniques can generate many false positives. To overcome this problem, a modified method, known as "**mirror orientation selection**" can be used to decrease the number of false-positive clones (Rebrikov et al., 2000).

16.4.3   SERIAL ANALYSIS OF GENE EXPRESSION (SAGE)

SAGE was the first sequencing-based method to quantify the abundance of thousands of transcripts simultaneously (Morozova et al., 2009). The basic principle is that a short DNA sequence of 9–11 bp, derived from a defined location within one transcript, contains enough information for the identification of that transcript, if the position of the sequence within the transcript is known. By counting the tags, SAGE provides an estimate of the abundance of the transcripts (Velculescu et al., 1995). Briefly, the method involves the generation of a library of clones containing concatenated short sequence tags, sequencing of clones using standard Sanger technique, matching of tag sequences in other databases to identify the transcripts, and finally counting of tags to estimate the relative abundance of the corresponding transcripts. The frequency of these transcripts in two or more SAGE libraries can be compared to distinguish differences in gene expression in the respective samples (Madden et al., 2000).

The main drawback of SAGE is the difficulty of identifying and annotating tags unambiguously, because short tags often match with multiple genes with similar coding sequences (Morozova et al., 2009). Some modifications in the original methodology were proposed, such as LongSAGE (Saha et al., 2002; Wei et al., 2004) and SuperSAGE (Matsumura et al., 2003), which result in tags with 21 and 26 bp, respectively, and DeepSAGE, which uses LongSAGE-derived ditags and replaces the Sanger sequencing with 454 pyrosequencing (Nielsen et al., 2006).

16.4.4 MICROARRAYS

Since their conception, microarrays have revolutionized the study of large-scale transcriptomes in different organisms and biological contexts (Schena et al., 1995). This technology consists of an array of single-stranded DNA molecules, called probes chemically linked onto a solid surface (glass slides), which are called chips. For comparison of different samples, mRNA is isolated from two plant populations (one with trait and other without trait) and used as templates for labeled cDNA synthesis with two different fluorescent properties (Figure 16.5).

Salazar et al. (2013) dissected genes and pathways underlying wood formation in three Eucalyptus species, and Rodrigues et al. (2013) dissected molecular mechanisms related to tolerance and susceptibility of citrus against *Xylella fastidiosa* by using RNA-seq.

Labeled cDNAs are hybridized with probes immobilized on the chip. Laser-attached fluorophores are excited to produce specific spectrums that are captured by a scanner. By using software, the fluorescence intensity is proportionally related to the level of gene expression, which enables the generation of the relative abundance of expression between samples (Brown, 1995).

16.4.5 RNA-SEQ

RNA-seq is a high-throughput sequencing of cDNA, which is based on the concept of direct sequencing of transcripts. This technique is more dynamic, reproducible, and provides a better estimate of the absolute expression levels (Nagalakshmi et al., 2008; Fu et al., 2009). RNA-seq allows us to identify isoforms of a gene, which are not easily detected using microarrays (Wang et al., 2009). RNA-seq uses NGS methods to sequence cDNA from RNA of biological samples, thus producing millions of short reads whose size depends on the platform used (Shendure and Ji, 2008).

16.5 PROTEOMICS

The study of the proteome involves the entire set of proteins expressed by the genome of a cell, but it can also be directed to cover just those proteins that are differentially expressed under specific conditions (Meire-

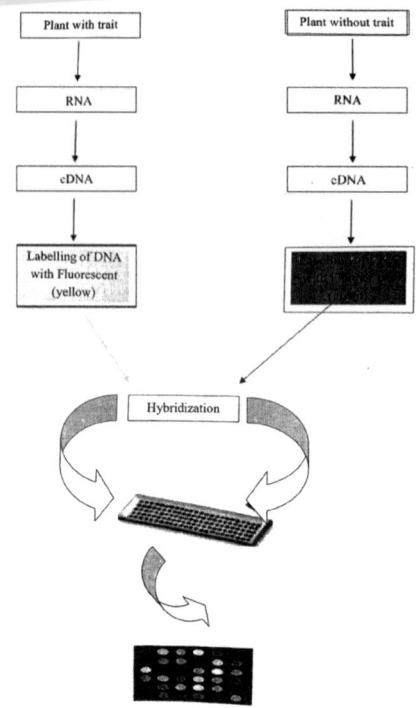

FIGURE 16.5   Gene expression analysis by using DNA microarray.

les, 2007). The first proteomics-related studies occurred in this context in the 1970s, when researchers began to map the proteins in a few organisms and created the first databases for proteins (Ofarrel, 1975; Klose, 1975; Scheele, 1975). The term "proteome," however, appeared only in 1994 to designate the group of proteins expressed by the genome of a cell in a given environment (Wilkins et al., 1995; Anderson and Anderson, 1996). At present, proteomics involves the functional analysis of gene products, a line of investigation conducted in a general area known as **"functional genomics"** which includes the large-scale identification, location, and compartmentalization of proteins, as well as the study and construction of protein interaction networks (Aebersold and Mann, 2003). The proteomic study in crop plants has several objectives, for example, to identify proteins that cause allergies in humans (Schenk et al., 2011) or to understand how the proteins are involved in the protection and response of plants subjected to biotic and abiotic stress (Kosova et al., 2011).

16.5.1 TECHNIQUES FOR EXTRACTION OF TOTAL PROTEINS

16.5.1.1 Extraction with TCA/Acetone

Damerval et al. (1986) developed this method to extract proteins from wheat seedlings. Because this method is simple and rapid, it is now used extensively to precipitate proteins from tissues or cells by treatment with trichloroacetic acid (TCA, 10% m/v) in acetone (–20°C) (Figure 16.6).

The method is based on protein denaturation in an acid and/or hydrophobic medium, which in turn helps the precipitation of proteins and the removal of interfering substances. Some protocols use direct precipitation of the protein extracts with TCA, but the use of TCA/acetone involves cleaning the pulverized tissue (Gorg et al., 1997). For further details on this protocol, please refer to Isaacson et al. (2006).

16.5.1.2 Phenolic Extraction

This method was originally developed by Hurkman and Tanaka (1986) for the isolation of membrane proteins, but it is used as an alternative to the extraction method with TCA/acetone for isolating total proteins of plants, especially when tissues are rich in phenolic compounds. The ground material is mixed with the extraction buffer containing phenol, and the phenol fraction is then precipitated with methanol or acetone. The pellet that is formed after the precipitation is normally white, but a yellow coloration may indicate that phenolic compounds have precipitated with the proteins (Figure 16.7).

When this occurs, maceration of the material with polyvinylpolypyrrolidone (PVPP) is recommended to help to remove the phenolic compounds (Wang et al., 2003). During the first extraction phase, all the material should be kept at 4°C; the recovery of the phenolic phase also requires a great deal of care (Faurobert et al., 2007). Furthermore, the phenol to be used should be equilibrated at a pH value of 8.0 by adjustment with Tris-HCl (Wang et al., 2008). As with TCA/acetone, the use of phenol helps to reduce protein degradation induced by endogenous proteolytic activity (Schuster and Davies, 1983). For the complete protocol, please refer to Isaacson et al. (2006).

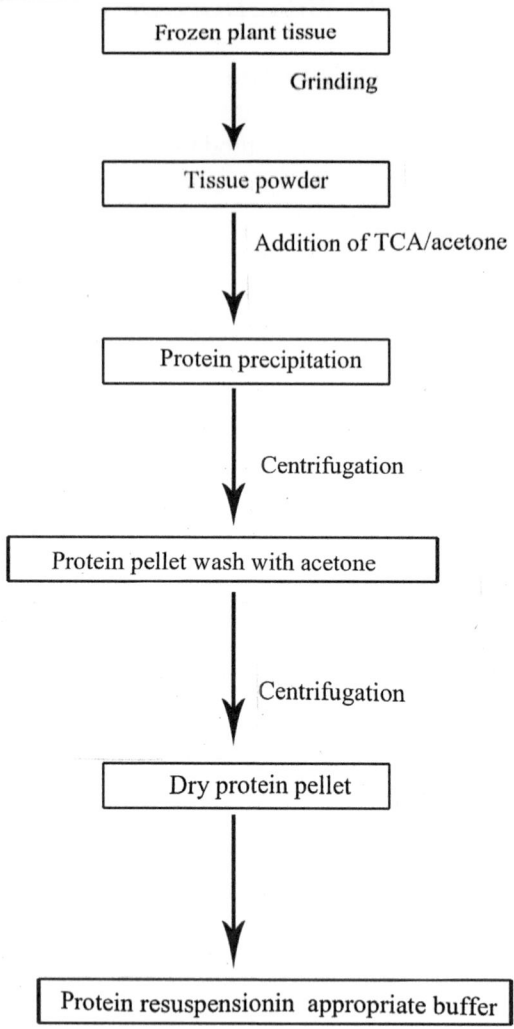

FIGURE 16.6 Extraction of total plant proteins with TCA/acetone.

16.5.2 *POST-TRANSLATIONAL MODIFICATIONS (PTMS)*

Post-translational modifications of proteins are covalent molecular changes that define the activity, location, turnover, and interaction of a protein with other proteins (Mann and Jensen, 2003). Thus, PTMs give the proteome an

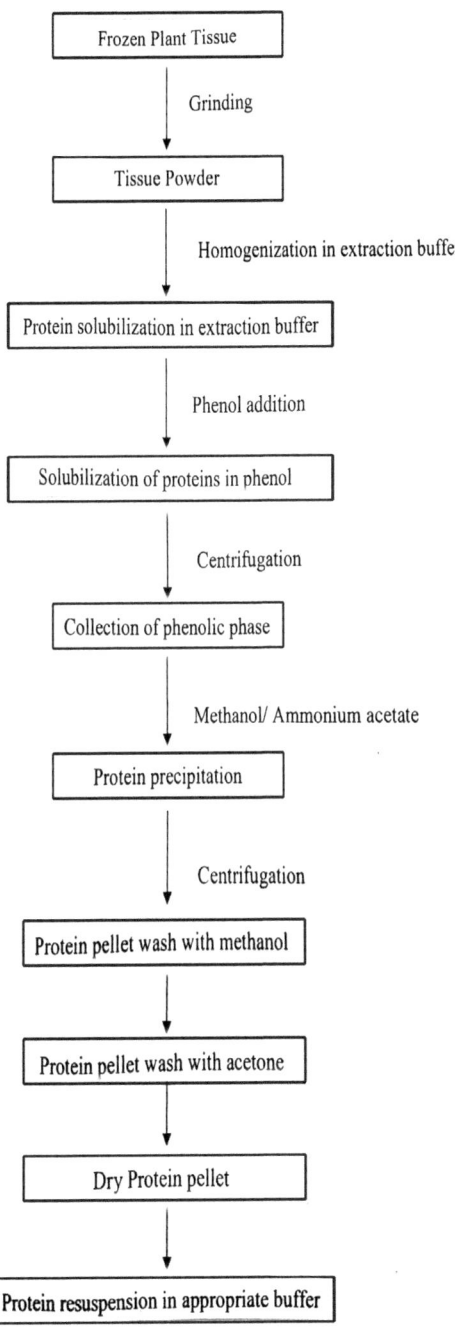

FIGURE 16.7   Extraction of total plant proteins with phenol.

enormous diversity, allowing the cell to respond flexibly to different stimuli (Howden and Huitema, 2012). Phosphorylation, glycosylation, ubiquitination, methylation, and acetylation are mostly responsible for PTMs. PTMs may change the chemical state of a protein statically or dynamically, and in such a subtle manner that it may not be easily detected by standard protein profiling techniques that use gel (Simon and Cravatt, 2008).

16.5.3 QUANTITATIVE PROTEOMICS

Proteomics based on mass spectrometry (Aebersold and Mann, 2003) play an important role in the understanding of fundamental biological processes, which has led to the development of methods that have capacity to quantify several thousands of proteins. The SILAC technique used for labeling amino acids with stable isotopes in cell culture (Ong et al., 2002) is the best-known method, particularly for its application to animal cells. Some modifications in SILAC are also present, which include pulsed-SILAC (Wu et al., 2012) and super-SILAC (Geiger et al., 2011). Direct coupling of reverse-phase liquid chromatography systems to mass spectrometers form the most sophisticated quantification technique presently in use. Protein identification is done by large-scale, multidimensional separation of peptides by using correlation of peptides to protein, which help in quantification of the protein without labeled with stable isotopes. The methods that use the intensity of the MS signal are more commonly employed for absolute quantifications (Schwanhausser et al., 2011). High3 (or Top3) is an alternative quantification method (Silva et al., 2006) for proteins.

16.6 METABOLOMICS

The actual technologies that make a qualitative and quantitative overview of metabolites present in an organism possible are called metabolomics (Hall, 2006). Plants metabolize more than 200,000 different molecules, which are designed for structuring and maintenance of tissues and organs as well as for the physiological processes associated with growth, development, reproduction, and defense (Weckwerth, 2003). Metabolic pathways are interconnected, dependent on other pathways by their substrates or products, and are also influenced by different environmental signals as well as being strongly dependent on genetic components and various levels of gene regulation.

Various techniques for separation and quantification of metabolites can be used simultaneously or in combination for the measurement of a particular group of metabolites (Dunn and Ellis, 2005). To identify and quantify compounds, it is necessary to use methodologies and devices according to the specific characteristics of each group of compounds (Table 16.2).

There are large number of applications for metabolomic technologies, including the study of metabolic pathways (Saito and Matsuda, 2010), metabolic engineering of plants (Rischer and Oksman, 2006), secondary metabolism of plants (Keurentjes, 2009), plant nutrition (Keurentjes, 2009), plant development (Peluffo et al., 2010), and plant phenotyping (Riedelsheimer et al., 2012). Metabolomics as well as other omics technologies such as transcriptomics and proteomics can be used to monitor genetic effects not only under normal growth conditions but also under stress, thereby allowing the estimation and evaluation of the potential risks associated with transgenic and other changes in the genome (Rischer and Oksman-Caldentey, 2006).

TABLE 16.2   Technologies Commonly Used in Analysis of Plant Metabolisms

Technology	Applications
GC-MS (Gas chromatography coupled to mass spectrometry)	The analysis of the lipophilic or polar compounds (e.g., sugars, organic acids, vitamins, tocopherols)
GC–GC–MS (Gas chromatography-Gas chromatography coupled to mass spectrometry)	Similar to GC-MS, but with a better separation of co-eluting compounds due to increased sensitivity of the combination GC–GC
SPME (solid phase micro extraction)	Analysis of volatile compounds (e.g., aromatic compounds, repellents)
CE-MS (Capillary electrophoresis coupled to mass spectrometry)	Analysis of polar compounds (fatty amines, Co-A derivatives, sugars, organic acids, vitamins, tocopherols)
LC-MS (Liquid chromatography coupled to mass spectrometry)	Analysis of secondary metabolites (mainly carotenoids, flavonoids, glucosinolates, vitamins)
FT-ICR-MS (Fourier transform ioncyclotron resonance mass spectrometry)	High resolution MS in combination with LC is highly potent. Identification of unknown metabolites by the mass/charge ratio
NMR (Nuclear magnetic resonance)	Non-destructive analysis of abundant metabolites in a sample
FAIMS (field asymmetric waveform ion mobility spectrometry)	Allows selection of specific ions, reducing ion suppression and matrix effects

(Source: Nejad, A. M., et al., 2013)

16.7 PHYSIONOMICS

Plant physiology integrates the entire knowledge of anatomy, morphology, and biochemistry to study functional processes ranging from seed germination to maturity. Plant physiology includes plant-water relations, light and shade, mineral nutrition, and biotic and abiotic stresses as well as the developmental process that make the plant body. Plant development is controlled by signaling molecules, which are present in little amount. Such molecules are known as plant hormones, and they work as signaling molecules; they participate in the plants' response to the environments, organogenesis, and growth and development.

The substance responsible for the growth of the oat (*Avena sativa*) coleoptile was named **Auxin** by the Dutch botanist Went (1928). In 1934, discovered the hormone gibberellin (GA) when studying rice (*Oryza sativa*) plants infected by the fungus *Gibberella fujikuroi* (*Fusarium moniliforme*) (Takahashi, 1998). In 1963, abscisic acid (ABA) was isolated from cotton (*Gossypium sp.*) and other plant species for the first time. Physiology is not a single technique but rather a combination of deep knowledge in itself, advanced by the appearance of new techniques, of which the **omics** is but the latest in a long series. Ultimately, because physiology aims to understand how plants function, the closest approach to **physionomics** would be **phenomics** (Berger et al., 2010; Furbank and Tester, 2011; Tisne et al., 2013). Phenomics may be defined as a series of high throughput techniques devised to automate and increase our capacity to identify and quantify phenotypes, eventually explaining them mechanistically through their genes, transcripts, proteins, and metabolites.

16.8 PHENOMICS

In conventional plant breeding, breeder's studies the nature of genetic variation in segregating populations is evaluated in multi-location/ years trials. Thus, with the use of appropriate experimental designs, researchers have obtained relevant information on the genetic diversity, heritability, environment influence, genetic value, and the best strategies to agronomically select superior crop varieties (Cobb et al., 2013). In general, accurate phenotyping is far more difficult than accurate genotyping for a variety of reasons, including the vast number of phenotypic traits and their sensitivity to the environmental factors (Table 16.3).

Plant phenomics is described as the study of growth, plant architecture, performance under different environmental condition and composition using high-throughput methods of data acquisition and analysis. Forward phenomics uses several methods to screen germplasm collections for specific traits. Reverse phenomics, on the other hand, consists analysis of a trait to unravel the various physiological, biochemical, and biophysical processes, and the genes involved in the regulation of the trait.

16.8.1 EXAMPLES OF LARGE-SCALE PHENOTYPING

In the genomics era, pattern has changed toward the evaluation of diversity at genetic level directly, because the pattern of the study of phenotype changed

TABLE 16.3 Comparisons between Various Features of Genotype and Phenotype

Sr. No.	Features	Genotype	Phenotype
1.	Complete set information	Genome	Phenome
2.	Study under discipline	Genomics	Phenomics
3.	Definition	Genetic makeup of an individual	Physical appearance
4.	Level of description	DNA sequence	Organism and even groups of individuals
5.	Identification	Difficult: require special techniques	Easy
6.	Environmental effect	No effect	Effected by environment
7.	Effect of growth stage	Not effected	Effected
8.	Expression effected by location	Not effected	Expression change with tissue and cell type
9.	Proper explanation of outcomes	Always possible	Extremely difficult to explain variation caused by environmental interaction with genotype
10.	Identification of complete set of components	Relatively simple	Complex

(Source: Singh and Singh, 2015)

due to various phenotyping tools instead of full season field evaluation. Due to the decreasing cost of genotyping technologies and whole genome sequencing, there is a strong motivation for the development of mapping populations in many species in order to understand the genetic control of complex traits and to discover useful genetics that can be easily deployed (McCouch et al., 2012). This effort has been a catalyst for new ideas about how to manipulate genetic variation in order to characterize and create relevant germplasm (Chen et al., 2011).

The phenotype of a plant is dynamic and unpredictable due to complex responses to endogenous and exogenous signals that are received cumulatively throughout the life of an organism. For understanding the genetic variation that underlies complex traits, it is necessary to carry out phenotyping activities with high accuracy and precision. Phenomics or next-generation phenotyping is understood to be the pursuit of accurate, precise, relevant, and economical phenotyping efforts (Cobb et al., 2013). The phenomics approaches, in general, evaluate many or all of the variable traits of relatively large populations that are grown in multiple environments (Table 16.4).

TABLE 16.4   Examples of Phenomics Approaches

Sr. No.	Name of instrument	Observation	References
1.	Digital cameras	Leaf area, rosette growth and the measurement of the characteristics of tissues, organs or individuals	Tisne et al. (2013)
2.	Infrared cameras	Visualize temperature gradients, which can indicate the degree of energy dissipation	Munns et al. (2010)
3.	Fluorescence detectors	Identify the differential responses of populations of seedlings, fruits or seeds to a stressor	Jansen et al. (2009)
4.	Light Detection and Ranging (LIDAR)	Growth rate through differences between small distances measured using a laser	Hosoi and Omasa (2009)

16.9   CONCLUDING REMARKS

The "omics" approach integrates genome, transcriptome, proteome, metabolome, physiognome, and phenome data into a single dataset and can lead to the identification of unknown genes and their regulatory networks involved in metabolic pathways of interest. The importance of determining the entire genome sequence of crop plant and other organisms was recognized more than three decades ago and was an important first step in ushering the field of genomics. DNA sequencing technology allows us to further study the interaction not only between nucleic acid and protein, but also between protein and protein. Transcriptome study helps the researchers in determining which sets of genes are turned on or off in a particular condition and quantify the changes in gene expression among different biological conditions. At present, proteomics involves the functional analysis of gene products, a line of investigation conducted in a general area known as **functional genomics** that includes the large-scale identification, location, and compartmentalization of proteins as well as the study and construction of protein interaction networks (Aebersold and Mann, 2003). Metabolomics as well as other omics technologies can be used to monitor genetic effects not only under normal growth conditions but also under stress, thus allowing the estimation and evaluation of the potential risks associated with transgenic changes in the genome. Plant physiology integrates all the knowledge of anatomy, morphology, and biochemistry to study functional processes ranging from seed germination to maturity. Phenomics or next-generation phenotyping is understood to be the pursuit of accurate, precise, relevant, and economical phenotyping efforts (Cobb et al., 2013).

KEYWORDS

- **genomics**
- **metabolomics**
- **phenomics**
- **physiognomics**
- **proteomics**
- **transcriptomics**

REFERENCES

Adams M. D., et al., (1991). Complementary DNA sequencing: expressed sequence tags and human genome project. *Science, 252*(5013), 1651–1656.

Aebersold, R. H., & Mann, M., (2003). Mass spectrometry-based proteomics. *Nature, 422*(6928), 198–207.

Anderson, N. G., & Anderson, N. L., (1996). Twenty years of two-dimensional electrophoresis: past, present and future. *Electrophoresis, 17*(3), 443–453.

Ansorge, W. J., (2009). Next-generation DNA sequencing techniques. *New Biotechnology, 25*(4), 195–203.

Berger, B., Parent, B., & Tester, M., (2010). High-throughput shoot imaging to study drought responses. *Journal of Experimental Botany, 61*(13), 3519–3528.

Borém, A., Fritsche-Neto, R., 2013. Biotecnologia aplicada ao melhoramento de plantas. 6th ed. Viscone do Rio Branco: Suprema Publishers, 336 pp.

Brown, O. P., (1995). Quantitative monitoring of gene expression patterns with a complementary DNA microarray. *Science, 270,* 467–470.

Chen, M., et al., (1997). Micro-colinearity in sh2-homologous regions of the maize, rice, and sorghum genomes. *Proceedings of the National Academy of Sciences USA, 94*(7), 3431–3435.

Chen, X., et al., (2011). Phosphoproteins regulated by heat stress in rice leaves. *Proteome Science, 9,* 37.

Cobb, J. N., et al., (2013). Next-generation phenotyping: requirements and strategies for enhancing our understanding of genotype–phenotype relationships and its relevance to crop improvement. *Theoretical and Applied Genetics, 126,* 867–887.

Damerval, C., et al., (1986). The technical improvements in two-dimensional electrophoresis increase the level of genetic variation detected in wheat-seedling proteins. *Electrophoresis, 7,* 52–54.

Desai, S., et al., (2001). Identification of differentially expressed genes by suppression subtractive hybridization. In: Stephen, H., & Levesey, R., (eds.). Func*tional Genomics: A Practical Approach.* Oxford: Oxford, University Press, pp. 45–80.

Diatchenko, L., et al., (1996). Suppression subtractive hybridization: a method for generating differentially regulated or tissue-specific cDNA probes and libraries. *Proceedings of the National Academy of Sciences USA, 93*(12), 6025–6030.

Dunn, W. B., & Ellis, D. I., (2005). Metabolomics: current analytical platforms and methodologies. *Trends in Analytical Chemistry, Amsterdam., 24,* 285–294.

Edwards, M., (2013). Whole-genome sequencing for marker discovery. In: Henry, R. J., (ed.) *Molecular Markers in Plants.* John Wiley & Sons, Inc., Ames, Iowa, USA. pp. 21–34.

Faurobert, M., Pelpoir, E., & Chaib, J., (2007). Phenol extraction of proteins for proteomic studies of recalcitrant plant tissues. *Methods in Molecular Biology, 355,* 9–14.

Feng, Y., et al., (2015). Nanopore-based Fourth-generation DNA Sequencing Technology. *Genomics Proteomics Bioinformatics., 13,* 4–16.

Fu, X., et al.,(2009). Estimating accuracy of RNA-Seq and microarrays with proteomics. *BMC Genomics., 10*(1), 161.

Furbank, R. T., & Tester, M., (2011). Phenomics – technologies to relieve the phenotyping bottleneck. *Trends in Plant Science, 16*(12), 635–644.

Geiger, T., et al., (2011). Use of stable isotope labeling by amino acids in cell culture as a spike-in standard in quantitative proteomics. *Nature Protocols*, *6*(2), 147–157.

Gorg, A., et al., (1997). Very alkaline immobilized pHgradients for two-dimensional electrophoresis of ribosomal and nuclearproteins. *Electrophoresis.*, *18*, 328–37.

Gurskaya, N. G., et al., (1996). Equalizing cDNA subtraction based on selective suppression of polymerase chain reaction: cloning of Jurkat cell transcripts induced by phytohemaglutinin and phorbol 12-myristate 13-acetate. *Annual Biochemistry*, *240*, 90–97.

Hall, R. D., (2006). Plant metabolomics: from holistic hope, to hype, to hot topic. *New Phytologist.*, *169*, 453–468.

Hosoi, F., & Omasa, K., (2009). Detecting seasonal change of broad-leaved woody canopy leaf area density profile using 3D portable LIDAR imaging. *Functional Plant Biology*, *36*(11), 998–1005.

Howden, A. J. M., & Huitema, E., (2012). Effector-triggered post-translational modifications and their role in suppression of plant immunity. *Frontiers in Plant Science*, *3*, 160.

Hurkman, W. J., & Tanaka, C. K., (1986). Solubilization of plant membrane proteins for analysis by two-dimensional gel electrophoresis. *Plant Physiology*, *81*, 802–806.

International Rice Genome Sequencing Project, (2005). The map-based sequence of the rice genome. *Nature*, *1*(7052), 793–800.

Isaacson, T., et al., (2006). Sample extraction techniques for enhanced proteomic analysis of plant tissues. *Nature Protocols*, *1*, 769–774.

Jansen, M., et al., (2009). Simultaneous phenotyping of leaf growth and chlorophyll fluorescence via Growscreen Fluoro allows detection of stress tolerance in Arabidopsis thaliana and other rosette plants. *Functional Plant Biology*, *36*, 902–914.

Keurentjes, J. B., (2009). Genetical metabolomics: Closing in on phenotypes. *Current Opinion inPlant Biology*, *12*, 223–230.

Kircher, M., & Kelso, J., (2010). High-throughput DNA sequencing–concepts and limitations. *BioEssays*, *32*(6), 524–536.

Klose, J., (1975). Protein mapping by combined isoeletric focusing and electrophoresis of mouse tissues. Anovel approach to testing for induced point mutations in mammals. *Humangenetik, Berlin.*, *26*(3), 231–243.

Kosova, K., et al., (2011). Plant proteome changes under abiotic stress – Contribution of proteomics studies to understanding plant stress response. *Review J. Proteomics.*, *74*, 1301–1322.

Lukianov, S. A., et al., (1994). Highly-effective subtractive hybridization of cDNA. *Bioorganicheskaya khimiya*, *20*, 701–704.

Lukyanov, S. A., Rebrikov, D., & Buzdin, A., A., (2007). Suppression subtractive hybridization. In: Buzdin, A. A. and Lukyanov S. A. (eds). Nucleic Acids Hybridization Modern Applications. *Dordrecht, The Netherlands:* Springer, pp. 53–84.

Madden, S. L., Wang, C. J., & Landes, G., (2000). Serial analysis of gene expression: from gene discovery to target identification. *Drug Discovery Today*, *5*(9), 415–425.

Mann, M., & Jensen, O. N., (2003). Proteomic analysis of post-translational modifications. *Nature Biotechnology*, *21*, 255–261.

Mardis, E. R., (2008). Next-generation DNA sequencing methods. *Annual Review of Genomics and Human Genetics*, *9*, 387–402.

Matsumura, H., et al., (2003). Gene expression analysis of plant host–pathogen interactions by SuperSAGE, *Proceedings of the National Academy of Sciences USA*, *100*, 15718–15723.

Maxam, A. M., & Gilbert, W., (1977). A new method for sequencing DNA. *Proceedings of the National Academy of Sciences USA, 74*(2), 560–564.

McCouch, S. R., et al., (2012). Genomics of gene banks: a case study in rice. *American Journal of Botany, 99,* 407–423.

Meireles, K. G. X., (2007). Aplicacoes da Proteomics na Pesquisa Vegetal. *Embrapa., 165,* 41.

Moore, G., et al., (1995). Grasses, line up and form a circle. *Current Biology, 5*(7), 737–739.

Morozova, O., Hirst, M., & Marra, M. A., (2009). Applications of new sequencing technologies for transcriptome analysis. *Annual Review of Genomics and Human Genetics, 10,* 135–151.

Munns, R., et al., (2010). New phenotyping methods for screening wheat and barley for beneficial responses to water deficit. *Journal of Experimental Botany, 61*(13), 3499–3507.

Mullis, K., et al., (1987). Specific enzymatic amplification of DNA *in vitro* : The polymerase chain reaction. *Cold Spring Harbor Symposia on Quantitative Biology, 51,* 263–273.

Nagalakshmi, U., et al., (2008). The transcriptional landscape of the yeast genome defined by RNA sequencing. *Science, 320,* 1344–1349.

Nagaraj, S. H., Gasser, R. B., & Ranganathan S., (2006). A hitchhiker's guide to expressed sequence tag (EST) analysis. *Briefings in Bioinformatics., 8*(1), 6–21.

Nejad, A. M., et al., (2013). *Next Generation Sequencing and Sequence Assembly.* Springer New York Heidelberg Dordrecht London, pp. 16.

Nielsen, K. L., Hogh, A. L., & Emmersen, J., (2006). Deep, SAGE–digital transcriptomics with high sensitivity, simple experimental protocol and multiplexing of samples. *Nucleic Acids Research, 34,* 133.

Nowrousian, M., (2010). Next-generation sequencing techniques for eukaryotic microorganisms: Sequencing-based solutions to biological problems. *Eukaryotic Cell, 9*(9), 1300–1310.

Ofarrel, P. H., (1975). High resolution two-dimensional electrophoresis of proteins. *The Journal of Biological Chemistry, 250*(10), 4007–4021.

Ong, S. E., et al., (2002). Stable isotope labeling by amino acids in cell culture, SILAC, as a simple and accurate approach to expression proteomics. *Molecular and Cellular Proteomics., 1*(5), 376–386.

Pandey, A., et al., (2008). Proteomics approach to identify dehydration responsive nuclear proteins from chickpea (*Cicer arietinum* L.). *Molecular and Cellular Proteomics., 7,* 88–107.

Paterson, A. H., et al., (2000). Comparative genomics of plant chromosomes. *Plant Cell, 12*(9), 1523–1540.

Peluffo, L., et al., (2010). Metabolic profiles of sunflower genotypes with contrasting response to *Sclerotinia sclerotiorum* infection. *Phytochemistry, 71,* 70–80.

Rafalski, J. A., Tingey, S. V., & Williams, J. G. K., (1990). RAPD markers, a new technology for genetic mapping and plant breeding. *AgBiotech News and Information, 3,* 645–648.

Rebrikov, D. V., et al., (2000). Mirror orientation selection (MOS): a method for eliminating false positive clones from libraries generated by suppression subtractive hybridization. *Nucleic Acids Research, 28,* 90.

Riedelsheimer, C., et al., (2012). Genomic and metabolic prediction of complex heterotic traits in hybrid maize. *Nature Genetics., 44,* 217–220.

Rischer, H., & Oksman-Caldentey, K. M., (2006). Unintended effects in genetically modified crops: revealed by metabolomics? *Trends in Biotechnology, 24,* 102–104.

Rodrigues, C. M., et al., (2013). RNA-Seq analysis of Citrus reticulata in the early stages of Xylella fastidiosa infection reveals auxin-related genes as a defense response. *BMC Genomics*, *14*(1), 676.

Saha, S., et al., (2002). Using the transcriptome to annotate the genome. *Nature Biotechnology*, *20*, 508–512.

Saito, K., & Matsuda, F., (2010). Metabolomics for functional genomics, systems biology, and biotechnology. *Annual Review of Plant Biology*, *61*, 463–489.

Salazar, M., M., et al., (2013). Xylem transcription profiles indicate potential metabolic responses for economically relevant characteristics of Eucalyptus species. *BMC Genomics*, *14*(1), 201.

Sanger, F., Nicklen, S., & Coulson, A. R., (1977). DNA sequencing with chain terminating inhibitors. *Proceedings of the National Academy of Sciences USA*, *74*, 5463–5467.

Schadt, E. E., Turner, S., & Kasarskis, A., (2010). A window into third-generation sequencing. *Hum. Mol. Genet.*, *19*, 227–240.

Scheele, G. A., (1975). Two-dimensional gel analysis of soluble proteins. Characterization of guinea pig exocrine pancreatic proteins. *The Journal of Biological Chemistry, Bethesda.*, *250*(14), 5375–5385.

Schena, M., et al., (1995). Quantitative monitoring of gene expression patterns with a complementary DNA microarray. *Science*, *270*(5235): 467–470.

Schenk, M. F., et al., (2011). Proteomic analysis of the major birch allergen Bet v 1 predicts allergenicity for 15 birch species. *Journal of Proteomics.*, *74*, 1290–1300.

Schuster, A. M., & Davies, E., (1983). Ribonucleic acid and protein metabolism in pea epicotyls. II. Response to wounding in aged tissue. *Plant Physiology*, *73*, 817–821.

Schwanhausser, B., et al., (2011). Global quantification of mammalian gene expression control. *Nature.*, *473*(7347), 337–342.

Shendure, J., & Ji, H., (2008). Next-generation DNA sequencing. *Nature Biotechnology*, *26*, 1135–1145.

Shendure, J., et al., (2005). Accurate multiplex polony sequencing of an evolved bacterial genome. *Science*, *309*(5741), 1728–1732.

Silva, J., C., et al., (2006). Absolute quantification of proteins by LCMSE – A virtue of parallel MS acquisition. *Molecular and Cellular Proteomics.*, *5*(1), 144–156.

Simon, G. M., & Cravatt, B. F., (2008). Challenges for the 'chemical-systems' biologist. *Natural Chemistry Biology*, *4*, 639–642.

Singh, B.D, and Singh, A. K. 2015. Marker-Assisted Plant Breeding: Principles and Practices. Springer India.

Takahashi, N., (1998). Discovery of gibberellin. *Discoveries in Plant Biology*, Singapore: *World Scientific* Publishing Co., *1*, 17–32.

Tester, M., & Langridge, P., (2010). Breeding technologies to increase crop production in a changing world. *Science*, *327*, 818–822.

The Arabidopsis Genome Initiative, (2000). Analysis of the genome sequence of the flowering plant Arabidopsis thaliana. *Nature*, *408*(6814), 796–815.

Tisne, S., et al., (2013). Phenoscope: an automated large-scale phenotyping platform offering high spatial homogeneity. *The Plant Journal*, *74*(3), 534–544.

Varshney, R. K., Graner, A., & Sorrells, M. E., (2005). Genomics-assisted breeding for crop improvement. *Trends in Plant Science*, *10*(12), 621–630.

Varshney, R. K., et al., (2009). Next-generation sequencing technologies and their implications for crop breeding. *Trends in Biotechnology*, *27*(9), 522–530.

Velculescu, V. E., et al., (1995). Serial analysis of gene expression. *Science-AAAS-Weekly Paper Edition. 270*(5235), 484–486.

Watson, J. D., & Crick, F. H. C., (1953). Genetical Implications of the Structure of Desoxyribosenucleic Acid. Nature, *171*, 964–967.

Wang, W., Tai, F., & Chen, S., (2008). Optimizing protein extraction from plant tissues for enhanced proteomics analysis. *Journal of Separation Science, 31*, 2032–2039.

Wang, Z., Gerstein, M., & Snyder, M., (2009). RNA-Seq: a revolutionary tool for transcriptomics. *Nature Reviews Genetics, 10*(1), 57–63.

Wang, W., et al.,(2003). Protein extraction for two dimensional electrophoresis from olive leaf, a plant tissue containing high levels of interfering compounds. *Electrophoresis, 24*, 2369–2375.

Weckwerth, W., (2003). Metabolomics in systems biology. *Annual Review of Plant Biology, 54*, 669–689.

Wei, C. L., et al., (2004). 5' Long serial analysis of gene expression (LongSAGE) and 3' LongSAGE for transcriptome characterization and genome annotation. *Proceedings of the National Academy of Sciences USA, 101*, 11701–11706.

Went, F. W., (1928). "Wuchsstoff und Wachstum." *Rec. Trav. Bot. Neerland., 24*, 1–116.

Wilkins, M. R., et al., (1995). Progress with proteome projects: why all proteins expressed by a genome should be identified and how to do it. *Biotechnology and Genetic Engineering Reviews, 13*, 19–50.

Wu, Z., Moghaddas Gholami, A., & Kuster, B., (2012). Systematic identification of the HSP90 regulated proteome. *Molecular and Cellular Proteomics, 11*, 1–14.

CHAPTER 17

RNA INTERFERENCE (RNAi) TECHNOLOGY IN BIOLOGY AND PLANT BREEDING

DINESH NARAYAN BHARADWAJ

Ex-Professor and Head, Department of Genetics and Plant breeding, C.S. Azad University of Agriculture and Technology, Kanpur–208002, India, Tel.: 7275380950, E-mail: prof.bharadwaj52@gmail.com

CONTENTS

17.1 INTRODUCTION

RNA interference (RNAi) is a naturally occurring mechanism that leads to the "silencing" of genes; consequently, the respective targeted protein is no longer synthesized. In nature, this mechanism is used for the regulation of specific genes and as defense mechanisms against viruses and transposons. Now, this technique can be used for loss-of-function studies where a gene is specifically silenced, and the impact of this loss is analyzed in cells or whole organisms. This can be performed under normal conditions or in the context of disease control.

RNAi is a biological process in which RNA molecules inhibit gene expression, typically by causing the destruction of specific mRNA molecules. RNAi is one of the most important technological breakthroughs in molecular biology, which allow biologists to directly observe the effects of the loss of function of specific genes in organisms (Figure 17.1). In the RNAi process, translation of a protein is prevented by selective degradation of its encoded mRNA. In nature, this mechanism likely evolved for cells to eliminate unwanted foreign gene expression as a defense mechanism against viruses. RNAi has had significant impact of gene function analysis in organisms. RNAi technology can validate target genes and functionally assess relevant disease genes. Thus, it leads to the development of effective therapeutics. RNAi was discovered by Andrew Fire and

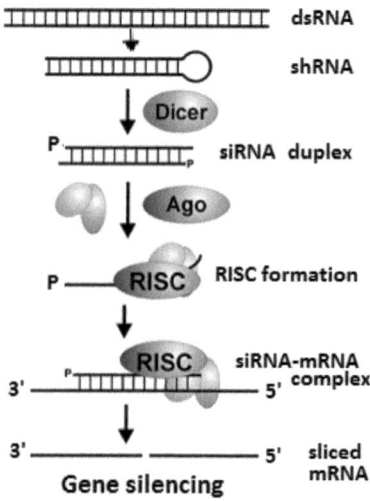

FIGURE 17.1 Process of gene silencing.

Craig Mello, for which they earned the Nobel Prize for Physiology in 2006.

During early 1990s, several scientists independently worked on inhibition of RNA protein expression in various plants and fungi. RNA interference (RNAi) is a natural process that cells use to "turn off" or silence unwanted or harmful genes. The initial discovery of this phenomenon was made in 1991, by scientists trying to deepen the color of petunias. Surprisingly, by introducing a gene for color, they found that the gene was turned off. Andrew Fire and Craig Mello (1998) in *Caenorhabditis elegans* observed that double-stranded RNA (dsRNA) was the source of sequence-specific inhibition of protein expression, which they called "RNA interference (RNAi)."

The well-studied outcome is posttranscriptional gene silencing (PTGS), which occurs when the guide strand pairs with a complementary sequence in a messenger RNA molecule and induces cleavage by Argonaute, the catalytic component of the RISC complex. In some organisms, this process spreads systemically, despite the initially limited molar concentrations of small interfering RNA (siRNA). RNAi is a valuable research tool, both in cell culture and living organisms, because synthetic dsRNA introduced into cells can selectively and robustly induce suppression of specific genes of interest. RNAi may be used for large-scale screens that systematically shut down each gene in the cell, which can help to identify the components necessary for a particular cellular process or an event such as cell division. The pathway is also used as a practical tool in biotechnology, medicine, and insecticides.

17.2 PRINCIPLES OF RNA INTERFERENCE

RNAi naturally evolved as a cell defense mechanism to eliminate effects of unwanted foreign genes that are often present in high copy numbers, such as viral genes, transposable elements, plasmids, and bacteria introduced in the cell. It has been observed that the expression of transgenes usually decreases as the number of copies increases in the cell. Similarly, the expression of endogenous homologous genes is also suppressed by the presence of the transgene (Napoli et al., 1990). Although such gene silencing can occur at the transcriptional level, the major mechanism of gene suppression or silencing occurs post-transcriptionally known as PTGS via RNAi (Pruss et al., 1997). This selective degradation of mRNAs is targeted by siRNAs (Van

Blokland et al., 1994). RNAi is defined as a phenomenon leading to PTGS after endogenous production or artificial introduction into a cell of siRNA with sequences complementary to the targeted gene (Bosher and Labouesse, 2000). Whereas the transcription of the gene is normal, the translation of the protein is prevented by selective degradation of its encoded mRNA. However, PTGS has emerged as a complex mechanism that involves several proteins and small RNAs. It is supposed that cells from prokaryotes to eukaryotes employ RNAi to firmly regulate protein levels in response to various environmental stimuli, and its fundamental importance in the selective suppression of protein translation by targeted degradation of the encoding mRNA.

RNAi technology is now used to identify and functionally assess the thousands of genes within the genome that potentially participate in disease development. This technology provides an efficient means in blocking of a specific gene expression and evaluating their response to chemical compounds and in signaling pathways of organisms. RNAi technology takes advantage of the cell's natural machinery, facilitated by siRNA molecules, to effectively knock down expression of a gene of interest. There are several ways to induce RNAi, such as by using synthetic molecules, RNAi vectors, and *in vitro* dicing. In mammalian cells, short pieces of dsRNA, siRNA about 20-25 nucleotides long, initiate the specific degradation of a targeted cellular mRNA. The antisense strand of the siRNA duplex becomes part of a multiprotein complex or RNA-induced silencing complex (RISC), which then identifies and separates the corresponding mRNA in two strands, i.e., passenger and guide strand, and cleaves it at a specific site. The passenger strand is degraded, while the RISC takes the guide strand to a specific mRNA site. Then, this cleaved message is targeted for degradation, which ultimately results in the loss of protein expression and unwanted target protein is not produced.

17.2.1 *POST-TRANSCRIPTIONAL GENE SILENCING (PTGS) AND THE DISCOVERY OF RNA INTERFERENCE*

In eukaryotic cells, PTGS and RNAi were discovered during genetic transformation. It was noticed that in plants (*Arabidopsis thaliana)* and the roundworm (*Caenorhabditis elegans),* the expression of mRNAs for the encoded transgene alone or together for homologous endogenous genes is very low or absent despite high levels of transcription (Fire, 1999; Marathe et al., 2000). This ability of RNAi allowed rapid identification of several genes that regulate the RNAi process (Figure 17.2).

The transgenes inserted into the genomes of plants by recombination, where the number of inserted copies, their chromosomal location and arrangement within the chromosome vary among transformants. The inverse correlation was observed between copy number and the level of gene expression, which indicated that an increased copy number of a particular gene results in silencing of that gene (Assaad et al., 1993). Primarily, it was thought that such gene silencing was due to reduced gene transcription and interactions between closely linked copies of gene that form secondary structures, which promotes methylation and inhibition of transcription (Ye and Signer, 1996). It was further observed that transcriptional gene silencing (TGS) could also occur in transgenes and one transgene can be silenced by another transgene. The presence of siRNAs and their cleavage into siRNAs of about 23 nucleotides their expression of dsRNA with corresponding sequences to open reading frames in plants results into PTGS (Hamilton and Baulconme, 1999). The expression of dsRNAs with sequences complementary to those of endogenous genes results in the selective silencing in *C. elegans* (Zamore et al., 2000) and *A. thaliana*. Further, it was found that injection of dsRNA was more effective to silence gene expression in *C. elegans* than single-stranded

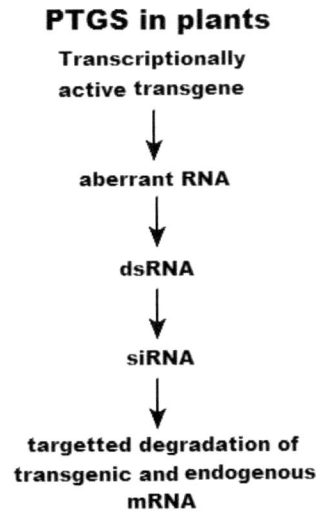

FIGURE 17.2   PTGS in plants.

antisense RNA (Fire et al., 1998). The suppression of the encoded protein by the targeted mRNA was found in many cycles of cell division. Cells possess a mechanism to amplify the RNAi process which is maintained by a trans-membrane protein called SID-1 within the cells of a common lineage function as a mediator of intercellular transfer of RNAi (Winston et al., 2002).

RNAi mechanism in *Drosophila* was first demonstrated by Kennerdell and coworkers (Kennerdell and Carthew, 1998) who showed the involvement of the *frizzled* and *frizzled2* genes in the wingless pathway after introduction of dsRNA into embryos where dsRNA as an extended hairpin-loop RNA was developed to induce heritable gene silencing (Kennerdell and Carthew, 2000), which allowed the identification of several endogenous genes roles in the RNAi process. An RNA nuclease, RISC (RNA-induced silencing complex), was responsible for the degradation of endogenous mRNA fragments (~25 nucleotides in length), could be used as guides by RISC (Hammond et al., 2000; Redfern et al., 2013) later RISC characterized as a ribonucleoprotein complex (Hammond et al., 2001), which incorporates one strand of a single-stranded RNA (ssRNA) fragment, such as microRNA (miRNA), which acts as a template for RISC to recognize complementary messenger RNA (mRNA) transcript. RNAi is an ATP-dependent and translation-independent event where the introduced dsRNA is processed into 21- to 23-nucleotide fragments that guide the cleavage of endogenous transcripts (Zamore et al., 2000). RNase III nuclease enzyme was found to be responsible for the processing of the dsRNA called Dicer, a protein with high homology to rde-1 in the *C. elegans* gene (Bernstein et al., 2001). Volpe et al. (2002) deleted argonaute, dicer, and RNA-dependent RNA polymerase homologs in yeast; this caused the accumulation of complementary transcripts from centromeric heterochromatic repeats and de-repression of transgenes integrated at the centromere and impairment of centromere function. The dsRNA that arise from centromeric repeats targets the formation and maintenance of heterochromatin through RNAi. Various components of PTGS are given in Table 17.1.

17.3   MECHANISM OF RNA INTERFERENCE

The knowledge of PTGS was gained by scientists (English et al., 1997) during observations of transcriptionally active genes and silencing of RNA viruses to a homologous endogenous gene and several molecules of ribonu-

TABLE 17.1 Components of Posttranscriptional Gene Silencing (PTGS)

Phenomenon	Organism	Mutation causing defective silencing	Gene function	Developmental defect
Posttranscriptional gene silencing	Plant (*A. thaliana*)	*sgs2/sde1*	RdRP	None
		sgs3	Unknown function	None
		sde3	Rec Q helicase	
		ago1	Translation initiation factor	Pleiotropic effects on development and fertility
		caf1	RNA helicase and RNase III	
Quelling	Fungus (*N. crassa*)	*qde-1*	RdRP	None
		qde-2	Translation initiation factor	None
		qde-3	RecQ DNA helicase	
RNA interference	Worm (*C. elegans*)	*ego-1*	RdRP	Gametogenetic defect and sterility
		rde-1	Translation initiation factor	None
		rde-2, rde-3, rde-4, mut-2	Unknown function	None
		K12H4.8 (*dcr-1*)	Dicer homologue RNA helicase & RNase III	Sterility
		mut-7	Helicase & RNaseD	None
		mut-14	DEAD box RNA helicase	
		smg-2	Upf1p helicase	
		smg-5	Unknown function	
		smg-6	Unknown function	
		sid-1	Transmembrane protein	
	Alga (*Chlamydomonas reinhardtii*)	*mut-6*	DEAH box RNA helicase	

cleoproteins that mediate RNAi, effects the selective degradation of targeted mRNAs. The production of dsRNA with sequence complementary to the mRNA being targeted is fundamental to the process of PTGS, and single-stranded RNA (ssRNA) is not sufficient to induce PTGS. Transgenes that synthesize dsRNA require only a few copies of dsRNA to achieve PTGS and can induce co-suppression. Such transgenes dsRNA include the synthesis of long hairpin mRNAs by transcription of an inverted repeat (Kennerdell and Carthew, 2000; Tavernarakis et al., 2000), and transcription of complementary sense and antisense strands by opposing promoters (Wang et al., 2000).

The proteins that mediate the RNAi process were identified through genetic screens of mutants resistant to RNAi in *C. elegans* and resistant to PTGS in Neurospora and Arabidopsis. Homologous genes are required in each species that encode proteins required for RNAi. Three homologous genes are rde-1 in *C. elegans* (Tabara et al., 1999), qde-2 in Neurospora (Cogoni et al., 1996) and ago-1 in Arabidopsis (Dalmay et al., 2000; Fagard et al., 2000). The proteins encoded by each of these genes share homologies with the eIF2C translation factors. In *C. elegans*, rde-1 plays a role in RNAi signal production. Another set of homologous genes involved in RNAi, which were identified in the screens for PTGS/RNAi mutants, include ego-1 in *C. elegans* (Tabara et al., 1999; Smardon et al., 2000), qde-1 in Neurospora (Cogoni and Macino, 1999), and sgs-2/sde-1 in Arabidopsis (Mourrain et al., 2000). Further studies in *C. elegans* identified the smg-2, smg-5, and smg-6 genes as being involved in PTGS. However, the products of these genes are not required for the initial silencing by dsRNA but are required for long-term maintenance of the gene suppression (Domeier et al., 2000).

The production of dsRNA from an ssRNA template is mediated by RNA-dependent RNA polymerases such as Ode-1, SDE-1, and Ego-1. The cleavage of dsRNA to produce siRNAs is mediated by Dicer and related proteins such as Ode-2, Ago-1, Rde-1, and Rde-4. The components of the RISC complex that mediate the recognition and degradation of the mRNA targeted by the siRNAs may include Rde-2, Mut-7, eIF2C1, and eIF2C2. Various components of PTGS for RNAi are presented in Table 17.1.

The first step is the production of dsRNA directed against an mRNA. The second step involves the recognition of dsRNA and its processing to produce 21- to 23-nucleotide siRNAs. The "effector step" is the recognition of the target mRNA by the siRNAs and the selective degradation of that mRNA. The introduction of dsRNA into cells, whether produced endogenously from exogenous plasmids or viral vectors, results in its recognition by an enzyme that cleaves the dsRNA into 21- to 23-nucleotide double-stranded fragments in an ATP-dependent, processive manner with a 2-nucleotide 3'-overhang and a 5'-phosphorylated end (Zamore et al., 2000; Elbashir et al., 2001b). This nuclease enzyme called Dicer is highly conserved in organisms (plants, fungi, worms, flies, and mammals); it is a member of the RNase III family of dsRNA-specific ribonucleases (Bernstein et al., 2001). Dicer enzymes recognize and process dsRNA (Bernstein et al., 2001; Ketting et al., 2001) and are essential for RNAi (Bernstein et al., 2001; Grishok et al., 2001; Ketting et al., 2001). Dicer not only processes dsRNA into siRNAs but also processes endogenous regulatory RNAs called microRNAs (miRNAs). Different domains of Dicer have been identified including a dsRNA binding domain; an RNase III activity domain; a helicase activity domain; and a PAZ domain (Piwi-Argonaut-Zwille domain), a region of a hundred amino acids, which could mediate interaction with argonaute proteins (Bernstein et al., 2001).

17.4   PATHWAYS OF RNAi

Cells produce ssRNA that provides a template for the formation of dsRNA, which involves the activity of RNA-dependent RNA polymerases. The dsRNA is then cleaved by a protein called Dicer to form small 21- to 23-nucleotide siRNAs. The siRNAs then associate with the specific mRNA targeted by their nucleotide sequence in a nucleic acid-protein complex called RISC, which includes RNase activity that degrades the mRNA at sites not bound by the siRNAs. The synthesis of the protein encoded by the mRNA targeted by the siRNAs is prevented, and that protein is selectively depleted from the cell.

The dsRNAs, which are cut into siRNAs and then (together with RISC complex) target and degrade an mRNA species. Cells contain large amounts of noncoding RNA including tRNAs, snRNAs, and rRNA. Among these, miRNA in particular is approximately 22 nucleotides long, is present in many different organisms from *C. elegans* to humans, and has received considerable attention; siRNAs and miRNAs can be used as a tool to suppress the expression of genes of interest (McManus et al., 2002).

The nonsense-mediated mRNA decay (NMD), although not strictly a PTGS phenomenon, is relevant to the RNAi. NMD is a process that appears to be a quality control mechanism that eliminates nonsense transcripts such as mRNAs with premature termination codons (Frischmeyer and Dietz, 1999). In yeast, this mechanism, coupled to mRNA splicing and pathway results in the degradation of aberrant mRNAs.

17.5   DESIGN AND SYNTHESIS OF SMALL INTERFERING RNAS (SiRNAS)

Double-stranded RNAs of 20–23 nucleotides (siRNAs) can be synthesized in large quantities and transfected into cells. siRNA-mediated silencing of gene expression is the interest of cDNA sequence of that gene. Targeting of siRNAs to any region of an mRNA induce degradation of the mRNA and stop production of the protein encoded by the mRNA. Several vectors are used for RNAi.

17.6   CONSTRUCTION OF PLASMIDS AND VIRAL VECTORS FOR RNA INTERFERENCE

Due to several reasons, expression plasmids and viral vectors are used in basic and applied RNAi research; these reasons include: (1) Expression vectors allow continuous production of siRNAs in cells and therefore, sustained depletion of the protein encoded by the targeted mRNA. (2) Particularly with viral vectors, the transfection efficiency of certain types of cells, especially postmitotic cells can be greatly increased. (3) Viral vectors are typically more effective in obtaining sustained expression (and gene silencing, in the case of RNAi *in vivo* (Figure 17.3). Lentiviral systems for shRNA delivery have also been developed; lentiviruses can infect noncycling and postmitotic cells and provide the advantage of not being silenced during development, thus allowing the generation of transgenic animals through infection of embryonic stem cells or embryos (Naldini, 1998; Lois et al., 2002; Pfeifer et al., 2002).

17.7   TRANSFECTION METHODS

Several transfection methods that are used to introduce oligodeoxynucleotides and DNA plasmids into cells are also used successfully to introduce

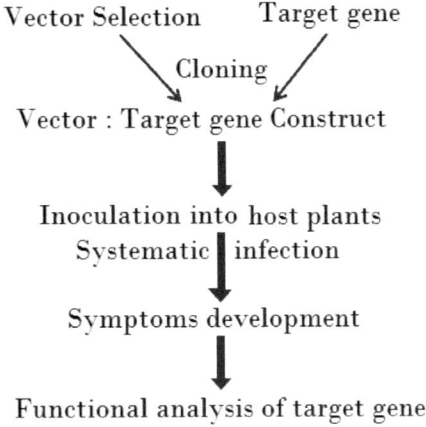

FIGURE 17.3   Vector used RNA interference.

siRNAs. It is therefore essential to optimize transfection conditions so that maximum gene silencing can be achieved. The following transfection parameters affect transfection and gene silencing efficacy:

i.   cell culture conditions (cell density and medium composition)
ii. the type and amount of transfection agent; the quality and amount of siRNA
iii. the length of time that the cells are exposed to the siRNA.

Postmitotic cells such as neurons and muscle cells tend to be more difficult to transfect using liposomes than mitotic cells such as stem cells, fibroblasts, and tumor cells.

17.8   APPLICATIONS OF RNA INTERFERENCE

17.8.1   RNAi TO ESTABLISH GENE FUNCTION

The most widely used RNAi technology is aimed to understand in cell culture and *in vivo* studies for the function of an individual or multiple proteins during various cellular processes. The cells can be transfected with different combinations of multiple siRNAs, each directed against a specific mRNA of interest, to elucidate the specific contributions of the proteins to a biological process involving a multiprotein complex. Wojcik and DeMartino (2002) studied RNAi methods to prove roles of different subunits of the proteasome in its assembly and function. Lee et al. (2002c) systematically inactivated 5690 genes in *C. elegans* using RNAi to identify genes that limit lifespan and identified a mitochondrial leucyl-tRNA synthetase gene; they showed that mutations of this gene impair mitochondrial function to increase lifespan.

17.8.2   SIGNAL TRANSDUCTION

Intercellular messenger molecules play important roles in the development and proper functioning of all organisms. These signals are growth factors like cytokines, cell adhesion molecules, neurotransmitters, steroids, and gasses (nitric oxide, etc.). Specific receptors located on the cell surface or within the cell transduce responses to the ligand via signaling cascades that form a complex involving kinases and transcription factors. RNAi methods provide powerful tools to establish the roles of individual proteins in the signal transduction pathway employed by a specific ligand. The efficacy of siRNA-mediated knockdown of signal transduction proteins has elucidated the role of biological signals and responses of cells. Adaptor proteins of the "Shc" family transduce signals from a diverse group of growth factors that signal through receptor tyrosine kinases. Liposome-mediated introduction of siRNA against a single isoform of "ShcA" into "HeLa" cells was used to selectively reduce levels of that "Shc," revealing its role in the regulation of cell proliferation (Kisielow et al., 2002). Neurotrophin receptors and integrins (receptors for extracellular matrix molecules such as laminin) are often coupled to a signaling pathway involving phosphatidylinositol 3-kinase and Akt kinase (Gary and Mattson, 2001; Cheng et al., 2003). Decreasing the amount of the 110 β subunit of phosphatidylinositol 3-kinase using siRNAs resulted in reduced growth and tissue invasiveness of tumor cells (Czauderna et al., 2003). RNAi has been successfully employed to unequivocally establish the roles of specific kinases in signal transduction processes. Calcium plays important roles as an intracellular signal pathway that mediates a variety of responses of cells to environmental stimuli. Calcium ions move across the plasma membrane as well as in and out of the endoplasmic reticulum and mitochondria. An inositol 1,4,5-trisphosphate (IP_3)-mediated release of intracellular calcium in the maturation of mouse oocytes.

17.8.3   CELL CYCLE REGULATION

In cancer development, tissue homeostasis and stem cell biology, genes that regulate the cell cycle have been in the center to employ RNAi and observed knockdown the expression of cell cycle genes, high levels of Mps1 contribute to tumorigenesis by attenuating the spindle assembly checkpoint requires. A centrosomal protein called "Cep135" in microtubule organization was observed by Ohta et al. (2002), and the role of "centrin-2" in centriole duplication was established (Salisbury et al., 2002). RNAi was used to show that the regulation of cell cycle progression in response to mitogens is controlled by the cyclin-dependent kinase inhibitor "p27" (Kip1) (Boehm et al., 2002). A novel human protein called Speedy, expressed only during the G_1/S phase of the cell cycle, was shown to enhance cell proliferation by enhancing the activity of Cdk2 (Porter et al., 2002).

17.8.4   DEVELOPMENT OF ORGANISMS

The development of the fertilized egg into an adult organism is a mystery, and RNAi technology can unravel the cellular and molecular events that regulate developmental processes. Methods for silencing single or multiple selected genes in developing embryos *in vivo* and stem cells in culture could reveal the functions of specific proteins in developmental processes. RNAi was used to demonstrate the transcriptional corepressor DRAP1 that inhibits the transcription factor FoxH1 and thereby regulates signaling by Nodal during mouse embryogenesis (Iratni et al., 2002). The transcription factor "myc" plays a primary role in the regulation of cell proliferation for the novel myc target gene mina53 in the regulation of cell proliferation (Tsuneoka et al., 2002).

17.8.5   MACROMOLECULAR SYNTHESIS AND DEGRADATION

Continuous synthesis of nucleic acids, proteins, lipids, and macromolecules is essential for cell survival and functions. RNAi technology is employed to understand the regulation of macromolecular synthesis and degradation.

17.8.6 CELL MOTILITY

The migration of cells within and between tissues, and extensions of cells such as the axons and dendrites of neurons, is a central process to the structural and functional organization of all tissues. RNAi methods to elucidate roles of specific proteins in regulating cell motility are now recognized. Depletion of the cytoskeletal linker protein "trypanin" (in trypanosomes) using siRNAs resulted in a loss of directional cell motility, which is caused by impaired coordination of flagellar beating (Hutchings et al., 2002). RNAi-mediated depletion of the integrin-interacting protein MIG-15 resulted in a dysregulation of commissural axon navigation in *C. elegans* (Poinat et al., 2002). Proteins critical for the function of the signaling and endocytic functions of these membrane domains are also identified using RNAi technology.

17.8.7 CELL DEATH (APOPTOSIS)

In the body, most cells and tissues have a fixed life span, and they then undergoes apoptosis, which is a form of programmed cell death wherein the cell dies in a well-controlled manner program; the dead cells are removed from the tissue without adversely affecting the adjacent healthy cells. Apoptosis also plays a key role in sculpting the cellular structure of tissues during embryonic and postnatal development (Baehrecke, 2002). The abnormal cell death is a major problem in a variety of diseases, including neurodegenerative disorders such as Alzheimer's diseases, Parkinson disease, and ischemic vascular conditions (Mattson, 2000) their study leads to targeted therapeutic drug development. Efforts are continued to prevent cell death of neurons and induce selective death of cancerous cells (Eldadah and Faden, 2000). RNAi has now been employed to establish roles for specific genes in apoptotic and anti-apoptotic pathways (Kartasheva et al., 2002) to deplete cells of apoptosis-inducing factor (AIF), thereby preventing apoptosis (Wang et al., 2002).

17.8.8 VIRAL INVASION/REPLICATION

RNAi mechanisms protect cells against infectious pathogens such as viruses. Treatment with siRNAs directed at genes *in vitro* has revealed roles of specific targeted proteins. Similarly, siRNAs directed against HIV-1 tat (reverse transcriptase genes) and against the NF-κB p65 subunit of the host cell revealed decreased expression of these viral and cellular proteins, and HIV-1 replication was inhibited; RNAi was involved in the control of HIV-1 gene expression (Surabhi and Gaynor, 2002). siRNAs could be used to completely eliminate the virus from infected cells.

17.9 THERAPEUTIC APPLICATIONS OF RNA INTERFERENCE

The RNAi mechanism is found to control diseases where selective depletion of one or a few specific proteins expected to slow or halt disease in the affected cells without tolerable side effects as discussed in the following subsections.

17.9.1 CANCER

RNAi technology has been rapidly developed in the inhibition of growth and progression of cancer cells by downregulation of the expression of a gene of interest. Generally, two abnormalities in cancer cells are observed: they exhibit deregulation of the cell cycle with uncontrolled growth and are resistant to death due to one or more proteins mediate apoptosis. RNAi is used to suppress the expression of a cell cycle gene in stopping tumor growth and killing the elective cancer cells, without damaging normal cells, and it is found to be more potent than antisense DNA in suppressing gene expression (Aoki et al., 2003). RNAi can also target multidrug resistance gene (MDR1) for re-sensitization to chemotherapy and silencing of double-strand break repair enzymes for enhanced effects of radio and chemotherapy.

17.9.2 INFECTIOUS DISEASES

Globally, viruses and bacteria are major causes of death worldwide and are also a cause of concern due to resistant strains and their potential use of infectious agents by terrorists (Tan et al., 2000; Franz and Zajtchuk, 2002). Other major infectious diseases include influenza, hepatitis, and West Nile virus. Bacterial infections also result in diseases like pneumonia and sepsis.

Several diseases can be treated by using RNAi and siRNA to inhibit cell infection and replication.

17.9.3   CARDIOVASCULAR AND CEREBROVASCULAR DISEASES

Globally, cardiovascular disease is the leading cause of death due to progressive occlusion of arteries by atherosclerosis that result in a myocardial stroke by death of cardiac muscle cells or brain cell neurons. The use of RNAi technology can intervene the process of atherosclerosis or reduce the damage to heart tissue and brain cells. The production of cell adhesion molecules can be selectively suppressed in cultured cells (Jarad et al., 2002). The drug mevastatin, an inhibitor of cholesterol synthesis, suppresses cell proliferation by inhibiting cyclin-dependent kinase-2 (Ukomadu and Dutta, 2003).

17.9.4   NEURODEGENERATIVE DISORDERS

The common age-related neurodegenerative disorders are Alzheimer's disease, Parkinsonism, Huntington's disease, and amyotrophic lateral sclerosis, which are caused due to dysfunction and death of specific cortical neurons that cause memory loss due to dopamine-producing neurons. These are genetic mutations in the huntingtin protein (Rubinsztein, 2002). The cultured neurons can be efficiently transfected with siRNAs and the targeted genes are effectively silenced to target synaptic proteins involved in the pathogenesis of neurodegenerative disorders.

17.9.5   RNAi AS A POTENTIAL THERAPEUTIC FOR HUMANS GENETIC DISEASES

The exciting therapeutic assurance of RNAi is possible by nucleases and viral vectors to target organs effected with dominant negative genetic disorders, where a mutant allele of a gene causes disease in the presence of a second normal copy of allele, have been posing a challenge since long in curing and treatments of this genetic disorder.

17.9.6 RNAi IN GENE REGULATION AND ANTIVIRAL RESPONSES

RNAi mechanisms play an important role in the regulation of cellular gene expression and provide natural antiviral immune responses; thus, they can be used as a natural defense mechanism against mobile endogenous transposons and invasion by exogenous viruses with dsRNA as an intermediate product. Specific-designed DNA molecules containing inverted repeat sequences can be transcribed into RNA molecules that form RNA hairpins, and correct sequences can be processed by the Dicer nuclease to form siRNAs. Thus, RNAi therapy for the prevention and treatment of viral infection is convenient (Figure 17.4).

17.9.7 RNA INTERFERENCE IN GENE KNOCKDOWN

The RNAi pathway is often used to study the gene function in cell culture *in vivo*. dsRNA is synthesized with a sequence complementary to the gene of interest and introduced into a cell or organism, where it is recognized as exogenous genetic material and activates the RNAi pathway, which decreases the expression of a targeted gene and gene product. This technique is known as gene knockdown.

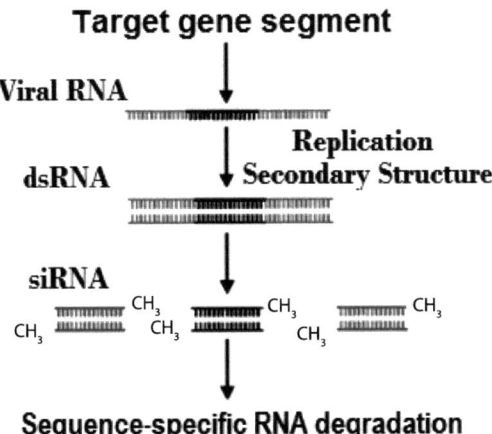

FIGURE 17.4 Antiviral mechanism of RNA interference.

17.9.8 RNA INTERFERENCE IN MEDICINE

The RNAi mechanism is used in selective gene silencing to establish the functions of genes and gene therapy approaches against treatment of specific diseases which has progressed potential gene targets for therapeutic intervention. The siRNAs or vector-mediated RNA expression methods suggested the potential of RNAi to block the disease process. Similarly, viral and bacterial genes are evident targets for RNAi-based therapeutic intervention for infectious diseases.

17.9.9 LIMITATIONS OF RNAI

Despite the proliferation of promising cell culture studies for RNAi-based drugs, some concern has been raised regarding the safety of RNAi, especially the potential for "off-target" effects, in which a gene with a coincidentally similar sequence to the targeted gene is also repressed and the error rate of off-target interactions is about 10%.

17.10 RNAi APPLICATIONS IN AGRICULTURE

Plant genome sequencing and expressed sequence tag (EST) have provided abundant sequence information for several plant species and gene function n(s). In *Arabidopsis thaliana*, the function of several genes is still unknown. In functional genomics, RNAi tools are widely used to analyze gene function to produce improved crop varieties (Table 17.2). The following three RNAi-based concepts have potential applications in plant functional genomics and agriculture:

a. Tissue-specific silencing, inducible silencing, and host delivered RNAi (hdRNAi) during plant-pest interaction. Tissue-specific promoters driving RNAi constructs can induce gene silencing in a particular organelle or tissue.
b. RNAi constructs with stress or chemical inducible promoters can be used to induce gene silencing only when required.

Both these concepts can be used together to achieve temporal and spatial control of gene silencing in plants.

TABLE 17.2   Applications of RNA interference in plants

Kingdom	Species	Stage tested	Delivery method
Plants	Monocots/dicots	Plant	Particle bombardment with siRNA/transgenics
Fungi	*Neurospora crassa*	Filamentous fungi	Transfection
	Schizosaccharomyces pombe	Filamentous fungi	Transgene
	Dictyostelium discoideum		Transgene
Algae	*Chlamydomonas reinhardtii*		Transfection

RNAi can be applied in two ways for crop improvement.

c.   In hdRNAi, dsRNA generated in an RNAi transgenic plant is delivered to interacting target organism (pest), thus activating gene silencing in the target organism. RNAi can be applied in agriculture, animal husbandry, and biofuel industry. As suppression of gene expression by RNAi is inheritable, this has been a tool for developing transgenic crop plants for resistance against disease, pests, drought, etc. in agriculture to develop biotic stress-tolerant crops.

RNAi can be applied in two ways for crop improvement

i.   Silencing of certain plant genes to improve agronomic traits or increase plant fitness against stress (Chen and Dixon, 2007; Shomura et al., 2008). The stable gene silencing method is used for host gene silencing hair pin RNAi (HGS-hpRNAi), where host plant is genetically engineered to alter its own gene expression for improving its agronomic superiority against biotic and abiotic stresses.

ii. By hdRNAi method, the gene(s) are silenced only in target organism (insect, pathogen and plant parasite) that damages crop known as hdRNAi. The hdRNAi mechanism can be subdivided into two categories. First is silencing of genes in biotic organisms (insect pests, nematodes, and parasitic weeds) that externally feed up on the plant and cause damage (hdRNAi-1). The second category is targeting gene silencing in viruses that enter the host plant cell (hdRNAi).

17.10.1 DISEASE RESISTANCE

Disease resistance by HGS-hpRNAi can be used against bacterial and fungal pathogens. In *Arabidopsis*, the bacterial component flagellin induces expression of a specific miRNA, which in turn leads to downregulation of signaling pathways that increases the plant's resistance (Fritz et al., 2006). Application of RNAi-mediated oncogene silencing to control crown gall disease was shown by Escobar et al. (2001). Tobacco plants gene silenced for glutathione S-transferase transcripts showed resistance to black shank disease (Hernández et al., 2009). In soybean, low lignin content increased plant fitness to fungal pathogen, cyst nematode resistance (Li X. et al., 2012), and provided resistance to *Sclerotinia sclerotiorum* (Peltier et al., 2009). Fatty acids and their derivatives play important signaling roles by negatively regulating plant bacterial disease resistance (Jiang et al., 2009; Li et al., 2008). The RNAi-mediated knockdown in rice homolog of gene, namely OsSSI2, enhanced resistance to leaf blight bacterium *Xanthomonas oryzae* and blast fungus *Magnaporthe grisea* (Jiang et al., 2009). This has potential for future applications for virus and other phyto-pathogens. RNAi is useful not only for functional genomic research in insects, but also for the control of insect pests where insect take up dsRNA through plant feeding (Huvenne and Smagghe, 2010) that effectively downregulate targeted genes in the insect. The hdRNAi-1 method-induced silencing in plant feeding insects opened the development of a new generation of insect-resistant crops. Transgenic corn plants engineered to express dsRNAs for vacuolar ATPase (tubulin genes) showed resistance to western corn rootworm (WCR) (Baum et al., 2007), wherein reduced plant damage, larval stunting, and larval mortality were observed (Figure 17.5). In cotton bollworm *Helicoverpa armigera*), larvae feeding on plant material expressing dsRNA specific to cytochrome P450 gene (CYP6AE14) showed retarded larvae growth (Mao et al., 2007). Thus hdRNAi-1 technology is also called as species-specific insecticide and is a potential alternative to chemical pesticides (Whyard et al., 2009).

Another important aspect of crop improvement is engineering of nematode resistance by two ways: (i) increasing plant resistance by silencing certain genes using HGS-hpRNAi and (ii) using hdRNAi-1 method (most effective). In this method, dsRNA or siRNAs is delivered from plant to nematode through ingestion of transgenic plant tissue (Gheysen and Van-

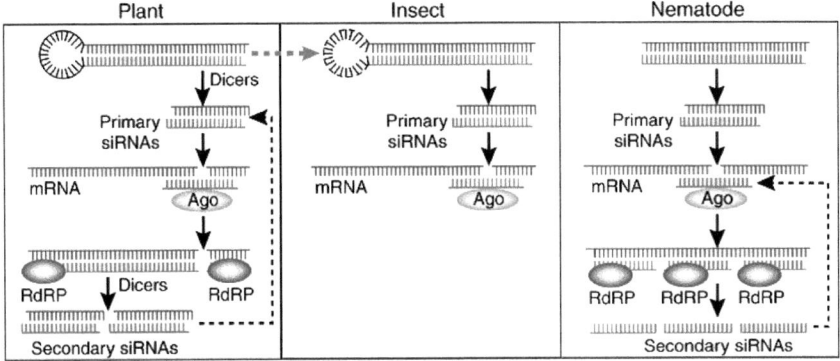

FIGURE 17.5   Mechanism of disease resistance in organisms.

holme, 2007). Transgenic plants with hdRNAi-1 were more successful with root-knot nematode (RKN) than cyst nematode (CN). Disease resistance created in various crops by RNAi is given in Table 17.3.

17.10.2   PEST CONTROL

The digestive system in an insect (nematode) is very different from that of mammals, and dsRNA designed to suppress specific genes in some pests can be provided in the diet to suppress or kill those pests. The sequence specificity of RNAi presents the opportunity to selectively target some pest species. Unlike some chemical pesticides, RNAi in plant is not expected to have any effect on non-target organisms (insects, nematodes, birds, reptiles, fish, and mammals). All transgenic-plant-mediated insect control technology is based on the plant-producing proteins derived from *Bacillus thuringiensis* (Bt). Combining Bt technology with a second, independent mode of insect control via RNAi would enhance to product performance and further guard against the development of resistance to Bt proteins (Goldstein et al., 2009).

17.10.3   DROUGHT TOLERANCE

The potential of hdRNAi-1 technology in plant parasite control and RNAi in Improved Drought Tolerance Gene silencing has been used to develop drought-tolerant plants (Ni et al., 2013). Conditional and specific downregulation of farnesyltransferase in canola using the AtHPR1 promoter driving

TABLE 17.3   Disease Resistance Created in Various Crops

Crop	Insect/Pathogen	Objective	Targeted genes
Arabidopsis thaliana	Meloidogyne sp.	Utilization of RNAi to silence the parasitism gene	16D10
Oryza sativa L	Magnaporthe grisea and Xanthomonas oryzae pv. oryzae	Functional analysis of a rice homolog SSI2 (OsSSI2) for disease resistance	OsSSI2
Prunus domestica L.	Plum pox virus (PPV)	To exploit the role of PTGS (RNAi) for virus resistance in a woody perennial species	PPV coat protein gene
Gossypium hirsutum	Helicoverpa armigera	Silencing a cotton bollworm P450 monooxy-genase gene by plant-mediated RNAi	Cytochrome P450 gene (CY-P6AE14)
Nicotiana rustica	Bemisia tabaci	Enhanced whitefly resistance via expressing double stranded RNA	v-ATPaseA
Zea mays	Diabrotica virgifera virgifera LeConte	Control of coleopteran insect pests through RNA interference	Genes encoding proteins
Medicago sativa	Acyrthosiphon pisum	RNAi knockdown of a salivary transcript	C002
Nicotiana benthamiana and Arabidopsis thaliana	Myzus persicae	To develop the plant-mediated RNAi technology for aphid resistance	M. persicae Rack1, M. persicae C002 (MpC002)
Nicotiana rustica	Helicoverpa armigera	Improvement of pest resistance in transgenic tobacco plants expressing dsRNA	EcR
Citrus aurantifolia	Citrus tristeza virus (CTV)	Transformation to generate transgenic plants carrying the coat protein gene of CTV	CTV-CP

Juglans regia L.	Agrobacterium tumefaciens	Application of oncogene silencing technology in the generation of crown gallresistant crops	Tryptophan mono-oxygenase (iaaM) and isopentenyl transferase (ipt)
Genus Malus	Agrobacterium tumefaciens	To provide effective method to produce crown gall resistant apple plants	iaaM, iaaH, and ipt
Oryza sativa L.	Nilaparvata lugens	Knockdown of midgut genes by dsRNA-transgenic plant-mediated RNA interference	NlHT1, Nlcar, Nltry
Genus Malus	Venturia inaequalis	To use to hairpin vector approach for resistance against V. inaequalis	GFP transgene and tri-hydroxy-naphthalene reductase gene (THN)

an RNAi construct showed yield protection under drought stress. Wang et al. (2000) observed the use of farnesyltransferase as an effective target for engineering drought tolerance in other plants. HGS-hpRNAi transgenic rice plants that cause silencing of the receptor for activated C-kinase 1 (RACK1) showed tolerance to drought stress. This approach is expected to produce more drought-tolerant crop plants in future.

17.10.4   RNAI FOR MALE STERILITY

RNAi technology has also been used to generate male sterility that is useful in the hybrid seed production program. Genes that are expressed solely in tissues involved in pollen production can be targeted through RNAi. Male sterile tobacco lines have been developed by inhibiting the expression of TA29, a gene necessary for pollen development. RNAi was also used to disrupt the expression of Msh1 in tobacco and tomato, resulting in rear-rangements in the mitochondrial DNA associated with naturally occurring cytoplasmic male sterility.

17.10.5   SEED GERMINATION

The plant seeds that have elevated lysine levels germinate poorly because the excess lysine levels in the seeds are not efficiently degraded during seed germination (Zhu et al., 2003), while the reduction of lysine level improves seed germination (Zhu et al., 2004; Tang et al., 2007). High-lysine maize obtained through RNAi generates quality and normal maize seeds with high levels of lysine-rich proteins (Tang et al., 2004).

17.10.6   RNAI IN PLANT FUNCTIONAL GENOMICS

A major challenge in the post-genomic era of plant biology is to determine the functions of all genes in the plant genome. RNAi offers specific and effective method in gene silencing. In addition, the expression of dsRNAs with inducible promoters can control the extent and timing of gene silencing; such essential genes are only silenced at the chosen growth stages or plant organs.

17.10.7 RNAI IN ENGINEERING PLANT METABOLIC PATHWAYS

RNAi has been used to modify plant metabolic pathways to enhance nutrient content and reduced toxin production (Table 17.4). The technique takes advantage of the heritable and stable RNAi phenotypes in plants.

17.10.8 IMPROVING NUTRITIONAL VALUE IN CROPS

Various crop improvements achieved by RNAi are presented in Table 17.5

17.10.8.1 Wheat

HGS-hpRNAi was applied to improve nutritional qualities of wheat grain. RNAi was successfully used to reduce γ-gliadins in bread wheat (Gil-Humanes et al., 2008; Gil-Humanes et al., 2012), and reduced levels of two starch-branching enzyme encoding genes (SBEIIa and SBEIIb) showed increased amylose content (Regina et al., 2006) that has health benefits.

17.10.8.2 Barley

Bayer Crop Science has developed barley plant varieties with the RNAi technology that are resistant to BYDV (barley yellow dwarf virus) (Wang et al., 2000).

17.10.8.3 Rice

Use of RNAi to improve rice plants (Kusaba et al., 2003) have made significant contribution by reduced level of glutenin and produced a variety LGC-1 (low glutenin content 1). The low glutenin content is a relief to the kidney patients who are unable to digest glutenin. Similarly, RNAi rice knockout transgenic lines for qSW5 (QTL for seed width on chromosome 5) showed increased seed weight. Further, semi-dwarf plants were generated from a taller rice variety QX1 by RNAi-mediated suppression of GA 20-oxidase (OsGA20ox2) gene expression, and increased panicle length, higher num-

TABLE 17.4   Examples of Novel Plant Traits Engineered through RNAi

Trait	Target Gene	Host	Application
Enhanced nutrient content	Lyc	Tomato	Increased concentration of lycopene (carotenoid antioxidant)
	DET1	Tomato	Higher flavonoid and b-carotene contents
	SBEII	Wheat, Sweet potato, Maize	Increased levels of amylose for glycemic management and digestive health
	FAD2	Canola, Peanut, Cotton	Increased oleic acid content
	SAD1	Cotton	Increased stearic acid content
	ZLKR/SDH	Maize	Lysine-fortified maize
Reduced alkaloid production	CaMXMT1	Coffee	Decaffeinated coffee
	COR	Opium poppy	Production of non-narcotic alkaloid, instead of morphine
	CYP82E4	Tobacco	Reduced levels of the carcinogen nornicotine in cured leaves
Heavy metal accumulation	ACR2	Arabidopsis	Arsenic hyperaccumulation for phytoremediation
Reduced polyphenol production	s-cadinene synthase gene	Cotton	Lower gossypol levels in cottonseeds, for safe consumption
Ethylene sensitivity	LeETR4	Tomato	Early ripening tomatoes
	ACC oxidase gene	Tomato	Longer shelf life because of slow ripening
Reduced allergenicity	Arah2	Peanut	Allergen-free peanuts
	Lolp1, Lolp2	Ryegrass	Hypo-allergenic ryegrass
Reduced production of lachrymatory factor synthase	lachrymatory factor synthase gene	Onion	Tearless onion

ber of seeds per panicle, and higher test (1000 grain) weight (Wang S. et al., 2012; Zhang Y. C. et al., 2013) were observed. Yang et al. (2013) enhanced cold tolerance in rice. RNAi trait can be introgressed with high-yielding target genotypes in backcross breeding program.

TABLE 17.5   Application of RNAi Interference and Crop Improvement

Application	Case study	Authors
Maize-Increased lysine level	Reduction of lysine catabolism and improving seed germination generating a dominant high-lysine maize variant by knocking out the expression of the 22-kD maize zein storage protein	Zhu et al., Tang et al.
Barley and Rice	Resistance of barley to BYDV and producing a rice variety called LGC-1 (low glutenin content 1) by RNAi technology	Wang et al., Kusaba et al., Williams et al.
Banana	Production of banana varieties resistant to the Banana Bract Mosaic Virus (BBrMV) by RNAi.	Rodoni et al.
Cotton	Transgenic cotton plants expressing a RNAi construct of the d-cadinene synthase gene of gossypol synthesis fused to a seed-specific promoter caused seed-specific reduction of Gossypol.	Sunilkumar et al.
Jute	Generating jute varieties with low lignin content by RNAi technology.	Williams et al.
Lathyrus sativus	RNAi construct designed to silence the genes encoding the two starch-branching isozymes of amylopectin synthesisR-NAi technology can be used to silence the gene(s) responsible for production of BOAA.	Regina et al.
Tomato	RNAi-mediated suppression of DET1 expression under fruit-specific promoters has recently shown to improve carotenoid and flavonoid levels in tomato fruits with minimal effects on plant growth.	Williams et al.
Coffee	RNAi technology has enabled the creation of varieties of Coffee that produces natural coffee with low or very low caffeine content.	Davuluri et al.
Flower color and modification		
Blue rose	Producing blue transgenic rose by knockdowning the cyanidin genes in rose and carnation by RNAi technology and introduce delphinidin genes.	Nishihara et al.
Flowering time		Van Uyen

TABLE 17.3 (Continued)

Application	Case study	Authors
Healthier oil	Using RNAi to silence the gene in cotton which codes for the enzyme that converts oleic acid into a different fatty acid	Waterhouse et al.
Pest control	Combining Bt technology with RNAi would both enhance product performance and further guard against the development of resistance to Bt proteins	Goldstein et al.
Wood and fruit quality	Down regulation of lignin biosynthesis pathways. -Producing transgenic hypoallergenic apples and a possible solution for the undesirable separation of juice into clear serum and particulate phase by using RNAi.	Hu et al., Teo et al., Amancio José de Souza et al.

17.10.8.4   Peanut

RNAi cause 70% increase in oleic acid content in transgenic peanut seeds (Yin et al., 2007) by silencing the oleate desaturase (FAD2) gene. Nutritional improvement was achieved by production of unsaturated fats without trans-fat in oil seed crops. RNAi silencing has been applied in genetic engineering of starches, oils and storage. The peanut allergy contributing gene Arah2 was knocked down by the RNAi method.

17.10.8.5   Cotton

Cotton is an important crop worldwide and is mainly used for fiber production. Malnutrition and starvation are a global problem (Sunilkumar et al., 2006). Cotton cake could be used as an extremely rich source of protein and calories but is not utilized because of a toxic compound Gossypol terpenoid. Gossypol is produced in vegetative parts of plants to protect damage by insects and pathogens. The RNAi transgenic cotton plants expressing the d-cadinene synthase gene of gossypol synthesis fused to a seed-specific promoter caused seed-specific reduction of this metabolite in seed tissues. These cotton plants are thus expected to have similar insect and pathogen resistance to that of wild type cotton and produce seeds with higher nutritional value (Tang et al., 2007).

17.10.8.6   Forage with Reduced Lignin

RNAi has been employed to produce transgenic forage with improved agronomic traits by manipulating plant architecture. The downregulation of specific cytochrome P450 enzymes involved in the lignin synthesis genes reduced lignin forage quality in alfalfa with increased digestibility of fodder to reduce bloating in cows. Reduced lignin-containing biomass also enhanced its application in paper pulp industry.

17.10.8.7   Banana

RNAi may be applied in the production of banana varieties resistant to the Banana Bract Mosaic Virus (BBrMV) in Southeast Asia and India (Rodoni et al., 1999). Sometimes, the entire banana crop is lost due to the attack of this virus that destroys the fruit by producing a bract region. Designing of RNAi vector to silence the coat protein (CP) region of the virus may develop a resistant banana variety to BBrMV, which is also safe to eat.

17.10.8.8   Jute

Application of RNAi involves the downregulation of an enzyme in the biosynthetic pathway of lignin in the two economically important Corchorus species: *C. capsularis* and *C. olitorius*. The enzyme 4-coumarate CoA ligase (4-Cl) is one of the key enzymes in the early stages of lignin biosynthesis. RNAi construct downregulates the expression of the 4-Cl gene to create a transgenic jute variety to reduce the lignin production. Thus, RNAi technology may prove to be a powerful molecular tool by generating jute varieties with low lignin content, are environmentally friendly, and enable cost-effective processing of fiber for the production of various economically important products such as high-quality paper and cloth (Williams et al., 2004).

17.10.8.9   *Latyrus sativus*

In several countries, people use leguminous crop leafy vegetable of *Lathyrus sativus* that contains a neurotoxin called β-oxalylamino-alanine (BOAA) (Spencer et al., 1986). These people suffer from a paralytic disease called lathyrism. The RNAi technology can silence the gene(s) responsible for the production of BOAA or reduce the levels of BOAA up to a safe level to use as food (Williams et al., 2004).

17.10.8.10   Tomato

Globally, tomato is used as a vegetable in a considerable amount of human diets. The fruits is rich in a number of vitamins with other health-promoting metabolites such as antioxidants, flavonoids, carotenoids, and lycopene, which provides the tomato fruit with its typical red color. Carotenoids are synthesized by the same biosynthetic pathway that synthesizes chlorophyll; this also influences tomato fruit quality by altering the levels of carotenoids and flavonoids (Adams-Phillips et al., 2004). Tomato has high pigment (hp-2) phenotype, which accumulates at elevated levels of carotenoids and flavonoids (Levin et al., 2003; Mustilli et al., 1999). The hp-2 mutants generally exhibit abnormal plant growth, and the DET1 gene is silenced in the transgenic plants (Mustilli et al., 1999; Levin et al., 2003; Davuluri et al., 2004). RNAi-mediated suppression of DET1 expression showed improvements in carotenoid and flavonoid levels in tomato fruits with minimal effects on plant growth and other fruit quality parameters (Davuluri et al., 2005).

17.10.8.10.1   Flavr Savr Tomato

Flavr Savr tomato, also known as CGN-89564-2, is a genetically modified tomato for shelf-life that was the first commercially grown food for human consumption. It was produced by the company Calgene and approved by the U.S. Food and Drug Administration (FDA) in 1992. (*Redenbaugh* et al., *1992*). It was first sold in 1994, and the production ceased in 1997. Calgene made history and was further acquired by Monsanto Company but could not succeed. Gupta et al. (2013) studied delayed ripening and improved fruit processing quality through RNAi in this variety.

17.10.8.11   Coffee

About 10% of the coffee in the world market is decaffeinate coffee (DECAF); decaf is required for caffeine-sensitive peoples. It is a stimulant of the central nervous system, the heart muscle, and the respiratory system and has a diuretic effect. Its adverse side-effects include insomnia, restlessness, and palpitations. A standard cup of filter coffee generally contains between 60–150 mg of caffeine, while decaffeinated coffee has about 2–4 mg of caffeine per cup. RNAi technology has enabled the creation of coffee varieties that produces natural coffee with low or very low caffeine content.

17.10.8.12   Healthy Oil

Silencing of genes by application of RNAi can be much helpful in seed-oil production where the health-related effect depends on the fatty acid composition of the oil. Palm oil is very high in palmitic acid, which makes it stable at high temperatures but also unhealthy for human consumption, as it raises LDL cholesterol levels. Olive oil, on the other hand, is high in linoleic acid, which is much healthier for human consumption, but it is not stable at high temperatures and therefore not good for frying. The best oil with no negative effects on cholesterol levels is high in oleic acid. RNAi is now used to silence the gene in cotton, which codes for the enzyme that converts oleic acid into a different fatty acid; this alters the seed oil composition from 10% to 75% oleic acid.

17.10.8.13   Improved Biofuel Production

Recalcitrance of plant material to a process called saccharification is a major limitation for conversion of lignocellulosic biomass to ethanol. Genetic reduction of lignin content can effectively overcome cell wall recalcitrance to bioconversion. Thus, it is possible to reduce or eliminate the costly pretreatment step by using biomass from low-lignin RNAi in plants. Applications of RNAi in transgenic plants can greatly reduce the cost of biofuel production. The major biofuel important crops are switchgrass and Miscanthus. Recently, HGS-hpRNAi was effectively used to reduce lignin content in switch grass (Fu et al., 2012).

17.10.8.14   Flower Color Modification in *Gentiana triflora* by RNAi-Mediated Gene Silencing

Gentiana scabra and their interspecific hybrids are one of the most popular floricultural plants in Japan and are used as ornamental cut flowers.

Florigene Ltd. and Suntory Ltd. have developed blue flowered carnations by genetic engineering, and these carnations are commercially produced in North America, Australia, and Japan (Tanaka et al., 2005). White-flowered transgenic gentians was produced by suppressing the chalcone synthase (CHS) gene using antisense technology (Nishihara et al., 2006). Gene silencing by RNAi was used to modify the flower color in petunia, torenia, and tobacco plants (Tsuda et al., 2004; Nishihara et al., 2005; Nakamura et al., 2006; Nakatsuka et al., 2007a, 2007b, 2008).

17.10.8.14.1 Blue rose

Cyanidin, pelargonidin, and delphinidin are anthocyanin precursors for all plant pigments. The cyanidin gene is responsible for a synthetic pathway that leads to the formation of red pigment, and a correspondent Delphinidin gene is the key gene for the formation of blue color. Scientists have successfully knocked down the cyanidin genes in rose and carnation by RNAi technology and introduce delphinidin genes, which, in natural condition, are absent in these two important cut flowers.

17.10.8.15 RNAi Fruit Quality

In some woody plants, self-incompatibility stands as a major problem in fruit set and breeding programs. The production of transgenic apple trees that are able to self-pollinate and develop into fruits was achieved by silencing the S-gene responsible for self-incompatibility. The self-compatible transgenic plants lacked the pistil S-RNase protein, which is the product of the S-gene. Fruit quality may also be enhanced by silencing experiments. Transgenic apple fruits silencing key enzyme ethylene production were significantly reduced displayed an increased shelf-life (Dandekar et al., 2004).

17.11 CONCLUDING REMARKS/FUTURE PERSPECTIVES OF RNAi TECHNOLOGY

In the past decade, RNAi technology has been an incredible functional genomics tool, and it will remain as a potential genomic tool for the next several decades to come as large scale knock down. Several lines are being developed in major crop plants. Application of RNA silencing in single cell organisms like *Chlamydomonas reinhardtii*, an algae, will enable high-throughput plant gene function identification (for abiotic stress tolerance). A stress-specific cDNA library cloned in a vector that can induce gene silencing (sense strand overexpression to induce RNAi) could be a potential source for developing knock down in *Chlamydomonas* lines for each gene to identify gene function. Molecular breeders can potentially use RNAi technology to incorporate insect or disease resistance into a high-yielding commercial hybrid or variety. With RNAi, it would be possible to target multiple genes for silencing using a thoroughly designed single transformation construct. Moreover, RNAi can also provide broad-spectrum resistance against pathogens with knock down in disease development and control.

KEYWORDS

- apoptosis
- dicer
- dsRNA
- mRNA
- PTGS
- quelling
- RISC
- RNA
- siRNA
- transcription
- transduction
- transfection
- translation

REFERENCES

Adams-Phillips, L., Barry, C., & Giovannoni, J., (2004). Signal transduction systems regulating fruit ripening. *Trends in Plant Science, 9,* 331–338.

Andrew, F., SiQun, Xu, Mary, K. M., Steven, A. K., Samuel, E. D., & Craig, C. M., (1998). Potent and specific genetic interference by double-stranded RNA in *Caenorhabditis elegan. Nature, 391,* 806–811.

Aoki, Y., Ciuca, D., Oidaira, H., Kamiya, T., & Kiyosawa, K., (2003). RNA interference may be more potent than antisense RNA in human cancer cell lines. *Clin. Exp. Pharmacol. Physiol., 30,* 96–102.

Assaad, F. F., Tucker, K. L., & Signer, E. R., (1993). Epigenetic repeat-induced gene silencing (RIGS) in Arabidopsis. *Plant. Mol. Biol., 22,* 1067–1085.

Baehrecke, E. H., (2002). How death shapes life during development. *Nat. Rev. Mol. Cell Biol., 3,* 779–787.

Baum, J. A., Bogaert, T., Clinton, W., Heck, G. R., Feldmann, P., Ilagan, O., Johnson, S., Plaetinck, G., Munyikwa, T., Pleau, M., Vaughn, T., & Roberts, J., (2007). Control of coleopteran insect pests through RNA interference. *Nature Biotechnology, 25,* 1322–1326.

Bernstein, E., Caudy, A. A., Hammond, S. M., & Hannon, G. J., (2001). Role for a bidentate ribonuclease in the initiation step of RNA interference. *Nature (London), 409,* 363–366.

Boehm, M., Yoshimoto, T., Crook, M. F., Nallamshetty, S., True, A., Nabel, G. J., & Nabel, E. G., (2002). A growth factor-dependent nuclear kinase phosphorylates p27(Kip1) and regulates cell cycle progression. *EMBO (Eur. Mol. Biol. Organ) J., 21,* 3390–3401.

Bosher, J. M., & Labouesse, M., (2000). RNA interference: genetic wand and genetic watchdog. *Nat. Cell Biol., 2,* E31–E36.

Chen, F., & Dixon, R. A., (2007). Lignin modification improves fermentable sugar yields for biofuel production. *Nature Biotechnology, 25,* 759–761.

Cheng, A., Wang, S., Yang, D., Xiao, R., & Mattson, M. P., (2003). Calmodulin mediates brain-derived neurotrophic factor cell survival signaling upstream of Akt kinase in embryonic neocortical neurons. *J. Biol. Chem., 278,* 7591–7599.

Cogoni, C., Irelan, J. T., Schumacher, M., Schmidhauser, T., Selker, E. U., & Macino, G., (1996). Transgene silencing of the al-1 gene in vegetative cells of Neurospora is mediated by a cytoplasmic effector and does not depend on DNA-DNA interactions or DNA methylation. *EMBO (Eur. Mol. Biol. Organ) J., 15,* 3153–3163.

Cogoni, C., & Macino, G., (1999). Gene silencing in Neurospora crassa requires a protein homologous to RNA-dependent RNA polymerase. *Nature (London), 399,* 166–169.

Colussi, P. A., Quinn, L. M., Huang, D. C., Coombe, M., Read, S. H., Richardson, H., & Kumar, S., (2000). Debcl, a proapoptotic Bcl-2 homologue, is a component of the Drosophila melanogaster cell death machinery. *J. Cell Biol., 148,* 703–714.

Czauderna, F., Fechtner, M., Aygun, H., Arnold, W., Klippel, A., Giese, K., & Kaufmann, J., (2003). Functional studies of the PI(3)-kinase signalling pathway employing synthetic and expressed siRNA. *Nucleic Acids Res., 31,* 670–682.

Dalmay, T., Hamilton, A., Rudd, S., Angell, S., & Baulcombe, D. C., (2000). An RNA-dependent RNA polymerase gene in Arabidopsis is required for posttranscriptional gene silencing mediated by a transgene but not by a virus. *Cell, 101,* 543–553.

Dandekar, A. M., Teo, G., Defilippi, B. G., Uratsu, S. L., Passey, A. J., Kader, A. A., Stow, J. R., Colgan, R. J., & James, D. J., (2004). Effect of down-regulation of ethylene biosynthesis on fruit flavor complex in apple fruit. *Transgenic Research, 13,* 373–384.

Davuluri, G. R., van Tuinen, A., Mustilli, A. C., Manfredonia, A., Newman, R., Burgess, D., Brummell, D. A., King, S. R., Palys, J., Uhlig, J., et al., (2004). Manipulation of DET1 expression in tomato results in photomorphogenic phenotypes caused by posttranscriptional gene silencing. *Plant Journal, 40,* 344–354.

Davuluri, G. R., van Tuinen, A., Fraser, P. D., Manfredonia, A., Newman, R., Burgess, D., Brummell, D. A., King, S. R., Palys, J., Uhlig, J., et al., (2005). Fruit-specific RNAi-mediated suppression of DET1 enhances carotenoid and flavonoid content in tomatoes. *Nature Biotechnology, 23,* 890–895.

Domeier, M. E., Morse, D. P., Knight, S. W., Portereiko, M., Bass, B. L., & Mango, S. E., (2000). A link between RNA interference and nonsense-mediated decay in *Caenorhabditis elegans*. *Science* (Washington DC, USA), *289*, 1928–1931.

Elbashir, S. M., Martinez, J., Patkaniowska, A., Lendeckel, W., & Tuschl, T., (2001b). Functional anatomy of siRNAs for mediating efficient RNAi in Drosophila melanogaster embryo lysate. *EMBO (Eur. Mol. Biol. Organ) J., 20*, 6877–6888.

Eldadah, B. A., & Faden, A. I., (2000). Caspase pathways, neuronal apoptosis and CNS injury. *J. Neurotrauma, 17*, 811–829.

Escobar, M. A., Civerolo, E. L., Summerfelt, K. R., & Dandekar, A. M., (2001). RNAi-mediated oncogene silencing confers resistance to crown gall tumorigenesis. *Proceedings of the National Academy of Sciences of the United States of America, 98*, 13437–13442.

Fagard, M., Boutet, S., Morel, J. B., Bellini, C., & Vaucheret, H., (2000). AGO1, QDE-2, and RDE-1 are related proteins required for post-transcriptional gene silencing in plants, quelling in fungi and RNA interference in animals. *Proc. Natl. Acad. Sci. USA, 97*, 11650–11654.

Fire, A., (1999). RNA-triggered gene silencing. *Trends Genet., 15*, 358–363.

Fire, A., Xu, S., Montgomery, M. K., Kostas, S. A., Driver, S. E., & Mello, C. C., (1998). Potent and specific genetic interference by double-stranded RNA in Caenorhabditis Elegans. *Nature* (London), *391*, 806–811.

Franz, D. R., & Zajtchuk, R., (2002). Biological terrorism: understanding the threat, preparation and medical response. *Dis. Mon., 48*, 493–564.

Frischmeyer, P. A., & Dietz, H. C., (1999). Nonsense-mediated mRNA decay in health and disease. *Hum. Mol. Genet., 8*, 1893–1900.

Fu, C., Sunkar, R., Zhou, C., Shen, H., Zhang, J. Y., Matts, J., et al., (2012). Overexpression of miR156 in switchgrass (*Panicum virgatum* L.) results in various morphological alterations and leads to improved biomass production. *Plant Biotechnol. J., 10*, 443–452.

Gary, D. S., & Mattson, M. P., (2001). Integrin signaling via the PI3-kinase-Akt pathway increases neuronal resistance to glutamate-induced apoptosis. *J. Neurochem., 76*, 1485–1496.

Gheysen, G., & Vanholme, B., (2007). RNAi from plants to nematodes. *Trends in Biotechnology, 25*, 89–92.

Gil-Humanes, J., Pistón, F., Hernando, A., Alvarez, J. B., Shewry, P. R., & Barro, F., (2008). Silencing of g-gliadins by RNA interference (RNAi) in bread wheat. *Journal of Cereal Science, 48*, 565–568.

Gil-Humanes, J., Pistón, F., Rosell, C. M., & Barro, F., (2012). Significant down-regulation of γ-gliadins has minor effect on gluten and starch properties of bread wheat. *J. Cereal Sci., 56*, 161–170.

Goldstein, D. A., Songstad, D., Sachs, E., & Petrick, J., (2009). RNA interference in plants. Monsanto. http://www.monsanto.com.

Grishok, A., Pasquinelli, A. E., Conte, D., Li, N., Parrish, S., Ha, I., Baillie, D. L., Fire, A., Ruvkun, G., & Mello, C. C., (2001). Genes and mechanisms related to RNA interference regulate expression of the small temporal RNAs that control *C. elegans* developmental timing. *Cell, 106*, 23–34.

Gupta, A., Pal, R. K., & Rajama, M. V., (2013). Delayed ripening and improved fruit processing quality in tomato by RNAi-mediated silencing of three homologs of 1-aminopropane-1-carboxylate synthase gene. *J. Plant Physiol., 170*, 987–995.

Hamilton, A. J., & Baulcombe, D. C., (1999). A species of small antisense RNA in post-transcriptional gene silencing in plants. *Science* (Washington DC), *286*, 950–952.

Hammond, S. M., Bernstein, E., Beach, D., & Hannon, G. J., (2000). An RNA-directed nucle ase mediates post-transcriptional gene silencing in Drosophila cells. *Nature* (London), *404*, 293–296.

Hammond, S. M., Boettcher, S., Caudy, A. A., Kobayashi, R., & Hannon, G. J., (2001). Argonaute2, a link between genetic and biochemical analyses of RNAi. *Science* (Washington DC), *293*, 1146–1150.

Hernández, I., Chacón, O., Rodriguez, R., Portieles, R., López, Y., Pujol, M., & Borrás-Hidalgo, O., (2009). Black shank resistant tobacco by silencing of glutathione S-transferase. *Biochemical and Biophysical Research Communications, 387*, 300–304.

Hutchings, N. R., Donelson, J. E., & Hill, K. L., (2002). Trypanin is a cytoskeletal linker protein and is required for cell motility in African trypanosomes. *J. Cell Biol., 156*, 867–877.

Huvenne, H., & Smagghe, G., (2010). Mechanisms of dsRNA uptake in insects and potential of RNAi for pest control: A review. *Journal of Insect Physiology, 56*, 227–235.

Iratni, R., Yan, Y. T., Chen, C., Ding, J., Zhang, Y., Price, S. M., Reinberg, D., & Shen, M. M., (2002). Inhibition of excess nodal signaling during mouse gastrulation by the transcriptional corepressor DRAP1. *Science* (Washington DC), *298*, 1996–1999.

Jiang, C. J., Shimono, M., Maeda, S., Inoue, H., Mori, M., Hasegawa, M., Sugano, S., & Takatsuji, H., (2009). Suppression of the rice fatty-acid desaturase gene OsSSI2 enhances resistance to blast and leaf blight diseases in rice. *Molecular Plant-Microbe Interactions, 22*, 820–829.

Kartasheva, N. N., Contente, A., Lenz-Stoppler, C., Roth, J., & Dobbelstein. M., (2002). p53 induces the expression of its antagonist p73 Delta-N establishing an autoregulatory feedback loop. *Oncogene, 21*, 4715–4727.

Kennerdell, J. R., & Carthew, R. W., (1998). Use of dsRNA-mediated genetic interference to demonstrate that frizzled and frizzled 2 Act in the wingless pathway. *Cell, 95*, 1017–1026.

Kennerdell, J. R., & Carthew, R. W., (2000). Heritable gene silencing in Drosophila using double-stranded RNA. *Nat. Biotechnol., 18*, 896–898.

Ketting, R. F., Fischer, S. E., Bernstein, E., Sijen, T., Hannon, G. J., & Plasterk, R. H., (2001). Dicer functions in RNA interference and in synthesis of small RNA involved in developmental timing in *C. elegans. Genes Dev., 15*, 2654–2659.

Kisielow, M., Kleiner, S., Nagasawa, M., Faisal, A., & Nagamine. Y., (2002). Isoform-specific knockdown and expression of adaptor protein ShcA using small interfering RNA. *Biochem. J., 363*, 1–5.

Krichevsky, A. M., & Kosik, K. S., (2002). RNAi functions in cultured mammalian neurons. *Proc. Natl. Acad. Sci. USA, 99*, 11926–11929.

Kusaba, M., Miyahara, K., Lida, S., Fukuoka, H., Takario, T., Sassa, H., Nishimura, M., & Nishio, T., (2003). Low glutenin content 1: a dominant mutation that suppresses the glutenin multigene family via RNA silencing in rice. *Plant Cell, 15*, 1455–1467.

Lee, S. S., Lee, R. Y., Fraser, A. G., Kamath, R. S., Ahringer, J., & Ruvkun, G., (2002c). A systematic RNAi screen identifies a critical role for mitochondria in *C. elegans* longevity. *Nat. Genet., 33*, 40–48.

Levin, I., Frankel, P., Gilboa, N., Tanny, S., & Lalazar, A., (2003). The tomato dark green mutation is a novel allele of the tomato homolog of the Deetiolated1 gene. *Theoretical and Applied Genetics, 106*, 454–460.

Li, K., Yuh-Shuh, W., Srinivasa Rao, U., Keri, W., Yuhong, T., Vatsala, V., Barney, J. V., Kent, D. C., Elison, B. B., & Kirankumar, S. M., (2008). Over expression of a fatty acid amide hydrolase compromises innate immunity in Arabidopsis. *The Plant Journal, 56,* 336–349.

Li, X., Wang, X., Zhang, S., Liu, D., Duan, Y., & Dong, W., (2012). Identification of soybean microRNAs involved in soybean cyst nematode infection by deep sequencing, *PLoS One. 7,* e3965010.1371/journal.pone.0039650.

Lois, C., Hong, E. J., Pease, S., Brown, E. J., & Baltimore, D., (2002). Germline transmission and tissue-specific expression of transgenes delivered by lentiviral vectors. *Science* (Washington DC), *295,* 868–872.

Mansoor, S., Amin, I., Hussain, M., Zafar, Y., & Briddon, R. W., (2006). Engineering novel traits in plants through RNA interference. *Trends in Plant Science, 11,* 559–565.

Mao, Y. B., Cai, W. J., Wang, J. W., Hong, G. J., Tao, X. Y., Wang, L. J., Huang, Y. P., & Chen, X. Y., (2007). Silencing a cotton bollworm P450 monooxygenase gene by plant-mediated RNAi impairs larval tolerance of gossypol. *Nature Biotechnology, 25,* 1307–1313.

Marathe, R., Anandalakshmi, R., Smith, T. H., Pruss, G. J., & Vance, V. B., (2000). RNA viruses as inducers, suppressors and targets of post-transcriptional gene silencing. *Plant Mol. Biol., 43,* 295–306.

Mattson, M. P., (2000). Apoptosis in neurodegenerative disorders. *Nat. Rev. Mol. Cell Biol., 1,* 120–129.

McManus, M. T., Petersen, C. P., Haines, B. B., Chen, J., & Sharp, P. A., (2002). Gene silencing using micro-RNA designed hairpins. *RNA, 8,* 842–850.

Mourrain, P., Beclin, C., Elmayan, T., Feuerbach, F., Godon, C., Morel, J. B., Jouette, D., Lacombe, A. M., Nikic, S., Picault, N., Remoue, K., Sanial, M., Vo, T. A., & Vaucheret, H., (2000). Arabidopsis SGS2 and SGS3 genes are required for posttranscriptional gene silencing and natural virus resistance. *Cell, 101,* 533–542.

Mustilli, A. C., Fenzi, F., Ciliento, R., Alfano, F., & Bowler, C., (1999). Phenotype of the tomato high pigment-2 mutant is caused by a mutation in the tomato homolog of Deetiolated1. *Plant Cell, 11,* 145–157.

Nakamura, N., Fukuchi-Mizutani, M., Miyazaki, K., Suzuki, K., & Tanaka, Y., (2006). RNAi suppression of the anthocyanidin synthase gene in Torenia hybrida yields white flowers with higher frequency and better stability than antisense and sense suppression. *Plant Biotechnol., 23,* 13–18.

Nakatsuka, T., Abe, Y., Kakizaki, Y., Yamamura, S., & Nishihara, M., (2007a). Production of red-flowered plants by genetic engineering of multiple flavonoid biosynthetic genes. *Plant Cell Rep., 26,* 1951–1959.

Nakatsuka, T., Pitaksutheepong, C., Yamamura, S., & Nishihara, M., (2007b). Induction of differential flower pigmentation patterns by RNAi using promoters with distinct tissue-specific activity. *Plant Biotechnol Rep., 1,* 251–257.

Nakatsuka, T., Mishibaa, K. I., Abe, Y., Kubota, A., Kakizaki, Y., Yamamura, S., & Nishihara, M., (2008). Plant Biotechnology, Flower color modification of gentian plants by RNAi-mediated gene silencing. *Plant Biotechnology., 25,* 61–68.

Naldini, L., (1998). Lentiviruses as gene transfer agents for delivery to non-dividing cells. *Curr. Opin. Biotechnol., 9,* 457–463.

Napoli, C., Lemieux, C., & Jorgensen, R., (1990). Introduction of a chimeric chalcone synthase gene into petunia results in reversible co-suppression of homologous genes in trans. *Plant Cell, 2,* 279–289.

Nishihara, M., Nakatsuka, T., & Yamamura, S., (2005). Flavonoid components and flower color change in transgenic tobacco plants by suppression of chalcone isomerase gene. *FEBS Lett., 579,* 6074–6078.

Nishihara, M., Nakatsuka, T., Hoaokawa, K., Yokoi, T., Abe, Y., Mishiba, K., & Yamamura, S., (2006). Dominant inheritance of white-flowered and herbicide-resistant traits in transgenic gentian plants. *Plant Biotechol., 23,* 25–31.

Ni, Z., Hu, Z., Jiang, Q., & Zhang, H., (2013). GmNFYA3 a target gene of miR169 is a positive regulator of plant tolerance to drought stress. *Plant Mol. Biol., 82,* 113–129.

Ohta, T., Essner, R., Ryu, J. H., Palazzo, R. E., Uetake, Y., & Kuriyama, R., (2002). Characterization of Cep135, a novel coiled-coil centrosomal protein involved in microtubule organization in mammalian cells. *J. Cell Biol., 156,* 87–99.

Peltier, A. J., Hatfield, R. D., & Grau, C. R. Soybean stem lignin concentration Price, D. R. G. and Gatehouse, J. A., (2009). RNAi-mediated crop protection against insects. *Trends in Biotechnology., 26,* 393–400.

Pfeifer, A., Ikawa, M., Dayn, Y., & Verma, I. M., (2002). Transgenesis by lentiviral vectors: lack of gene silencing in mammalian embryonic stem cells and preimplantation embryos. *Proc. Natl. Acad. Sci. USA, 99,* 140–145.

Pilate, G., Guiney, E., Holt, K., Petit-Conil, M., Lapierre, C., Leple, J., Pollet, B., Mila, I., Webster, E. A., Marstorp, H. G., Hopkins, D. W., Jouanin, L., Boerjan, W., Schuch, W., Cornu, D., & Halpin, C., (2002). Field and pulping performances of transgenic trees with altered lignification. *Nature Biotechnology, 20,* 607–612.

Poinat, P., De Arcangelis, A., Sookhareea, S., Zhu, X., Hedgecock, E. M., Labouesse, M., & Georges-Labouesse, E., (2002). A conserved interaction between beta1 integrin/PAT-3 and Nck-interacting kinase/MIG-15 that mediates commissural axon navigation in *C. elegans. Curr. Biol., 12,* 622–631.

Porter, L. A., Dellinger, R. W., Tynan, J. A., Barnes, E. A., Kong, M., Lenormand, J. L., & Donoghue, D. J., (2002). Human Speedy: a novel cell cycle regulator that enhances proliferation through activation of *Cdk2. J. Cell Biol., 157,* 357–366.

Pruss, G., Ge, X., Shi, X. M., Carrnington, J. C., & Vance, V. B., (1997). Plant viral synergism: the potyviral genome encodes a broad-range pathogenicity enhancer that transactivates replication of heterologous viruses. *Plant Cell, 9,* 1749–1759.

Redenbaugh, K., Bill, H., Belinda, M., Matthew, K., Ray, S., Rick, S., Cathy, H., & Don, E., (1992). *Safety Assessment of Genetically Engineered Fruits and Vegetables: A Case Study of the Flavr Savr Tomato.* CRC Press, pp. 288.

Redfern, A. D., Colley, S. M., Beveridge, D. J., Ikeda, N., Epis, M. R., Li, X., et al., (2013). RNA-induced silencing complex (RISC) proteins PACT, TRBP, and Dicer are SRA binding nuclear receptor coregulators. *Proc. Natl. Acad. Sci. USA, 110,* 6536–6541.

Regina, A., Bird, A., Topping, D., Bowden, S., Freeman, J., Barsby, T., Kosar-Hashemi, B., Li, Z., Rahman, S., & Morell, M., (2006). High-amylose wheat generated by RNA interference improves indices of large-bowel health in rats. *Proceedings of the National Academy of Sciences of the USA, 103,* 3546–3551.

Rodoni, B. C., Dale, J. L., & Harding, R. M., (1999). Characterization and expression of the coat protein-coding region of the banana bract mosaic potyvirus, development of diagnostic assays and detection of the virus in banana plants from five countries in Southeast Asia. *Archives of Virology, 144,* 1725–1737.

Rubinson, D. A., Dillon, C. P., Kwiatkowski, A. V., Sievers, C., Yang, L., Kopinja, J., Zhang, M., McManus, M. T., Gertler, F. B., Scott, M. L., & Van Parijs, L., A lentivirus-based Tang, G., Galili G., (2003). Using RNAi to improve plant nutritional value: from mechanism to application. *TRENDS in Biotechnology, 22*(9) 463–469.

Rubinsztein, D. C., (2002). Lessons from animal models of Huntington's disease. *Trends Genet., 18,* 202–209.

Salisbury, J. L., Suino, K. M., Busby, R., & Springett, M., (2002). Centrin-2 is required for centriole duplication in mammalian cells. *Curr. Biol., 12,* 1287–1292.

Smardon, A., Spoerke, J. M., Stacey, S. C., Klein, M. E., Mackin, N., & Maine, E. M., (2000). EGO-1 is related to RNA-directed RNA polymerase and functions in germ-line development and RNA interference in *C. elegans. Curr. Biol., 10,* 169–178.

Shomura, A., Izawa, T., Ebana, K., Ebitani, T., Kanegae, H., Konishi, S., & Yano, M., (2008). Deletion in a gene associated with grain size increased yields during rice domestication. *Nature Genetics, 40,* 1023–1028.

Spencer, P. S., Roy, D. N., Ludolph, A., Hugon, J., Dwivedi, M. P., & Schaumburg, H. H., (1986). Lathyrism: evidence for role of the neuroexcitatory aminoacid BOAA. *Lancet, 2*(8515), 1066–1067.

Sunilkumar, G., Campbell, L. M., Puckhaber, L., Stipanovic, R. D., & Rathore, K. S., (2006). Engineering cottonseed for use in human nutrition by tissue specific reduction of toxic gossypol. *Proceedings of the National Academy of Sciences, 103,* 18054–18059.

Surabhi, R. M., & Gaynor, R. B., (2002). RNA interference directed against viral and cellular targets inhibits human immunodeficiency virus type 1 replication. *J. Virol., 76,* 12963–12973.

Tabara, H., Sarkissian, M., Kelly, W. G., Fleenor, J., Grishok, A., Timmons, L., Fire, A., & Mello, C. C., (1999). The Rde-1 gene, RNA interference and transposon silencing in *C. Elegans. Cell, 99,* 123–132.

Tabara, H., Yigit, E., Siomi, H., & Mello, C. C., (2002). The dsRNA binding protein RDE-4 interacts with RDE-1, DCR-1 and a DExH-box helicase to direct RNAi in *C. elegans. Cell, 109,* 861–871.

Tan, Y. T., Tillett, D. J., & McKay, I. A., (2000). Molecular strategies for overcoming antibiotic resistance in bacteria. *Mol. Med. Today, 6,* 309–314.

Tanaka, Y., Katsumoto, Y., Brugliera, F., & Mason, J., (2005). Genetic engineering in floriculture. *Plant Cell Tiss. Org. Cult., 80,* 1–24.

Tang, G., & Galili, G., (2004). Using RNAi to improve plant nutritional value: from mechanism to application. *TRENDS in Biotechnology, 22*(9), 463–469.

Tang, G., Galili, G., & Zhuang, X., (2007). RNAi and microRNA: breakthrough technologies for the improvement of plant nutritional value and metabolic engineering. *Metabolomics, 3,* 357–369.

Tavernarakis, N., Wang, S. L., Dorovkov, M., Ryazanov, A., & Driscoll, M., (2000). Heritable and inducible genetic interference by double-stranded RNA encoded by transgenes. *Nat. Genet., 24,* 180–183.

Tsuda, S., Fukui, Y., Nakamura, N., Katsumoto, Y., Yonekura-Sakakibara, K., Fukuchi-Mizutani, M., Ohira, K., Ueyama, Y., Ohkawa, H., Holton, T. A., Kusumi, T., & Tanaka, Y., (2004). Flower color modification of Petunia 83ybrid commercial varieties by metabolic engineering. *Plant Biotechnol., 21,* 377–386.

Ukomadu, C., & Dutta, A., (2003). Inhibition of cdk2 activating phosphorylation by mevastatin. *J. Biol. Chem., 278,* 4840–4846.

Van Blokland, R., Van Der Geest, N., Mol, J. M. N., & Kooter, J. M., (1994). Transgene-mediated expression of chalcone synthase expression in Petunia hybrida results from an increase in RNA turnover. *Plant J., 6,* 861–877.

Van Uyen, N., (2006). Novel approaches in plants breeding RNAi technology. *Proc. of Intern. Workshop on Biotechnol. in Agri.,* 12–16.

Volpe, T. A., Kidner, C., Hall, I. M., Teng, G., Grewal, S. I., & Martienssen, R. A., (2002). Regulation of heterochromatic silencing and histone H3 lysine-9 methylation by RNAi. *Science* (Washington DC), *297,* 1833–1837.

Wang, X., Tang, C., Chai, J., Shi, Y., & Xue, D., (2002). Mechanisms of AIF-mediated apoptotic DNA degradation in *Caenorhabditis elegans*. *Science* (Washington DC), *298*, 1587–1592.

Wang, Z., Morris, J. C., Drew, M. E., & Englund, P. T., (2000). Inhibition of Trypanosoma brucei gene expression by RNA interference using an integratable vector with opposing T7 promoters. *J. Biol. Chem.*, *275*, 40174–40179.

Wang, S., Wu, K., Yuan, Q., Liu, X., Liu, Z., Lin, X., et al., (2012). Control of grain size, shape and quality by OsSPL16 in rice. *Nat. Genet.*, *44*, 950–954.

Waterhouse, P. M., Graham, M. ;lW. and Wang, M.B. (1998). Virus resistance and gene silencing can be induced by simultaneous expression of sense and antisense RNA. *Proc. of Natl Acad. Sci.*, USA 95, 13959–13964.

Whyard, S., Singh, A. D., & Wong, S., (2009). Ingested double-stranded RNAs can act as species-specific insecticides. *Insect Biochemistry and Molecular Biology, 39*, 824–832.

Williams, M., Clark, G., Sathasivan, K., & Islam, A. S., (2004). RNA Interference and its Application in Crop Improvement. *Plant tissue culture and Biotechnology*, 1–18.

Winston, W. M., Molodowitch, C., & Hunter, C. P., (2002). Systemic RNAi in *C. elegans* requires the putative transmembrane protein SID-1. *Science* (Washington DC), *295*, 2456–2459.

Wojcik, C., & DeMartino, G. N., (2002). Analysis of Drosophila 26 S proteasome using RNA interference. *J. Biol. Chem.*, *277*, 6188–6197.

Yang, C., Li, D., Mao, D., Liu, X., Li, C., Li, X., et al., (2013). Overexpression of microRNA319 impacts leaf morphogenesis and leads to epce. 12130 [inhanced cold tolerance in rice] (*Oryza sativa* L.). *Plant Cell Environ.*, *36*, 2207–2218.

Ye, F., & Signer, E. R., (1996). RIGS (repeat-induced gene silencing) in Arabidopsis is transcriptional and alters chromatin configuration. *Proc. Natl. Acad. Sci. USA*, *93*, 10881–10886.

Yin, D., Deng, S., Zhan, K., & Cui, D., (2007). High-oleic peanut oils produced by HpRNA-mediated gene silencing of oleate desaturase. *Plant Molecular Biology Reporter, 25*, 154–163.

Zamore, P. D., Tuschl, T., Sharp, P. A., & Bartel, D. P., (2000). RNAi: Double-stranded RNA directs the ATP-dependent cleavage of mRNA at 21 to 23 nucleotide intervals. *Cell*, *101*, 25–33.

Zhang, Y. C., Yu, Y., Wang C. Y., Li, Z. Y., Liu, Q., Xu, J., et al., (2013). Overexpression of microRNA OsmiR397 improves rice yield by increasing grain size and promoting panicle branching. *Nat. Biotechnol.*, *31*, 848–852.

Zhu, X., & Galili, G., (2003). Increased lysine synthesis coupled with a knockout of its catabolism synergistically boosts lysine content and also trans-regulates the metabolism of other amino acids in Arabidopsis seeds. *Plant Cell, 15*, 845–853.

Zhu, X., & Galili, G., (2004). Lysinemetabolism is concurrently regulated by synthesis and catabolism in both reproductive and vegetative tissues. *Plant Physiology, 135*, 129–136.

CHAPTER 18

GENOME EDITING IN PLANT BREEDING

D. R. MEHTA[1] and A. K. NANDHA[2]

[1]Associate Professor, Department of Genetics and Plant Breeding, Junagadh Agricultural University, Junagadh–362001, Gujarat, India

[2]PhD Scholar, Department of Biotechnology, Junagadh Agricultural University, Junagadh–362001, Gujarat, India

CONTENTS

18.1 INTRODUCTION

In the mid of the 21st century, overall agricultural productivity will need to increase by at least 70% to fulfill the requirements of growing population. At the same time, agricultural production will need to become more resilient to biotic and may inadvertently generate varieties that are ill-equipped to cope up with the biotic as well as abiotic stresses in future (Madesis et al., 2016). A broad range and multifaceted solution is certainly required for such problems. To comply with such essentials, plant breeding must produce new varieties. For example, research is in progress to develop plants with ability to grow in higher CO_2 and higher temperature, so that they can survive in such global warming era. But the process of such gene transformation via simple plant breeding techniques needs more than 20 years. Moreover, such breeding methods are also accompanied by other challenges (Jones, 2016).

With the help of newly emerged techniques known as genome editing, scientists are seeking to develop new crop varieties that are more potent than before in various criteria. Such genomic techniques and tools have converted tedious and time-consuming process into a more effective, more efficient, and much potential process for the plant breeders.

Genome editing is the process of editing an organism's DNA by altering, removing, or adding nucleotides to the genome. This process is accomplished by using engineered nucleases that can make a double-stranded "cut" or a single-stranded "nick" in an organism's DNA. These sites are altered and then repaired by either homologous recombination or non-homologous end joining. This type of genetic engineering is useful for a number of applications, including targeted gene mutations, chromosomal rearrangement, and creation of transgenic plants and animals.

18.2 DIFFERENCE BETWEEN GENOME-EDITED PLANTS AND GENETICALLY MODIFIED PLANTS

Genome-edited plants are not classified as genetically modified plants. With traditional genetic engineering, genes are often introduced into a plant's DNA that does not arise naturally in the species, for example, genes for resistance to a specific herbicide. Different processes exist for this: for example, the genes can be "shot" into the plant cells by using a kind of "gene gun."

With genome editing, one can cut the DNA with a protein at a predefined location. The genome editing method known as CRISPR/Cas9 has become the most common method (Weigel, 2016). One can then modify the DNA at the interface or insert new sections. So, genome editing should be viewed as a variant of mutation breeding, with the difference that the generation of particular mutations is targeted.

The major advantage of these modifications can be obtained in the same way as they are made in traditional breeding and crossing experiments. For example, individual letters of the genetic code can be exchanged. This corresponds to a modification that can also arise through natural mutation. Short sections of DNA can also be inserted, and in this way, genes from a species can be replaced with genes from its other varieties or from closely related species–something that is also done in traditional cross-breeding (Weigel, 2016).

18.3   CROP IMPROVEMENT USING GENOME EDITING TECHNIQUES

One of the major goals of conventional plant breeding is to remove or to add certain traits of crops to enhance their nutritional values or resistance to diverse biotic and abiotic stress (Allard, 1999; Moose and Mumm, 2008). For instance, the high level of erucic acids and glucosinolates in rapeseed was successfully removed by conventional breeding in 1970s, and now, rapeseed has become the third most important source of vegetable oils in the world (Gupta et al., 2007). While this conventional breeding relies chiefly on natural variation in a gene of interest, physical or chemical mutagens have been used to generate random crop variants. In addition, the development of RNAi methods enables target genes to be silenced in specific tissues or at certain times, which results in the removal of unwanted traits from crops (Kusaba, 2004; Tang and Galili, 2004). However, RNAi-mediated gene silencing had to overcome challenges from incomplete gene silencing, the co-silencing of unintended genes (off-targets), and the random integration of foreign DNA into plant genomes (if T-DNA harboring RNAi construct is transformed). Genome editing technologies can overcome some of these limitations of conventional breeding and RNAi-based approach to be used to generate improved crops indistinguishable from naturally occurring mutant crops.

Genome editing technologies rely on engineered endonucleases (EENs) that cleave DNA in a sequence-specific manner because of the presence of a sequence-specific DNA-binding domain or RNA sequence. Through recognition of the specific DNA sequence, these nucleases can efficiently and precisely cleave the targeted genes. The double strand breaks (DSBs) of DNA consequently result in cellular DNA repair mechanisms, including homology-directed repair (HDR) and error-prone non-homologous end joining breaks (NHEJ), leading to gene modification at the target sites

18.4   DSB IS A KEY MECHANISM IN GENOME EDITING

Traditionally, natural or induced mutagenesis has been used in plant breeding to generate genetic variation that increases crop yields. In recent decades, mutagens such as ethyl methyl sulfonate (EMS) and γ-rays have been used to induce new traits by randomly introducing mutations into the genome (Sikora et al., 2011). To obtain superior mutant organisms using these strategies, mutant populations must be screened to identify individual plants that have the desired phenotype. Following natural or induced DSBs, DSB repair systems are responsible for the generation of mutations in the genome, and NHEJ and HDR are the main pathways by which this occurs (Osakabe et al., 2012). NHEJ uses various repair enzymes to join the ends of the DNA following a DSB. This repair system has two sub-pathways: the Ku-dependent NHEJ pathway and the Ku-independent NHEJ pathway (Deriano and Roth, 2013). In Ku-dependent NHEJ, the DNA end protection factors—Ku70/80 proteins—bind to the end of the DNA strand at the break site and recruit the repair enzyme ligase IV and its cofactor (Deriano and Roth, 2013). This Ku-dependent pathway can sometimes generate several nucleotide insertions or deletions at the break site. When the factors involved in Ku-dependent NHEJ are absent, the alternative end-joining (Ku-independent) pathway can repair DSBs in eukaryotes (Frit et al., 2014). Microhomology-mediated end joining (MMEJ) is a major Ku-independent NHEJ pathway. A recent study of the MMEJ pathway proposed that the microhomology (MM) regions are resected, annealed at the MM regions, filled in by DNA polymerase, and finally re-ligated by DNA ligase (Crespan et al., 2012). Thus, MMEJ produces a longer deletion at the DSB site compared with Ku-dependent NHEJ.

In contrast to error-prone NHEJ pathways, HDR is usually an error-free DSB repair pathway, because HDR uses a DNA template to replace the DNA sequence at the break point accurately. HDR is restricted to late S/G2 phases in the cell cycle, whereas NHEJ functions throughout the entire cell cycle. Therefore, NHEJ is the major DSB repair pathway in eukaryotes (Sonoda et al., 2006). Both NHEJ and HDR repair pathways can be exploited for nuclease-based genome editing. Based on these DSB repair mechanisms, site-directed mutagenesis mediated via NHEJ- or HDR-directed gene knock-in/correction can be performed at specific locations in the genome. In addition, recent studies have shown that targeting cleavage to specific loci can induce chromosomal rearrangements such as deletions of a few megabase pairs (Lee et al., 2010; Xiao et al., 2013), duplications (Lee et al., 2012), inversions (Lee et al., 2012; Xiao et al., 2013), and translocations (Brunet et al., 2009).

18.5 TYPES OF GENOME EDITING

The words "genome editing" define the use of a suite of site-directed nucleases (SDN) capable of cutting or otherwise modifying predetermined DNA sequences in the genome. Examples of SDNs are: zinc-finger nucleases (ZFN), transcription activator-like effector nuclease (TALEN), and clustered regularly interspaced short palindromic repeats (CRISPRs). Others such as meganucleases (MN) and oligonucleotide-directed mutagenesis (ODM) also exist as yet unpublished molecules with genome-editing potential (Tables 18.1 and 18.2). Regardless of the specific tool, they have one thing in common: to cut or otherwise modify a predetermined sequence in the target genome and generate novel phenotypes in animals and plants (Jones, 2016).

18.5.1 MEGANUCLEASES

Meganucleases or homing endonuclease, endodeoxyribonucleases characterized by a large recognition site (double-stranded DNA sequences of 12 to 40 base pairs), were the first class of sequence-specific nucleases used in plants, and they continue to be deployed to achieve complex genome modifications. The term *meganuclease* reflects the name "Omega" that was coined for the first known homing system as well as for designating sequence specific endonucleases proteins that could recognize large targets (from 12 to 40 bp) (Thierry and Dujon, 1992). An advantage of mega-

TABLE 18.1 Comparative Attributes of Plant Genome Editing Techniques

Sr. No.	Features	Types of genome editing technique			
		Mega nuclease	**ZFN**	**TALEN**	**CRISPR-Cas**
1.	Components	Target recognition domain Nuclease domain	Zinc-finger domains Non-specific *Fok* I nuclease domain	TALE domains Non-specific *Fok* I nuclease domain	CrRNA Cas9 protein
2.	Length of target Sequence (bp)	14-40	18-24	24-59	20-22
3.	Target recognition efficiency	Low	High	High	High
4.	Cleavage efficiency	High	High	High	High with possibility of multiplexing
5.	Level of experiment setup	Complicated procedure of redesigning for each new target site and need expertise in protein engineering	Complicated procedure of redesigning for each new target site and need expertise in protein engineering	Relatively easy procedure of designing for each new target site	Easy and very fast procedure of designing for each new target site
6.	Off-target effect	Detected	Detected	Detected	Detected
7.	Requirement for cloning	Require cloning	Cloning is necessary	Require cloning	No need for cloning
8.	Structural protein domain	I-DmoI and I-CreI	ZFNs work as dimeric and only protein component required	Work as dimeric and required protein components	Consist single monomeric protein and chimeric RNA
9.	Target reorganization specificity	Moderate	High	High	High

TABLE 18.2 Various Crop Plants Targeted for Gene Editing

No.	Species	Nuclease	Gene	References
1.	Arabi-dopsis	Meganuclease	*RAD51C* and *XRCC3*	Roth et al. (2012)
		Oligonucle-otide-Directed Mutagenesis -ODM	*EPSPS*	Sauer et al. (2016a)
		ZFN	*HSP18.2, ABI4, ubi10, ADH1, TT4, At1g53430-At1g53440, PPO, TYLCCNV*	Lloyd et al. (2005); Osakabe et al. (2010); Zhang et al. (2010); Petolino et al. (2010); Even-faiteson et al. (2011); Qi et al. (2013a); de Pater et al. (2013); Chen et al. (2014)
		TAL effector nuclease-TALEN	*ADH1, TT4, MAP-KKK1, DSK2B,* and *NATA2*	Cermak et al. (2011); Christian et al. (2013)
		CRISPR/Cas9	*BRI1, JAZ1, GA1, CHLI1, CHLI2, FAD2-1A and FAD2-1B, ADH1, RTEL1, NAC052, NAC050*	Li et al. (2013); Jiang et al. (2013); Feng et al. (2013); Mao et al. (2013); Feng et al. (2014); Schiml et al. (2014); Jiang et al. (2014a,b); Haun et al. (2014); Johnson et al. (2015); Gao et al. (2015a,b,c); Ning et al. (2015); Fauser et al. (2014).
2..	Canola	Zinc finger Nuclease (ZFN)	*KASII*	Gupta et al. (2012)
3	Cotton	Meganuclease	*cry2Ae, bar*	D'Halluin et al. (2013)
4.	Potato	TALEN	*VInv, ALS*	Nicolia et al. (2015); Clasen et al. (2015)
		CRISPR/Cas9	*StIAA2*	Wang et al. (2015)
5.	Soya-bean	ZFN	*DCL4a and DCL4b,*	Curtin et al. (2011, 2013)
		TALEN	*FAD2-1A and FAD2-1B*	Haun et al. (2014)
		CRISPR/Cas9	*GmFEI2* and *GmSHR, Glyma07g14530, GN20GG, Glyma06 g14180, Glyma08g0 2290* and *Glyma12g 37050*	Jacobs et al. (2015); Cai et al. (2015) Michno et al. (2015) Sun et al. (2015)

TABLE 10.2   (Continued)

No.	Species	Nuclease	Gene	References
6.	Tobacco	ODM	*SuRA, SuRB, ALS SuRA and SuRB*	Beetham et al. (1999); Kochevenko and Willmitzer (2004); Ruiter et al. (2003)
		ZFN	*ALS SuRA* and *SuRB, uidA, NPTII, GUS*	Wright et al. (2005); Maeder et al. (2008); Townsend et al. (2009); Cai et al. (2009); Petolino et al. (2010); Baltes et al. (2014)
		TALEN	*Hax3, dHax3, ALS*	Cermak et al. (2011); Mahfouz et al. (2011); Zhang et al. (2013)
		CRISPR/Cas9	*CYCD3, AtCRU3, HLA3, ABP1, NtPDS, NtPDR6*	Li et al. (2013); Jiang et al. (2013); Nekrasov et al. (2013); Gao et al. (2015a,b,c); Johnson et al. (2015)
7.	Tomato	TALEN	*PROCERA*	Lor et al. (2014)
		CRISPR/Cas9	*SlAGO7, SHORT-ROOT, SCARECROW*	Brooks et al. (2014); Ron et al. (2014)
8.	Barley	TALEN	*HvPAPhy_a*	Wendt et al. (2013); Gurushidze et al. (2014)
9.	Maize	Meganuclease	*AmCyan1, DsRed2, MS26, LG1*	Gao et al. (2010); Djukanovic et al. (2013); Martin-Ortigosa et al. (2014)
		ODM	*AHAS*	Zhu et al. (1999)
		ZFN	*IPK1, PAT, AAD1*	Shukla et al. (2009); Ainley et al. (2013)
		TALEN	*ZmIPK, ZmIPK1, ZmMRP4, ZmPDS, gl2*	Liang et al. (2014); Char et al. (2015); Beetham et al. (2014)
		CRISPR/Cas9	*ZmIPK, ZmIPK1, ZmMRP4, ZmPDS, LIG1, Ms26, MS45, ALS1 and ALS2*	Liang et al. (2014); Svitashev et al. (2015)
10.	Rice	ODM	*ALS*	Okuzaki and Toriyma (2004)
		TALEN	*OsBADH2, Hax3, dHax3.*	Li et al. (2012); Shan et al. (2013a, 2015); Mahfouz et al. (2011)

TABLE 18.2 (Continued)

No.	Species	Nuclease	Gene	References
		CRISPR/Cas9	*OsSWEET14, Os-SWEET11, INOX, PDS, ROC5, SPP, YSA, CAO1, LAZY1,OsMYB1, OsYSA, Os-ROC5, OsDERF1, OsPDS,OsMSH1, OsSPP, OsEPSPS, Os06g0133000, OsFTL11, Os07g0261200, Os02g0700600*	Jiang et al. (2013); Shan et al. (2013); Feng et al. (2013); Miao et al. (2013); Zhang et al. (2014); Zhou et al. (2014); Xu et al. (2014); Endo et al. (2015); Xie et al. (2015); Shimatani et al. (2015); Zhang et al. (2016); Ma et al. (2015)
11.	Sorghum	CRISPR/Cas9	*OsSWEET14, Os-SWEET11*	Jiang et al. (2013)
12.	Wheat	TALEN	*TaMLO-A1*	Wang et al. (2014)
		CRISPR/Cas9	*INOX, PDS, TaMLO-A1*	Shan et al. (2013b); Upadhyay et al. (2013); Wang et al. (2014)
13.	*Brassica napus*	ODM	*AHAS*	Gocal, (2015); Gocal et al. (2015); Ruiter et al., (2003)
14.	Flax	ODM	*EPSPS*	Beetham et al. (2014); Sauer et al. (2016b)
15.	Sweet orange	CRISPR/Cas9	*CsPDS*	Jia and Wang (2014)
16.	Liverwort	CRISPR/Cas9	*ARF1*	Sugano et al. (2014)
17.	*Populus tomentosa*	CRISPR/Cas9	*PtoPDS, PtPDS1 and PtPDS2, 4CL1 and 4CL2*	Fan et al., (2015); Tingting et al. (2015); Zhou et al. (2015)
18.	*Medicago trancatula*	CRISPR/Cas9	*GN20GG*	Michno et al., 2015
19.	Black cottonwood	CRISPR/Cas9	*PtoPDS, 4CL1 and 4CL2*	Fan et al., (2015); Zhou et al. (2015)
20.	Wild cabbage (*B. oleracea*)	CRISPR/Cas9	*HvPm19*	Lawrenson et al. (2015)

(Source: Weeks et al., 2016).

nucleases is their size. Meganucleases are, therefore, considered to be the most specific naturally occurring restriction enzymes. Meganucleases can be divided into five families based on sequence and structure motifs: (i) LAGLIDADG, (ii) GIY-YIG, (iii) HNH, (iv) His-Cys box, and (v) PD-(D/E) XK. Among the LAGLIDADG family, one of the six families of homing endonucleases, meganuclease has become a valuable tool for the study of genomes and genome engineering from last 15–20 years. Meganucleases are "naturally occurring molecular DNA scissors" that can be used to replace, eliminate, or modify sequences. It has high sequence specificity. By modifying their recognition sequence through protein engineering, the targeted sequence can be changed. Meganucleases are used to modify all genome types, whether bacterial, plant, or animal. They are among the smallest nucleases – comprising only around 165 amino acids (aa), which makes them amenable to most delivery methods, including vectors with limited cargo capacities, such as plant RNA viruses.

Meganucleases are mainly represented by two main enzyme families collectively known as homing endonucleases: intron endonucleases and intein endonucleases. Introns propagate by intervening at a precise location in the DNA, where the expression of the meganuclease produces a break in the complementary intron- or intein-free allele. For inteins and group I introns, this break leads to the duplication of the intron or intein at the cutting site by means of the homologous recombination repair for double-stranded DNA breaks.

To date, a purposeful role within the host has not been identified for these proteins, and they tend to be classified as **selfish genetic elements**. With few exceptions, LAGLIDADG proteins exhibit one of two primary activities: (a) they function as RNA maturases involved in facilitating the splicing of their own intron or (b) they function as highly specific endonucleases capable of recognizing and cleaving the exonexon junction sequence wherein their intron resides, thus giving rise to the moniker **homing endonuclease**.

Relative to other sequence-specific nucleases, however, meganucleases are challenging to re-design for new target specificity. Redesign is hindered by the non-modular nature of the protein. For example, within the LAGL-IDADG family of meganucleases, the amino acids responsible for binding DNA overlap with those for DNA cleavage; therefore, attempting to alter the DNA-binding domain can affect the enzyme's catalytic activity.

The use of meganucleases in plants has been limited to naturally occurring meganucleases (e.g., I-SceI, I-CreI) or to redesigned nucleases made by groups with expertise in structure-based design or the capacity to carry out high-throughput *in vitro* screens to identify active nucleases from libraries of variants (Pâques and Duchateau, 2007).

It has been hypothesized that homing endonucleases have a so called "life-cycle" (i) first, they start out as invasive endonucleases capable of mobilizing their coding sequence; (ii) upon "invasion," they acquire a concomitant RNA maturase activity to help ensure proper splicing of their intron; (iii) over time, the "invasive" nuclease activity is lost, leaving only the RNA maturase function, and (iv) finally, upon losing the maturase activity, propagation of the intron becomes unviable and the intron is lost. It is thus inferred that the functionality of a given LAGLIDADG protein (endonuclease, maturase, or both) represents a snapshot into the current state of its life cycle.

18.5.2 OLIGONUCLEOTIDE-DIRECTED MUTAGENESIS

Oligonucleotide-directed mutagenesis (ODM) or chimeraplasty is based on the site-specific correction or directed mutation (base substitution, addition, or deletion) of a target gene through the introduction of a chemically synthesized complementary chimeric DNA and/or RNA oligo-nucleotide designed to have one or more bases that do not pair with the endogenous gene sequence (Cole-Strauss et al., 1996). The DNA and 2'-O-methyl RNA residues form a double-stranded molecule capped at both ends by sequences that fold into a hairpin. The template oligonucleotide is typically approximately 20–80 nucleotides in size. The resulting helical distortion induced by the mismatch is subsequently recognized by the DNA repair machinery of the cell and corrected using the DNA sequence of the chimera as a template (Sargent et al., 2011).

Moerschell et al. (1988) was the first to show that single-stranded DNA oligonucleotides could direct *in vivo* base changes in *Saccharomyces cerevisiae*. To date, numerous publications report that synthetic oligonucleotides can induce small sequence changes in nuclear DNA of eukaryotic cells. The genetic changes that can be obtained using oligonucleotide-directed mutagenesis include the introduction of a new mutation (replacement of one or a few base pairs), the reversal of an existing mutation (gene repair), or the induction of short deletions in native genomic sequences (Lusser et al., 2011).

At present, the potential applicability of gene repair by ODM seems to be limited by the lack of knowledge on how these molecules are functioning and, consequently, by low efficiency of correction. For most ODM-specialized researchers, mechanistic studies to identify critical proteins involved in the gene correction process will undoubtedly play an important role for designing rational approaches toward increasing the targeting efficiency, efficacy, and safety.

18.5.3   ZINC-FINGER NUCLEASES

ZNF is a protein motif that contains a bound Zn ion and a protein "finger." It was first discovered in *Xenopus laevis* (African clawed frog) as the DNA binding domain of transcription factor (Li et al., 1992). A study of the transcription of a particular RNA sequence revealed that the binding strength of a small transcription factor was due to the presence of zinc-coordinating finger-like structures. Amino acid sequencing of TFIIIA revealed nine tandem sequences of 30 amino acids, including two invariant pairs of cysteine and histidine residues. Extended X-ray absorption fine structure confirmed the identity of the zinc ligands: two cysteines and two histidines. The DNA-binding loop formed by the coordination of these ligands by zinc were thought to resemble fingers; hence, it was named as zinc finger nucleases. More recent work on the characterization of proteins in various organisms has revealed the importance of zinc ions in polypeptide stabilization.

Zinc finger nucleases (ZFNs) are artificial proteins consisting of an engineered zinc finger DNA-binding domain fused to the cleavage domain of the *FokI* restriction endonuclease. ZFNs can be used to induce double-stranded breaks (DSBs) in specific DNA sequences and that makes ZFNs enable to promote site-specific homologous recombination and targeted manipulation of genomic loci in a variety of different cell types (Kim et al., 1997). A long-term goal of the Zinc Finger Consortium is to develop ZFNs as broadly applicable and readily accessible molecular tools for performing targeted genetic alterations. The ability to alter the sequence or structure of any gene of interest would be enormously useful for biological research and molecular therapeutics.

Like the meganucleases, zinc-finger nucleases are relatively small (around 300 aa per monomer; around 600 aa per nuclease pair), making them amenable to most delivery methods. DNA targeting by zinc-finger nucleases is achieved by arrays of zinc fingers, each of which typically binds to triplet of nucleotide. The modular structure of zinc finger (ZF) motifs and modular recognition by ZF domains make them the most versatile of the DNA recognition motifs for designing artificial DNA-binding proteins (Pabo et al., 2001; Beerli and Barbas, 2002). Each ZF motif consists of ~30 amino acids and folds into $\beta\beta\alpha$ structure, which is stabilized by chelation of a zinc ion by the conserved Cys_2His_2 residues. The ZF motifs bind DNA by inserting the α-helix into the major groove of the DNA double helix (Pavletich and Pabo, 1991). Each finger primarily binds to a triplet within the DNA substrate. Key amino acid residues at positions -1, $+1$, $+2$, $+3$, $+4$, $+5$, and $+6$ relative to the start of the α-helix of each ZF motifs contribute to most of the sequence-specific interactions with the DNA site (Pavletich and Pabo, 1991; Shi and Berg, 1995; Elrod-Erickson and Pabo, 1999). These amino acids can be changed, while maintaining the remaining amino acids as a consensus backbone to generate ZF motifs with different triplet sequence-specificities (Desjarlais and Berg, 1993).

Several of such ZF motifs bind to longer sequences of DNA in tandem to form ZFPs (Liu et al., 1997; Beerli et al., 1998; Kim and Pabo, 1998). The designed ZFPs provide a powerful platform technology as other functionalities like non-specific FokI cleavage domain (N), transcription activator domains (A), transcription repressor domains (R), and methylases (M) can be fused to obtain various forms of ZFNs (Kim et al., 1996; Smith et al., 2000; Mani et al., 2005; Alwin et al., 2005) as zinc finger transcription activators (ZFA) (Zhang et al., 2000; Liu et al., 2001; Rebar et al., 2002; Bartsevich et al., 2003; Dai et al., 2004; Rebar, 2004), zinc finger transcription repressors (ZFR) (Bartsevich and Juliano, 2000; Ren et al., 2002; Snowden et al., 2003), and zinc finger methylases (ZFM) (Xu and Bestor, 1997; McNamara et al., 2002), respectively.

18.5.3.1 Classification of Zinc Finger Nucleases

Initially, the term zinc finger was solely used to describe DNA-binding motif found in *Xenopus laevis*; however, it is now used to refer to any number of structures related by their coordination of a zinc ion. In general, zinc fingers coordinate zinc ions with a combination of cysteine and histidine residues. Originally, the number and order of these residues was used to classify different types of zinc fingers. More recently, a more systematic method has been used to classify zinc finger proteins. This method classifies zinc finger proteins into "fold groups" based on the overall shape of the protein backbone in the folded domain. The most common "fold groups" of zinc fingers are the Cys$_2$His$_2$-like, treble clef, and zinc ribbon.

18.5.3.1.1 *Two Ways of Classification of ZFN*

1) **Numbers of cysteines and histidines:** Typical zinc finger motifs are composed of two cysteines followed by two histidines

2) **Structure:** Zinc fingers are structurally diverse and are present among proteins that perform a broad range of functions in various cellular processes. Each available zinc finger structure can be placed into one of the eight fold groups (Table 18.3) that are defined based on the structural properties in the vicinity of the zinc-binding site. Evolutionary relatedness of proteins within fold groups is not implied, but each group is divided into families of potential homologues. Table 18.3 shows the different structures and their key features (Krishna et al., 2003):

   a) **Cys$_2$His$_2$:** The Cys2His2-like fold group is by far the best-characterized class of zinc fingers and is extremely common in mammalian transcription factors. These domains adopt a simple ββα fold and have the amino acid sequence motif: **X2-Cys-X2,4-Cys-X12-His-X3,4,5-His.** This class of zinc fingers can have a variety of functions such as binding RNA and mediating protein-protein interactions, but it is best known for its role in sequence-specific DNA-binding proteins such as Zif268. In such proteins, individual zinc finger domains typically occur as tandem repeats with two, three, or more fingers comprising the DNA-binding domain of the protein. These tandem arrays can bind in the major groove of DNA and are typically spaced at 3-bp intervals. The α-helix of each domain can make sequence-specific contacts to DNA bases; residues from a single recognition helix can contact 4 or more bases to yield an overlapping pattern of contacts with adjacent zinc fingers.

TABLE 18.3   Structure-Based Classification of ZFNs (Krishna et al., 2003)

Sr. No.	Fold group	Representative structure	Ligand placement
1.	C_2H_2		Two ligands from a knuckle and two more from the C terminus of a helix
2.	Gag knuckle		Two ligands from a knuckle and two from a short helix or loop
3.	Treble clef		Two ligands from a knuckle and two more from the N terminus of a helix
4.	Zinc ribbon		Two ligands each from two knuckles
5.	Zn_2/Cys_6		Two ligands from N terminus of a helix and two more from a loop
6.	TAZ$_2$ domain like		Two ligands, each from the termini of two helices
7.	Zinc binding loops		Four ligands from a loop

TABLE 10.3 (Continued)

Sr. No.	Fold group	Representative structure	Ligand placement
8.	Metallo-thionein		Cysteine rich metal binding loop

b) **Gag-knuckle:** This fold group is defined by two short β-strands connected by a turn followed by a short helix or loop and resembles the classical Cys_2His_2 motif with a large portion of the helix and β-hairpin truncated. The retroviral nucleocapsid protein from HIV and other related retroviruses are examples of proteins possessing these motifs. The gag-knuckle zinc finger in the HIV NC protein is the target of a class of drugs known as zinc finger inhibitors.

c) **Treble-clef:** The treble-clef motif consists of a β-hairpin at the N-terminus and an α-helix at the C-terminus; each contributes two ligands for zinc binding, although a loop and a second β-hairpin of varying length and conformation can be present between the N-terminal β-hairpin and the C-terminal α-helix. These fingers are present in a diverse group of proteins that frequently do not share sequence or functional similarity with each other. The best-characterized proteins containing treble-clef zinc fingers are the nuclear hormone receptors.

d) **Zinc ribbon:** The zinc ribbon fold is characterized by two beta-hairpins forming two structurally similar zinc-binding sub-sites.

e) **Zn_2/Cys_6:** The canonical members of this class contain a binuclear zinc cluster in which two zinc ions are bound by six cysteine residues. These zinc fingers can be found in several transcription factors including the yeast Gal4 protein.

f) **Miscellaneous:** Metallothioneins are cysteine-rich loops of about 60–70 residues that bind a variety of metals. No clearly defined regular secondary structural elements can be detected in metallothioneins, and metal-binding sites in them do not appear to be similar to those of other proteins. Although the precise biological function of metallothioneins is not clear, they are known to sequester excess metal ions from the cellular environment and possibly protect from metal toxicity.

18.5.3.2 Mechanism of the ZFN System

Zinc finger molecule bind with 3 specific nucleotides in specific sequence, while the nuclease enzyme binds at the spacer region, which cause cleavage at the specific site. If donor is present and insertion of specific gene sequence is required, then this cleaved sequence followed homologous recombination or in the absence (Carroll, 2011) of donor/for mutation, it follows NHEJ.

Different strategies can potentially be used when using ZFNs to modify the genome of plant species. Targeting a given native genome sequence (which is not typically composed of a palindrome-like sequence) requires the delivery and expression of at least two ZFN monomers in the same cell. ZFN expression can lead not only to site-specific mutagenesis, but also to gene stacking and/or gene replacement, depending on the presence and structure of the donor DNA and the plant DNA-repair machinery. The use of ZFNs as site-specific mutagens was initially tested in transgenic *Arabidopsis* plants, which were engineered to carry a semi-palindromic target site for a well-defined QQR ZFN.

The initial success in plant genome engineering with ZFNs was reported in *Arabidopsis thaliana* (Lloyd et al., 2005) and tobacco (*Nicotiana tabacum*) (Wright et al., 2005). Further developments included a small number of crop plants such as maize and soybean, and introduced important agronomic traits. Herbicide-resistance mutations were introduced into tobacco endogenous acetolactate synthase genes (ALS SuRA and SuRB) loci by the ZFN-mediated gene targeting at frequencies exceeding 2% of transformed cells for mutations as far as 1.3 kilobases from the ZFN cleavage site (Townsend et al., 2009). In addition, gene disruption and repair by the NHEJ pathway was observed in the tobacco protoplasts with more than 40% of recombinant plants having modifications in multiple SuR alleles.

These authors have also tested for HR-based gene replacement by electro-poration of ZFN plasmids into tobacco protoplasts along with SurB donor templates modified at positions known to confer herbicide resistance. Successful gene replacement was observed at frequencies ranging from 0.2% to 4% (Townsend et al., 2009). Another example of gene editing using ZFNs was published by Shukla et al. (2009) in which the authors generated ZFNs that targeted the maize inositol-1,3,4,5,6-pentakisphosphate enzyme gene (IPK1). The disruption of the IPK1 gene reduces the level of phytate, a known antinutritional compound found in seeds of many cereals. The IPK1 locus was disrupted by ZFN-mediated DSBs and repaired by HR using donor templates with an herbicide resistance gene and short, locus-specific homology arms (Shukla et al., 2009). However, the requirement of the two binding events with the correct orientation and proper spacing permits specific targeting of the unique recognition sites. A site-specific DSB into any locus of interest can be subsequently repaired by the NHEJ pathway, which leads to occasional loss or gain of the genetic information and thereby, could be used to introduce small insertions or deletions (Urnov et al., 2010). The outcome of the repair process also could disrupt the gene function. Apart from that, ZFNs are difficult to design and are expensive as compared to TALENs.

18.5.4 TALENS

TALENs are a recent addition to the arsenal of sequence-specific nucleases, and they quickly became adopted for plant genome engineering. TALENs are similar to ZFNs in that they have a DNA-binding domain derived from TALE proteins fused to Fok1 cleavage domain. TALE proteins are transcription factors from the plant bacterial pathogen *Xanthomonas*. These bacteria are pathogens of crop plants such as rice, pepper, and tomato, and they cause significant economic damage to agriculture, which motivated their thorough study. The bacteria were found to secrete effector proteins (TALEs) to the cytoplasm of plant cells, which affect processes in the plant cell and increase its susceptibility to the pathogen. Further investigation of the effector protein action mechanisms revealed that they are capable of DNA binding and activating the expression of their target genes via mimicking the eukaryotic transcription factors. (Nemudryi et al., 2014). The bacteria produce such proteins to target and control the expression of specific plant gene promoters in order to mimic host transcription factors and thus achieve successful infection (Bogdanove et al., 2010). One advantage of TALENs, compared to meganucleases and zinc-finger nucleases, is their modular DNA binding domain. The TALE DNA binding domain is composed of direct repeats consisting of 33–35 amino acids. Two amino acids within these repeats, termed *repeat-variable diresidues* (RVDs), recognize a target nucleotide (e.g., the most widely used RVDs and their nucleotide targets are HD, cytosine; NG, thymine; NI, adenine; and NN, guanine and adenine).

This one-to-one correspondence of a single RVD to a single DNA base, together with effective methods for cloning arrays of the DNA binding motif, have nearly eliminated the design challenges encountered with zinc-finger nucleases and meganucleases. Another advantage of TALENs is their target specificity. TALEN monomers are typically designed with 15–20 RVDs, and, as a result, a TALEN target site is frequently >30 bp. This relatively large target site makes TALENs the most specific of all the nucleases, and it may contribute to reduced toxicity compared to zinc-finger nucleases.

The only real drawback to the use of TALENs is their large size (about 950 aa; around1900 aa per pair) and repetitive nature, making delivery to plant cells a challenge. TALENs are typically delivered to plant cells by direct delivery of DNA to protoplasts or by stable integration of TALEN-encoding constructs into plant genomes.

18.5.4.1   Structure of TALEN

DNA recognition by TALENs is achieved through arrays of TAL effector motifs. The fundamental building block of DNA binding region of TALENs, i.e., TALEs, typically consists of a 16 to 20 single repeat monomers, with each monomer of 34 amino acid residues in length. (Ackerveken et al., 1996; Zhu et al., 1998; Schornack et al., 2006). Each repeat targets a specific base of DNA, the identity of which is determined by a hypervariable pair of amino acid typically found at position 13 and 14 of the repeat. Each of these pair of amino acids binds to specific nucleotide in the target DNA (Herbers et al., 1992; Yang et al., 2005; Boch et al., 2009; Moscou and Bogdanove, 2009). The remaining 34 amino acids in the repeats are nearly always the same. Thus, the overall number of repeats matches an equal number of bases. In 2010, first nucleases were attached to TALE proteins for genome engineering.

Although highly conserved, the exception is the hypervariable amino acid residues at positions 12 and 13, called repeat-variable di-residues (RVDs). Different RVDs associate preferentially with different nucleotides, with the four most common RVDs (HD, NG, NI, and NN) accounting for each of the four nucleotides (C, T, A, and G, respectively). RVD is highly variable and show a strong correlation with specific nucleotide recognition. By fusing the TALE sequence to a *FokI* nuclease, the synthetic compound is now called a **TALEN.** However, *FokI* works only in a **dimerized form,** and hence, TALENs are always designed as pairs binding opposing strands of the DNA to allow **dimerization of *FokI* in a spacer region** that is bridging the two TALE binding sites. The treatment of genome with TALEN leads to specific DNA binding and subsequent DSB (double strand break) in which the genome repairs through a NHEJ or HDR.

18.5.4.2   Mechanism of the TALEN System

TALEN can be used to edit genomes by inducing double-strand breaks (DSB), which cells respond to with repair mechanisms. NHEJ directly ligates DNA from either side of a DSB where there is very little or no sequence overlap for annealing. This repair mechanism induces errors in the genome via indels (insertion or deletion) or chromosomal rearrangement; any such errors may render the gene products coded at that location non-functional. Because this activity can vary depending on the species, cell type, target gene, and nuclease used, it should be monitored when designing new systems. A simple heteroduplex cleavage assay can be run which detects any difference between two alleles amplified by PCR. Cleavage products can be visualized on simple agarose gels or slab gel systems.

Alternatively, DNA can be introduced into a genome through NHEJ in the presence of exogenous double-stranded DNA fragments. HDR can also introduce foreign DNA at the DSB as the transfected double-stranded sequences are used as templates for the repair enzymes (Figure 18.1).

The DNA-binding specificity of TALE repeats is solely based on the nature of the RVDs in a typical 34-aa repeat (Boch and Bonas, 2010) suggesting that the amino acids at positions 12 and 13 probably interact directly with the DNA bases. The consecutive array of repeats binds to a consecutive DNA sequence, and rearranging repeat units generates novel custom-designed DNA-binding specificities with high potential for biotechnology

(Curtin et al., 2012). Computational and molecular biological analyses have enabled to decipher the TALE code for DNA recognition (Boch et al., 2009; Moscou and Bogdanove, 2009). Likewise, the number of repeats and the sequence of the RVD region will determine the length and sequence of the target DNA. DNA-binding domains of individual TALENs typically contain between 15 and 30 RVDs and can then target 15 to 30 nucleotides in the host genome accordingly (Cermak et al., 2011). However, in contrast to zinc finger proteins, the re-engineering of the linkage between repeats is not necessary to construct long arrays of TALEs (Christian et al., 2010) because of the lack of interaction between every three nucleotides.

TALENs have been used to generate targeted modifications in a variety of plant species, such as *A. thaliana* (Cermak et al., 2011), tobacco (Mahfouz and Li, 2011; Mahfouz et al., 2011), and rice (Li et al., 2012). Li et al. (2012) have presented one of the first successful demonstrations of a nuclease-mediated modification of agronomic importance. These authors have exploited TALEN technology to edit a specific susceptibility gene *Os11N3* (also called *OsSWEET14*) in rice to thwart the virulence strategy of *Xanthomonas oryzae* and thereby engineer heritable genome modifications for resistance to bacterial blight. During infection, the bacteria secrete effector proteins that target DNA sequences in the promoter region of the rice sucrose-efflux transporter gene (OsSWEET14) (Curtin et al., 2012). It is interesting to note that OsSWEET14 plays a crucial role in the development of the plant and obtaining a knockout mutant to circumvent the effects of the pathogen was not feasible.

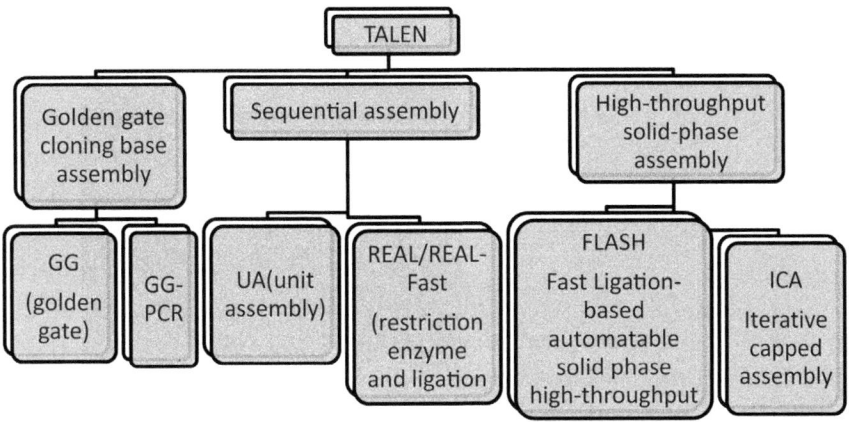

FIGURE 18.1   Construction of TALEN using various techniques.

The pathogen's virulence was successfully reduced by disrupting the promoter sequence bound by the pathogen effector.

18.5.5   CRISPR/CAS

The last evolution of genetic tools came from the study of the defense system of bacteria, more specifically its reaction toward phages. First CRISPR cluster repeats were reported by Ishino et al. in 1987, but the function remained a mystery. In the early 2000, it was discovered that some bacterial genomes featured short regularly spaced repeats (SRSR), sequences that were later renamed clustered regularly interspaced short palindromic repeats (CRISPR). The most recent addition to the sequence-specific nuclease family, CRISPR/Cas, is proving to be the nuclease-of-choice for plant genome engineering. Unlike the other three nuclease classes, which target DNA through protein/DNA interactions, CRISPR/Cas uses a guide RNA molecule (gRNA) to direct an endonuclease, Cas9, to a target DNA sequence. As a result, redirecting CRISPR/Cas is extremely simple, requiring only the cloning of a 20-nt sequence (complementary to a target DNA sequence) within a gRNA expression construct. One limitation of the CRISPR/Cas system may be off-target cleavage. Whereas 20 nucleotides are used to direct Cas9 binding and cleavage, the system tolerates mismatches, with a higher tolerance for mismatches at the 50 end of the targeting-RNA sequence.

To reduce the likelihood of off-target cleavage, alternative CRISPR/Cas reagents have been developed, including paired Cas9 nickases, fusion of catalytically dead Cas9 to FokI, and shortening of the gRNA targeting sequence. Possibly, the simplest approach to minimize off-target cleavage is to design gRNAs that have minimum sequence homology to other sites within the plant genome. In addition to potential off-targeting, another limitation of the CRISPR/Cas system is size. Cas9 is around 1400 aa, making it one of the largest sequence-specific nucleases. However, for vectors that are unable to deliver Cas9, it may be possible to generate plant lines that constitutively express Cas9; therefore, only the delivery of gRNA(s) is required.

An example of genome editing of crop plants was performed by Shan et al. (2013b) who designed several sgRNAs and codon-optimized Cas9, which target different DNA strands of four rice endogenous genes and one wheat gene. They demonstrated different levels of mutagenesis (frequency of insertions and deletions) and the possibility of homology-directed repair by co-transformation of Cas9, sgRNA and single-stranded DNA oligos into plant cells. Another report demonstrated the use of a multiplex CRISPR/Cas9 technique for gene simultaneously edition of two sites (CHLOROPHYLL A OXYGENASE1 and LAZY1) in the *Arabidopsis* genome by applying a single CRISPR/Cas9 construct harboring two sgRNA expression cassettes (Mao et al., 2013). In addition, it has been demonstrated that the CRISPR/ Cas9 system can provide homozygous gene editing within one reproductive generation (Zhang et al., 2014).

18.5.5.1 Mechanism of the CRISPR–Cas System

The key step in editing an organism's genome is selective targeting of a specific sequence of DNA. Two biological macromolecules, the Cas9 protein and guide RNA, interact to form a complex that can identify target sequences with high selectivity. The Cas9 protein is responsible for locating and cleaving target DNA in both natural and artificial CRISPR/Cas systems. The Cas9 protein has six domains, REC I, REC II, Bridge Helix, PAM Interacting, HNH, and RuvC. (Jinek et al., 2014; Nishimasu et al., 2014).

The Rec I domain is the largest and is responsible for binding guide RNA. The role of the REC II domain is not yet well understood. The arginine-rich bridge helix is crucial for initiating cleavage activity upon binding of target DNA (Nishimasu et al., 2014). The PAM-interacting domain confers PAM specificity and is therefore responsible for initiating binding to target DNA (Anders et al., 2014; Jinek et al., 2014; Nishimasu et al., 2014; Sternberg et al., 2014). The HNH and RuvC domains are nuclease domains that cut single-stranded DNA. They are highly homologous to HNH and RuvC domains found in other proteins (Jinek et al., 2014; Nishimasu et al., 2014).

The Cas9 protein remains inactive in the absence of guide RNA (Jinek et al., 2014). In engineered CRISPR systems, guide RNA is composed of a single strand of RNA that forms a T-shape comprising one tetraloop and two or three stem loops (Jinek et al., 2012; Nishimasu et al., 2014). The guide RNA is engineered to have a 5' end that is complimentary to the target DNA sequence. This artificial guide RNA binds to the Cas9 protein and, upon binding, induces a conformational change in the protein. The conformational change converts the inactive protein into its active form. The mechanism of the conformational change is not completely understood, but Jinek and colleagues hypothesize that steric interactions or weak binding between protein side chains and RNA bases may induce the change (Jinek et al., 2014).

Once the Cas9 protein is activated, it stochastically searches for target DNA by binding with sequences that match its protospacer adjacent motif (PAM) sequence (Sternberg et al., 2014). A PAM is a two- or three-base sequence located within one nucleotide downstream of the region complementary to the guide RNA. PAMs have been identified in all CRISPR systems, and the specific nucleotides that define PAMs are specific to the particular category of the CRISPR system (Mojica et al., 2009). The PAM in *Streptococcus pyogenes* is 5′-NGG-3′ (Jinek et al., 2012). When the Cas9 protein finds a potential target sequence with the appropriate PAM, the protein will melt the bases immediately upstream of the PAM and pair them with the complementary region on the guide RNA (Sternberg et al., 2014). If the complementary region and the target region pair properly, the RuvC and HNH nuclease domains will cut the target DNA after the third nucleotide base upstream of the PAM (Anders et al., 2014)

Among three types of the CRISPR/Cas system (Table 18.4), type II is used because of its nuclease activity due to the Cas9 protein. The type II CRISPR/Cas system recognizes and targets the genetic material of pathogens via three stepwise processes, namely acquisition, expression, and interference.

The type II CRISPR/Cas system has been exploited for genome editing in various organisms (Table 18.5). The design and construction of CRISPR/Cas9 constructs are relatively straightforward, cheap, and devoid of intellectual property barriers (Van Der Oost et al., 2009; Garneau et al., 2010; Horvath and Barrangou, 2010; Marraffini and Sontheimer, 2010). The acquisition process involves recognition and integration of foreign DNA as spacer within the CRISPR locus. Generally, the protospacer contains a short stretch (2–5 bp) of conserved nucleotides (PAMs) that act as a recognition motif for acquisition of the DNA fragment (spacer). The insertion of a single copy of spacer of approximately 30 bp occurs at the leader side of the CRISPR array and is followed by its duplication (Garneau et al., 2010). Mutations in the PAMs of the viral genome can thwart CRISPR-mediated immunity against pathogen attacks (Garneau et al., 2010). In the second step of the CRISPR/Cas system function, i.e., expression, the long pre-crRNA is actively transcribed from the CRISPR locus and processed into crRNAs with the help of Cas proteins (Cas1,

TABLE 18.4 Different Classes of the CRISPR/Cas System and Their Unique Features

Type	Organism	Cas protein associate with CRISPR	Cleavage domain	Function	Cleavage Target
Type I	Bacteria (*E. coli*) and Archaea (*P. aeruginosa*)	Cas1, Cas2, Cas3, Cas5, Cas6 and Cas7 protein	HD nuclease domain of Cas3	Recognize PAM sequence in DNA, Act as ATP-dependent helicase	DNA
Type II	Bacteria (*S. thermophilus*)	Cas1, Cas2, Cas9 and Cas4/Csn2 protein	RuvC-like Nuclease domain near the N terminus and HNH (McrA-like) nuclease in the middle of Cas9	crRNA biogenesis and 5' trimming of the crRNA	DNA
Type III	Archaea (*S. epidermis, L. lactis* and *P. furiosus*)	Cas1, Cas2, Cas10 and Cas6 protein	Catalytic triad of Cas6 and Csm/Cmr complex	Palm domain of nucleic acid polymerases and nucleotide cyclases, sequence-specific processing step and trimming	DNA/RNA
Type V	Bacteria (*Alicyclobacillus, Oleiphilus sp.*)	Cas1, Cas2, Cas4, Cas12a,b,c and cpf1 protein, C2C3	RuvC like Nuclease. Lacks HNH nuclease	Form DNA recognition competent structure, crRNA maturation	DNA
Type VI	Bacteria (*Leptotrichia buccalis, Fusobacterium perfoetens* and *Prevotella buccae*)	Cas13a and Cas13b, evolve from HEPN domain	Unclear	ssRNA inactivation, editing, modification or localization	RNA

(Source: Adapted from Kuma & Jain 2015).

Cas2, Cas9, and Cas4/Casn2) and the tracrRNA molecule. Recently, the tracrRNA has also been reported to participate in the processing of pre-crRNA in *S. pyogenes* (Karvelis et al., 2013). The tracrRNA pairs with

TABLE 18.5 List of Targeted Gene(s) via the CRISPR/Cas9 System in Different Plant Species

No.	Plant species	Target gene(s)	Delivery method	References
1.	*Arabidopsis Thaliana*	AtPDS3,AtFLS2, AtRACK1b and AtRACK1c	Protoplast co-transfection and *Agrobacterium* infiltration	Li et al. (2013)
		GFP	*Agrobacterium*-mediated transfor-mation	Jiang et al. (2014b)
		CHL1, CHL2 and TT4i	*Agrobacterium*-mediated transfor-mation	Mao et al. (2013)
		BRI1, JAZ1 and YFP	*Agrobacterium*-mediated transfor-mation	Feng et al. (2013)
2.	*Nicotiana benthamiana*	NbPDS3	*Agrobacterium*-mediated transfor-mation	Li et al. (2013)
		NbPDS	*Agrobacterium*-mediated transfor-mation	Nekrasov et al. (2013)
		Nbpds	*Agrobacterium*-mediated transfor-mation	Upadhyay et al. (2013)
		NbPDS	*Agrobacterium*-mediated transfor-mation	Belhaj et al. (2013)
		GFP	*Agrobacterium*-mediated transfor-mation	Jiang et al. (2013)
3.	*Oryza sativa*	OsPDS, OsBADH2,Os02g23823 and OsMPK2	Transformation using particle bom-bardment	Shan et al. (2013b)
		OsSWEET11 and Os-SWEET14	*Agrobacterium*-mediated transfor-mation	Jiang et al. (2013)
		OsMYB1	*Agrobacterium*-mediated transfor-mation	Mao et al. (2013)
		ROC5, SPP and YSA	*Agrobacterium*-mediated transfor-mation	Feng et al. (2013)

TABLE 18.5 (Continued)

No.	Plant species	Target gene(s)	Delivery method	References
		OsMPK5	*Agrobacterium*-mediated transformation	Xie and Yang (2013)
		CAO1 and LAZY1	*Agrobacterium*-mediated transformation	Mao et al. (2013)
4.	*Triticum aestivum*	TaMLO	Protoplast transformation	Shan et al. (2013b)
		Tainox and TaPDS	*Agrobacterium*-mediated transformation	Upadhyay et al. (2013)
		TaMLO-A1	Transformation using particle bombardment	Wang et al. (2015)
5.	*Sorghum bicolor*	DsRED2	*Agrobacterium*-mediated transformation	Jiang et al. (2013b)
6.	*Marchantia polumorpha*	MpARF1	*Agrobacterium*-mediated transformation	Sugano et al. (2014)

the repeat region of crRNA via base complementarity and facilitates the processing of pre-crRNA into crRNA (Deltcheva et al., 2011). The processed crRNAs enter the CRISPR-associated complex for antiviral defense (CASCADE) and help to recognize as well as base pair with a specific target region of the foreign DNA (Deltcheva et al., 2011). During the interference step, the crRNA guides the Cas protein complex to the specific target region of the foreign DNA for cleavage and provides immunity against pathogen attacks (Garneau et al., 2010; Marraffini and Sontheimer, 2010). The components of the CRISPR/Cas system, crRNA and tracrRNA, can be fused into sgRNA to direct Cas9 for introducing target-specific DSBs (Jinek et al., 2012). The designing of sgRNAs is also simple and thus may be preferred for genome editing. Initially, the type II CRISPR/Cas system was programmed to induce cleavage at various sites in DNA *in vitro* (Jinek et al., 2012). Recently, the CRISPR/Cas technology has been adopted for genome editing in bacteria, yeast, and other organisms for efficient targeted mutagenesis (Cong et al., 2013; Dicarlo et al., 2010, Gratz et al., 2013; Hwang et al., 2013; Jiang et al., 2013; Qi et al., 2013; Wang et al., 2013; Yang et al., 2013). The target mutation efficiency of the CRISPR/Cas9 system was found to be similar to that of ZFNs and TALENs *in vivo* (Hwang et al., 2013).

Thus, the CRISPR/Cas9 system has successfully been applied to various crop species, including rice, maize, wheat, barley, sorghum, soybean, *Brassica oleracea*, tomato, potato, lettuce, sweet orange, poplar, and grapevine (Lawrenson et al., 2015; Ren et al., 2016; Sovova et al., 2016). The first edited agronomic traits reported in the literature relate to disease tolerance (powdery mildew in wheat (Wang et al., 2014), bacterial blight in rice (Jiang et al., 2013), potyviruses in cucumber (Chandrasekaran et al., 2016), modified ripening profiles in tomato (Ito et al., 2015), male sterility (Li et al., 2016b) and factors involved in yield in rice (Li et al., 2016a, Xu et al., 2016), and drought tolerance in maize (Shi et al., 2016). Chabannes et al. (2013) studied the first CRISPR varieties that will be marketed and probably present well characterized traits inserted in optimized backgrounds (waxy starch, tolerance to herbicides), but in the long term, innovative properties could be developed in relation to biotic and abiotic stresses, yield, quality (nutritional compounds), physiology (improved biomass, accelerated flowering time), molecular pharming (therapeutic molecules), or even genome remodeling (e.g., bananas without endogenous banana streak virus sequences).

18.6 WHY CRISPR-CAS9 IS MORE TRUSTWORTHY THAN TALENs AND ZFNs?

For the assessment of any genome editing tool, the percentage of achieving desired mutation known as targeting efficiency (TE) is regarded as the most reliable attribute. The success ratio of Cas9 TE can be compared with other techniques like TALENs or ZFNs (Wendt et al., 2013; Ma et al., 2014) (Figure 18.2). For example, custom-designed ZFNs and TALENs could only achieve 1–50% efficiencies (Maeder et al., 2008; Miller et al., 2011; Wolt et al., 2016). Conversely, TE of Cas9 in animals and plants, i.e., zebra-fish and maize, respectively, was observed up to 70% and it ranges between 2 and 5% in case of induced pluripotent stem cells (Fu et al., 2013; Hsu et al., 2013). Later, CRISPR/Cas9 efficiency was recorded up to 9.2% as compared to ZFN efficiency that was lower than 1% in the case of pig IGF2 (Insulin- like growth factor 2). Reports are available that broadly describe better genome targeting of single cell mouse embryo up to 78% and successful effectual germline transmission by using dual sgRNAs (Zhang et al., 2013; Zhou et al., 2014; Wolt et al., 2016). Moreover, the incidence of off-target mutations is also an effective parameter for the assessment of genome editor's performance. Such mutations may be observed in sites that have dissimilarity of small number of nucleotides in comparison with original sequence till they are neighbors of protospacer adjacent motif (PAM) sequence. The DNA sequences are used to transcribe crRNA-targeting sequences known as protospacers. These consist of short sequences and found clustered in bacterial genome in form of a group called CRISPR array. The PAM sequence is absolute need of Cas9 for binding its target. Cas9 do not cleave the protospacer sequence in the absence of adjacent PAM sequence. This favors the stance that Cas9 can endure mismatches up to five bases within the protospacer region (Fu et al., 2014; Sander and Joung, 2014) or one base divergence in the PAM sequence (Hsu et al., 2013). Other than facilitation activity for genome alterations, the wild-type Cas9 nuclease has capacity to be transformed into Cas9 after inactivation of catalytic domains. Furthermore, effector fusion usage can enhance the range of genome engineering modalities attainable by adopting Cas9. Normally, off target mutations are bit difficult to detect because these require full genome sequencing to completely rule the mount. So, unanimous opinion is that CRISPR/Cas9 facilitates plant genome interrogation, as it enables high efficiency generation of mutants

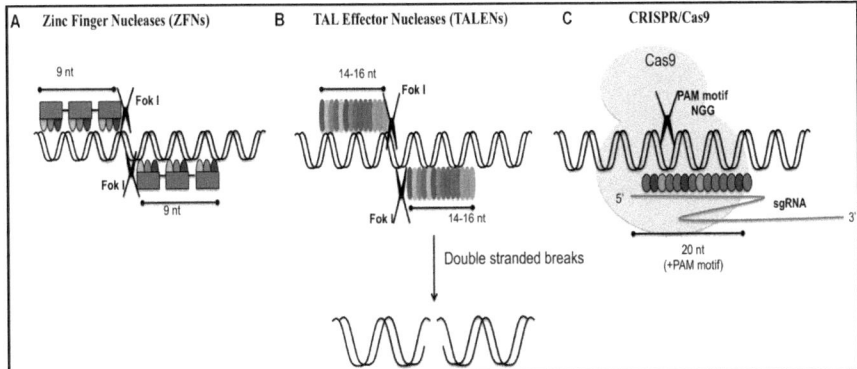

FIGURE 18.2 Comparison of CRISPR/Cas9 with ZFNs and TALENs. (Source: Adapted from Weeks et al., 2016).

bearing multiple gene mutations. This effective approach endorses high specificity of wide range of genome editing applications.

18.7   IS GENOME EDITING POSSIBLE FROM A LEGAL POINT OF VIEW OR WOULD IT REQUIRE CHANGE IN THE LAW?

The German Genetic Engineering Act states that the descendants of a genetically modified plant must also be classified as genetically modified. The fact that genome-edited plants temporarily contained the gene for cutting protein would make them and their descendants genetically modified plants forever – despite the fact that the foreign gene was removed without trace. This was certainly not the intention of the legislator as genome engineering did not yet exist when the Genetic Engineering Act was passed. Thus, it is suggested that the Genetic Engineering Act should not be applied to genome-edited plants.

The new genome editing techniques with RNP-based transformation could pave the way for solving problems in food security, by developing cultivars by which desired traits from a gene pool of wild species. Thus, novel and valuable plants generated by genome editing techniques can regain useful traits overlooked during domestication; these traits help plants to survive in unpredictable global environmental changes.

18.8   CONCLUDING REMARKS

Genome editing tools provide new strategies for genetic manipulation in plants and are likely to assist in engineering desired plant traits by modifying endogenous genes. Such technology will have a major impact on applied crop improvement and commercial product development. All techniques will no doubt be revolutionized by virtue of being able to make targeted DNA sequence modifications rather than random changes. These technologies involve simple point mutation, which is indistinguishable from native biological process and would seem to have no basis for regulatory scrutiny. On the other hand, others may lead to plant genome variations, which are similar to GMOs. Such technology offers scientists and plant breeders a flexible and relatively easy approach to accelerate breeding practices in a wide variety of crop species, providing another tool that we can use to improve food security in the future. In gene modification, these targetable nucleases have potential applications to become alternatives to standard breeding methods to identify novel traits in economically important plants and more valuable in plant breeding and biotechnology as modifying a specific site rather than the whole gene.

KEYWORDS

- **CRISPR/Cas**
- **genome editing**
- **meganucleases**
- **oligonucleotide-directed mutagenesis**
- **TALENs**
- **zinc-finger nucleases**

REFERENCES

Ackerveken, G. V. D., Marois, E., & Bonas, U., (1996). Recognition of the bacterial avirulence protein avrbs3 occurs inside the host plant cell. *Cell, 87,* 1307–1316.

Ainley, W. M., et al., (2013). Trait stacking via targeted genome editing. *Plant Biotechnol. J., 11,* 1126–1134.

Allard, R. W., (1999). *Principles of Plant Breeding.* 2nd (ed.). John Wiley & Sons, New York.

Alwin, S., et al., (2005). Custom zinc-finger nucleases for use in human cells. *Mol. Ther., 12,* 610–617.

Anders, C., et al., (2014). Structural basis of PAM-dependent target DNA recognition by the Cas9 endonuclease. *Nature, 513,* 569–573.

Baltes, N. J., et al., (2014). DNA replicons for plant genome engineering. *Plant Cell, 26,* 151–163.

Bartsevich, V. V., & Juliano, R. L., (2000). Regulation of the MDR1 gene by transcriptional repressors selected using peptide combinatorial libraries. *Mol. Pharmacol., 58,* 1–10.

Bartsevich, V. V., et al., (2003). Engineered zinc finger proteins for controlling stem cell fate. *Stem Cells, 21,* 632–637.

Beerli, R. R., & Barbas, C. F., (2002). III, Engineering polydactyl zinc-finger transcription factors. *Nat. Biotechnol., 20,* 135–141.

Beerli, R. R., et al., (1998). Toward controlling gene expression at will: specific regulation of the erbB-2/ HER-2 promoter by using polydactyl zinc finger proteins constructed from modular building blocks. *Proc. Natl. Acad. Sci. USA, 95,* 14628–14633.

Beetham, P. R., et al., (1999). A tool for functional plant genomics: Chimeric RNA/DNA oligonucleotides cause *in vivo* gene-specific mutations. *Proc. Natl. Acad. Sci. USA*, *96*, 8774–8778.

Beetham, P. R., et al., (2014). Methods and compositions for increasing efficiency of targeted gene modification using oligonucleotide-mediated gene repair. *US Patent Application WO* 2014144987 A2.

Belhaj, K., et al., (2013). Plant genome editing made easy: targeted mutagenesis in model and crop plants using the CRISPR/Cas system. *Plant Methods*, *9*, 39.

Boch, J., & Bonas, U., (2010). Xanthomonas AvrBs3 family-type III effectors: discovery and function. *Annu. Rev. Phytopathol.*, *48*, 419–436.

Boch, J., et al., (2009). Breaking the code of DNA binding specificity of TAL-type III effectors. *Sci.*, *326*, 1509–1512.

Bogdanove, A. J., Schornack, S., & Lahaye, T., (2010). TAL effectors: finding plant genes for disease and defense. *Curr. Opin. Pl. Biol.*, *13*, 394–401.

Brooks, C., et al., (2014). Efficient gene editing in tomato in the first generation using the clustered regularly interspaced short palindromic repeats/CRISPR-associated9 system. *Plant Physiol.*, *166*, 1292–1297.

Brunet, E., et al., (2009). Chromosomal translocations induced at specified loci in human stem cells. *Proc. Natl Acad. Sci. USA*, *106*, 10620–10625.

Cai, C. Q., et al., (2009). Targeted transgene integration in plant cells using designed zinc finger nucleases. *Plant Mol. Biol.*, *69*, 699–709.

Cai, Y., et al., (2015). CRISPR/Cas9- mediated genome editing in soybean hairy roots. *PLoS ONE*, *10*, e0136064. doi: 10.1371/journal.pone.0136064.

Carroll, D., (2011). Genome engineering with zinc-finger nucleases. *Genetics.*, *188*, 773–782.

Cermak, T., et al., (2011). Efficient design and assembly of custom TALEN and other TAL effector-based constructs for DNA targeting. *Nucleic Acids Res.*, *39*, e82.

Chabannes, M., et al., (2013). Three infectious viral species lying in wait in the banana genome. *J. Virol.*, *87*, 8624–8637.

Chandrasekaran, J., et al., (2016). Development of broad virus resistance in non-transgenic cucumber using CRISPR/Cas9 technology. *Mol. Plant Pathol.*, *17*, 1140–1153.

Char, S. N., et al., (2015). Heritable site-specific mutagenesis using TALENs in maize. *Plant Biotechnol. J.*, doi:10.1111/ pbi.12344.

Chen, W., et al., (2014). Inhibiting replication of begomoviruses using artificial zinc finger nucleases that target viral-conserved nucleotide motif. *Virus Genes.*, *48*, 494–501.

Christian, M., et al., (2013). Targeted mutagenesis of *Arabidopsis thaliana* using engineered TAL effector nucleases. *G3 (Bethesda)*, *3*, 1697–1705.

Christian, M., et al., (2010). Targeting DNA double-strand breaks with TAL effector nucleases. *Genetics.*, *186*, 757–761.

Clasen, B. M., et al., (2015). Improving cold storage and processing traits in potato through targeted gene knockout. *Plant Biotechnol. J.*, doi:10.1111/pbi.12370.

Cole-Strauss, A., et al., (1996). Correction of the mutations responsible for sickle cell anemia by an RNA-DNA oligonucleotide. *Sci.*, *273*, 1386–1389.

Cong, L., et al., (2013). Multiplex genome engineering using CRISPR/Cas systems. *Sci.*, *339*, 819–823.

Crespan, E., et al., (2012). Microhomology-mediated DNA strand annealing and elongation by human DNA polymerases k and b on normal and repetitive DNA sequences. *Nucleic Acids Res.*, *40*, 5577–5590.

Curtin, S. J., et al., (2013). Targeted mutagenesis for functional analysis of gene duplication in legumes. *Methods Mol. Biol.*, *1069*, 25–42.

Curtin, S. J., et al., (2011). Targeted mutagenesis of duplicated genes in soybean with zinc-finger nucleases. *Plant Physiol.*, *156*, 466–473.

Curtin, S. J., Voytas, D. F., & Stupar, R. M., (2012). Genome engineering of crops with designer nucleases. *Plant Genome J.*, *5*, 42.

D'Halluin, K., et al., (2013). Targeted molecular trait stacking in cotton through targeted double-strand break induction. *Plant Biotechnol. J.*, *11*, 933–941.

Dai, Q., et al., (2004). Engineered zinc finger-activating vascular endothelial growth factor transcription factor plasmid DNA induces therapeutic angiogenesis in rabbits with hind limb ischemia. *Circulation*, *110*, 2467–2475.

De Pater, S., et al., (2013). ZFN mediated gene targeting of the Arabidopsis protoporphyrinogen oxidase gene through Agrobacterium-mediated floral dip transformation. *Plant Biotechnol. J.*, *11*, 510–515.

Deltcheva, E., et al., (2011). CRISPR RNA maturation by *trans-encoded* small RNA and host factor RNase III. *Nature*, *471*, 602–607.

Deriano, L., & Roth, D. B., (2013). Modernizing the nonhomologous end-joining repertoire: alternative and classical NHEJ share the stage. *Annu. Rev. Genet.*, *47*, 433–455.

Desjarlais, J. R., & Berg, J. M., (1993). Use of a zinc-finger consensus sequence framework and specificity rules to design specific DNA-binding proteins. *Proc. Natl Acad. Sci. USA*, *90*, 2256–2260.

Dicarlo, J. E., et al., (2013). Genome engineering in *Saccharomyces cerevisiae* using CRISPR–Cas systems. *Nucleic Acids Res.*, *47*, 4336–4343.

Djukanovic, V., et al., (2013). Male-sterile maize plants produced by targeted mutagenesis of the cytochrome P450-like gene (MS26) using a re-designed I-CreI homing endonuclease. *Plant J.*, *76*, 888–899.

Elrod-Erickson, M., & Pabo, C. O., (1999). Binding studies with mutants of Zif268. Contribution of individual side chains to binding affinity and specificity in the Zif268 zinc finger–DNA complex. *J. Biol. Chem.*, *274*, 19281–19285.

Endo, M., Mikami, M., & Toki, S., (2015). Multigene knockout utilizing off target mutations of the CRISPR/Cas9 system in rice. *Plant Cell Physiol.*, *56*, 41–47.

Even-Faitelson, L., et al., (2011). Localized egg-cell expression of effector proteins for targeted modification of the Arabidopsis genome. *Plant J.*, *68*, 929–937.

Fan, D., et al., (2015). Efficient CRISPR/Cas9- mediated targeted mutagenesis in *Populus* in the first generation. *Sci. Rep.*, *5*, 12217.

Fauser, F., et al., (2012). *In planta* gene targeting. *Proc. Natl. Acad. Sci. USA*, *109*, 7535–7540.

Feng, Z., et al., (2013). Efficient genome editing in plants using a CRISPR/Cas system. *Cell Res.*, *23*, 1229–1232.

Feng, Z., et al., (2014). Multigeneration analysis reveals the inheritance, specificity, and patterns of CRISPR/Cas induced gene modifications in Arabidopsis. *Proc. Natl Acad. Sci. USA*, *111*, 4632–4637.

Frit, P., et al., (2014). Alternative end-joining pathway(s): bricolage at DNA breaks. *DNA Repair*, *17*, 81–97.

Fu, Y., et al., (2013). High-frequency off-target mutagenesis induced by CRISPR-Cas nucleases in human cells. *Nat. Biotechnol.*, *31*, 822–826.

Fu, Y., et al., (2014). Improving CRISPR-Cas nuclease specificity using truncated guide RNAs. *Nat. Biotechnol.*, *32*, 279–284.

Gao, H., et al., (2015a). Expression activation and functional analysis of HLA3, a putative inorganic carbon transporter in *Chlamydomonas reinhardtii. Plant J., 82,* 1–11.

Gao, H., et al., (2010). Heritable targeted mutagenesis in maize using a designed endonuclease. *Plant J., 61,* 176–187.

Gao, J., et al., (2015b). CRISPR/Cas9-mediated targeted mutagenesis in *Nicotiana tabacum. Plant Mol. Biol., 87,* 99–110.

Gao, Y., et al., (2015c). Auxin binding protein 1 (ABP1) is not required for either auxin signaling or Arabidopsis development. *Proc. Natl Acad. Sci. USA, 112,* 2275–2280.

Garneau, J. E., et al., (2010). The CRISPR/Cas bacterial immune system cleaves bacteriophage and plasmid DNA. *Nature, 468,* 67–71.

Gocal, G., (2015). Non-transgenic trait development in crop plants using oligo-directed mutagenesis: Cibus' Rapid Trait Development System. In: NABC Report 26. *New DNA-Editing Approaches: Methods, Applications and Policy for Agriculture.* North American Agricultural Biotechnology Council, Ithaca, NY., pp. 97–105.

Gocal, G. F. W., Schopke, C., & Beetham, P. R., (2015). Oligo mediated targeting gene editing. In: *Advances in New Technology for Targeted Modification of Plant Genomes.* Zhang, F., Puchta, H., & Thompson, J. G., (eds.). Springer, New York, N. Y., pp. 73–89.

Gratz, S. J., et al., (2013). Genome engineering of Drosophila with the CRISPR RNA-guided Cas9 nuclease. *Genetics, 194,* 1029–1035.

Gupta, M., et al., (2007). History, origin, and evolution. *Adv. Bot. Res., 45,* 1–20.

Gupta, M., et al., (2012). Transcriptional activation of Brassica napus beta-ketoacyl-ACP synthase II with an engineered zinc finger protein transcription factor. *Plant Biotechnol. J., 10,* 783–791.

Gurushidze, M., et al., (2014). True-breeding targeted gene knock-out in barley using designer TALE nuclease in haploid cells. *PLoS One, 9,* e92046.

Haun, W., et al., (2014). Improved soybean oil quality by targeted mutagenesis of the fatty acid desaturase 2 gene family. *Plant Biotechnol. J., 12,* 934–940.

Herbers, K., Conrads-Strauch, J., & Bonas, U., (1992). Race-specificity of plant resistance to bacterial spot disease determined by repetitive motifs in a bacterial avirulence protein. *Nature, 356,* 172–174.

Horvathm, P., & Barrangoum, R., (2010). CRISPR/Cas, the immune system of bacteria and archaea. *Sci., 327,* 167–170.

Hsu, P. D., et al., (2013). DNA targeting specificity of RNA-guided Cas9 nucleases. *Nat. Biotechnol., 31,* 827–832.

Hwang, W. Y., et al., (2013). Efficient genome editing in zebrafish using a CRISPR–Cas system. *Nature Biotech., 31,* 227–229.

Ishino, Y., et al., (1987). Nucleotide sequence of the *iap* gene, responsible for alkaline phosphatase isozyme conversion in *Escherichia coli* and identification of the gene product. *J. Bacteriol., 169,* 5429–5433.

Ito, Y., et al., (2015). Biochemical and Biophysical Research Communications CRISPR/Cas9-mediated mutagenesis of the RIN locus that regulates tomato fruit ripening. *Biochem. Biophys. Res. Commun., 467*(1), 76–82.

Jacobs, T. B., et al., (2015). Targeted genome modifications in soybean with CRISPR/Cas9. *BMC Biotechnol., 15,* 16.

Jia, H., & Wang, N., (2014). Targeted genome editing of sweet orange using Cas9/sgRNA. *PLoS One, 9:*e93806.

Jiang, W., et al., (2013). Demonstration of CRISPR/Cas9/ sgRNA-mediated targeted gene modification in Arabidopsis, tobacco, sorghum and rice. *Nucleic Acids Res., 41*(20), e188.

Jiang, W., et al., (2014a). Successful transient expression of Cas9 and single guide RNA genes in *Chlamydomonas reinhardtii. Eukaryot. Cell, 13,* 1465–1469.

Jiang, W., Yang, B., & Weeks, D. P., (2014b). Efficient CRISPR/Cas9-mediated gene editing in *Arabidopsis thaliana* and inheritance of modified genes in the T2 and T3 generations. *PLoS One, 9,* e99225.

Jinek, M., et al., (2012). A programmable dual RNA guided DNA endonuclease in adaptive bacterial immunity. *Sci., 337,* 816–821.

Jinek, M., et al., (2014). Structures of Cas9 endonucleases reveal RNA-mediated conformational activation. *Sci., 343,* 1247997.

Johnson, R. A., et al., (2015). Comparative assessments of CRISPR-Cas nucleases' cleavage efficiency in planta. *Plant Mol. Biol., 87,* 143–156.

Jones, H. D., (2016). Are plants engineered with CRISPR technology genetically modified organisms? *Biochem. Soc.,* pp. 14–17.

Karvelis, T., et al., (2013). crRNA and tracrRNA guide Cas9-mediated DNA interference in *Streptococcus thermophilus. RNA Biol., 10,* 841–851.

Kim, J. S., & Pabo, C. O., (1998). Getting a handhold on DNA: design of poly-zinc finger proteins with femtomolar dissociation constants. *Proc. Natl Acad. Sci. USA, 95,* 2812–2817.

Kim, Y. G., Cha, J., & Chandrasegaran, S., (1996). Hybrid restriction enzymes: zinc finger fusions to FokI cleavage domain. *Proc. Natl Acad. Sci. USA, 93,* 1156–1160.

Kim, Y. G., et al., (1997). Site-specific cleavage of DNA-RNA hybrids by zinc finger/FokI cleavage domain fusions. *Gene, 203,* 43–49.

Kochevenko, A. and Willmitzer, L., (2004). Chimeric RNA/DNA oligonucleotide-based site-specific modification of the tobacco acetolactate synthase gene. *Plant Physiol., 132,* 174–184.

Krishna, S. S., Majumdar, I., & Grishin, N. V., (2003). Survey and summary: Structural classification of zinc fingers. *Nucleic Acids Res., 31*(2), 532–550.

Kumar, V., & Jain, M., (2015). The CRISPR–Cas system for plant genome editing: advances and opportunities. *J. Exp. Bot., 66*(1), 47–57.

Kusaba, M., (2004). RNA interference in crop plants. *Curr. Opin. Biotechnol., 15,* 139–143.

Lawrenson, T., et al., (2015). Induction of targeted, heritable mutations in barley and *Brassica oleracea* using RNA-guided Cas9 nuclease. *Genome Biol., 16,* 258.

Lee, H. J., et al., (2012). Targeted chromosomal duplications and inversions in the human genome using zinc finger nucleases. *Genome Res., 22,* 539–548.

Lee, H. J., Kim, E., & Kim, J. S., (2010). Targeted chromosomal deletions in human cells using zinc finger nucleases. *Genome Res., 20,* 81–89.

Li, J. F., et al., (2013). Multiplex and homologous recombination mediated genome editing in Arabidopsis and *Nicotiana benthamiana* using guide RNA and Cas9. *Nat. Biotechnol., 31,* 688–691.

Li, L., Wu, L. P., & Chandrasegaran, S., (1992). Functional domains in FokI restriction endonuclease. *Proc. Natl. Acad. Sci. USA, 89,* 4275–4279.

Li, M., et al., (2016a). Reassessment of the four yield-related genes Gn1a, IPA1, DEP1 and GS3 in rice using a CRISPR/cas9 system. *Front. Plant Sci., 7,* 377.

Li, Q., et al., (2016b). Development of japonica Photo-Sensitive Genic Male Sterile Rice Lines by Editing Carbon Starved Anther Using CRISPR/Cas9. *J. Genet. Genomics, 43,* 415–419.

Li, T., et al., (2012). High-efficiency TALEN-based gene editing produces disease-resistant rice. *Nat. Biotechnol., 30,* 390–392.

Liang, Z., (2014). Targeted mutagenesis in *Zea mays* using TALENs and the CRISPR/Cas system. *J. Genet. Genomics., 41,* 63–68.

Liu, P. Q., et al., (2001). Regulation of an endogenous locus using a panel of designed zinc finger proteins targeted to accessible chromatin regions. Activation of vascular endothelial growth factor A. *J. Biol. Chem., 276,* 11323–11334.

Liu, Q., et al., (1997). Design of polydactyl zinc-finger proteins for unique addressing within complex genomes. *Proc. Natl Acad. Sci. USA, 94,* 5525–5530.

Lloyd, A., et al., (2005). Targeted mutagenesis using zinc-finger nucleases in Arabidopsis. *Proc. Natl. Acad. Sci., 102,* 2232–2237.

Lor, V. S., (2014). Targeted mutagenesis of the tomato PROCERA gene using TALENS. *Plant Physiol., 166,* 1288–1291.

Lusser, M., (2011). New plant breeding techniques: State-of-the-art and prospects for commercial development. *JRC Scientific and Technical Rep.,* 1–219.

Ma, X. L., (2015). A robust CRISPR/Cas9 system for convenient, high-efficiency multiplex genome editing in monocot and dicot plants. *Mol. Plant,* doi: 10.1016/j.molp.04.007.

Ma, Y., Zhang, L., & Huang, X., (2014). Genome modification by CRISPR. *FEBS J., 281,* 5186–5193.

Madesis, P., et al., (2016). Perspective of genome editing in plant breeding. *Adv. Pl. Agri. Res., 3*(6), 1.

Maeder, M. L., et al., (2008). Rapid "open-source" engineering of customized zinc-finger nucleases for highly efficient gene modification. *Mol. Cell, 31,* 294–301.

Mahfouz, M. M., & Li, L., (2011). TALE nucleases and next generation GM crops. *GM Crops., 2,* 99–103.

Mahfouz, M. M., et al., (2011). *De novo*-engineered transcription activator-like effector (TALE) hybrid nuclease with novel DNA binding specificity creates double-strand breaks. *Proc. Natl. Acad. Sci. USA, 108,* 2623–2628.

Mani, M., et al., (2005). Design, engineering and characterization of zinc finger nucleases. *Biochem. Biophys. Res. Commun., 335,* 447–457.

Mao, Y., et al., (2013). Application of the CRISPR-Cas system for efficient genome engineering in plants. *Mol. Pl., 6,* 2008–2011.

Marraffini, L. A., & Sontheimer, E. J., (2010). CRISPR interference: RNA directed adaptive immunity in bacteria and archaea. *Nature Rev. Gen., 11,* 181–190.

Martin-Ortigosa, S., et al., (2014). Mesoporous silica nanoparticle-mediated intracellular cre protein delivery for maize genome editing via loxP site excision. *Plant Physiol., 164,* 537–547.

McNamara, A. R., et al., (2002). Characterisation of site-biased DNA methyltransferases: specificity, affinity and subsite relationships. *Nucleic Acids Res., 30,* 3818–3830.

Miao, J., et al., (2013). Targeted mutagenesis in rice using CRISPR-Cas system. *Cell Res., 23,* 1233–1236.

Michno, J. M., et al., (2015). CRISPR/Cas mutagenesis of soybean and *Medicago truncatula* using a new webtool and a modified Cas9 enzyme. *GM Crops Food, 6,* 243–252. doi: 10.1080/21645698. 2015. 1106063.

Miller, J. C., et al., (2011). A TALE nuclease architecture for efficient genome editing. *Nat. Biotechnol.*, *29*, 143–148.

Moerschell, R. P., Tsunasawa, S., & Sherman, F., (1988). Transformation of yeast with synthetic oligonucleotides. *Proc. Natl. Acad. Sci. USA*, *85*, 524–528.

Mojica, F. J. M., et al., (2009). Short motif sequences determine the targets of the prokaryotic CRISPR defence system. *Microbiol.*, *155*, 733–740.

Moose, S. P., & Mumm, R. H., (2008). Molecular plant breeding as the foundation for 21st century crop improvement. *Plant Physiol.*, *147*, 969–977.

Moscou, M. J., & Bogdanove, A. J., (2009). A simple cipher governs DNA recognition by TAL effectors. *Sci.*, *26*, 1501.

Nekrasov, V., et al., (2013). Targeted mutagenesis in the model plant *Nicotiana benthamiana* using Cas9 RNA-guided endonuclease. *Nat. Biotechnol.*, *31*, 691–693.

Nemudryi, A. A., et al., (2014). TALEN and CRISPR/Cas genome editing systems: tools of discovery. *Acta. Naturae*, *6*(3), 19–40.

Nicolia, A., et al., (2015). Targeted gene mutation in tetraploid potato through transient TALEN expression in protoplasts. *J. Biotechnol.*, *204*, 17–24.

Ning, Y. Q., et al., (2015). Two novel NAC transcription factors regulate gene expression and flowering time by associating with the *histone demethylase* JMJ14. *Nucleic Acids Res.*, *43*, 1469–1484.

Nishimasu, H., et al., (2014). Crystal structure of Cas9 in complex with guide RNA and target DNA. *Cell*, *156*(5), 935–949.

Okuzaki, A., & Toriyama, K., (2004). Chimeric RNA/DNA oligonucleotide-directed gene targeting in rice. *Plant Cell Report*, *22*, 509–512.

Osakabe, K., Endo, M., & Toki, S., (2012). DNA double-strand breaks and homologous recombination in higher plants. In Plant Mutagenesis — Principles and Applications. Edited by Quingyao, S., pp. 71–80. The Smiling Hippo, Greece.

Osakabe, K., Osakabe, Y., & Toki, S., (2010). Site-directed mutagenesis in *Arabidopsis* using custom-designed zinc finger nucleases. *Proc. Natl Acad. Sci. USA.*, *107*, 12034–12039.

Pabo, C. O., Peisach, E., & Grant, R. A., (2001). Design and selection of novel Cys2His2 zinc finger proteins. *Annu. Rev. Biochem.*, *70*, 313–340.

Pâques, F., & Duchateau, P., (2007). Meganucleases and DNA double-strand break-induced recombination perspectives for gene therapy. *Current Gene Therapy*, *7*, 49–66.

Pavletich, N. P., & Pabo, C. O., (1991). Zinc finger-DNA recognition: crystal structure of a Zif2608–*DNA complex at 2. 1 A. Sci.*, *252*, 809–817.

Petolino, J. F., et al., (2010). Zinc finger nuclease-mediated transgene deletion. *Plant Mol. Biol.*, *73*, 617–628.

Qi, L. S., et al., (2013). Repurposing CRISPR as an RNA-guided platform for sequence-specific control of gene expression. *Cell.*, *5*, 1173–1183.

Qi, Y., (2013a). Targeted deletion and inversion of tandemly arrayed genes in *Arabidopsis thaliana* using zinc finger nucleases. *G3 (Bethesda)*, *3*, 1707–1715.

Rebar, E. J., (2004). Development of pro-angiogenic engineered transcription factors for the treatment of cardiovascular disease. *Expert Opin. Investig. Drugs*, *13*, 829–839.

Rebar, E. J., et al., (2002). Induction of angiogenesis in a mouse model using engineered transcription factors. *Nature Med.*, *8*, 1427–1432.

Ren, C., et al., (2016). CRISPR/Cas9-mediated efficient targeted mutagenesis in Chardonnay (*Vitis vinifera* L.). *Sci. Rep.*, *6*, 32289.

Reii, D., et al., (2002). PPARgamma knockdown by engineered transcription factors: exogenous PPAR gamma2 but not PPAR gamma1 reactivates adipogenesis. *Genes Dev.*, *16*, 27–32.

Ron, M., et al., (2014). Hairy root transformation using *Agrobacterium rhizogenes* as a tool for exploring cell type-specific gene expression and function using tomato as a model. *Plant Physiol.*, *166*, 455–469.

Roth, N., et al., (2012). The requirement for recombination factors differs considerably between different pathways of homologous double-strand break repair in somatic plant cells. *Plant J.*, *72*, 781–790.

Ruiter, R., et al., (2003). Spontaneous mutation frequency in plants obscures the effect of chimeraplasty. *Plant Mol. Biol.*, *53*, 675–689.

Sander, J. D., & Joung, J. K., (2014). CRISPR-Cas systems for editing, regulating and targeting genomes. *Nat. Biotechnol.*, *32*, 347–355.

Sargent, R. G., Kim, S., & Gruenert, D. C., (2011). Oligo/polynucleotide-based gene modification: strategies and therapeutic potential. *Oligonucleo.*, *21*, 55–75.

Sauer, N. J., et al., (2016a). Oligonucleotide-directed mutagenesis for precision gene editing. *Plant Biotechnol. J.*, *14*, 496–502. doi: 10.1111/ pbi. 12496.

Sauer, N. J., et al., (2016b). Oligonucleotide-mediated genome editing provides precision and function to engineered nucleases and antibiotics in plants. *Plant Physiol.*, *170*, 1917–1928.

Schiml, S., Fauser, F., & Puchta, H., (2014). The CRISPR/Cas system can be used as nuclease for in planta gene targeting and as paired nickases for directed mutagenesis in Arabidopsis resulting in heritable progeny. *Plant J.*, *80*, 1139–1150.

Schornack, S., et al., (2006). Gene-for-gene-mediated recognition of nuclear-targeted AvrBs3-like bacterial effector proteins. *J. Plant Physiol.*, *163*, 256–272.

Shan, Q., et al., (2015). Creation of fragrant rice by targeted knockout of the OsBADH2 gene using TALEN technology. *Plant Biotechnol. J.*, *13*, 791–800. doi: 10.1111/pbi. 12312.

Shan, Q., et al., (2013). Targeted genome modification of crop plants using a CRISPR-Cas system. *Nat. Biotechnol.*, *31*, 686–688.

Shi, J., et al., (2016). ARGOS8 variants generated by CRISPR-Cas9 improve maize grain yield under field drought stress conditions. *Plant Biotechnol. J.*, 15(2), *207–216*

Shi, Y., & Berg, J. M., (1995). A direct comparison of the properties of natural and designed zinc-finger proteins. *Chem. Biol.*, *2*, 83–89.

Shimatani, Z., et al., (2015). Positive-negative-selection-mediated gene targeting in rice. *Front. Plant Sci.*, *5*, 748.

Shukla, V. K., et al., (2009). Precise genome modification in the crop species *Zea mays* using zinc-finger nucleases. *Nature*, *459*, 437–441.

Sikora, P., et al., (2011). Mutagenesis as a tool in plant genetics, functional genomics, and breeding. *Int. J. Plant Genomics*, *2011*, 314829.

Smith, J., et al., (2000). Requirements for double-strand cleavage by chimeric restriction enzymes with zinc finger DNA-recognition domains. *Nucleic Acids Res.*, *28*, 3361–3369.

Snowden, A. W., et al., (2003). Repression of vascular endothelial growth factor A in glioblastoma cells using engineered zinc finger transcription factors. *Cancer Res.*, *63*, 8968–8976.

Sonoda, E., et al., (2006). Differential usage of non-homologous end-joining and homologous recombination in double strand break repair. *DNA Repair*, *5*, 1021–1029.

Sovová, T., et al., (2016). Genome Editing with Engineered Nucleases in Economically Important Animals and Plants: State of the Art in the Research Pipeline. *Curr. Issues Mol. Biol., 21,* 41–62.

Sternberg, S. H., et al., (2014). DNA interrogation by the CRISPR RNA-guided endonuclease Cas9. *Nature, 507*(7490), 62–67.

Sugano, S. S., et al., (2014). CRISPR/Cas9 mediated targeted mutagenesis in the liverwort *Marchantia polymorpha* L. *Plant Cell Physiol., 55,* 475–481.

Sun, X., et al., (2015). Targeted mutagenesis in soybean using the CRISPR-Cas9 system. *Sci. Rep., 5,* 10342. doi: 10.1038/srep10342.

Svitashev, S., et al., (2015). Targeted mutagenesis, precise gene editing and site-specific gene insertion in maize using Cas9 and guide RNA. *Plant Physiol., 169,* 931–945.

Tang, G., & Galili, G., (2004). Using RNAi to improve plant nutritional value: from mechanism to application. *Trends Biotechnol., 22,* 463–469.

Thierry, A., & Dujon, B., (1992). Nested chromosomal fragmentation in yeast using the meganuclease I-Sce I: a new method for physical mapping of eukaryotic genomes. *Nucleic Acids Res., 20,* 5625–5631.

Tingting, L., et al., (2015). Highly efficient CRISPR/Cas9 mediated targeted mutagenesis of multiple genes in *Populus. Yi. Chuan, 37,* 1044–1052.

Townsend, J. A., et al., (2009). High-frequency modification of plant genes using engineered zinc-finger nucleases. *Nature, 459,* 442–5.

Upadhyay, S. K., et al., (2013). RNA-guided genome editing for target gene mutations in wheat. *G3 (Bethesda), 3,* 2233–2238.

Urnov, F. D., et al., (2010). Genome editing with engineered zinc finger nucleases. *Nat. Rev. Genet., 11,* 636–646.

Van der Oost. J., et al., (2009). CRISPR-based adaptive and heritable immunity in prokaryotes. *Trends Biochem. Sci., 34,* 401–407.

Wang, H., et al., (2013). One-step generation of mice carrying mutations in multiple genes by CRISPR/Cas-mediated genome engineering. *Cell, 153,* 910–918.

Wang, S., et al., (2015). Efficient targeted mutagenesis in potato by the CRISPR/Cas9 system. *Plant Cell Rep.,* doi: 10.1007/s00299–15–1816–7.

Wang, Y., et al., (2014). Simultaneous editing of three homoeoalleles in hexaploid bread wheat confers heritable resistance to powdery mildew. *Nat. Biotechnol., 32*(9), 947–951.

Weeks, D. P., Spalding, M. H. and Yang, B., (2016). Use of designer nucleases for targeted gene and genome editing in plants. *Plant Biotech. J., 14,* 483–495.

Weigel, D., (2016). The end product is what matters: Research report. *Developmental Biol.,* https://www.mpg.de/en.

Wendt, T., et al., (2013). TAL effector nucleases induce mutations at a pre-selected location in the genome of primary barley transformants. *Plant Mol. Biol., 83,* 279–285.

Wolt, J. D., Wang, K., & Yang, B., (2016). The regulatory status of genome-edited crops. *Plant Biotechnol. J., 14,* 510–518.

Wright, D. A., et al., (2005). High- frequency homologous recombination in plants mediated by zinc-finger nucleases. *Pl. J., 44,* 693–705.

Xiao, A., et al., (2013). Chromosomal deletions andinversions mediated by TALENs and CRISPR/Cas in zebrafish. *Nucleic Acids Res., 41,* e141.

Xie, K., & Yang, Y., (2013). RNA-guided genome editing in plants using a CRISPR–Cas system. *Molecular Plant, 6,* 1975–1983.

Xie, K., Minkenberg, B., & Yang, Y., (2015). Boosting CRISPR/Cas9 multiplex editing capability with the endogenous tRNA-processing system. *Proc. Natl Acad. Sci. USA, 112,* 3570–3575.

Xu, G. L., & Bestor, T. H., (1997). Cytosine methylation targetted to pre-determined sequences. *Nature Genet., 17,* 376–378.

Xu, R., et al., (2014). Gene targeting using the *Agrobacterium* tumefaciens-mediated CRISPR-Cas system in rice. *Rice (N Y), 7,* 5.

Xu, R., et al., (2016). Rapid improvement of grain weight via highly efficient CRISPR/Cas9-mediated multiplex genome editing in rice. *J. Genet. Genom., 43*(9), 529–532.

Yang, B., Sugio, A. and White, F. F., (2005). Avoidance of Host Recognition by Alterations in the Repetitive and C-Terminal Regions of AvrXa7, a Type III Effector of Xanthomonas oryzae pv. Oryzae. Mol. *Plant Microbe Interact., 18,* 142–149.

Yang, H., et al., (2013). One-step generation of mice-carrying reported and conditional alleles by CRISPR/Cas-mediated genome engineering. *Cell, 154,* 1370–1379.

Zhang, F., et al., (2010). High frequency targeted mutagenesis in *Arabidopsis thaliana* using zinc finger nucleases. *Proc. Natl Acad. Sci. USA, 107,* 12028–12033.

Zhang, H., et al., (2016). TALEN mediated targeted mutagenesis produces a large variety of heritable mutations in rice. *Plant Biotechnol. J., 14*(1), 186–194.

Zhang, H., et al., (2014). The CRISPR/Cas9 system produces specific and homozygous targeted gene editing in rice in one generation. *Plant Biotechnol J., 12,* 797–807.

Zhang, L., et al., (2000). Synthetic zinc finger transcription factor action at an endogenous chromosomal site. Activation of the human erythropoietin gene. *J. Biol. Chem., 275,* 33850–33860.

Zhang, Y., et al., (2013). Transcription activator-like effector nucleases enable efficient plant genome engineering. *Plant Physiol., 161,* 20–27.

Zhou, H., et al., (2014). Large chromosomal deletions and heritable small genetic changes induced by CRISPR/Cas9 in rice. *Nucleic Acids Res., 42,* 10903–10914.

Zhou, X., et al., (2015). Exploiting SNPs for biallelic CRISPR mutations in the outcrossing woody perennial *Populus* reveals 4-coumarate: CoA ligase specificity and redundancy. *New Phytol., 208,* 298–301.

Zhu, T., et al., (1999). Targeted manipulation of maize genes *in vivo* using chimeric RNA/DNA oligonucleotides. *PNAS, 96,* 8768–8773.

Zhu, W., et al., (1998). AvrXa10 contains an acidic transcriptional activation domain in the functionally conserved c terminus. *Mol. Plant Microbe Interact., 11,* 824–832.

CHAPTER 19

NANOBIOTECHNOLOGY IN AGRICULTURE

D. N. BHARADWAJ and SANJAY SINGH

C. S. Azad University of Agriculture & Technology, Kanpur–208002, India

CONTENTS

ABSTRACT

Nanotechnology is a promising and new field of interdisciplinary research. It opens up a wide opportunity in various fields of human welfare such as pharmaceuticals, medicine, electronics, and agriculture. Agriculture is an essential source of food, feed, and shelter to human sustenance. The increasing global population in near future will be around 10 billion in 2050, making it hard for the agricultural sector to produce more agricultural commodities with limiting natural resources. A large population in developing and developed countries is still facing food insecurity in one or another way. To meet challenges like deforestation, water crisis, global warming, and climate change, there is a need to create drought- and pest-resistant crops coupled with high production and yield. Currently, nanotechnology seems to provide assurance in various fields of human needs to revolutionize the areas of agriculture, pharmaceuticals, medicines, textile, energy, communication, and information technology. Therefore, the application of nanotechnology to all these sectors is fetching attention of scientists and policy makers. Specifically, the application of nanotechnologies to agriculture and food industry carries potential benefits to reduce not only agricultural inputs but also improve food safety and nutrition quality with improved processing. The benefits of nanotechnology are widespread in agriculture to make need-based strategies for the management of inputs like nanoencapsulated seeds, nanofertilizers, nanoinsecticides, nanopesticides, and nanoherbicides by the applications of nanomaterials in sustainable agriculture management.

19.1   NANOTECHNOLOGY AND NANOSCALE

- Manipulating materials and systems at the scale of atoms and molecules.
- *Nanomaterials* measure a few hundred nanometers or less.
- A nanometer (nm) is one billionth of one meter (i.e., 1/1,000,000,000 meter), smaller than the wavelength of visible light and a hundred-thousandth the width of a human hair.
- Nanotechnology is the ability to create and manipulate atoms and molecules on the smallest of scales. The word "Nano" in Greek means dwarf.
- Nanotechnology deals with anything that measures between 1 and 100 nm.

19.2   NANOBIOTECHNOLOGY

Nanobiotechnology joins the breakthroughs in nanotechnology to those in molecular biology. The DNA molecule has interesting features for its use in nanotechnology, namely its mini size of about 2 nm diameter; short structural repeat (helical pitch) of about 3.4–3.6 nm; and its "stiffness," with a persistence length 50 nm. Therefore, chemically, DNA will be a key player in bottom-up nanotechnology. *In vitro* genetic manipulation was first performed in 1970 due to its "sticky ends" that are short single-stranded overhang protruding from the end of a double-stranded helical DNA molecules have complementary arrangements of the nucleotide bases these are adenine, cytosine, guanine, and thymine which cohere to form a molecular complex which is the best example of programmable molecular recognition. Molecular biologists can help nanotechnologists to understand and access the nanostructures and nanomachines with extraordinary properties of biological molecules and cell processes, and nanotechnologists can then accomplish many difficult goals rather than build silicon scaffolding for nanostructures. The ladder structure of DNA provides nanotechnologists with a natural framework for assembling nanostructures, and its highly specific bonding properties bring atoms together in a predictable pattern to create a nanostructure. DNA has been used not only to build nanostructures but also as an essential component of nanomachines. DNA is the excellent information storage molecule, and it may serve as the basis for the next generation of computers. As microprocessors and microcircuits shrink to nanoprocessors and nanocircuits, DNA molecules mounted onto silicon chips may replace microchips with electron flow-channels etched in silicon. Such biochips are DNA-based processors that use the extraordinary information of DNA.

Other biological molecules relate to all applications of genomics including mammalian, plant, and microbes. Genomics provides the basic tools and subsequently the technology for gathering sequence information of all genes of an organism and designing innovative devices to probe the biological information and its application in diverse fields such as medicine and agriculture.

Research in nanobiotechnology is advancing toward the ability of sequencing the DNA in nanofabricated gel-free systems, which would allow significantly more rapid DNA sequencing, coupled with powerful genetic association analysis, DNA sequencing data of the crop germplasm, gene pool, and wild relatives, which potentially can provide highly useful information about molecular markers associated with agronomically and economically important traits. Thus, nanobiotechnology can enhance the pace of progress in molecular marker-assisted breeding for crop improvement.

The term "nanotechnology" was first used by Norio Taniguchi in 1974. The smallness of nanometer can be observed as the nanometer (nm) is one-billionth of a meter, smaller than the wavelength of visible light and a hundred-thousandth the width of a human hair and Nanotechnology measuring between 1-100 nm side by side to match the width of a human hair. **Nanotech** is manipulation of matter at the molecular, atomic, and supramolecular scale. Nanotechnology is a scientific approach that manipulates physical as well as chemical properties of a material or substance at the molecular level. DNA double helix has a diameter of around 2 nm. On the other hand, the smallest cellular life forms, the bacteria of the genus *Mycoplasma*, are around 200 nm in length. Nanotechnology may be able to create many new materials and devices with a vast range of applications, such as in nanoelectronics, biomaterials, nanomedicine, and energy production. Biotechnology involves manipulating knowledge of cellular processes at the genetic and molecular level to engineer new products and services in diverse fields of sciences including agriculture and medicine (Fakruddin et al., 2012). Globally, agriculture is the only resource (food, feed, fuel, and shelter) of human sustenance, Above 60% world's population depends on agricultural activities for their livelihood (Brock et al., 2011).

Nanotechnology has proved to have the potential to revolutionize the agricultural and food industry by manipulating with novel tools and molecular understanding in the management of crop diseases by detection and control, increased absorption of nutrients, and improved nutritional composition with enhanced yields. It can also modify the plants' capability to combat adverse environmental conditions and show better response to fertilizers, pesticides, and insecticides (Tarafdar et al., 2013). Nanotechnology has envisaged new challenges and demand for safe and healthy food, minimizing the risk of disease pathogens, and environmental threats to agricultural crop production (Biswal et al., 2012). The potential detection of pathogens can be done even before the symptoms develop by integrated sensing and monitoring and controlling systems containing bioactive systems such as drugs, pesticides, nutrients, probiotics, nutraceuticals, and implantable cell bioreactors. New nanostructured catalysts are available to enhance the efficiency of herbicides and pesticides at very low and safe doses, which will also protect against the environmental hazards by reducing pollution and cleaning up existing pollutants in soil and water.

Nanobiotechnology is competent to use the alternative renewable energy sources, develop genetically modified crops, animals, and precision farming techniques (Prasad et al., 2012a). Nanotechnology occupies a prominent position in transforming agriculture and food production through the development of nanomaterials to open novel applications in plant biotechnology and agriculture. Currently, nanochemical pesticides are already in use with several other applications that are still in early stages. The intensive farming systems includes the fine tuning and more precise micromanagement of soils; the most efficient and targeted use of inputs, new toxin formulations for pest control, developing new crop and animal traits, and the diversification and differentiation of farming practices and products to revolutionize sustainable agriculture management are concerns to modern approaches of nanotechnology.

19.3 APPLICATIONS OF NANOTECHNOLOGY IN AGRICULTURAL PRODUCTION

Nanotechnology helps agricultural sciences to reduce environmental pollution by production of nanoparticles and nanocapsules of pesticides and fertilizers which have the ability to control its delivery with effective absorption at the time of its requirement. These nanoparticles have higher efficiency at lower doses as compared to traditional pesticides used by farmers and are safer for environment. These nanoparticles of drugs can also be used to alter the kinetic profiles of drug release, leading to more sustained release of drugs with a minimum dose requirement (Sharon et al., 2010). The nano-based diagnostic kits detect problems rapidly (Prasanna, 2007) because of nano sensors used which also protect the environment through the use of alternative or renewable energy sources and clean-up pollutants. The demand is growing for healthy, safe food due to changing weather patterns (Hager, 2011). There are various applications of nano-technology in agriculture as discussed below:

Agriculture	Processing	Products	Nutrition
New pesticides	Microencapsulation of flavors and aromas	UV protection	Neutraceuticals
Genetic engineering	Gelation agents	Antimicrobials	Nutrient delivery
Identity preservation	Nanoemulsions	Condition and misuse monitor	Fortification of mineral/vitamin
Sensors for soil condition measurement	Anti-caking	High barrier plastics	Purification of drinking water
	Sanitation of equipment	Security/anti counterfeiting	Supplements
		Contaminant sensors	

19.4 PRECISION FARMING

Precision farming has been a long-awaited goal of plant breeders to maximize crop yields, with minimum inputs of fertilizers, insecticides, pesticides, herbicides, etc., through environmental monitoring and using targeted action. Precision farming makes the best use of global satellite positioning systems (GPS), computers, and remote sensing devices to measure localized environmental conditions for real-time monitoring. Thus, crops are grown precisely by identifying the nature and local problems at the maximum efficiency. The centralized soil conditions and crop's seedling developmental stages, fertilizer, agrochemicals, and water use efficiency are fine-tuned to low input costs and potentially high yield to farmers. Precision farming can also help to reduce agricultural waste and reduce environmental pollution. Therefore, sensors and monitoring systems developed with nanotechnology will have a great impact on precision farming.

The nanosensors can be distributed throughout the field to monitor soil conditions, crop growth, etc. Nanotechnology-based biosensors have increased sensitivity for earlier response to environmental changes such as:

- Nanosensors utilizing tiny carbon nanotubes (nanocantilevers) trap and measure individual proteins or even small molecules.
- Nanoparticles (nanosurfaces) are capable to trigger an electrical or chemical impulse to signal the presence of any contaminant such as bacteria.
- Some other nanosensors can trigger an enzymatic reaction by using nanodendrimers to target the presence of chemicals and proteins.
- Consequently, precision farming with the help of smart sensors will permit farmers to make better decisions for enhanced agricultural productivity by using accurate information.

19.5   APPLICATION OF NANOTECHNOLOGY IN SEED SCIENCE

Seed is a natural nanogift to human; it is a self-perpetuating biological entity that can survive even in harsh environmental conditions. Nanotechnology can be applied to harness the good quality of seeds. In cross or wind-pollinated crops, seed production is a tedious process because pollen load may cause contamination and affect genetic purity. The use of bionanosensors can help to reduce pollen contamination, even for pollens from genetically modified (GM) crops.

Stored seeds are often damaged by pathogens, but nanocoating of seeds with Zn, Mn, Pt, Au, and Ag can protect seeds for longer periods. The encapsulation and controlled release techniques have revolutionized the use of pesticides and herbicides. These seeds can be imbibed with nanoencapsulations with bacterial strain, known as "Smart seed," which also reduces the seed use rate and improves crop performance.

Khodakovskaya et al. (2009) have reported the use of carbon nanotubes (CNTs) to improve tomato seed germination through better permeation of moisture where CNTs serve as new pores for water permeation by penetration of seed coat to channelize the water from the substrate into the seeds; this is a good option of seed germination rainfed agricultural system.

19.6 NANOTECHNOLOGY IN FOOD PROCESSING

Nanotechnology-based food processing is making impact on functional foods, which respond to the body's requirements and can deliver nutrients more efficiently. These foods will remain dormant in the body and deliver nutrients to cells when required, i.e., food "on demand." Other types of food processing includes adding nanoparticles to existing foods to enable increased nutrient absorption. An example is tuna fish oil (source of omega 3 fatty acids), a product of Australia. Another nutraceutical produced by Israel, known as Nutralease, includes lycopene, beta-carotene, lutein, phytosterols, CoQ10, and docosahexaenoic acid (DHA)/ eicosapentaenoic acid (EPA). These Nutralease particles allow these compounds to enter the bloodstream from the gut more easily, thus increasing their bioavailability. Similarly, Shemen Industries produced Canola Activa oil, which reduces cholesterol intake into the body by 14%, through bile solubilization and has potential applications in the pharmaceutical industry. Unilever is developing low-fat ice creams by decreasing the size of emulsion particles that give ice-cream its texture; this will reduce 90% emulsion and decrease fat content from 16% to about 1%. All these new developments in food processing will make the concept of super foodstuff a reality.

19.7 NANOTECHNOLOGY IN FOOD PACKAGING

Consumers are demanding fresh food to be kept for long time, and the packaging materials should be easy for handling, safe, and healthy with an effective packaging material. By using nanoparticle technology, Bayer (2005) has developed plastic packaging that will keep food fresh for a longer time with the help of silicate nanoparticles. Nanoparticles such as titanium dioxide, zinc oxide, and magnesium oxide as well as their combination can be efficient in killing microorganisms and are cheaper and safer instead of metal-based nanoparticles. In food packaging, oxygen spoils the fat in meat and cheese and turns them pale, but nanoparticles in Durethan® block air to penetrate inside the meat. Silver nanoparticles have been used in refrigerator manufacturing to inhibit bacterial growth and eliminate odors.

19.8 COSMETIC INDUSTRY

Nanotechnology is now also being used in cosmetic industry for producing transparent creams. Royal BodyCare, a company has marketed a new product called NanoCeuticals, which is a colloid (or emulsion) of particles of less than 5 nm in diameter; the company claims that the product will scavenge free radicals, increase hydration, and balance the body's pH. This company also developed NanoClusters™, a nanosize powder combined with nutritional supplements; it enhances the absorption of nutrients and vitamins directly to the skin, in addition to providing UV protection. Some other companies are manufacturing anti-aging formulations by using nanoparticles.

19.9 MICROARRAYS AND EXPRESSION PROFILING

Microarray-based hybridization methods allow to simultaneous measure the expression level of thousands of genes. This provides information about many different aspects of gene regulation and function with novel formats for sequence determination, and patterns of genomic expression can have significantly higher throughput than current technologies. Thousands of DNA or protein molecules are arrayed on glass slides to create DNA chips and protein chips, respectively. These developments in microarray technology use customized beads in place of glass slides; so, nanofabrication techniques can be used in pattern surface chemistry for the production of various biosensor and biomedical applications in the area of:

i. Determination of new genomic sequences.
ii. Scanning of genes for polymorphism that might have an impact on phenotype.
iii. Comprehensive survey of the pattern of gene(s) expression in organisms when exposed to biotic/abiotic stress.
iv. Detect mutations in disease-related genes.
v. Monitor gene activity.
vi. Identify important genes to crop productivity.
vii. Improve screening for microbes used in bioremediation.

19.10   NANOSENSORS FOR MONITORING SOIL CONDITIONS, GROWTH HORMONE AND NUTRIENTS UPTAKE

The efficient use of natural agricultural resources like water, nutrients chemicals like DHA /EPA (fatty acids are omega-3 fats, which are found in cold water fish) during farming acts as nanosensors are user friendly. The application of GPS with the satellite imaging system in crop fields might enable farmers to detect stresses, drought, and crop pests through nanosensors in the field, which can provide knowledge of diseases like existence of plant viruses and soil nutrient uptake (Ingale and Chaudhari, 2013). These nanosensors also minimize fertilizer consumption and environmental pollution. Nanoencapsulated slow-release fertilizers have been widely used (DeRosa et al., 2010) with information of the manufacture, nanobarcodes. etc. Li et al. (2005) gave the idea of barcodes for economical, proficient, effortless decoding and recognition of diseases by using fluorescent based tools. Nanotechnology can enable to study plant's regulation of hormones such as auxin; this will help scientists to know how plant roots acclimatize to their environment, particularly to marginal soils (McLamore et al., 2010).

19.11   NANOBIOSENSORS

The novel sensor devices can be incorporated into microbial cells that can be used in the diagnosis of Minerals (MN) deficiency and soil toxicity. Such GM microbial sensors were constructed and used to evaluate the immobilization and bioavailability of Zn in different soils (Liu et al., 2012a; Maderova and Paton, 2013). The Zn-specific biosensor, *Pseudomonas putida* X4 (pczcR3GFP), was constructed by fusing a promoterless enhanced green fluorescent protein (egfp) gene with the czcR3 promoter in the chromosome of *P. putida* X4. The fluorescent reporter strain detected about 90% of the Zn content in soil-water extracts of soil samples amended with Zn; this is an alternative system for the convenient evaluation of Zn toxicity in the environment (Liu et al., 2012a).

Engineered nanobiosensors and nanoprobes such as DNA or RNA-based aptamers (synthetic nucleic acids) are highly specific and sensitive devices that allow the detection of very low quantities of analytes in individual living cells or fluids. These aptamers are capable of binding tightly to a target of interest of monoclonal antibodies (Bunka and Stockley,

2006). These nanobiosensors are sensitive to recognize individual chemical species and metabolites at specific locations of tissues, organs, or fluids (Vo-Dinh et al., 2006) in the terrestrial ecosystems through interactions, chemical signaling and communications established between plant roots and soil microorganisms.

These microcapsules may have potential for applications in targeted delivery systems for the controlled release of drugs, pesticides, or other payloads such as plant nutrients (Zhang et al., 2013).

The incorporation of nanodevices in plant nutrition may, therefore, allow for the development of efficient technological platforms to detect and treat nutrient deficiencies in soils and plants and in real time. Intelligent nanodevices or biosensors may help deliver macronutrients and MNs according to the temporal and spatial MN requirements of crops during the growing season.

Nanosensors are sophisticated instruments that respond to physicochemical and biological aspects and transfer that response into a signal that can be used for crop production (NNCO, 2009). These nanosensors allow the detection of contaminants such as microbes, pests, nutrient content, and plant stress due to drought, temperature, insect or pathogen pressure, or lack of nutrients; this helps farmers in potential and efficient use of inputs by indicating the nutrient or water status of crop plants to apply nutrients, water, or crop protectants (insecticide, fungicide, or herbicide) only at requirement in real time monitoring; this is linked with autonomous sensors to a GPS system. These nanosensors could be distributed throughout the field where they can monitor soil conditions and crop growth to trigger an electrical or chemical signal in the presence of a contaminant such as bacteria and viruses. Therefore, precision farming with the help of smart sensors leads to enhanced agriculture productivity by providing accurate information to farmers to make better decisions.

19.12   NANOTECHNOLOGY IN PURIFICATION OF IRRIGATED WATER

Instead using costly chemicals and UV light to purify water, several nanomaterials developed economically can be used effectively in purification of irrigation water, such as nano-enabled water treatment techniques based on membrane filters derived from CNTs, nanoporous ceramics, and magnetic nanoparticles (Hillie and Hlophe, 2007). These CNT filters can remove contaminants (heavy metals like lead, uranium, and arsenic) and toxicants and pathogens from potable water. Nanoceramic charged filters can trap bacteria and viruses with negative charge. These filters are capable to remove microbial endotoxins, genetic materials, pathogenic viruses, and micro-sized particles (Argonide, 2005). The magnetic nanoparticles and magnetic separators are produced at very low magnetic field gradients. These nanocrystals of monodisperse magnetite (Fe3O4) have a strong and irreversible interaction with arsenic while retaining their magnetic properties (Yavuz et al., 2006). The treated water could be used for irrigation purpose.

19.13   DETOXIFICATION OR REMEDIATION OF HARMFUL POLLUTANTS

Nanomineral synthetic clay candles of hydrotalcite can easily remove arsenic from water (Gilman, 2006), and the technology is currently used in many developing countries to filter organisms from drinking water. Nanoscale zero-valent iron is the most widely used nanomaterial that could be used to remediate pollutants in soil or groundwater. Other nanomaterials that are used in remediation include nanoscale zeolites, metal oxides, CNTs and fibers, enzymes, various noble metals, and titanium dioxide. Nanoparticle filters can also be used to remove organic particles and pesticides like DDT (dichorodiphenyl-trichloroethane), endosulfan, malathion, and chlor-pyrifos from water. Several other nanoparticle filters have been used in developed countries for remediation at waste sites (Karn et al., 2009).

19.14   NANOCAPSULES IN EFFICIENT DELIVERY OF AGROCHEMICALS

Nanoencapsulation is currently the most promising technology for the protection of host plants against insect pests. The efficient delivery of agrochemicals like fertilizers, pesticides, fungicides, etc. is possible by nanoencapsulation for nutrition and against any particular host for insect pest control, which allows absorption of these agrochemicals into the plants (Scrinis and Lyons, 2007) at the required time. This process is also successful to deliver DNA and other desired chemicals into plant tissues for the protection of host plants against insect pests (Torney, 2009). The nanoencapsulation release mechanisms include diffusion, dissolution, biodegradation, and osmotic pressure at specific pH (Ding and Shah, 2009; Vidhyalakshmi et al., 2009) into the target host tissue through nanoencapsulation.

After the second green revolution, fertilizers have increased crop production and now play a pivotal role. Nanofertilizers have enhanced nutrient use efficiency and have overcome the problem of eutrophication; these fertilizers release nutrients very efficiently at a specific time of requirement to the crop as compared to ordinary fertilizers (Liu et al., 2006a). Nanofertilizers also reduce nitrogen loss due to leaching and emissions and enhance long-term incorporation by soil microorganisms. The slow controlled release of fertilizers may also improve soil health by decreasing toxic effects associated with fertilizer's overapplication (Suman et al., 2010).

Nanoencapsulation process-controlled chemical methods have revolutionized the use of agrochemicals. The pesticides and herbicides inside nanoparticles are being developed to release when proper environment triggers for their requirement; this has facilitated greater production of crops with less injury to plants and workers. The nanoparticles of agrochemicals are produced within the size of 100–250 nm, and they dissolve in water more effectively. Some other companies are producing water or oil-based nanosuspensions of pesticides or herbicides within the range of 200–400 nm, in the form of gels, creams, liquids, etc. Syngenta, the world's largest agrochemical corporation, is using nanoemulsions in its pesticide products like Primo MAXX®, a plant growth regulator, which if applied prior to the onset of stress such as heat, drought, disease, or traffic can strengthen the physical structure of turf grass and allow it to withstand ongoing stresses throughout the growing season. Similarly, the encapsulated product from Syngenta delivers a broad-spectrum control of primary and secondary insect pests of cotton, rice, peanuts, and soybeans. Marketed under the name Karate® ZEON, this is a quick-release micro-encapsulated product containing the active compound lambda-cyhalothrin (a synthetic insecticide based on the structure of natural pyrethrins), which breaks open on contact with leaves. In contrast, the encapsulated product "gutbuster" only breaks open to release its content when it comes into contact with alkaline environments such as the stomach of certain insects.

19.15   NANO-BASED SMART DRUG DELIVERY SYSTEMS

Smart drug delivery systems can detect and treat an infection and nutrient deficiency and provide timed-release drugs and micronutrients in the organisms. The use of pesticides in crops has tremendously increased in the second half of the 20th century, as DDT is one of the most effective and widespread chemical that is highly toxic and persistent, thereby affecting human and animal health and ecosystems by its residual effects; it was later banned. Now, integrated pest management (IPM) systems is used during crop rotation along with biological pest control methods. IPM is one of the very effective methods and is used in several countries including India. Nanoscale devices with novel properties are being developed to be used for agricultural crop production. These devices can identify plant health issues before they appear and become severe. These nanosensors or smart devices will act as both a preventive and an early warning system to deliver chemicals in a controlled and targeted manner and are capable to respond with remedial action or alert the farmers for proper action against problem.

19.16   ZEOLITES FOR WATER RETENTION

Zeolites are naturally occurring crystalline aluminum silicates that enable water infiltration and retention in the soil due to its very porous properties and capillary suction. They act as a natural wetting agent for nonwetting sands and assist water distribution through soils. This can improve significantly the water retention of sandy soils and increase porosity in clay soils. Improving water-retention capacity of soils, consequently increased crop production in drought areas. Further, the application of zeolite will further improve soil's ability to retain water and nutrients to produce higher yields.

19.17   NANOCOATINGS AND NANOFEED ADDITIVES

In poultry houses, self-sanitizing photocatalyst coating with nanotitanium dioxide (TiO_2) could be used to destroy and inhibit bacterial growth through oxidation in light and humidity. The nanoparticles-coated poultry feed binds pathogenic bacteria that helps to reduce food-borne pathogens. In presence of light and humidity, the unique photocatalytic of nano TiO_2 is activated when exposed to UV light. TiO_2 then oxidizes and destroys bacteria. This unique coating is approved by the Canadian Food Inspection Agency (CFIA). In Denmark, the Chicken and Hen Infection Program (CHIP) involves self-cleaning and disinfection by nanocoatings (Clemants, 2009). The nanoscale smooth surface provides effective disinfection and cleaning. Currently, Danish researchers are trying to develop coatings incorporating nanosilver, which may work without UV light for activation. Modified nanoclays (montmorillonite nanocomposite) can ameliorate the harmful effects of aflatoxin on poultry (Shi et al., 2006). Research on nanoparticles and insect control is geared toward introduction of faster and eco-friendly pesticides in near future (Bhattacharyya et al., 2010). Surface-modified hydrophobic as well as lipophilic nanosilica can be significantly used as a new drug for treating nuclear polyhedrosis virus (BmNPV) infection, a major problem in silkworm industry. A study on *Bombyx mori* (silk worm) has clearly shown that nanoparticles could stimulate more fibroin protein production, which can assist in the future in producing CNT (Bhattacharyya et al., 2008; Bhattacharyya, 2009).

19.18 NANOHERBICIDES

Nanoherbicides blend with the soil, eradicate weeds in an eco-friendly way without leaving their toxic residues, and prevent the growth of weeds that have become resistant to conventional herbicides. This will easily destroy weeds seed in the soil and prevent them from germinating when weather and soil conditions become favorable for their growth. Most weeds survive and spread their underground tubers and deep roots; removal of weeds manually can infect and spread into uninfected areas. The application of nanosized active ingredient formulation through the use of an adjuvant can be a good solution if the active ingredient is combined with a smart delivery system; the herbicides will be applied only when they are required in the field. Crop yield is highly affected if the soil is contaminated with weeds and weed seeds. The herbicide efficacy can be enormously improved if used with nanotechnology, resulting in more crop production without any harm.

19.19 NANOTECHNOLOGY IN ORGANIC FARMING

The success of second green revolution was mainly due to high input of irrigation, fertilizers, insecticides, pesticides, herbicides, etc. But now, organic farming has goal to increase crop productivity with low inputs through monitoring environmental variables and applying targeted action. Organic farming is sophisticated farming that makes use of computers, GPS systems, and remote sensing devices to measure highly localized environmental and weather conditions; thus, the crops grow at minimum inputs and maximum crop harvesting efficiency is obtained. Organic farming uses centralized weather data to determine soil conditions, plant development, seeding, fertilizer, chemical and water use efficiency with lower production costs and potentially higher production return to agriculturists. Such precision farming can also help to reduce agricultural waste and help to control environment pollution.

19.20   NANOPARTICLES AND PLANT DISEASE CONTROL

Some of the nanomaterials like silver, carbon, silica, and alumino-silicates show antimicrobial activity and thus enable to make more economic agricultural production. Young (2009) stated that silver displays diverse modes of inhibitory action on microorganisms; therefore, it can be used relatively safer to control various plant pathogens than commercially used fungicides. Silver influences various biochemical processes in microorganisms in their cells' routine functions and plasma membrane (Pal et al., 2007). Silver nanoparticles prevent biomolecule inhibition by the expression of ATP and associated proteins (Yamanka et al., 2005).

As reported by Kumar and Yadav (2009), Prasad et al. (2011), Swamy and Prasad (2012), and Prasad and Swamy (2013), the use of nanoparticles can be considered as eco-friendly and cost-effective alternate and effective approach in controlling pathogenic microbes. So, the nanoparticles have a great potential in the management of plant diseases instead of synthetic fungicides (Park et al., 2006). Zinc oxide (ZnO) and magnesium oxide (MgO) nanoparticles are found to be effective antibacterial and antiodor agents (Shah and Towkeer, 2010), and their nanostructures are used as an attractive antibacterial ingredient in many products like anti-microbial preservative for wood or food products (Huang et al., 2005; Aruoja et al., 2009; Sharma et al., 2009). As their nanocapsules provide better penetration through cuticle and allow slow and controlled release of active ingredients on the target weed at the same time, they are safe for plants and cause less environmental pollution hazards (Barik et al., 2008). Besides, they have strong inhibitory, antibactericidal, antimicrobial, and broad-spectrum effects (Swamy and Prasad, 2012; Prasad et al., 2012b; Prasad and Swamy, 2013).

19.21 NANOPARTICLES AS PESTICIDES

Nanoparticles against insects and pests are also very effective and can be used in the preparation of new pesticides, insecticides, and insect repellants (Barik et al., 2008; Gajbhiye et al., 2009).

19.22 NANOPARTICLE GENE-MEDIATED DNA TRANSFER METHOD

Nanotechnology has promising applications in nanoparticle gene-mediated DNA transfer (Torney, 2009). Therefore, it can be used to deliver DNA fragments and other desired chemicals into plant tissues for the protection of host plants against insect pests. Porous hollow silica nanoparticles (PHSNs) loaded with validamycin (pesticide) can be used as an efficient delivery system of water-soluble pesticide for its controlled release; this makes it a promising carrier in agriculture, especially for controlled delivery of pesticides whose immediate as well as prolonged release is needed to plants (Liu et al., 2006b). The engineered nanoparticles may be better to evade the body's defense mechanism because of their nano size or protective coatings.

Yet, there is no standardization for the use and testing of nanotechnology and products incorporating the nanomaterials are being produced without any check. The ability of nanomaterials to infiltrate the human body is well known, but to date, there is no information of its effects on human health. Nanotechnology is a new and emerging science, before its mass scale application, scientists must study its harmful effects on human, animal and environment. Agriculture scientists are exploring a new technology and looking at every possibility to improve current methods in all fields of agriculture. Therefore, extensive studies are recommended to understand the mechanism for nanoparticles materials, their toxicity, and their impact on natural environment.

19.23 CONCLUSIONS AND FUTURE PERSPECTIVES

New advances in nanotechnology can play a major factor in modern agriculture. The successful development and application of nanoplatforms in medicine *in vitro* have generated interest in agri-nanotechnology (Nair et al., 2010). The technological innovations in nanobiotechnology will occupy a prominent position in transforming agricultural systems and food production worldwide.

The applications of nanotechnology has the potential to change the existing pattern of agricultural production by way of better management and conservation of inputs to plant production. Nanotechnology application in agriculture and food systems can do much more (Sugunan and Dutta, 2004). Introduction of a new technology always has an ethical and social responsibilities associated with unforeseen risks that may arise abruptly along with the tremendous positive potential.

Nanotechnology application in agriculture food systems is in infancy, and many more applications are expected to come in the near future. We expect to have several new processes for biosynthesis of nanoparticles in the area of electrochemical sensor, biosensors, medicine, healthcare, and agriculture to reduce pollution and to make agriculture more environmental friendly (Suman et al., 2010).

Thus, nanotechnology will revolutionize agriculture including pest management in the near future. Over the next two decades, the green revolution would be accelerated by means of nanotechnology. Nanoparticles would help to produce new pesticides, insecticides, and insect repellents (Owolade et al., 2008).

The outlook of nanoscience in agriculture is vague because many issues like GM crops need through attention in agricultural research and technologies. Agriculture technology should take advantage of the powerful tools of nanotechnology for the benefit of mankind.

KEYWORDS

- agriculture
- nanobiotechnology
- nano-encapsulaed seeds
- nano-fertilizer
- nano-herbicides
- nano-insecticides
- nano-pesticides

REFERENCES

Argonide, (2013). NanoCeram filters. Argonide Corporation. <http://sbir.nasa.gov/SBIR/successes/ss/9–072text.html>. Accessed on December 10.

Aruoja, V., Dubourguier. H., Kasamets, C., & Kahru, K. A., (2009). Toxicity of nanoparticles of CuO, ZnO and TiO2 to microalgae, *Pseudokirchneriella subcapitata*. *Sci. Total Environ.*, *407,* 1461–1468.

Barik, T. K., Sahu, B., & Swain, V., (2008). Nanosilica-from medicine to pest control. *Parasitolol. Res.*, *103,* 253–258.

Bhattacharyya, A., (2009). Nanoparticles from drug delivery to insect pest control. *Akshar.*, *1*(1), 1–7.

Bhattacharyya, A., Bhaumik, A., Pathipati, U. R., Mandal, S., & Epidi, T. T., (2010). Nanoparticles – A recent approach to insect pest control. *Afr. J. Biotechnol.*, *9*(24), 3489–3493.

Bhattacharyya, A., Gosh, M., Chinnaswamy, K. P., Sen, P., Barik, B., Kundu, P., & Mandal, S., (2008). Nanoparticle (allelochemicals) and silkworm physiology. In: *Recent Trends in Seribiotechnology* (Chinnaswamy, K. P., & Vijaya Bhaskar Rao, A. eds.) Bangalore, India., pp. 58–63.

Biswal, S. K., Nayak, A. K., Parida, U. K., & Nayak, P. L., (2012). Applications of nanotechnology in agriculture and food sciences. *Int. J. Sci. Innovat. Discov.*, *2*(1), 21–36.

Brock, D. A., Douglas, T. E., & Queller, D. C., (2011). Strassmann JE. Primitive agriculture in a social amoeba. *Nature.*, *469,* 393–396.

Bunka, D. H. J., & Stockley, P. G., (2006). Aptamers come of age – at last. *Nat. Rev. Microbiol. 4,* 588–596.

Clemants, M., (2009). Pullet production gets silver lining. Poultry International, April. <http://www.wattagnet.com/Poultry_International/4166.html>.

DeRosa, M. C., Monreal, C., Schnitzer, M., Walsh, R., & Sultan, Y., (2010). Nanotechnology in fertilizers. *Nat. Nanotechnol.*, *5,* 91–94.

Ding, W. K., & Shah, N. P., (2009). Effect of various encapsulating materials on the stability of probiotic bacteria. *J. Food Sci.*, *74*(2), M100–M107.

Fakruddin, Md, Hossain, Z., & Afroz, H., (2012). Prospects and applications of nanobiotechnology: a medical perspective. *J. Nanobiotechnol.*, *10,* 31.

Gajbhiye, M., Kesharwani, J., Ingle, A., Gade, A., & Rai, M., (2009). Fungus mediated synthesis of silver nanoparticles and its activity against pathogenic fungi in combination of fluconazole. *Nanomedicine.*, *5*(4), 282–286.

Gilman, G. P., (2006). A simple device for arsenic removal from drinking water using hydrotalcite. *Sci. Total Environ.*, *366,* 926–931.

Hillie, T., & Hlophe, M., (2007). Nanotechnology and the challenge of clean water. *Nat. Nanotechnol.*, *2,* 663–664.

Huang, L., Dian-Qing, L., Yan-Jun, W., Min David, G., & Xue, E. D., (2005). Controllable preparation of nano-MgO and investigation of its bactericidal properties. *J. Inorg. Biochem.*, *99,* 986–993.

Ingale, A. G., & Chaudhari, A. N., (2013). Biogenic synthesis of nanoparticles and potential applications: An eco-friendly approach. *J. Nanomed. Nanotechol.*, *4,* 165. doi: 10.4172/2157–7439. 1000165.

Karn, B., Kuiken, T., & Otto, M., (2009). Nanotechnology and in situ remediation: A review of benefits and potential risks. *Environ. Health Persp.*, *117*(12), 1823–1831.

Kumar, V., & Yadav, S. K., (2009). Plant-mediated synthesis of silver and gold nanoparticles and their applications, *J. Chem. Technol. Biotechnol., 84*, 151–157.

Li, Y., Cu, Y. T., & Luo, D., (2005). Multiplexed detection of pathogen DNA with DNA based fluorescence nanobarcodes. *Nat. Biotechnol., 23*, 885–889.

Liu, F., Wen, L. X., Li, Z. Z., Yu, W., Sun, H. Y., & Chen, J. F., (2006b). Porous hollow silica nanoparticles as controlled delivery system for water soluble pesticide. *Mat. Res. Bull., 41*, 2268–2275.

Liu, X., Feng, Z., Zhang, S., Zhang, J., Xiao, Q., & Wang, Y., (2006a). Preparation and testing of cementing nano-subnano composites of slow- or controlled release of fertilizers. *Scientia Agricultura Sinica., 39*, 1598–1604.

Liu, P., Huang, Q., & Chen, W., (2012a). Construction and application of a zinc-specific biosensor for assessing the immobilization and bioavailability of zinc in different soils. *Envir. Pollut., 164*, 66–72.

Maderova, L., & Paton, G. I., (2013). Deployment of microbial sensors to assess zinc bioavailability and toxicity in soils. *Soil Boil. Biochem., 66*, 222–2.

Zhang, Y., Chan, H. F., & Leong, K. W., (2013). Advanced materials and processing for drug delivery: The past and the future. *Adv. Drug Deliv. Rev., 65*, 104–120.

McLamore, E. S., Diggs, A., CalvoMarzal, P., Shi, J., Blakeslee, J. J., Peer, W. A., Murphy, A. S., & Porterfield, D. M., (2010). Noninvasive quantification of endogenous root auxin transport using an integrated flux microsensor technique. *Plant J., 63*, 1004–1016.

Nair, R., Varghese, S. H., Nair, B. G., Maekawa, T., Yoshida, Y., & Kumar, D. S., (2010). Nanoparticle material delivery to plants. *Plant Sci., 179*, 154–163.

Owolade, O. F., Ogunleti, D. O., & Adenekan, M. O., (2008). Titanium dioxide affects disease development and yield of edible cowpea. *EJEAF Chem., 7*(50), 2942–2947.

Pal, S., Tak, Y. K., & Song, J. M., (2007). Does the antibacterial activity of silver nanoparticles depend on the shape of the nanoparticle? A study of Gram negative bacterium *Escherichia coli. Appl. Environ. Microbiol., 73*, 1712–1720.

Park, H. J., Kim, S. H., Kim, H. J., & Choi, S. H., (2006). A new composition of nanosized silica-silver for control of various plant diseases. *Plant Pathol. J., 22*, 25–34.

Prasad, K. S., Pathak, D., Patel, A., Dalwadi, P., Prasad, R., & Patel, P., Kaliaperumal S. K., (2011). Biogenic synthesis of silver nanoparticles using *Nicotiana tobaccum* leaf extract and study of their antibacterial effect. *Afr. J. Biotechnol., 9*(54), 8122–8130.

Prasad, R., Bagde, U. S., & Varma, A., (2012a). Intellectual property rights and agricultural biotechnology: an overview. *Afr. J. Biotechnol., 11*(73), 13746–13752.

Prasad, R., & Swamy, V. S., (2013). Antibacterial activity of silver nanoparticles synthesized by bark extract of *Syzygium cumini. J. Nanopart.* http://dx.doi.org/., 10. 1155/2013/431218.

Prasad, R., Swamy, V. S., & Varma, A., (2012b). Biogenic synthesis of silver nanoparticles from the leaf extract of *Syzygium cumini* (L.) and its antibacterial activity. *Int. J. Pharm. Bio Sci., 3*(4), 745–752.

Prasanna, B. M., (2007). Nanotechnology in agriculture. *ICAR National Fellow, Division of Genetics, I. A. R. I.*, New Delhi, India. pp. 111–118.

Sharma, V. K., Yngard, R. A., & Lin, Y., (2009). Silver nanoparticles: green synthesis and their antimicrobial activities. *Adv. Colloid Interface Sci., 145*, 83–96.

Sharon, M., Choudhary, A. K., & Kumar, R., (2010). Nanotechnology in agricultural diseases and food safety. *J. Phytol. 2*, 83–92.

Shah, M. A., & Towkeer, A., (2010). *Principles of Nanosciences and Nanotechnology*. Narosa Publishing House, New Delhi.

Shi, Y. H., Xu, Z. R., Feng, J. L., & Wang, C. Z., (2006). Efficacy of modified montmorillonite nanocomposite to reduce the toxicity of aflatoxin in broiler chicks. *Anim. Feed Sci. Technol.*, *129*, 138–148.

Sugunan, A., & Dutta, J., (2008). Nanotechnology vol. *2*: *Environmental Aspects* (Krug Harald (ed.), Wiley-VCH, Weinheim.

Suman, P. R., Jain, V. K., & Varma, A., (2010). Role of nanomaterials in symbiotic fungus growth enhancement. *Curr. Sci.*, *99*, 1189–1191.

Swamy, V. S., & Prasad, R., (2012). Green synthesis of silver nanoparticles from the leaf extract of *Santalum album* and its antimicrobial activity. *J. Optoelectron. Biomed. Mater.*, *4*(3), 53–59.

Tarafdar, J. C., Sharma, S., & Raliya, R., (2013). Nanotechnology: Interdisciplinary science of applications. *Afr. J. Biotechnol.*, *12*(3), 219–226.

Torney, F., (2009). Nanoparticle mediated plant transformation. Emerging technologies in plant science research. *Interdepartmental Plant Physiology Major Fall Seminar Series. Phys.*, pp. 696.

Vidhyalakshmi, R., Bhakyaraj, R., & Subhasree, R. S., (2009). Encapsulation the future of probiotics-A Review. *Adv. Biol. Res.*, *3*(3–4), 96–103.

Vo-Dinh, T., Kasili, P., & Wabuyele, M., (2006). Nanoprobes and nanobiosensors for monitoring and imaging individual living cells. *Nanomedicine. 2,* 22–30.

Yamanka, M., Hara, K., & Kudo, J., (2005). Bactericidal actions of silver ions solution on *Escherichia coli* studying by energy filtering transmission electron microscopy and proteomic analysis. *Appl. Environ. Microbiol.*, *71*, 7589–7593.

Yavuz, C. T., Mayo, J. T., Yu, W., W., Prakash, A., Falkner, J. C., Yean, S., Cong, L., Shipley, H. J., Kan, A., Tomson, M., Natelson, D., & Colvin, V. L., (2006). Low-field magnetic separation of monodisperse Fe_3O_4 nanocrystals. *Science*, *314*, 964–967.

Young, K. J., (2009). Antifungal activity of silver ions and nanoparticles on phytopathogenic fungi. *Plant Dis.*, *93*(10), 1037–1043.

Zhang, Y., Chan, H. F., & Leong. K. W., (2013). Advanced materials and processing for drug delivery: The past and the future. *Adv. Drug Deliv. Rev.*, *65*, 104–120.

INDEX